AF411450

Greening Cities Shaping Cities: Pinpointing Nature-Based Solutions in Cities between Shared Governance and Citizen Participation

Greening Cities Shaping Cities: Pinpointing Nature-Based Solutions in Cities between Shared Governance and Citizen Participation

Editors

Israa H. Mahmoud
Eugenio Morello
Giuseppe Salvia
Emma Puerari

MDPI • Basel • Beijing • Wuhan • Barcelona • Belgrade • Manchester • Tokyo • Cluj • Tianjin

Editors

Israa H. Mahmoud
Laboratorio di Simulazione
Urbana Fausto Curti,
Department of Architecture
and Urban Studies (DAStU),
Politecnico di Milano,
20133 Milan, Italy

Eugenio Morello
Laboratorio di Simulazione
Urbana Fausto Curti,
Department of Architecture
and Urban Studies (DAStU),
Politecnico di Milano,
20133 Milan, Italy

Giuseppe Salvia
UCL – University College
of London
UK

Emma Puerari
University of Groningen
The Netherlands

Editorial Office
MDPI
St. Alban-Anlage 66
4052 Basel, Switzerland

This is a reprint of articles from the Special Issue published online in the open access journal *Sustainability* (ISSN 2071-1050) (available at: https://www.mdpi.com/journal/sustainability/special_issues/greening_cities).

For citation purposes, cite each article independently as indicated on the article page online and as indicated below:

LastName, A.A.; LastName, B.B.; LastName, C.C. Article Title. *Journal Name* **Year**, *Volume Number*, Page Range.

ISBN 978-3-0365-4703-9 (Hbk)
ISBN 978-3-0365-4704-6 (PDF)

Cover image courtesy of Eugenio Morello

Contents

About the Editors . . . . **vii**

Israa H. Mahmoud, Eugenio Morello, Giuseppe Salvia and Emma Puerari
Greening Cities, Shaping Cities: Pinpointing Nature-Based Solutions in Cities between Shared
Governance and Citizen Participation
Reprinted from: *Sustainability* **2022**, *14*, 7011, doi:10.3390/su14127011 . . . . **1**

Rob Roggema
From Nature-Based to Nature-Driven: Landscape First for the Design of Moeder Zernike in
Groningen
Reprinted from: *Sustainability* **2021**, *13*, 2368, doi:10.3390/su13042368 . . . . **9**

Alessandro Arlati, Anne Rödl, Sopho Kanjaria-Christian and Jörg Knieling
Stakeholder Participation in the Planning and Design of Nature-Based Solutions. Insights from
CLEVER Cities Project in Hamburg
Reprinted from: *Sustainability* **2021**, *13*, 2572, doi:10.3390/su13052572 . . . . **31**

Sophie Mok, Ernesta Mačiulytė, Pieter Hein Bult and Tom Hawxwell
Valuing the Invaluable(?)—A Framework to Facilitate Stakeholder Engagement in the Planning
of Nature-Based Solutions
Reprinted from: *Sustainability* **2021**, *13*, 2657, doi:10.3390/su13052657 . . . . **49**

Maria Beatrice Andreucci, Angela Loder, Martin Brown and Jelena Brajković
Exploring Challenges and Opportunities of Biophilic Urban Design: Evidence from Research
and Experimentation
Reprinted from: *Sustainability* **2021**, *13*, 4323, doi:10.3390/su13084323 . . . . **65**

Didem Kara and Gülden Demet Oruç
Evaluating the Relationship between Park Features and Ecotherapeutic Environment: A
Comparative Study of Two Parks in Istanbul, Beylikdüzü
Reprinted from: *Sustainability* **2021**, *13*, 4600, doi:10.3390/su13094600 . . . . **89**

Wendel Henrique Baumgartner
Parque Augusta (São Paulo/Brazil): From the Struggles of a Social Movement to Its
Appropriation in the Real Estate Market and the Right to Nature in the City
Reprinted from: *Sustainability* **2021**, *13*, 5150, doi:10.3390/su13095150 . . . . **113**

**Giulio Senes, Paolo Stefano Ferrario, Gianpaolo Cirone, Natalia Fumagalli, Paolo Frattini,
Giovanna Sacchi and Giorgio Valè**
Nature-Based Solutions for Storm Water Management—Creation of a Green Infrastructure
Suitability Map as a Tool for Land-Use Planning at the Municipal Level in the Province of
Monza-Brianza (Italy)
Reprinted from: *Sustainability* **2021**, *13*, 6124, doi:10.3390/su13116124 . . . . **135**

**Rembrandt Koppelaar, Antonino Marvuglia, Lisanne Havinga, Jelena Brajković and
Benedetto Rugani**
Is Agent-Based Simulation a Valid Tool for Studying the Impact of Nature-Based Solutions on
Local Economy? A Case Study of Four European Cities
Reprinted from: *Sustainability* **2021**, *13*, 7466, doi:10.3390/su13137466 . . . . **153**

Ana Mitić-Radulović and Ksenija Lalović
Multi-Level Perspective on Sustainability Transition towards Nature-Based Solutions and
Co-Creation in Urban Planning of Belgrade, Serbia
Reprinted from: *Sustainability* **2021**, *13*, 7576, doi:10.3390/su13147576 . . . . **171**

**Israa H. Mahmoud, Eugenio Morello, Chiara Vona, Maria Benciolini, Iliriana Sejdullahu,
Marina Trentin and Karmele Herranz Pascual**
Setting the Social Monitoring Framework for Nature-Based Solutions Impact: Methodological
Approach and Pre-Greening Measurements in the Case Study from CLEVER Cities Milan
Reprinted from: *Sustainability* **2021**, *13*, 9672, doi:10.3390/su13179672 . . . . **193**

Peter Brokking, Ulla Mörtberg and Berit Balfors
Municipal Practices for Integrated Planning of Nature-Based Solutions in Urban Development
in the Stockholm Region
Reprinted from: *Sustainability* **2021**, *13*, 10389, doi:10.3390/su131810389 . . . . **221**

Patricia Sanches, Fabiano Lemes de Oliveira and Gabriela Celani
Green and Compact: A Spatial Planning Model for Knowledge-Based Urban Development in
Peri-Urban Areas
Reprinted from: *Sustainability* **2021**, *13*, 13365, doi:10.3390/su132313365 . . . . **241**

Nathalie Nunes, Emma Björner and Knud Erik Hilding-Hamann
Guidelines for Citizen Engagement and the Co-Creation of Nature-Based Solutions: Living
Knowledge in the URBiNAT Project
Reprinted from: *Sustainability* **2021**, *13*, 13378, doi:10.3390/su132313378 . . . . **263**

Barbara Ester Adele Piga, Gabriele Stancato, Nicola Rainisio and Marco Boffi
How Do Nature-Based Solutions' Color Tones Influence People's Emotional Reaction? An
Assessment via Virtual and Augmented Reality in a Participatory Process
Reprinted from: *Sustainability* **2021**, *13*, 13388, doi:10.3390/su132313388 . . . . **301**

Chuloh Jung, Nahla Al Qassimi, Mohammad Arar and Jihad Awad
The Improvement of User Satisfaction for Two Urban Parks in Dubai, UAE: Bay Avenue Park
and Al Ittihad Park
Reprinted from: *Sustainability* **2022**, *14*, 3460, doi:10.3390/su14063460 . . . . **327**

Roberto Falanga, Jessica Verheij and Olivia Bina
Green(er) Cities and Their Citizens: Insights from the Participatory Budget of Lisbon [†]
Reprinted from: *Sustainability* **2021**, *13*, 8243, doi:10.3390/su13158243 . . . . **351**

About the Editors

Israa H. Mahmoud

Israa H. Mahmoud, Ph.D., is an Architect and Urban Planner by education. She holds a Ph.D. in urban regeneration and economic development. Since 2018, she has been a Post-Doc research fellow at the Laboratorio di Simulazione Urbana Fausto Curti, Department of Architecture and Urban Studies (DAStU), at the Politecnico di Milano. Now she is the research team leader together with Prof. Eugenio Morello on Clever Cities, a European Commission—Horizon 2020 Funded Project, as an expert of co-creation guidance for cities to implement nature-based solutions in socially inclusive urban regeneration processes. Between 2016 and 2017, she took part in a European Commission—MARIE-CURIE Action funded Horizon 2020 Project—MAPS-LED Project as an early-stage researcher (ESR), and later as an experienced researcher (ER) at San Diego State University, CA, USA. At Politecnico di Milano, she also lectures about nature-based solutions in the Master of Science in "Urban Planning and Policy Design", as well as in the Master of "Sustainable Architecture and Landscape Design". Recently, she co-led the "Greening Cities, Shaping Cities" International Symposium at the Politecnico di Milano in October 2020. In 2022, she co-edited a book on "Nature-based Solutions for Sustainable Urban Planning" https://link.springer.com/book/10.1007/978-3-030-89525-9.

Eugenio Morello

Eugenio Morello, Ph.D. An architect by education, Eugenio Morello is Associate Professor of Urban Planning and Design at the Politecnico di Milano, Department of Architecture and Urban Studies (DAStU). He is a coordinator and research scientist at the Laboratorio di Simulazione Urbana Fausto Curti (since 2010) and the Climate Change Risk and Resilience Lab (since 2020). Since 2017, he has been the rector's delegate for environmental sustainability. He is the principal investigator of the European Horizon 2020 projects 'Clever Cities' (2018–2023), 'Sharing Cities' (2016–2021), and the Urban Innovative Action (UIA) Air-Break (2020–2023). His research interest is situated in the interplay between urban design and environmental quality, climate design, resilience, and adaptation to climate change. He investigates the integration of environmental aspects and energy transition solutions for the design of sustainable neighborhoods towards the closing of energy and environmental cycles. More recently, his research work has opened new insights on the topic of collaborative consumption and sharing society and how these new paradigms can inform urban planning, urban design, and co-creation approaches.

Giuseppe Salvia

Giuseppe Salvia, Ph.D., is a Senior Lecturer in Design and Innovation at Brunel University London. His core expertise is design strategies and approaches for sustainable change. Giuseppe has a track record of research projects conducted in globally established universities, including University College London and Politecnico di Milano. The focus on projects regard socio-technical innovation and implications for social, environmental, and economic sustainability, applied to a number of sectors; these include green spaces and nature-based solutions, urban dynamics and health, thermal comfort and heating, product lifetimes, and material efficiency. Giuseppe has developed an interest in understanding behaviours and social practices, for delivering strategies of change, through systems-thinking and codesign methods. For these, he has collaborated with recognized private and public organizations.

Emma Puerari

Emma Puerari, Ph.D., is an Assistant Professor in Urban Design and Planning at the Faculty of Spatial Sciences, Department of Planning and Environment at University of Groningen. The core of her research focuses on the role of design in enabling spatial and institutional innovation within urban environments and territories, addressing the interplay of planning and design domains in sustainability transition processes. In this respect, the territorial impact of citizens' engagement in co-creation and co-production processes is a crucial part of her research interest. Within her current institution, she is member of the "Centre for Advanced Studies in Urban Science and Design" (CASUS) at the "Urban and Regional Studies Institute (URSI) and member of the Graduate School of Spatial Sciences as PhD supervisor. Emma has been a Post-doc researcher at TU Delft, Faculty of Industrial Design Engineering (IDE) and visiting researcher at the Dutch Research Institute for Transitions (DRIFT) (2016–2018). She has been active as a researcher and manager on several national and European research projects: Participatory City Making, NWO-STW, for TU Delft and DRIFT (2016–2018); Community Participation in Planning, Erasmus +, for DAStU POLIMI (2015–2016); MyNeighbourhood I MyCity, EU-CIP PSP Grant Agreement no 325227 for DAStU POLIMI (2013–2015); Periphèria, EU-CIP PSP Grant Agreement no 271015, for DAStU POLIMI (2012–2013).

 *sustainability*

Editorial

Greening Cities, Shaping Cities: Pinpointing Nature-Based Solutions in Cities between Shared Governance and Citizen Participation

Israa H. Mahmoud [1,*], **Eugenio Morello** [1,*], **Giuseppe Salvia** [2] **and Emma Puerari** [3]

[1] Laboratorio di Simulazione Urbana Fausto Curti, Department of Architecture and Urban Studies (DASTU), Politecnico of Milan, Via Bonardi, 3, 20133 Milan, Italy
[2] Institute for Environmental Design and Engineering-The Bartlett School of Environment, Energy and Resources, UCL–University College of London, London WC1E 6BT, UK; g.salvia@ucl.ac.uk
[3] Urban Design and Planning, Department of Planning and Environment, Faculty of Spatial Sciences, University of Groningen, Mercator Building, Room 3.38 Landleven 1, 9747 AD Groningen, The Netherlands; e.puerari@rug.nl
*   Correspondence: israa.mahmoud@polimi.it (I.H.M.); eugenio.morello@polimi.it (E.M.)

Citation: Mahmoud, I.H.; Morello, E.; Salvia, G.; Puerari, E. Greening Cities, Shaping Cities: Pinpointing Nature-Based Solutions in Cities between Shared Governance and Citizen Participation. *Sustainability* **2022**, *14*, 7011. https://doi.org/10.3390/su14127011

Received: 20 May 2022
Accepted: 5 June 2022
Published: 8 June 2022

Publisher's Note: MDPI stays neutral with regard to jurisdictional claims in published maps and institutional affiliations.

The topic of pinpointing Nature-Based Solutions (NBS) in the urban context has been cultivating interests lately from different scholars, urban planning practitioners and policy-makers. This Special Issue originates from the Greening Cities Shaping Cities Symposium held at the Politecnico di Milano (12–13 October 2020), aiming at bridging the gap between the science and practice of implementing NBS in the built environment [1], as well as highlighting the importance of citizen participation in shared governance and policy making. The Special Issue was also made open to other contributions from outside the symposium in order to allow for contributions from a major scientific and practical audience wherever possible. Indeed, we have gathered contributions from Italy, Germany, the Netherlands, Turkey, Brazil, Portugal, Denmark, France, Bulgaria, Sweden, Hungary, Spain, the UAE, the UK, and the USA.

In particular, a specific focus in this Special Issue is given to investigations on how NBS and urban greening strategies are re-shaping the built environment and the whole imagery of cities, both from a spatial and a governance perspective [2,3]. The intended result is a set of contributions providing insights and food for thought to urban debates on design and planning theory, policy and practice around NBS. Nowadays, cities are making use of nature as a solution to many challenges, without radically and critically addressing the full potential of interpreting green planning as a powerful urban design instrument and governance feature [4]. For instance, how will vegetation infilling strategies affect planners' toolkits and decision-making procedures? How can we get citizens involved in the design and management process around NBS?

Hence, within this Special Issue, an attentive selection of contributions mainly looked at addressing the procedural gaps in greening city strategies that are nowadays at the forefront of re-shaping many urban fabrics, specifically by investigating governance and citizen participation.

A strong emphasis on the viability NBS for implementation often encounters hindrances on the governance scale and lacks a strong functional governance model in order to "make it work". A big tranche of this pitfall is due to the lack of capacities and communication between municipal departments, as well as the need to raise awareness on how NBS operate on a day-to-day activity. Hence, the capacity building and awareness activities result as one major need in cities' decision-making processes to make the implementation of NBS more inclusive and their management shared among more stakeholders within a sustainable urban planning approach [5].

NBS are living and dynamic systems and require specific attention in design and maintenance. Hence, the engagement and the active role of citizens is crucial [6]. One important aspect of innovation in NBS implementation nowadays is its inclusivity and its relatedness to citizen-centred approaches for implementation in Urban Living Labs (ULLs). The notion is that ULLs allow a flexible structural pathway and include a variety of sleeve tools to bring everyone on board [7]. Lessons in this section would mainly address successful case studies from physical or digital ULLs experiences in implementing NBS in urban regeneration processes.

Lastly, some of the Special Issue's contributions also address whether the embeddedness of NBS in cities tangibly affects urban morphologies and radically impacts our approach to urban planning, urban design strategies and, consequently, urban governance models [8]. Integrating nature-based greening plans and NBS seems to be happening more and more frequently in city strategic planning and city visions; however, a deep recognition of the role of greening in shaping the overall imagery of cities and renovating the role of green planning as a quintessential element of design and planning seems to be lacking a deeper and conscious debate.

In sum, in this Special Issue, we aimed to touch base on many aspects related to NBS conceptualization, public acceptance, implementation and upscaling in cities. The finality of our exploration is to find clues towards a critical understanding and interpretation of how greening cities is affecting urban shaping, both from morphological and governance point of views. Other related questions on biodiversity and citizens engagement [9] or more technical ones on climate change mitigation and adaptation using remote sensing methods and GIS were not alluring to the authors of this Special Issue; however, they remain a starting point for further scientific investigations.

Article 1: From Nature-Based to Nature-Driven: Landscape First for the Design of Moeder Zernike in Groningen.

In this first article, Roggema gives a fresh perspective on climate change adaptation using a nature-driven approach [10]. He methodologically applies a research-through-design-process on a case study, namely, "Moeder Zernike campus in Groningen, Netherlands". Roggema integrates food systems, coastal and water shortage dynamics as well as urban agriculture in one visionary future plan for the area using NBS. This research article looks at the tensions between short-term practices, adaptive climate change management relying only on data availability and, lastly, on a longer-term view working towards the unknown impact of future climate change. the main takeaway from this article is how embracing a nature-driven perspective to urban design increases the adaptive capacity, the ecological diversity and the range of healthy food grown on a university campus using a co-creative design-led approach as a way to take nature as the basis for urban transformation.

Article 2: Stakeholder Participation in the Planning and Design of Nature-Based Solutions. Insights from CLEVER Cities Project in Hamburg.

Arlati et al. [11] present reflections on the co-creation practices of NBS deployed in the frame of the Horizon 2020 project CLEVER Cities by analyzing and discussing the case study of Hamburg, Germany. The focus of the article is based on an analysis of the stakeholder engagement methods called to collaborate in the environment of Urban Living Labs (ULLs) for the co-creation of NBS and the role those stakeholders played along the process. The potential of NBS to foster participation and support sustainability transitions was recognized in the Hamburg case study under the circumstances granted by the Horizon 2020 Programme. The paper argues that the current governance mechanisms should undergo structural changes in order to allow a broader collaboration and steer the transition process.

Article 3: Valuing the Invaluable(?)—A Framework to Facilitate Stakeholder Engagement in the Planning of Nature-Based Solutions.

In this article, Mok et al. [12] present and discuss the logic of a framework aiming at facilitating stakeholder engagement in the planning of NBS from the project UNaLab—a Horizon 2020 Research and Innovation Action. They exploit the challenges and trade-offs in approaches of NBS valuation with the goal of identifying key values and engaging beneficiaries from the public, private and civil society sectors in the development of NBS. Applied methods such as focus groups, interviews and surveys were used to assess different framework components and their interlinkages, as well as to test their applicability in urban planning. The authors develop a case for 'softer' approaches to NBS value assessment tools in order to encourage awareness-raising, stakeholder engagement and mobilize local actors around NBS to complement 'harder' valuation mechanisms. Through a survey with experts from the projects and several workshops, authors further developed their framework based on providing a structured approach, which can be used in multiple contexts to facilitate navigation through the complexity of a common understanding between actors from different backgrounds and thus support the formation of new alliances for NBS planning and implementation.

Article 4: Exploring Challenges and Opportunities of Biophilic Urban Design: Evidence from Research and Experimentation.

Andreucci et al. [13] explore how the benefits of nature are understood for different environments and multiple scales, ranging from a building (e.g., workplace) to the neighborhood (e.g., arts and conference complex) and up to the citywide scale. For this aim, the authors embrace biophilic design theory and make a case for the importance of deepening the understanding and application of this approach, which is often considered of secondary priority. The multi-scale examples of NBS implemented in both London and Chicago and articulated in the article reinforce the importance of a systems-thinking approach, as the authors also infer in the conclusions. Diverse are the perceptions, experiences and feelings that people may develop while interacting with NBS while in the Shard or at the Barbican; however, they are both components of the same city, in which people accomplish their micro-mundane routines and co-exist in different ways.

Article 5: Evaluating the Relationship between Park Features and Eco therapeutic Environment: A Comparative Study of Two Parks in Istanbul, Beylikdüzü.

Kara and Oruç [14] address the therapeutical benefits of (re-)establishing a connection with nature, especially in the urban environment. Informed by a literature review on eco-psychology and eco-therapy, as well as a case study carried out in a district of Istanbul, the authors explore how the physical attributes of the space affect the user experience of being in a park, the connection to nature and therefore the therapeutic benefits deriving from this. The results suggest that the experience of and connectedness to nature is complex, with several factors and determinants, as it may be sensible to expect for some. In our view, the main takeaway from the study is the importance of adopting a user-centered approach to landscape designing and policy making in order to unleash the potential psychological benefits that NBS can provide. This reflection may encourage stakeholders to reflect on how ready and equipped they are for this approach to be operationalized.

Article 6: Parque Augusta (São Paulo/Brazil): From the Struggles of a Social Movement to Its Appropriation in the Real Estate Market and the Right to Nature in the City.

In his article, Baumgartner [15] reports on the narratives of the implementation from Parque Augusta in the center of São Paulo, Brazil. After years of struggle with the city and the real estate developers carried out by an organized social movement and citizens to avoid building speculation on a precious green space, the collaborative co-design of an urban park, enriched by green solutions (NBS), has followed. However, during the park's construction, the pressure of the properties surronded by high-density buildings and the reduction in the implementation of the previously agreed-on green solutions opened reflections on urban greening processes, on appropriaton dynamics of green areas and on the right to nature in the city. The natural elements play a key role and represent a powerful medium in activating citizens in safeguarding and enhancing left-over spaces in cities. The article proves how such an experience can inform local governments in deploying such civil society engagement around nature, to improve democracy and support decision-making processes and the planning of a city's green space system.

Article 7: Nature-Based Solutions for Storm Water Management—Creation of a Green Infrastructure Suitability Map as a Tool for Land-Use Planning at the Municipal Level in the Province of Monza-Brianza (Italy).

Senes et al. [16] develop a methodology to define Green Infrastructure for stormwater management at the municipal level, with an application in the Province of Monza-Brianza, Italy. NBS diffused in the city, in combination with the sewer infrastructure, will help see improvements arising from reductions in stormwater quantity and reduced sewage overflows. The goal of this study is to support cities in setting up Green Infrastructure Suitability Maps as a tool for land-use planning. Hence, aiming at identifying non-urbanized areas where rainwater can potentially infiltrate, considering also site-specific soil characteristics, the proposed methodology is defined based on three phases, namely: the definition of the territorial information needed, the production of base maps and the production of a Suitability Map. The authors demonstrate how the spatial mapping of NBS proves to be an effective tool to support the decision-making process for spatial planning.

Article 8: Is Agent-Based Simulation a Valid Tool for Studying the Impact of Nature-Based Solutions on Local Economy? A Case Study of Four European Cities.

Koppelaar et al. [17] describe an agent-based model which reveals the potential interconnection between the assessment of the wealth of the commercial urban fabric and the development of wide NBS (e.g., parks). The reflections are drawn from longitudinal case studies in three different countries. Despite the limitations of the work finely acknowledged in their discussion, the authors make the case for the added value of the model, which supports the decision-making process of urban developments by calculating the indirect financial benefit of implementing NBS. The article may also raise reflections for the reader about the wider system of places and practices that the NBS belong to and should be considered with in order to assess and foster the benefits associated to them.

Article 9: Multi-Level Perspective on Sustainability Transition towards Nature-Based Solutions and Co-Creation in Urban Planning of Belgrade, Serbia.

With this paper, Mitić-Radulović and Lalović [18] explore the challenge of achieving clear, coherent and ambitious urban greening strategies embedded in urban planning and developed in a co-creative, participatory and inclusive manner within the European context. The work, using the Multi-Level Perspective (MLP) on sustainability transitions, observes the urban planning system in Belgrade, Serbia, as a socio-technical regime with a focus on two recent urban development initiatives in Belgrade, the Capital of Serbia, as the specific context of analysis. In particular, the article examines informal urban planning instruments that can be implemented by the practitioners of niche innovations to engage constructively and appropriately in co-creation, supporting urban planners and NBS advocates in the Serbian and EU enlargement context.

Article 10: Setting the Social Monitoring Framework for Nature-Based Solutions Impact: Methodological Approach and Pre-Greening Measurements in the Case Study from CLEVER Cities Milan.

In this article, Mahmoud et al. [19] set a new methodological approach for monitoring the social impacts of NBS on human health and wellbeing, social cohesion and environmental justice, as well as citizens' perception about safety and security related to the NBS implementation process. Their methodological approach relies on a co-creation process using several steps of scoping and gathering information based on the case study of the Milanese context from CLEVER Cities Horizon 2020 project. The authors examined the relevance of using NBS in addressing social co-benefits by analyzing data from questionnaires submitted to citizens and participants of activities during pre-greening interventions against a set of five major indicators: (1) place, use of space and relationship with nature; (2) perceived ownership and sense of belonging; (3) psychosocial issues, social interactions and social cohesion; (4) citizen perception about safety and security; and lastly, (5) knowledge about CLEVER interventions and NBS benefits in relation to the socio-demographics of the questionnaires' respondents. Lastly, the results are cross-compared within the three areas of interventions of the project Urban Living Labs (so called CLEVER Action Labs).

The article hence pinpoints the importance of co-producing social monitoring methods with citizens to set the boundaries for NBS place-based interventions and accentuate citizens' perceptions about their wellbeing, general health and strong sense of neighborhood belonging. A wider interest is noted towards civic participation in co-management and becoming informed about NBS interventions in the Milanese context.

Article 11: Municipal Practices for Integrated Planning of Nature-Based Solutions in Urban Development in the Stockholm Region.

Brokking, Mörtberg and Balfors [20] explore how NBS are addressed in urban development processes. The authors propose a study of municipal planning practices related to NBS and their contribution to regional green infrastructures and social and ecological qualities. They run their analysis on three case studies in the Stockholm region of Sweden. They run a mixed method approach using focus groups, interviews and through the study of official documents. The results of their study highlight that, while the institutional conditions play a fundamental role in shaping the planning processes that can challenge the ability to enhance social and ecological qualities, the planning and the design of urban green spaces play a key role in the engagement of the communities. Co-creation sessions are fundamental for the development of specific competences for the development of innovative solutions on private and public green areas. Despite the differences between the different case studies, the paper concludes that a knowledge-driven and integrative planning process can foster the potential of NBS for green and sustainable cities.

Article 12: Green and Compact: A Spatial Planning Model for Knowledge-Based Urban Development in Peri-Urban Areas.

Sanches, Lemes de Oliveira and Celani [21] define a multi-scalar spatial planning model for peri-urban areas and urban voids able to reconcile medium-to-high building densities with the provision of ecosystem services. They employ a three-scale spatial planning model: micro, meso and macro. Subsequently, the model is applied to the case of the International Hub for Sustainable Development (HIDS) in Campinas, Brazil. An urban design proposal was developed during an international workshop in July 2020 and was secondly completed with experts' workshops and planning professionals. Lastly, in 2021, the model was evaluated and validated through a series of workshops looking at evidence-based solutions and the evaluation of their results in real-time. This research puts a mark on the practical application of modelling in design exercises towards reducing the gap between theory and practice, which is beneficial to the approach of NBS.

Article 13: Guidelines for Citizen Engagement and the Co-Creation of Nature-Based Solutions: Living Knowledge in the URBiNAT Project.

Nunes, Björner and Hilding-Hamann [22] focus on citizens' participation within the context of urban regeneration projects. Their work aims to develop specific guidelines for the development of co-creation of NBS. The work was developed within the framework of the Horizon 2020 project URBiNAT that focuses on the regeneration of underserved urban districts. The article describes the processes followed within such a project: the collection of scientific and practical input from both researchers and practitioners first, followed by a deeper analysis of selected participants. The results highlight what the authors described as an 'ecology of knowledges' based on a 'living' framework, addressing the needs of a broad set of citizens and contexts. The paper includes a discussion on the implementation of co-creation practices in the development of NBS. The conclusions broaden the research context to include the refinement of the NBS approach, with participation being seen as both a means and an end to it.

Article 14: How Do Nature-Based Solutions' Color Tones Influence People's Emotional Reaction? An Assessment via Virtual and Augmented Reality in a Participatory Process.

Piga et al. [23] examine the effects of NBS on people's emotions, focusing on the reliability of Augmented Reality (AR) and Virtual Reality (VR) simulations as means for engaging citizens in participatory processes. Their case studies explore the reaction to existing and designed NBS, showing that some color tones of NBS, namely green and lime, reduce the unpleasantness experienced while viewing the urban environment. Such effect is confirmed both in AR and VR, suggesting that increasing urban greenery can have a positive effect. The results of VR are fully consistent with the previous literature, whereas in AR, some variables show a different pattern. The authors suggest that available digital tools are a valuable support for envisioning sustainable urban transformations with diverse stakeholders, although further interdisciplinary studies are needed to tackle the technical and ethical implications of such technologies.

Article 15: The Improvement of User Satisfaction for Two Urban Parks in 2 Dubai, UAE: Bay Avenue Park and Al Ittihad Park.

Jung et al. [24] develop a conventional user satisfaction method and questionnaires for analyzing users' satisfaction in two urban parks in Dubai, Bay Avenue Park and Al Ittihad Park. The authors, using a comparative analysis, expose different park users' behavior, satisfaction level (based on park environment and accessibility) and users' demographic information. Following a descriptive statistics and frequency method, the authors perform a multiple regression analysis to better understand the physical environment factors affecting each of the two parks' satisfaction level. Both parks, being the green structures of neighborhoods and located within walking distances of residential areas, prove to be highly satisfying, in particular thanks to the presence of natural elements such as green spaces, trees and trails. This research can be used as basic data for improving the future planning of urban parks in Dubai, towards a more greening approach to urban planning, including governmental policy, vision and implementation. This could be possible in the future by conducting research on more diverse types of parks, other greening strategies and detailing accessibility-related environmental factors, such as health, community wellbeing and other physical characteristics, such as the width of sidewalks and types of pavement materials.

Review Article 16: Green(er) Cities and Their Citizens: Insights from the Participatory Budget of Lisbon.

In this article, Falanga, Verheij and Bina [25] examine the role of the Participatory Budget (PB) as a potential driver of urban sustainability. The experience of Lisbon, in Portugal, a city recognized internationally as a leader in participatory budgeting the early 2000s, is analyzed and discussed. The authors propose a multimethod approach in the analysis of data on PB calls in Lisbon, investigating emerging trends and variations in citizen proposals, projects, votes and public funding. Emerging key topics show links and trade-offs between locally embedded participation and the international discourse on urban sustainability. A growing interest of project proposals focusing on NBS, involving citizens and businesses, is emerging. Thoroughly analyzing PB data as an expression of citizens' interests and priorities is key to enabling cities to better integrate them into urban planning strategies and—as argued by the authors—to counteract the dominant engineered approach towards sustainability, mainly focused on green growth and innovation.

Author Contributions: Conceptualization, I.H.M. and E.M.; methodology, I.H.M.; formal analysis, I.H.M. and E.M.; investigation, I.H.M. and E.M.; writing—original draft preparation, I.H.M.; writing—review and editing, E.M.; visualization, G.S. and E.P.; supervision, I.H.M. and E.M.; project administration, I.H.M. and E.M. All authors have read and agreed to the published version of the manuscript.

Funding: The 'Greening Cities: Shaping Cities' symposium was co-funded by the Department of Architecture and Urban Studies (DASTU), Politecnico di Milano, Milan, Italy, under the Grant number MPO0DOTA02 and the CLEVER Cities project. This project has received funding from the European Union's Horizon 2020 innovation action program under grant agreement no. 776604. The sole responsibility for the content of this publication lies with the authors. It does not necessarily represent the opinion of the European Union. Neither the EASME nor the European Commission are responsible for any use that may be made of the information contained herein.

Acknowledgments: The editors would like to thank all authors for their excellent contributions, as well as the scientific external guest editors (especially for articles 10, 12 and 14) and reviewers who helped achieve this Special Issue in its final form, they are all listed on the Special issue webpage here https://www.mdpi.com/journal/sustainability/special_issues/greening_cities.

Conflicts of Interest: The authors declare no conflict of interest.

References

1. Greening Cities Shaping Cities: An International Research Symposium. Available online: https://www.greeningcities-shapingcities.polimi.it/ (accessed on 20 May 2022).
2. Albert, C.; Schröter, B.; Haase, D.; Brillinger, M.; Henze, J.; Herrmann, S.; Gottwald, S.; Guerrero, P.; Nicolas, C.; Matzdorf, B. Addressing Societal Challenges through Nature-Based Solutions: How Can Landscape Planning and Governance Research Contribute? *Landsc. Urban Plan.* **2019**, *182*, 12–21. [CrossRef]
3. Zingraff-Hamed, A.; Hüesker, F.; Albert, C.; Brillinger, M.; Huang, J.; Lupp, G.; Scheuer, S.; Schlätel, M.; Schröter, B. Governance Models for Nature-Based Solutions: Seventeen Cases from Germany. *Ambio* **2021**, *50*, 1610–1627. [CrossRef] [PubMed]
4. Grace, M.; Scott, A.J.; Sadler, J.P.; Proverbs, D.G.; Grayson, N. Exploring the Smart-Natural City Interface; Re-Imagining and Re-Integrating Urban Planning and Governance. *Emerald Open Res.* **2020**, *2*, 7. [CrossRef]
5. Mahmoud, I.H.; Morello, E.; Lemes de Oliveira, F.; Geneletti, D. *Nature-Based Solutions for Sustainable Urban Planning*, 1st ed.; Contemporary Urban Design Thinking; Mahmoud, I.H., Morello, E., Lemes de Oliveira, F., Geneletti, D., Eds.; Springer International Publishing: Cham, Switzerland, 2022; ISBN 978-3-030-89525-9. [CrossRef]
6. Mahmoud, I.; Morello, E. Co-Creation Pathway for Urban Nature-Based Solutions: Testing a Shared-Governance Approach in Three Cities and Nine Action Labs. In *Smart and Sustainable Planning for Cities and Regions*; Bisello, A., Vettorato, D., Ludlow, D., Baranzelli, C., Eds.; Springer International Publishing: Cham, Switzerland, 2021; pp. 259–276. ISBN 9783030577643.
7. Mahmoud, I.H.; Morello, E.; Ludlow, D.; Salvia, G. Co-Creation Pathways to Inform Shared Governance of Urban Living Labs in Practice: Lessons From Three European Projects. *Front. Sustain. Cities* **2021**, *3*, 690458. [CrossRef]
8. Boros, J.; Mahmoud, I. Urban Design and the Role of Placemaking in Mainstreaming Nature-Based Solutions. Learning From the Biblioteca Degli Alberi Case Study in Milan. *Front. Sustain. Cities* **2021**, *3*, 635610. [CrossRef]
9. Vona, C.; Mahmoud, I.; Benciolini, M.; Belardi, M.; Trentin, M.; Sejdullahu, I. Il Coinvolgimento Dei Cittadini per La Biodiversità Urbana Attraverso Le NBS: L'esperienza CLEVER Cities. *Reticula* **2021**, *28*, 95–107.
10. Roggema, R. From Nature-Based to Nature-Driven: Landscape First for the Design of Moeder Zernike in Groningen. *Sustainability* **2021**, *13*, 2368. [CrossRef]
11. Arlati, A.; Rödl, A.; Kanjaria-Christian, S.; Knieling, J. Stakeholder Participation in the Planning and Design of Nature-Based Solutions. Insights from CLEVER Cities Project in Hamburg. *Sustainability* **2021**, *13*, 2572. [CrossRef]
12. Mok, S.; Mačiulytė, E.; Bult, P.H.; Hawxwell, T. Valuing the Invaluable(?)—A Framework to Facilitate Stakeholder Engagement in the Planning of Nature-Based Solutions. *Sustainability* **2021**, *13*, 2657. [CrossRef]
13. Andreucci, M.B.; Loder, A.; Brown, M.; Brajković, J. Exploring Challenges and Opportunities of Biophilic Urban Design: Evidence from Research and Experimentation. *Sustainability* **2021**, *13*, 4323. [CrossRef]
14. Kara, D.; Oruç, G.D. Evaluating the Relationship between Park Features and Ecotherapeutic Environment: A Comparative Study of Two Parks in Istanbul, Beylikdüzü. *Sustainability* **2021**, *13*, 4600. [CrossRef]
15. Baumgartner, W.H. Parque Augusta (São Paulo/Brazil): From the Struggles of a Social Movement to Its Appropriation in the Real Estate Market and the Right to Nature in the City. *Sustainability* **2021**, *13*, 5150. [CrossRef]
16. Senes, G.; Ferrario, P.S.; Cirone, G.; Fumagalli, N.; Frattini, P.; Sacchi, G.; Valè, G. Nature-Based Solutions for Storm Water Management—Creation of a Green Infrastructure Suitability Map as a Tool for Land-Use Planning at the Municipal Level in the Province of Monza-Brianza (Italy). *Sustainability* **2021**, *13*, 6124. [CrossRef]
17. Koppelaar, R.; Marvuglia, A.; Havinga, L.; Brajković, J.; Rugani, B. Is Agent-Based Simulation a Valid Tool for Studying the Impact of Nature-Based Solutions on Local Economy? A Case Study of Four European Cities. *Sustainability* **2021**, *13*, 7466. [CrossRef]
18. Mitić-Radulović, A.; Lalović, K. Multi-Level Perspective on Sustainability Transition towards Nature-Based Solutions and Co-Creation in Urban Planning of Belgrade, Serbia. *Sustainability* **2021**, *13*, 7576. [CrossRef]
19. Mahmoud, I.H.; Morello, E.; Vona, C.; Benciolini, M.; Sejdullahu, I.; Trentin, M.; Pascual, K.H. Setting the Social Monitoring Framework for Nature-Based Solutions Impact: Methodological Approach and Pre-Greening Measurements in the Case Study from CLEVER Cities Milan. *Sustainability* **2021**, *13*, 9672. [CrossRef]
20. Brokking, P.; Mörtberg, U.; Balfors, B. Municipal Practices for Integrated Planning of Nature-Based Solutions in Urban Development in the Stockholm Region. *Sustainability* **2021**, *13*, 10389. [CrossRef]
21. Sanches, P.; Lemes de Oliveira, F.; Celani, G. Green and Compact: A Spatial Planning Model for Knowledge-Based Urban Development in Peri-Urban Areas. *Sustainability* **2021**, *13*, 13365. [CrossRef]
22. Nunes, N.; Björner, E.; Hilding-Hamann, K.E. Guidelines for Citizen Engagement and the Co-Creation of Nature-Based Solutions: Living Knowledge in the URBiNAT Project. *Sustainability* **2021**, *13*, 13378. [CrossRef]
23. Piga, B.E.A.; Stancato, G.; Rainisio, N.; Boffi, M. How Do Nature-Based Solutions' Color Tones Influence People's Emotional Reaction? An Assessment via Virtual and Augmented Reality in a Participatory Process. *Sustainability* **2021**, *13*, 13388. [CrossRef]
24. Jung, C.; Al Qassimi, N.; Arar, M.; Awad, J. The Improvement of User Satisfaction for Two Urban Parks in Dubai, UAE: Bay Avenue Park and Al Ittihad Park. *Sustainability* **2022**, *14*, 3460. [CrossRef]
25. Falanga, R.; Verheij, J.; Bina, O. Green(Er) Cities and Their Citizens: Insights from the Participatory Budget of Lisbon. *Sustainability* **2021**, *13*, 8243. [CrossRef]

Article

From Nature-Based to Nature-Driven: Landscape First for the Design of Moeder Zernike in Groningen

Rob Roggema [1,2]

[1]   Cittaideale, 6706LC Wageningen, The Netherlands; rob@cittaideale.eu; Tel.: +31-6-1534-9807
[2]   Institute for Culture and Society, Western Sydney University, Parramatta, NSW 2150, Australia

Abstract: Global climate change impacts the future of urbanism. The future is increasingly uncertain, and current responses in urban planning practice are often human-centered. In general, this is a way to respond to change that is oriented towards improving the life of people in the short term, often extracting resources from the environment at dangerous levels. This impacts the entire ecological system, and turns out to be negative for biodiversity, resilience, and, ultimately, human life as well. Adaptation to climatic impacts requires a long-term perspective based in the understanding of nature. The objective of the presented research is to find explorative ways to respond to the unknown unknowns through designing and planning holistically for the Zernike campus in Groningen, the Netherlands. The methods used in this study comprise co-creative design-led approaches which are capable of integrating sectoral problems into a visionary future plan. The research findings show how embracing a nature-driven perspective to urban design increases the adaptive capacity, the ecological diversity, and the range of healthy food grown on a university campus. This study responds to questions of food safety, and growing conditions, of which the water availability is the most pressing. Considering the spatial concept, this has led to the necessity to establish a novel water connection between the site and the sea.

Keywords: nature-based solutions; landscape and urban design; urban agriculture and food systems; coastal dynamics; Groningen

Citation: Roggema, R. From Nature-Based to Nature-Driven: Landscape First for the Design of Moeder Zernike in Groningen. *Sustainability* **2021**, *13*, 2368. https://doi.org/10.3390/su13042368

Academic Editor: Israa H. Mahmoud

Received: 16 January 2021
Accepted: 18 February 2021
Published: 22 February 2021

Publisher's Note: MDPI stays neutral with regard to jurisdictional claims in published maps and institutional affiliations.

1. Introduction

Climate change is one of humanity's biggest problems [1], and this is largely part related to the ways in which humans live, even causing a new geological era, the anthropocene [2]. The climate's impact on land use, productivity, and food security [3], ecology [4], livability [5–7], and safety, which is under pressure of accelerated sea level rise [8,9], is moving beyond planetary boundaries [10]. The question of whether policy responses can deal with these uncertainties is investigated in this article. It is clear that adaptation is inescapable [11,12].

Current spatial planning focuses on the past, reiterating former policies for novel problems [13,14]. New problems, however, cannot be solved with solutions derived from the actions that caused them. Contemporary policies in the Groningen political arena tend to 'muddle through' [15], and are focused on the near future and on well-understood problems [16,17].

Current achievable policy outcomes are rooted in an existing context of political negotiations and compromises in governance. Wicked problems cannot be dealt with using linear answers—a common mistake. A 'negotiated average' provides solutions for an already changed problem the moment the solution is brought forward, contradicting the long-term larger scale [18], forming an adaptation gap [19,20]. Planning that responds to emergencies is continuously 'muddling through' [15], while uncertainty and wicked problems [21] require 'unsafe planning' [22], bridging this gap (Figure 1).

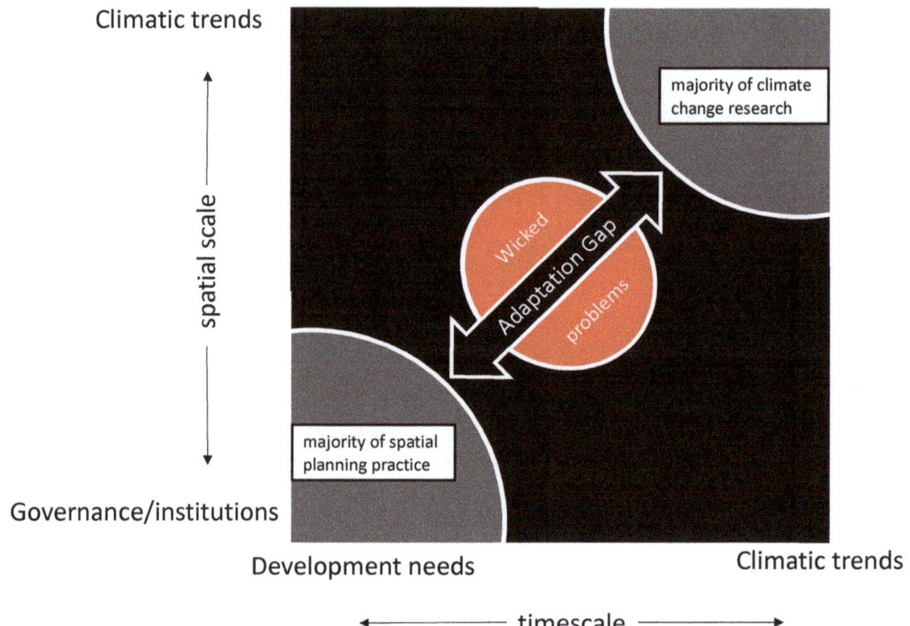

Figure 1. The adaptation gap and wicked problems [19].

Instead, climate adaptation requires innovation in spatial planning that bypasses short-term path-dependency. Therefore, coping with uncertainty [23] is essential, and short-term spatial policymaking needs to be replaced by applying the art of the long view [24] and unsafe planning [22], allowing us to leapfrog current policies.

Climate adaptation is viewed as a spatial challenge [25], positioning 'design' to discover holistic solutions [26–28]. This way, adaptation is designed for the Zernike campus, located on the northern fringe of the city of Groningen in the Netherlands (Figure 2). This university campus is home to two universities, several research institutes, and many enterprises and start-ups. It hosts approximately 30,000 people, including students and staff. The Moeder Zernike project is part of the design-manifestation of the Climate Adaptation Week Groningen, in which the task is to propose solutions for a 100-year future.

This article takes nature-based solutions (NBS) as the point of departure and investigates the potentialities of how to anticipate climate impacts.

Figure 2. Zernike campus in the northern fringe of the City of Groningen.

2. Research Problem

2.1. Problem Definition

Climate adaptation is often practiced with a human-centred objective. In many cases, this leads to increased deterioration of ecosystems and biodiversity loss, puts pressure on agricultural systems, and leads to economic problems and foodborne diseases. Moreover, the lack of green space decreases the quality of life of humans, both physically as mentally.

2.1.1. Ecology

The worldwide decrease in biodiversity, up to 68% [29,30], has a negative impact on soil and water quality. This impacts the functioning of all kinds of natural processes. In 2007, populations of the black-tailed godwit, redshank, oystercatcher, and lapwing were 10–60% lower than in 1990 [31], and this decline in open farmland birds has not stopped (Figure 3). The numbers of the black-tailed godwit decreased by 40%, which is internationally important since the major breeding area in Europe is in the Netherlands [32]. As a result of intensified farmland and cattle farming, this trend has continued to decline since 1960 [31]. Biodiversity loss since the second half of last century is very high. This umbilical cord cannot be severed, as human life depends on nature.

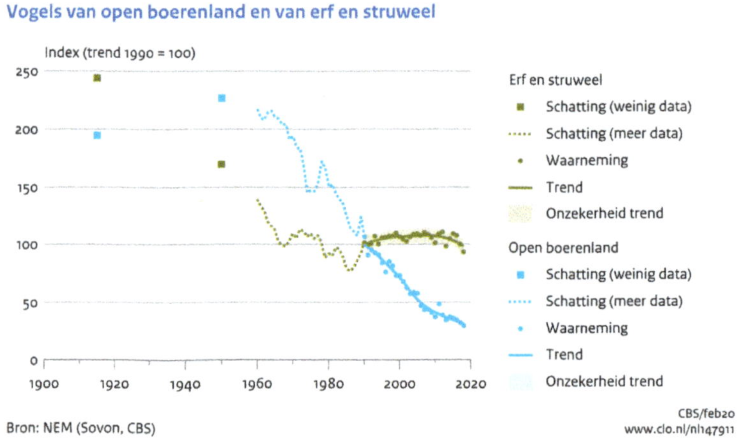

Figure 3. Trends in bird populations open farmland in the Netherlands (in blue) [31].

2.1.2. Agriculture

A few crops, including potatoes, wheat, and sugar beet, and grassland dominate the current agriculture in Groningen [33], and these land uses lead to the dewatering of peatlands, causing carbon emissions constituting approximately 40% of the total Groningen emissions from agriculture [34]. The EU subsidizes agricultural production, and due to future reduction in these subsidies [35], farmer's incomes are increasingly under pressure. Globally, food security concerns rise [36]. Groningen's agriculture is not capable of feeding its own population, and it must therefore import food from around the globe. Food safety poses another serious risk [37], as it may cause illnesses, epidemics, or pandemics. Dewatering in combination with relative sea level rise causes soil subsidence and increasing salinity [38]. This problem has increased through recent droughts [39], creating additional risks for the fresh water-dependent crops. In Groningen, agriculture has undergone a transformation. As part of general developments in the Netherlands [40], land consolidation in the 20th century led to larger parcels (Figure 4), increasing the vulnerability to illnesses and economic change.

Figure 4. Landscape change in the Marne area as a result of land consolidation, from 1925 to 1975.

Agriculture finds itself at an intersection: will it follow existing pathways of increasing efficiency of production methods, using up all the natural resources and emaciating the landscape, or growing crops that enrich the soil, keeping nature healthy?

2.1.3. The Soil-Salinity Complex

The soil in the landscape of the Groningen area is subjected to a toxic mix of increasing salinity [41–43], soil subsidence resulting from dewatering the agricultural lands in conjunction with peat oxidation [44–47] and gas extractions [48–50], and the desiccation of the land [51,52], mutually exaggerating each other's impacts. Dewatering of the land causes soil subsidence, increasing the influence of seepage from the sea, which is even more detrimental due to the land drying out as result of climatic droughts. Increasing differences between the level of the land and the rising sea strengthens salinification, which causes further dewatering, which, in turn, causes additional soil subsidence, subsequently exaggerating salinity. This vicious circle of land degradation leads to ever more compromised conditions for growing food, induces a strictly controlled and managed water system, causes further loss of biodiversity, and instigates an (assumed) need to raise dikes to prevent the land from flooding. The impact of this process on traditional forms of agriculture in the northern Netherlands makes it difficult for these farming types to stay economically viable. Salinity alone is leading to an economic loss of EUR 1166/ha for potato fields [53].

2.1.4. Health-Problems

Resulting from the human-centred way in which we grow food and have organised our settlements, human and ecological health has come under increasing pressure. Urbanization is considered to be one of the most important health challenges of the 21st century [54], being associated with an increase in chronic and non-communicable conditions such as obesity, stress, poor mental health, and a decline in physical activity [55]. When a population becomes more urbanized, green space has a positive influence on mental health, social cohesion, and physical behavior [56,57]. In urban environments with a lack of reasonable amounts of accessible green spaces, health problems tend to increase, as the opportunities to exercise are limited, leading to higher chances of obesity [58–61]. More children living in these precincts suffer from attention disorder at school [62–65] or encounter ADHD and similar illnesses [66–69]. The influence of green space on children's spatial working memory and their cognitive functioning is positive and strongly related to academic achievement in children's performance [62,63]. Psychological problems amongst adults [70–72] cause higher levels of stress and domestic violence [73]. People living in areas without access to nature were 1.27 times more likely to experience symptoms of depression [74], and these areas were found to have higher crime levels compared to other areas [75]. Meanwhile, the health-related outcomes of living close to natural areas [76–83] and being able to undertake physical activity in nature [84–88] include reduced levels of morbidity [89] by reducing cardiovascular disease [90]. Patients with views of trees and greenery out their windows heal faster and need less medication [57], indicating the restorative influence of gardens in many different urban contexts [91–99]. Furthermore, green space provides cleaner air and mitigates heat, hence creating a healthier atmosphere in the city [100–103].

2.2. Research Objective and Question

The objective of this research is to explore whether taking a nature-driven approach to planning and design for climate adaptation brings benefits for liveability, productivity and ecology on the Zernike campus. The research question is: how can a nature-driven campus be designed that offers positive impacts in the face of the threat of future changes caused by climate change?

3. Methodology

The applied methodology is design-led [27–29]. The design results are continuously valued in a research-through-design process [104–110]. By applying 'designerly' explorations [111,112], out-of-the-box thinking and creativity are enhanced. This overarching methodological view is elaborated in detail in specific methods in every stage of the research process (Figure 5). In practice, these methods are intertwined:

1. The analysis of current impacts and threats resulting from climate change is undertaken through a literature review of the most recent academic results in the Netherlands, such as the climate scenario [113], the national delta-program [114], recent insights into accelerated sea level rise [9], and the novel risk of droughts [115].

2. During the second stage of the research, a comparative analysis [116,117] is undertaken into three specific viewpoints, defining the way to guide climate adaptation.

3. After the preferred viewpoint is chosen, the core question of how to design a campus that could be self-reliant is investigated. Through literature review of the concepts of self-reliant areas [118–121] and self-sufficiency [122–124], the programmatic contours for the design are formulated.

4. The program of demand for the growth of food, its spatial implications, and the required amounts and spaces for water are quantitatively analysed [125–127].

5. Once the quantitative consequences of a self-reliant campus become clear, the quest for a holistic design intervention, responding to this long-term view [25], is explored via futuring [128,129] and spatial visioning [130,131]. A design that responds to uncertainties regarding the future and leapfrogs current policy constraint backtracking [132], oriented towards creating a (spatial) tipping point [133], is applied to change path-dependency. Understanding the theory of complexity [19,134] and its processes of self-organisation and emergence [135] in an urban design context is essential. In order to include these concepts, a co-creative working method is chosen, through which collaborative design work can take place in a design charrette [136–139] approach.

6. The final stage of the research takes the integrated spatial view as the starting point for thematic spatial explorations of food-, eco- and waterscapes. The creative process is here mingled with analytical interactions, typical for a research by design method. The spatial propositions are permanently assessed, and new design questions are raised, which in turn lead to adjusted design explorations. In a cyclic process, the designs are tested and modified until satisfied. In the Moeder Zernike project, the thematic aspects are separated from each other using a layered mapping method [140], making it possible to quantitatively and qualitatively investigate the consequences of spatial choices.

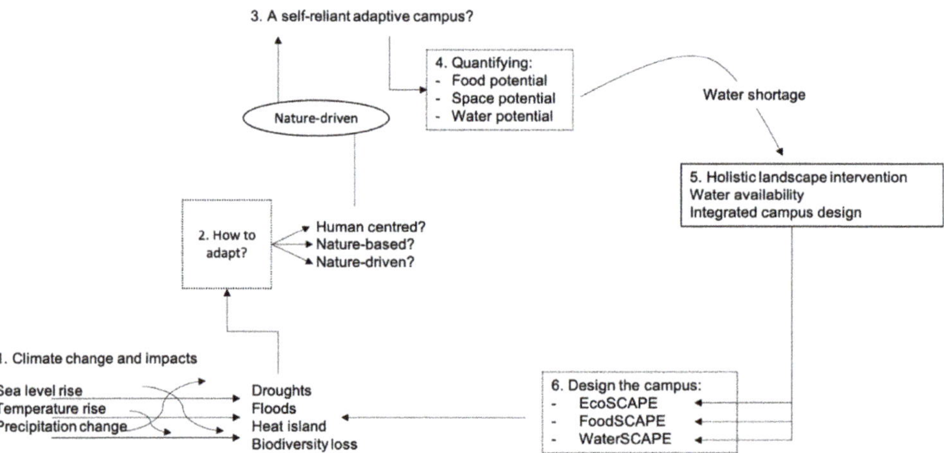

Figure 5. Methodology.

4. Results

4.1. Ways to Respond

The main objective is to take nature as the basis for the urban transformation. In this sense, the city is seen as being part of nature. This could be a bold statement; however, the city of Rotterdam recently presented itself as a wilderness park because its urban wildlife is so abundant [141]. Being part of nature enhances the health not only of the urban wilderness, but also of its human inhabitants when living close to green areas [142]. Developing urban areas need to achieve balance in the exchange of materials, resources, and the potential to allow co-existential living systems, urban and natural, to emerge and evolve by creating regenerative cities [143–146]. However, cities, having extracted natural resources at a large scale ever since the industrial revolution, should become reciprocal [147]. Three gradations of responsiveness are distinguished (Figure 6): searching for human-centred contrast, establishing nature-driven contact, and striving for nature-driven contract [148].

	CONTRAST	CONTACT	CONTRACT
IMAGE OF NATURE	wilderness	accessible nature	ecosystem services
Formal Interaction	city and nature have sharp boundaries, protected areas	city and nature intertwine	city and nature take each others form
commentary	*bring the city to nature 'satellites' and 'garden cities'*	*insert nature into the city 'green wedges' and 'parks'*	*go for a complete mix, 'reweaving the urban tapestry' and 'broadacre city'*
Functional Interaction	city and nature are each others jungle	city and nature come to each others rescue	city and nature take on each others form
commentary	*'places to get lost'*	*regulated leisure in nature*	*produce food on your own gardenlot*
Physical Interaction	city and nature keep their distance	city and nature exchange information	city and nature take on each others construction
commentary	*natural expression of the city 'non human' outside*	*natural expression of the city 'well tempered' environment outside*	*expression of city and agriculture 'new hybrids' in- and outside the city*
VISION OF THE CITY	from 'Cabanes' to 'Metropolis'	'Green-Blue infrastructure' to 'Lobe city'	from 'Subtopia' to 'Metabolic City'

Figure 6. Relationships of nature and the city [148].

When a perspective of the city is taken where urban life is seen in contrast with nature, the wilderness is the opposite of human life. This human-centered view aims to enhance human ability, overcome human limitations, and human preferences and concerns are explicitly considered in the design [149,150]. In such a human-centered view, nature has instrumental value [151], and according to Aristotle, "nature has made all things specifically for the sake of man" (cited in [152]). The human-centered approach considers basic human needs, motivations, and meaningful experiences in relation to green areas [153].

Nature-based solutions bring nature and the city into contact to improve urban sustainability [154]. 'Nature-based solutions aim to help societies address a variety of environmental, social and economic challenges in sustainable ways. Nature-based solutions use the features and complex system processes of nature in order to achieve desired outcomes that ideally are resilient to change' [155]. They are 'solutions that are inspired and supported by nature, which are cost-effective, simultaneously provide environmental, social and economic benefits and help build resilience. Such solutions bring more, and more diverse, nature and natural processes into cities, landscapes and seascapes, through locally adapted, resource-efficient and systemic interventions' [155]. Hence, nature-based solutions are seen as deliberate interventions seeking to use the properties of nature to address societal challenges.

In contrast with the human-centered view on nature, a biocentric perspective considers humans as members of an interconnected 'web of life' [156,157]; nature and the city have a moral 'contract', with people seen as an integral part of nature, rather than its master or steward [151]. Current practice often starts with an economically driven program, after which green spaces are fitted in. A nature-driven approach [142,158,159] takes the ecological system as the foundation for the design within which other (urban) functions are embedded. The resilience and self-organising power of ecosystems safeguard their own existence, in which humans but also non-human organisms survive in the context of uncertain climatic conditions.

Choosing for a nature-driven future places humans within the wider natural system and allows the system to be guided by its own resilience and synergies. It opens the pathway towards an autonomously operating symbiosis of nature and man, but it also implies that all resources should and need to be supplied, used, and treated within those same boundaries.

4.2. Self-Reliance

A self-reliant area, in this case the Zernike campus, should therefore be able to function without substantial support from outside the system, e.g., the city, neighbourhood, or broader area. This means that all that is generated and wasted should stay and be used within the system boundaries. On top of this, the area, as part of its natural environment, needs to stay within its own 'planetary boundaries' [10], downscaled from the global level to the area of observation. The goal to design a nature-driven campus implies that natural systems determine the productivity on campus. In addition to preventing decreasing biodiversity [29], a nature-driven campus grows its own food, provides a cooling environment, generates sufficient renewable energy, and the water system, with its in- and outgoing flows, is bound to the area's boundaries. The key question is whether a nature-driven, self-reliant campus, in the context of climate change, can grow sufficient healthy food for its consumers (students, staff, visitors and residents) and if enough space and water is available. Moeder Zernike transforms from a parasite, fetching its resource lifelines from outside, to an amoeba (Figure 7), self-sufficiently generating resources from within.

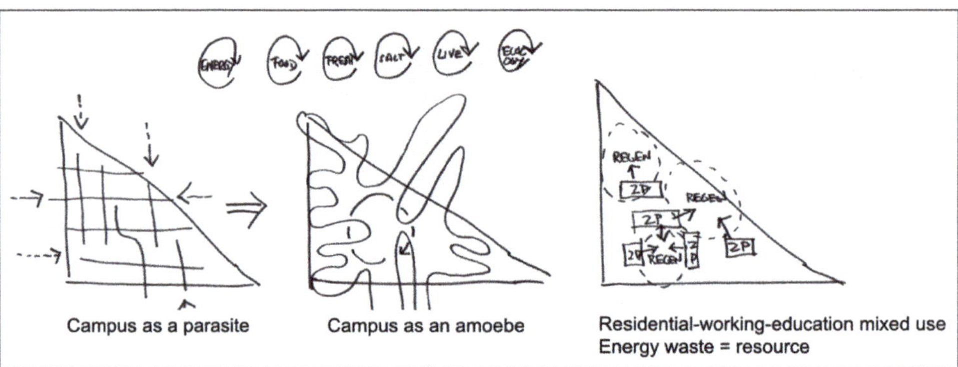

Figure 7. The Zernike amoeba, regenerating its material flows.

4.3. Quantifying Potentials

4.3.1. Food

Firstly, the amount of food required to provide the new diet [160], based on a per person estimation of food categories (Figure 8) applied to all consumers on campus (30,000 students and staff and an additional 10,000 of new residents), is calculated. These amounts are combined with the number of meals the different groups, students, staff, and residents, consume on campus, taking into account holiday periods. The total amount of food needed is almost 5.5 million kg per year.

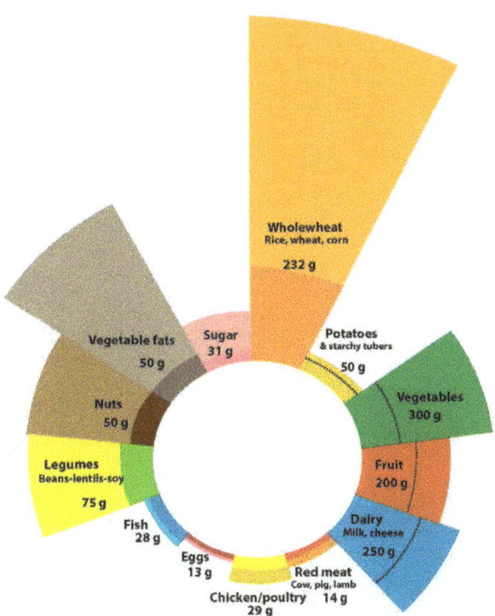

Figure 8. Amounts of food for a healthy diet, recalculated for the Dutch context. Based on [160].

4.3.2. Space

The area needed for growing all crops (Figure 9) adds up to an area of approximately 0.1 km^2 (or 100,000 m^2). Assuming the largest portion of these crops will grow inside, on rooftops, or clinging to the facades of campus buildings, using novel multiple harvest technologies, such as aqua- and aeroponics, the useable spaces of current buildings is calculated (Figure 10). Potentially 140,000 m^2 of rooftop area and 90,000 m^2 indoors is available. Additionally, inside existing buildings, almost 10,000 homes can be realised. The total area for growing food on Zernike is 230,000 m^2, more than twice the required space.

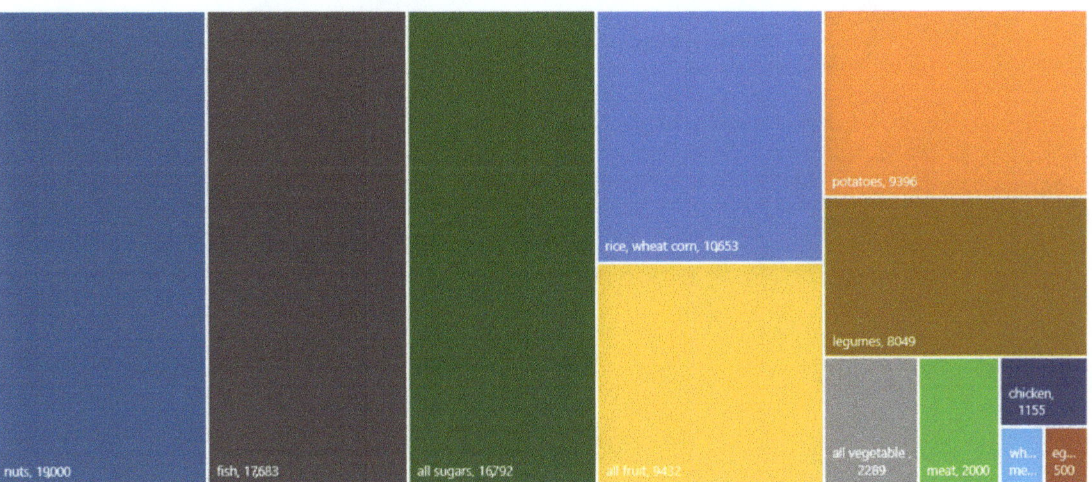

Figure 9. Area needed to produce food in m^2.

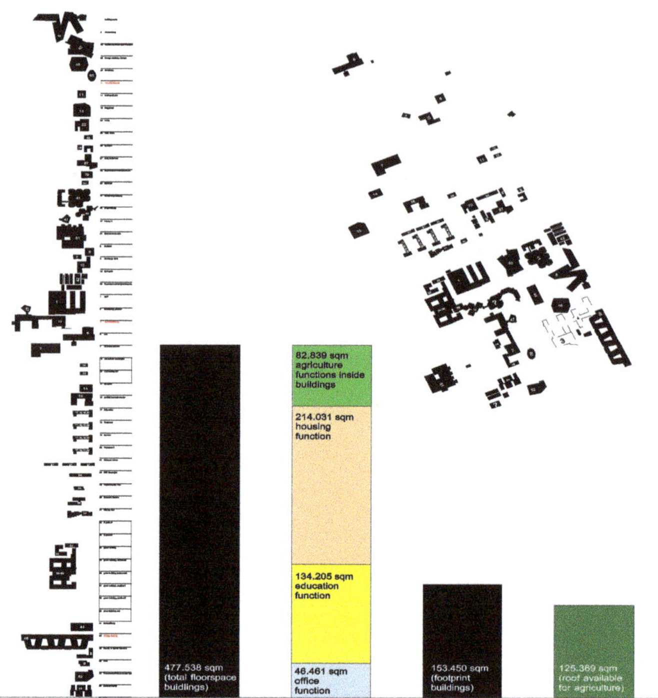

Figure 10. Area available for food production in and on buildings.

4.3.3. Water

The total amount of water needed to grow all crops is almost 3.5 billion litres/year, equalling nearly 1400 Olympic swimming pools. On top of this, nearly 400 Olympic swimming pools (approx. 1 billion litres) of drinking water are required for daily use, totalling 1800 pools/year. Analysis of the expected amounts of precipitation on Zernike (Table 1) shows that, according the driest climate scenario, a little more than 2500 Olympic pools are available. Around 50% of this water evaporates [161], hence only 1250 will remain for usage. This implies that on a yearly basis, there is a shortage of 550 pools (1.4 billion litres) of water. Especially in the context of increasing droughts in the Netherlands [115], current rainfall will not suffice to grow all the crops on Zernike, and water from outside the campus is needed. In order to avoid extracting water from other users in neighbourhoods across the city, only one option remains: retrieve water from the sea.

Table 1. Calculation of rainwater amounts for the Zernike campus.

Month	Current Climate (1981–2010)		2085 Wl Scenario Based on 2019 Weather (+3.5 Degrees/No Changes in Currents)		2085 Wl Scenario Based on 2019 Weather (+3.5 Degrees/No Changes in Currents)		2085 Wh Scenario Based on Climate Data (+3.5 Degrees, +Changes in Currents)		2085 Wh Scenario Based on 2019 Weather (+3.5 Degrees, +Changes in Currents)	
	Rain (In Olympic Pools)	Cumulative								
Jan	710	710	580	807.2	660.3	660,3	937	937	768	768
Feb	400	1110	210	1269.7	247.8	908.1	550	1487	303	1071
Mar	390	1500	880	1732.8	1016.8	1924.9	571	2058	1208	2279
Apr	−130	1370	−280	1635.5	−269.8	1655.1	−134	1924	−302	1977
May	−240	1130	−560	1436	−567.5	1087.6	−251.8	1672.2	−610.2	13,668
Jun	−110	1020	−200	1389.6	−149.9	937.7	−105.4	1566.8	−206.2	1160.6
Jul	−140	880	−750	1123.6	−845.5	92.2	−468.8	1098	−938.5	222.1
Aug	10	890	−140	1022.5	−243.6	−151.4	−292.5	805.5	−408	−185.9
Sep	340	1230	1020	1275.5	899	747.6	71.8	877.3	595.4	409.5
Oct	550	1780	820	1862.6	874.65	1622.25	621.4	1498.7	923.8	1333.3
Nov	740	2520	550	2651.25	586.3	2208.55	831	2329.7	618.2	1951.5
Dec	760	3280	520	3460.9	554.05	2762.6	852.2	3181.9	583.4	2534.9
		3280		3460.9		2762.6		3181.9		2534.9

4.4. Holistic Intervention

The crucial factor in the design for Moeder Zernike therefore is the supply of enough water to feed agriculture on campus. Out of the co-creative design process and analyses, one impactful proposition emerged, to establish a lifeline between campus and the infinite water source of the Wadden Sea. This is a large-scale long-term intervention benefitting a multitude of aspects: ecology, safety, food, and water. By establishing this tipping point, Moeder Zernike is suddenly placed in a new ecological context, where fresh water and saline influences collide. The saline influence also induces a spatial novelty in the form of an inversed wierde (a cultural relic found everywhere in the northern landscape), creating a freshwater reservoir (Figure 11).

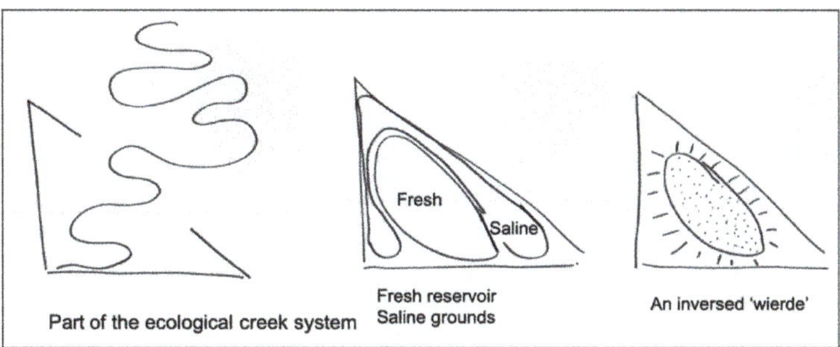

Figure 11. Moeder Zernike embraces fresh and saltwater, creating an inversed wierde for freshwater storage.

As saltwater flows around the freshwater reservoir (Figure 12), it brings nutrients and sediment, leaving behind fertile soils, enriching agricultural potential and providing the dynamic environment for a steep increase in biodiversity.

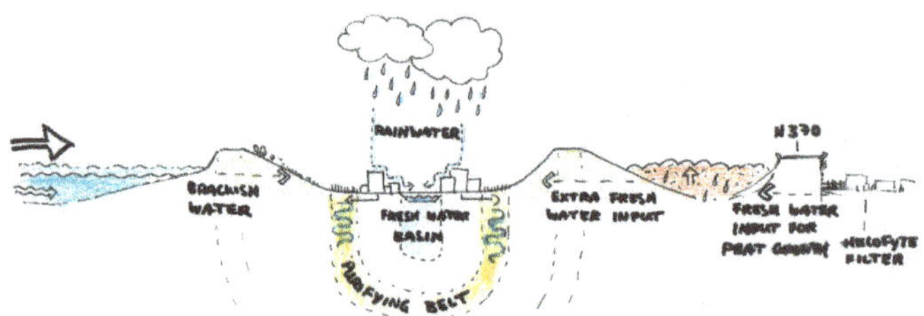

Figure 12. Cross-section, showing the different environments and water features.

The inversed wierde protects Moeder Zernike against outside influences, while a sandy membrane simultaneously filters saline water for use on campus. This provides the urgently needed water for growing food. Spatially, a coherent inner world emerges, in which the experimental life within Moeder Zernike takes place, while outwardly, the campus presents itself as a spatial entity within the surrounding landscape (Figure 13).

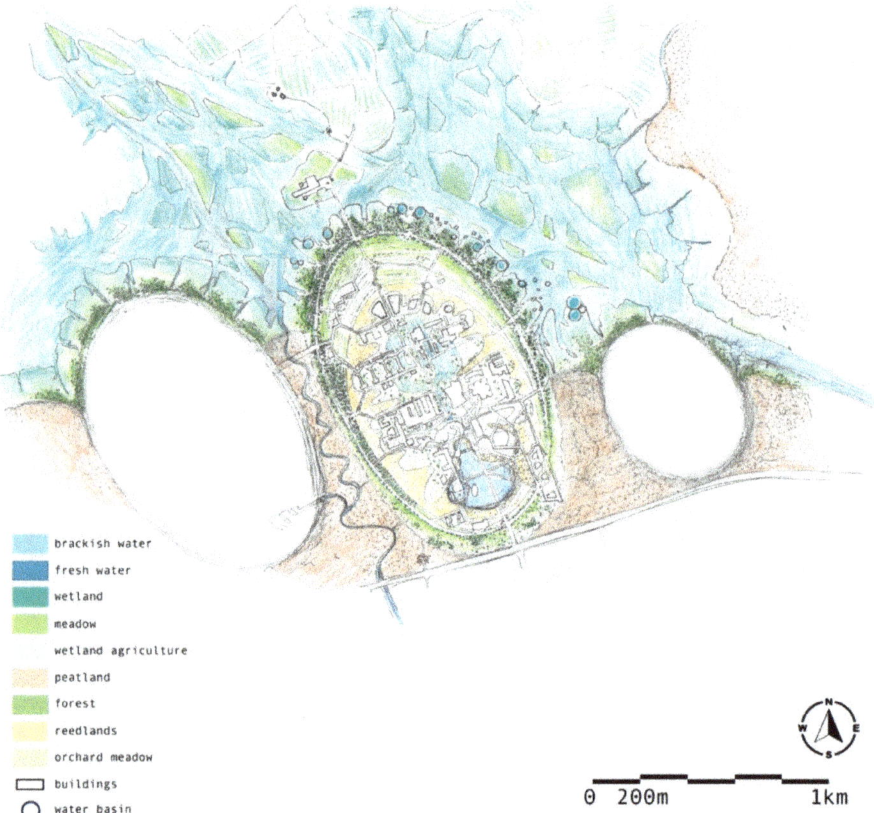

Figure 13. Moeder Zernike as a freshwater reservoir in a saline landscape.

4.5. Design of Scapes

The fundamental choice of bringing water from the sea inland to provide the conditions to become self-reliant impacts ecology, food-, and waterscapes.

4.5.1. Foodscape

A rich diversity of growing conditions emerges (Figure 14), producing everything for the consumption of the new diet. Apart from the food production in existing buildings, the campus will contain fishing grounds in open water with salmon, eel, and sturgeon, have saline aquafarms at the campus edges for prawns and lobster, free-ranging cattle wandering the slopes of the wierde, while inside orchards nuts will be grown. In the terraced landscape, water is used multiple times, trickling down through fruit and berry plantations and rice paddies. Freshwater fish, carp and tilapia, live in the water reservoir, whilst the southern mound is home to caves for chicory and fungi, mushrooms and insects. On top of these, a publicly accessible picking garden is foreseen so urban residents can freely gather their lunch and dinner ingredients.

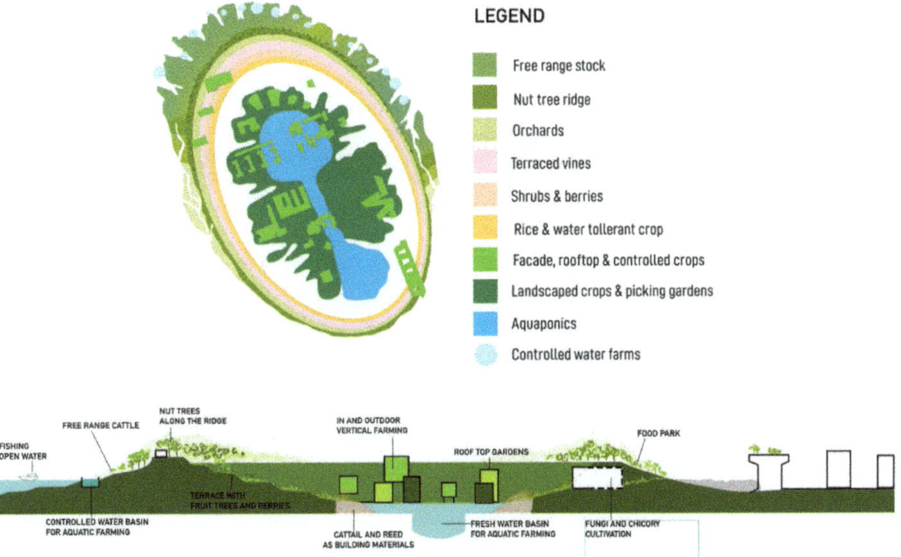

Figure 14. Foodscape of Zernike.

4.5.2. Waterscape

By connecting the campus with the sea, saline, brackish, and fresh water are all part of the waterscape (Figure 15). The protective sand edge purifies the saline water before it enters campus, at the same time protecting Moeder Zernike against the spring tide. All wastewater from buildings is filtered and cleaned in a helophyte system, making it usable for growing crops. During periods of heavy rainfall, the central lake fills up, is home to fish, provides the water for indoor crop growth, and cools the environment on hot days. The pond also functions as a heat exchanger, generating energy. Slowly growing peat is supplied with pre-purified household water from urban neighborhoods.

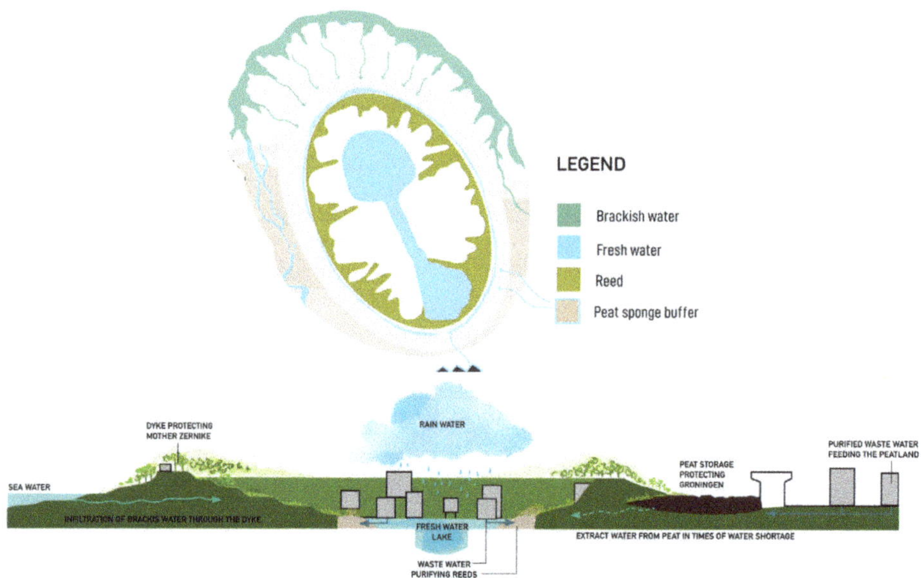

Figure 15. Waterscape of Zernike.

4.5.3. Ecoscape

Ecological qualities are enhanced, bringing brackish and saline waters together with freshwater conditions, kindling new gradients. A diverse range of ecotypes emerge (Figure 16), such as new islands and wetlands, occasionally flooding, while other parts permanently rise above the water level. The reed-lands of the helophytes are home to insects, small fish, and reptiles, in turn attracting all kinds of birds. An abundant range of species inhabit the inner side of the ridge. Built structures will offer a unique rock-biotope for specific plants, butterflies, bees, and nesting places for birds and bats. Finally, peatlands offer a habitat to water birds, insects, and a range of reed plants.

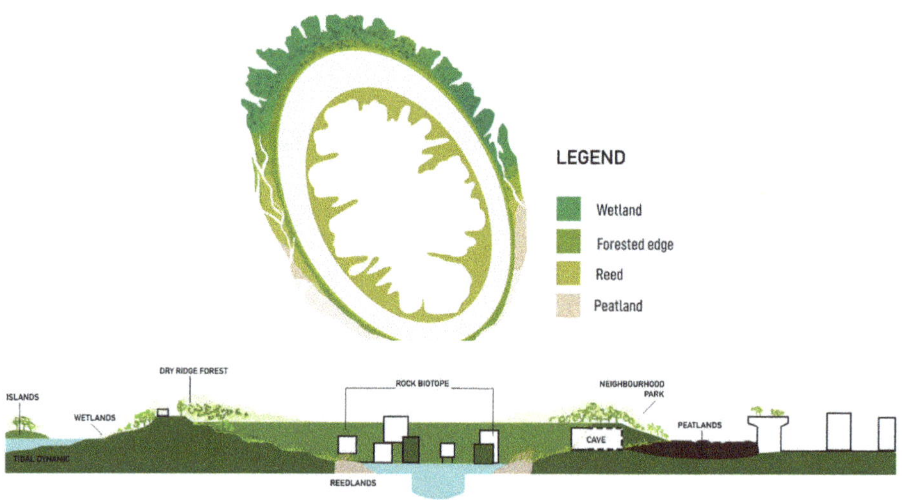

Figure 16. Ecoscape of Moeder Zernike.

5. Discussion

The research presented in this article illuminates tensions between three forms of future planning. First, current short-term practices often plan for the known knowns, and decide on that basis, even if there is a chance of making bad decisions. The second way of acting, dubbed adaptive management, delays decisions until more knowledge or data are available. The third way proposes developing a long-term view then working towards that view with the understanding that the future is full of unknown unknowns. In this case, it is well-understood that the unknown unknowns are difficult to comprehend, especially regarding long-term or deep uncertainties, and imagining a far future could then guide the way forward.

In this article, a pledge is made for using imagination to plan for a desired future, as opposed to potentially making the wrong decision or postponing decisions until we know more. Therefore, the Zernike plan takes on the largest quests of our time—biodiversity, food supply and climate impacts—and unites them around an imagined long-term future, on the basis of which implications for current decisions can be derived. This not only gives direction—it also brings coherence and inspiration, and offers comprehensive spatial thinking for the campus grounds.

The question, however, is how policy making and political decision making may be diverted from the current practice of responsiveness, short-term orientation or the ever-apparent quest for more knowledge, data, and research. Indeed, more information is not always needed; instead, larger insights must emerge so that decision making can be based on wisdom rather than rationality.

Moreover, the process of creating an inspirational design is, in a way, magical, and can be used more profoundly. The magic happens when out of a set of problems and questions, at a certain moment, a vision comes forward, resulting from irreducible co-creative ways of drawing, building, talking, and exchanging ideas. This approach, in which interaction delivers tangible results, is often underestimated in planning processes. In practice, seemingly estimated policy boundaries limit approaches that explore the unexplored. Often, these are seen to undermine the current culture, put the intangible hidden agreements out in the open, or overhaul just adopted plans and policies. The fear of discovering something new, which is essential for an unknown future, is paralyzing the involved bureaucrats, planners, and policymakers. This rusty cultural constraint prohibits free and novel ideas from emerging. This could be dangerous, as continuing on a familiar pathway will hardly ever offer solutions for the unknowns. Three typical options remain:

1. Breaking the barriers in a way that is acceptable, by means of inspiration, future thinking, and offering a pleasant and plausible future that differs from the known world.
2. Experimenting on and developing small novelties in a controlled context that guide the way to what could be possible in the future.
3. Waiting for something to go terribly wrong, causing a tipping point for changing course. A disaster could overcome the fear of change, as it becomes clear to everyone existing approaches have caused the devastation.

Naturally, it is preferable to anticipate such disastrous events before they happen and be quick enough to change course before the worst occurs.

6. Conclusions

The Moeder Zernike research has illustrated how a nature-driven approach can be applied for designing a future oriented plan. By doing so, the plan is able to incorporate nature as the driving force and allows the emergence of a self-reliant and biodiverse urban precinct.

The distinction between human-centred, nature-based, and nature-driven may sound artificial, but it brings crucial questions to the debate. Do we, as humans, design for our own sake, to get better lives for ourselves, as the human species? Or do we offer nature-based solutions, that adjust urban environments so there is also place and space for natural processes? Does this, looking through human spectacles at nature, suffice for our survival in increasing constraining futures? Or do we need to start with nature and let nature drive the human behaviour of urban dwellers? This poses the question as to where humans fit in the natural environment surrounding them.

The plan for Moeder Zernike has shown that a plan that is driven by nature starts with understanding the landscape and its ecological, systemic features and characteristics. Starting the design process by firstly looking at the landscape guarantees a development of cities and urban contexts that are embedded in and embraced by nature. This offers humans the best proposition for sustainably surviving on the planet.

Author Contributions: Conceptualization, methodology, investigation, writing—original draft preparation, writing—review and editing, R.R. The author has read and agreed to the published version of the manuscript.

Funding: This research received no external funding.

Institutional Review Board Statement: Not applicable.

Informed Consent Statement: Not applicable.

Data Availability Statement: Not applicable.

Conflicts of Interest: The author declares no conflict of interest. This work has been presented on Greening cities shaping cities symposium in October 2020, https://www.greeningcities-shapingcities.polimi.it/ (accessed on 1 October 2020).

References

1. Pachauri, R.K.; Allen, M.R.; Barros, V.R.; Broome, J.; Cramer, W.; Christ, R.; Church, J.A.; Clarke, L.; Dahe, Q.; Dasgupta, P.; et al. *Climate Change 2014: Synthesis Report. Contribution of Working Groups I, II and III to the Fifth Assessment Report of the Intergovernmental Panel on Climate Change*; IPCC: Geneva, Switzerland, 2014; p. 151.
2. Crutzen, P.J. Geology of mankind. *Nat. Cell Biol.* **2002**, *415*, 23. [CrossRef] [PubMed]
3. *IPCC Special Report on Climate Change, Desertification, Land Degradation, Sustainable Land Management, Food Security, and Greenhouse Gas Fluxes in Terrestrial Ecosystems*; Summary for Policymakers; Approved Draft. IPCC: Geneva, Switzerland, 2019. Available online: https://www.ipcc.ch/report/srccl/ (accessed on 10 August 2020).
4. Díaz, S.; Settele, J.; Brondizio, E.; Ngo, H.T.; Guèze, M.; Agard, J.; Arneth, A.; Balvanera, P.; Brauman, K.A.; Butchart, S.; et al. *Summary for Policymakers of the Global Assessment Report on Biodiversity and Ecosystem Services of the Intergovernmental Science-Policy Platform on Biodiversity and Ecosystem Services*; IPBES-Secretariat: Bonn, Switzerland, 2019.
5. Roberts, W.O. Climate Change and the Quality of Life for the Earth's New Millions. *Proc. Am. Philos. Soc.* **1976**, *120*, 230–232. Available online: https://www.jstor.org/stable/986562 (accessed on 21 December 2020).
6. Mupedziswa, R.; Kubanga, K.P. Climate change, urban settlements and quality of life: The case of the Southern African Development Community region. *Dev. S. Afr.* **2016**, *34*, 196–209. [CrossRef]
7. Liang, L.; Deng, X.; Wang, P.; Wang, Z.; Wang, L. Assessment of the impact of climate change on cities livability in China. *Sci. Total Environ.* **2020**, *726*, 138339. [CrossRef]
8. Lindsey, R. Climate Change: Global Sea Level. Climate.gov, 2018. Available online: https://www.climate.gov/news-features/understanding-climate/climate-change-global-sea-level (accessed on 27 July 2020).
9. Haasnoot, M.; Bouwer, L.; Diermanse, F.; Kwadijk, J.; Van der Spek, A.; Oude Essink, G.; Delsman, J.; Weiler, O.; Mens, M.; Ter Maat, J.; et al. *Mogelijke Gevolgen van Versnelde Zeespiegelstijging Voor het Deltaprogramma. Een Verkenning*; Deltares: Delft, The Netherlands, 2018.
10. Rockström, J.; Steffen, W.; Noone, K.; Persson, Å.; Chapin, F.S.; Lambin, E.F.; Lenton, T.M.; Scheffer, M.; Folke, C.; Schellnhuber, H.J.; et al. A safe operating space for humanity. *Nature* **2009**, *461*, 472–475. [CrossRef] [PubMed]
11. Schipper, E.L.F.; Burton, I. (Eds.) *Adaptation to Climate Change*; Earthscan: London, UK, 2009.
12. Global Commission on Adaptation. *Adapt Now: A Global Call for Leadership on Climate Resilience*; Global Center on Adaptation: Gronningen/Rotterdam, The Netherlands; Washington, DC, USA; World Resources Institute: Washington, DC, USA, 2019.
13. Schubert, D. Cities and plans—The past defines the future. *Plan. Perspect.* **2019**, *34*, 3–23. [CrossRef]
14. Molotch, H.; Freudenburg, W.; Paulsen, K.E. History Repeats Itself, But How? City Character, Urban Tradition, and the Accomplishment of Place. *Am. Sociol. Rev.* **2000**, *65*, 791–823. [CrossRef]
15. Lindblom, C.E. The Science of "Muddling Through". *Public Adm. Rev.* **1959**, *19*, 79–88. [CrossRef]
16. Roggema, R.; Genel Altinkaya, Ö.; Psarra, I. *De Toukomst is al Lang Begonnen*; Hanze University of Applied Sciences Groningen: Groningen, The Netherlands, 2020.
17. Roggema, R. Bypassing the Obvious: Implementing Cutting Edge Ideas for Futuring Urban Landscapes. *Urban Reg. Plan.* **2021**, *6*, 1–14. [CrossRef]

18. Horstmann, B. *Framing Adaptation to Climate Change: A Challenge for Building Institutions*; Deutsches Institut für Entwicklungspolitik: Bonn, Germany, 2008.
19. Roggema, R. Swarm Planning: The Development of a Methodology to Deal with Climate Adaptation. Ph.D. Thesis, Delft University of Technology and Wageningen University and Research Centre, Delft, The Netherlands, 2012.
20. United Nations Environment Programme. *Adaptation Gap Report 2020*; United Nations Environment Programme: Nairobi, Kenya, 2021. Available online: https://www.unep.org/adaptation-gap-report-2020 (accessed on 21 December 2020).
21. Rittel, H.W.J.; Webber, M.M. Dilemmas in a General Theory of Planning. *Policy Sci.* **1973**, *4*, 155–169. [CrossRef]
22. Davy, B. Plan It Without a Condom! *Plan. Theory* **2008**, *7*, 301–317. [CrossRef]
23. Merry, U. *Coping with Uncertainty. Insights from the New Sciences of Chaos, Self-Organisation and Complexity*; Praeger Publishers: Westport, CT, USA, 1995.
24. Schwarz, P. *The Art of the Long View. Planning for the Future in an Uncertain World*; Bantam Doubleday Dell Publishing Group Inc.: New York, NY, USA, 1991.
25. Roggema, R. *Adaptation to Climate Change: A Spatial Challenge*; Springer International Publishing: Berlin/Heidelberg, Germany, 2009.
26. Wrigley, C. Principles and practices of a design-led approach to innovation. *Int. J. Des. Creativity Innov.* **2017**, *5*, 235–255. [CrossRef]
27. Balz, V.E. Regional design: Discretionary approaches to regional planning in The Netherlands. *Plan. Theory* **2017**, *17*, 332–354. [CrossRef]
28. Yan, W.; Roggema, R. Developing a Design-Led Approach for the Food-Energy-Water Nexus in Cities. *Urban Plan.* **2019**, *4*, 123–138. [CrossRef]
29. Almond, R.E.A.; Grooten, M.; Petersen, T. *Living Planet Report 2020—Bending the Curve of Biodiversity Loss*; WWF: Gland, Switzerland, 2020.
30. Provincie Groningen. De Toestand van Natuur en Landschap in de Provincie Groningen Achtergronddocuent. 2017. Available online: https://destaatvangroningen.nl/uploads/toestand-van-natuur-en-landschap-achtergronddocument.pdf (accessed on 15 June 2020).
31. CBS; PBL; RIVM; WUR. Boerenlandvogels, 1915–2018 (Indicator 1479, Versie 11, 5 Februari 2020). Centraal Bureau voor de Statistiek (CBS), Den Haag; PBL Planbureau voor de Leefomgeving, Den Haag; RIVM Rijksinstituut voor Volksgezondheid en Milieu, Bilthoven; en Wageningen University and Research, Wageningen. 2020. Available online: www.clo.nl (accessed on 21 December 2020).
32. Teunissen, W.; Soldaat, L. *Indexen en Trends van een Aantal Weidevogelsoorten uit het Weidevogelmeetnet. Periode 1990–2004*; SOVON-Informatie 2005/13; SOVON Vogelonderzoek Nederland: Nijmegen, The Netherlands, 2005.
33. Available online: www.statline.cbs.nl (accessed on 21 December 2020).
34. Van Well, E.; Rougoor, C. *Landbouw en Klimaatverandering in Groningen*; CLM: Culemborg, The Netherlands, 2016.
35. Ministerie van LNV. *Europese Landbouw-Beleid. Toelichting op de Betalingen in het Kader van het Gemeenschappelijk Landbouwbeleid in het Boekjaar 2019*; Ministerie van LNV: Den Haag, The Netherlands, 2019.
36. Rosegrant, M.W.; Cline, S.A. Global food security: Challenges and policies. *Science* **2003**, *302*, 1917–1919. [CrossRef] [PubMed]
37. World Health Organization. *Food Safety Risk Analysis: A Guide for National Food Safety Authorities*; WHO; FAO: Rome, Italy, 2006.
38. Van Staveren, G.; Velstra, J. *Verzilting van Landbouwgronden in Noord-Nederland in het Perspectief van de Effecten van klimaatverandering*; Onderzoeksprogramma Klimaat voor Ruimte/Climate Changes Spatial Planning: Den Haag, The Netherlands, 2012.
39. De Vries, S. Increasing Drought: A New Enemy for The Netherlands. 20 August 2020. Available online: https://www.the-low-countries.com/article/increasing-drought-a-new-enemy-for-the-netherlands (accessed on 21 December 2020).
40. Van der Woud, A. *Het Landschap de Mensen. Nederland 1850–1940*; Prometeus: Amsterdam, The Netherlands, 2020.
41. Velstra, J.; Groen, J.; De Jong, K. Observations of salinity patterns in shallow groundwater and drainage water from agricultural land in the northern part of The Netherlands. *Irrig. Drain.* **2011**, *60*, 51–58. [CrossRef]
42. Kroes, J.; Supit, I. Impact analysis of drought, water excess and salinity on grass production in The Netherlands using historical and future climate data. *Agric. Ecosyst. Environ.* **2011**, *144*, 370–381. [CrossRef]
43. Raats, P.A. Salinity management in the coastal region of the Netherlands: A historical perspective. *Agric. Water Manag.* **2015**, *157*, 12–30. [CrossRef]
44. Heuff, F.; Mulder, G.; Van Leijen, F.; Samiei Esfahany, S.; Hanssen, R. Dynamics of peat soils in the 'Green Heart' of the Netherlands measured by satellite radar interferometry. In Proceedings of the 20th EGU General Assembly, Vienna, Austria, 4–13 April 2018; p. 17443.
45. Stuyfzand, P.J. The impact of land reclamation on groundwater quality and future drinking water supply in the Netherlands. *Water Sci. Technol.* **1995**, *31*, 47–57. [CrossRef]
46. De Waal, J.; Roest, J.; Fokker, P.; Kroon, I.; Breunese, J.; Muntendam-Bos, A.; Oost, A.; Van Wirdum, G. The effective subsidence capacity concept: How to assure that subsidence in the Wadden Sea remains within defined limits? *Neth. J. Geosci.* **2012**, *91*, 385–399. [CrossRef]
47. Erkens, G.; Van der Meulen, M.; Middelkoop, H. Historical land use caused carbon release in the Dutch coastal peatlands. In Proceedings of the EGU General Assembly 2010, Vienna, Austria, 2–7 May 2010; p. 12386.

48. Schoonbeek, J.B. Land subsidence as a result of natural gas extraction in the province of Groningen. In Proceedings of the SPE European Spring Meeting, Amsterdam, The Netherlands, 8–9 April 1976.

49. Van Thienen-Visser, K.; Fokker, P.A. The future of subsidence modelling: Compaction and subsidence due to gas depletion of the Groningen gas field in the Netherlands. *Neth. J. Geosci.* **2017**, *96*, s105–s116. [CrossRef]

50. Pottgens, J.J.; Brouwer, F.J. Land subsidence due to gas extraction in the northern part of the Netherlands. *Land Subsid.* **1991**, *200*, 99–108.

51. Suykerbuyk, W.; Govers, L.L.; Van Oven, W.; Giesen, K.; Giesen, W.B.; De Jong, D.J.; Bouma, T.J.; Van Katwijk, M.M. Living in the intertidal: Desiccation and shading reduce seagrass growth, but high salinity or population of origin have no additional effect. *PeerJ* **2018**, *6*, e5234. [CrossRef]

52. Boogerd, A.; Groenewegen, P.; Hisschemöller, M. Knowledge utilization in water management in The Netherlands related to desiccation. *JAWRA J. Am. Water Resour. Assoc.* **1997**, *33*, 731–740. [CrossRef]

53. Tzemi, D.; Ruto, E.; Gould, I.; Bosworth, G. Economic Impacts of Salinity Induced Soil Degradation. In *Baseline Study Final Report*; University of Lincoln: Lincoln, UK, 2020. Available online: https://northsearegion.eu/media/14789/chap2-economic-analysis-of-salinization.pdf (accessed on 21 December 2020).

54. World Health Organisation. Urban Health. 2015. Available online: www.who.int/topics/urban_health/en/ (accessed on 11 November 2016).

55. Dye, C. Health and Urban Living. *Science* **2008**, *319*, 766–769. [CrossRef]

56. Cox, D.T.C.; Shanahan, D.F.; Hudson, H.L.; Plummer, K.E.; Siriwardena, G.M.; Fuller, R.A.; Anderson, K.; Hancock, S.; Gaston, K.J. Doses of Neighborhood Nature: The Benefits for Mental Health of Living with Nature. *BioScience* **2017**, *67*, 147–155. [CrossRef]

57. Cox, D.T.; Shanahan, D.F.; Hudson, H.L.; Fuller, R.A.; Gaston, K.J. The impact of urbanisation on nature dose and the implications for human health. *Landsc. Urban Plan.* **2018**, *179*, 72–80. [CrossRef]

58. Epstein, L.H.; Wing, R.R.; Penner, B.C.; Kress, M.J. Effect of diet and controlled exercise on weight loss in obese children. *J. Pediatr.* **1985**, *107*, 358–361. [CrossRef]

59. Epstein, L.H.; Paluch, R.A.; Consalvi, A.; Riordan, K.; Scholl, T. Effects of manipulating sedentary behavior on physical activity and food intake. *J. Pediatr.* **2002**, *140*, 334–339. [CrossRef] [PubMed]

60. Epstein, L.; Raja, S.; Gold, S.; Paluch, R.; Pak, Y.; Roemmich, J. Reducing sedentary behavior the relationship between park area and the physical activity of youth. *Psychol. Sci.* **2006**, *17*, 654–659. [CrossRef] [PubMed]

61. Epstein, L.H.; Paluch, R.A.; Roemmich, J.N.; Beecher, M.D. Family-based obesity treatment, then and now: Twen-ty-five years of pediatric obesity treatment. *Health Psychol.* **2007**, *26*, 381. [CrossRef] [PubMed]

62. Dovey, R. Can Parks Make Kids Better at Math? Next City, 11 September 2018. Available online: https://nextcity.org/daily/entry/can-parks-make-kids-better-at-math (accessed on 21 December 2020).

63. Flouri, E.; Papachristou, E.; Midouhas, E. The role of neighbourhood greenspace in children's spatial working memory. *Br. J. Educ. Psychol.* **2018**, *89*, 359–373. [CrossRef]

64. Louv, R. *Vitamin, N. The Essential Guide to a Nature-Rich Life: 500 Ways to Enrich Your Family's Health & Happiness*; Atlantic Books: London, UK, 2016.

65. Shore, R. Kids Need Access to Nature for Mental Health. Vancouver Sun. 17 April 2017. Available online: https://vancouversun.com/health/local-health/kids-need-access-to-nature-for-mental-health (accessed on 21 December 2020).

66. Li, D.; Sullivan, W.C. Impact of views to school landscapes on recovery from stress and mental fatigue. *Landsc. Urban Plan.* **2016**, *148*, 149–158. [CrossRef]

67. Sullivan, W.; Chang, C. Mental Health and the Built Environment. In *Making Healthy Places*; Dannenberg, A., Frumkin, H., Jackson, R., Eds.; Island Press/Center for Resource Economics: Washington, DC, USA, 2011; pp. 106–116.

68. Taylor, A.F.; Kuo, F.E.; Sullivan, W.C. Views of nature and self-discipline: evidence from inner city children. *J. Environ. Psychol.* **2002**, *22*, 49–63. [CrossRef]

69. Van Dijk-Wesselius, J.; Maas, J.; Hovinga, D.; Van Vugt, M.; Van den Berg, A.E. The impact of greening schoolyards on the appreciation, and physical, cognitive and social-emotional well-being of schoolchildren: A prospective intervention study. *Landsc. Urban Plan.* **2018**, *180*, 15–26. [CrossRef]

70. Husqvarna Group. *Global Green Space Report 2013. Exploring Our Relationship to Forests, Parks and Gardens around the Globe*; Husqvarna AB: Stockholm, Sweden, 2013.

71. Mennis, J.; Mason, M.; Ambrus, A. Urban greenspace is associated with reduced psychological stress among adolescents: A Geographic Ecological Momentary Assessment (GEMA) analysis of activity space. *Landsc. Urban Plan.* **2018**, *174*, 1–9. [CrossRef] [PubMed]

72. Thompson, C.W.; Roe, J.; Aspinall, P.; Mitchell, R.; Clow, A.; Miller, D. More green space is linked to less stress in deprived communities: Evidence from salivary cortisol patterns. *Landsc. Urban Plan.* **2012**, *105*, 221–229. [CrossRef]

73. Bureau of Crime Statistics and Research. Mapping Rates of Apprehended Domestic Violence Orders (ADVOs). NSW Government. Available online: www.bocsar.nsw.gov.au/Pages/bocsar_news/ADVO-Rates.aspx (accessed on 21 December 2020).

74. Min, K.-B.; Kim, H.-J.; Kim, H.-J.; Min, J.-Y. Parks and green areas and the risk for depression and suicidal indicators. *Int. J. Public Health* **2017**, *62*, 647–656. [CrossRef] [PubMed]

75. Wright, T. *Western Sydney, Where Pollies Would have you Think Crime Control is at Sea*; Sydney Morning Herald: Sydney, Australia, 2013.

76. Jackson, L.E. The relationship of urban design to human health and condition. *Landsc. Urban Plan.* **2003**, *64*, 191–200. [CrossRef]
77. Gidlöf-Gunnarsson, A.; Öhrström, E. Noise and well-being in urban residential environments: The potential role of perceived availability to nearby green areas. *Landsc. Urban Plan.* **2007**, *83*, 115–126. [CrossRef]
78. Kaplan, S. The restorative effects of nature—Toward an integrative framework. *J. Environ. Psychol.* **1995**, *15*, 169–182. [CrossRef]
79. Kaplan, R.; Kaplan, S. *The Experience of Nature: A Psychological Perspective*; Cambridge University Press: Cambridge, UK, 1989.
80. Maas, J.; Verheij, R.; Groenewegen, P.; De Vries, S.; Spreeuwenberg, P. Green space urbanity, and health: How strong is the relation? *J. Epidemiol. Community Health* **2006**, *60*, 587–592. [CrossRef]
81. Maas, J.; Verheij, R.; De Vries, S.; Spreeuwenberg, P.; Schellevis, F.G.; Groenewegen, P.P. Morbidity is related to a green living environment. *J. Epidemiol. Community Health* **2009**, *63*, 967–973. [CrossRef] [PubMed]
82. Maller, C.; Townsend, M.; Pryor, A.; Brown, P.; St Leger, L. Healthy nature healthy people: 'contact with nature' as an upstream health promotion intervention for populations. *Health Promot. Int.* **2006**, *21*, 45–54. [CrossRef] [PubMed]
83. Thompson, C.W.; Aspinall, P.; Roe, J. Access to Green Space in Disadvantaged Urban Communities: Evidence of Salutogenic Effects Based on Biomarker and Self-report Measures of Wellbeing. *Procedia Soc. Behav. Sci.* **2014**, *153*, 10–22. [CrossRef]
84. Bird, W. *Natural Thinking, Investigating the Links between the Natural Environment, Biodiversity and Mental Health*; Royal Society for the Protection of Birds: Sandy, UK, 2007.
85. Carrus, G.; Scopelliti, M.; Lafortezza, R.; Colangelo, G.; Ferrini, F.; Salbitano, F.; Agrimi, M.; Portoghesi, L.; Semenzato, P.; Sanesi, G. Go greener, feel better? The positive effects of biodiversity on the well-being of individuals visiting urban and peri-urban green areas. *Landsc. Urban Plan.* **2015**, *134*, 221–228. [CrossRef]
86. Marselle, M.; Irvine, K.; Warber, S. Examining group walks in nature and multiple aspects of well-Being: A large-scale study. *Ecopsychology* **2014**, *6*, 134–147.
87. Pretty, J.; Peacock, J.; Sellens, M.; Griffin, M. The mental and physical health outcomes of green exercise. *Int. J. Environ. Health Res.* **2005**, *15*, 319–337. [CrossRef] [PubMed]
88. Tzoulas, K.; Korpela, K.; Venn, S.; Yli-Pelkonen, V.; Kaźmierczak, A.; Niemela, J.; James, P. Promoting ecosystem and human health in urban areas using Green Infrastructure: A literature review. *Landsc. Urban Plan.* **2007**, *81*, 167–178. [CrossRef]
89. Mitchell, R.; Popham, F. Greenspace, urbanity and health: Relationships in England. *J. Epidemiol. Community Health* **2007**, *61*, 681–683. [CrossRef]
90. Gascon, M.; Triguero-Mas, M.; Martínez, D.; Dadvand, P.; Rojas-Rueda, D.; Plasència, A.; Nieuwenhuijsen, M.J. Residential green spaces and mortality: A systematic review. *Environ. Int.* **2016**, *86*, 60–67. [CrossRef]
91. Söderback, I.; Söderström, M.; Schälander, E. Horticultural therapy: The 'healing garden' and gardening in reha-bilitation measures at Danderyd Hospital Rehabilitation Clinic, Sweden. *Pediatric Rehabil.* **2004**, *7*, 245–260. [CrossRef] [PubMed]
92. Marcus, C. Healing gardens in hospitals. *Interdiscip. Des. Res. e-J.* **2007**, *1*, 1–27.
93. Lau, S.S.Y.; Yang, F. Introducing Healing Gardens into a Compact University Campus: Design Natural Space to Create Healthy and Sustainable Campuses. *Landsc. Res.* **2009**, *34*, 55–81. [CrossRef]
94. Lottrup, L.; Grahn, P.; Stigsdotter, U.K. Workplace greenery and perceived level of stress: Benefits of access to a green outdoor environment at the workplace. *Landsc. Urban Plan.* **2013**, *110*, 5–11. [CrossRef]
95. Krasny, M. *and Tidball, K. Civic Ecology: Adaptation and Transformation from the Ground Up*; MIT Press: Cambridge, MA, USA, 2015.
96. Pudup, M.B. It takes a garden: Cultivating citizen-subjects in organized garden projects. *Geoforum* **2008**, *39*, 1228–1240. [CrossRef]
97. Kuo, F.; Sullivan, W. Environment and Crime in the Inner City: Does Vegetation Reduce Crime? *Environ. Behav.* **2001**, *33*, 343–367. [CrossRef]
98. Kuo, F.; Bacaicoa, M.; Sullivan, W. Transforming inner-city landscapes: Trees, sense of safety and preference. *Environ. Behav.* **1998**, *30*, 28–59. [CrossRef]
99. Okvat, H.; Zautra, A. Community gardening: A parsimonious path to individual, community, and environmental resilience. *Am. J. Community Psychol.* **2011**, *47*, 374–387. [CrossRef] [PubMed]
100. Abhijith, K.; Kumar, P.; Gallagher, J.; McNabola, A.; Baldauf, R.; Pilla, F.; Broderick, B.; Di Sabatino, S.; Pulvirenti, B. Air pollution abatement performances of green infrastructure in open road and built-up street canyon environments—A review. *Atmos. Environ.* **2017**, *162*, 71–86. [CrossRef]
101. Kumar, P.; Druckman, A.; Gallagher, J.; Gatersleben, B.; Allison, S.; Eisenman, T.S.; Hoang, U.; Hama, S.; Tiwari, A.; Sharma, A.; et al. The nexus between air pollution, green infrastructure and human health. *Environ. Int.* **2019**, *133*, 105181. [CrossRef] [PubMed]
102. Zupancic, T.; Westmacott, C.; Bulthuis, M. *The Impact of Green Space on Heat and Air Pollution in Urban Communities: A Meta-Narrative Systemic Review*; David Suzuki Foundation: Vancouver, ON, Canada, 2015.
103. Hewitt, C.N.; Ashworth, K.; MacKenzie, A.R. Using green infrastructure to improve urban air quality (GI4AQ). *Ambio* **2020**, *49*, 62–73. [CrossRef] [PubMed]
104. Conversano, I.; del Conte, L.; Mulder, I. Research through Design for accounting values in design. In Proceedings of the 4th Biennial Research through Design Conference, Delft, Rotterdam, The Netherlands, 19–22 March 2019.
105. De Jong, T.M. *Kleine Methodologie voor Ontwerpend Onderzoek*; BOOM Uitgevers: Amsterdam, The Netherlands, 1992.
106. Hauberg, J. Research by Design—A research strategy. *Rev. Lusófona Arquit. Educ. Archit. Educ. J.* **2011**, *5*, 46–56.
107. Milburn, L.-A.S.; Brown, R.D. The relationship between research and design in landscape architecture. *Landsc. Urban Plan.* **2003**, *64*, 47–66. [CrossRef]

108. Roggema, R. Research by Design: Proposition for a Methodological Approach. *Urban Sci.* **2016**, *1*, 2. [CrossRef]
109. Rosemann, J. The Conditions of Research by Design in Practice. In *Proceedings of the Research by Design: International Conference Faculy of Architecture Delft University of Technology in Co-Operation with the EAASE/AEEA, Delft, The Netherlands, 1–3 November 2000*; Delft University Press: Delft, The Netherlands, 2001; pp. 63–68.
110. Swann, C. Action Research and the Practice of Design. *Des. Issues* **2002**, *18*, 49–61. [CrossRef]
111. Chamberlain, P.; Bonsiepe, G.; Cross, N.; Keller, I.; Frens, J.; Buchanan, R.; Meroni, A.; Krippendorff, K.; Stappers, P.J.; Jonas, W.; et al. Design Research Now. In *Design Research Now*; Springer International Publishing: Berlin/Heidelberg, Germany, 2007; pp. 41–54.
112. Hocking, V.T. Designerly ways of knowing: What does design have to offer. In *Tackling Wicked Problems through the Transdisciplinary Imagination*; Brown, V.A., Harris, J.A., Russell, J.Y., Eds.; Earthscan: London, UK, 2010; pp. 242–250.
113. Van den Hurk, B.; Siegmund, P.; Klein Tank, A. *KNMI'14: Climate Change Scenarios for the 21st Century—A Netherlands Perspective*; Scientific Report WR2014-01; KNMI: De Bilt, The Netherlands, 2014.
114. Ministerie van Infrastructuur en Milieu. Deltaporgramma 2021. 2020. Available online: https://dp2021.deltaprogramma.nl (accessed on 21 December 2020).
115. INFRAM. *Rapport Eerste Fase Beleidstafel Droogte*; Ministerie van Infrastructuur en Milieu: Den Haag, The Netherlands, 2019.
116. Van de Vijver, F.; Leung, K. *Methods and Data Analysis of Comparative Research*; Allyn & Bacon: Boston MA, USA, 1997.
117. Denk, T. Comparative multilevel analysis: Proposal for a methodology. *Int. J. Soc. Res. Methodol.* **2010**, *13*, 29–39. [CrossRef]
118. Barton, H. Eco-neighbourhoods: A review of projects. *Local Environ.* **1998**, *3*, 159–177. [CrossRef]
119. Mougeot, L.J.A. For self-reliant cities: Urban food production in a globalizing South. In *For Hunger-Proof Cities: Sustainable Urban Food Systems*; Koc, M., MacRae, R., Mougeot, L.J.A., Welsh, J., Eds.; International Development Research Centre: Ottawa, ON, Canada, 1999; pp. 11–25.
120. Morris, D. *Self-Reliant Cities*; The New Rules Project: Minneapolis, MN, USA, 2008.
121. Shuman, M. *Going Local: Creating Self-Reliant Communities in a Global Age*; Routledge: London, UK, 2013.
122. Storey, J.B.; Baird, G. Towards the Self-Sufficient City Building. *Trans. Inst. Prof. Eng. N. Z.* **1999**, *26*, 1–8.
123. Tait, M. Urban villages as self-sufficient, integrated communities: A case study in London's Docklands. *Urban Des. Int.* **2003**, *8*, 37–52. [CrossRef]
124. Núñez-Ríos, J.E.; Aguilar-Gallegos, N.; Sánchez-García, J.Y.; Cardoso-Castro, P.P. Systemic Design for Food Self-Sufficiency in Urban Areas. *Sustainability* **2020**, *12*, 7558. [CrossRef]
125. Karnib, A. A Quantitative Assessment Framework for Water, Energy and Food Nexus. *Comput. Water Energy Environ. Eng.* **2017**, *6*, 11–23. [CrossRef]
126. Yang, Y.J.; Goodrich, J.A. Toward quantitative analysis of water-energy-urban-climate nexus for urban adaptation planning. *Curr. Opin. Chem. Eng.* **2014**, *5*, 22–28. [CrossRef]
127. Soininen, J. Spatial structure in ecological communities–a quantitative analysis. *Oikos* **2016**, *125*, 160–166. [CrossRef]
128. Fry, T. *Design Futuring*; Zed Books: London, UK, 2009; pp. 71–77.
129. Cornish, E. *Futuring: The Exploration of the Future*; World Future Society: Chicago IL, USA, 2004.
130. Shipley, R.; Newkirk, R. Vision and visioning in planning: What do these terms really mean? *Environ. Plan. B Plan. Des.* **1999**, *26*, 573–591. [CrossRef]
131. Gaffikin, F.; Sterrett, K. New Visions for Old Cities: The Role of Visioning in Planning. *Plan. Theory Pract.* **2006**, *7*, 159–178. [CrossRef]
132. Roggema, R. Adaptation to climate change: Does spatial planning help? Swarm planning does! *Manag. Nat. Resour. Sustain. Dev. Ecol. Hazards II* **2009**, *127*, 161–172. [CrossRef]
133. Gladwell, M. *The Tipping Point: How Little Things Can Make a Big Difference*; Little, Brown and Company: New York, NY, USA, 2000.
134. Roggema, R. *Swarming Landscapes: The Art of Designing for Climate Adaptation*; Springer: Berlin/Heidelberg, Germany, 2012; p. 260.
135. Portugali, J. *Self-Organisation and the City*; Springer: Berlin/Heidelberg, Germany, 2012.
136. Condon, P.M. *Design Charrettes for Sustainable Communities*; Island Press: Washington, DC, USA, 2008.
137. Lennertz, B.; Lutzenhiser, A. *The Charrette Handbook. The Essential Guide for Accelerated Collaborative Community Planning*; The American Planning Association: Chicago, IL, USA, 2006.
138. Roggema, R.; Martin, J.; Horne, R. Sharing the climate adaptive dream: The benefits of the charrette approach. In Proceedings of the ANZRSAI Conference, Canberra, Australia, 6–9 December 2011; AERU Research Unit: Canterburry, UK, 2011.
139. Roggema, R. *The Design Charrette: Ways to Envision Sustainable Futures*; Springer: Berlin/Heidelberg, Germany, 2013; p. 335.
140. McHarg, I.L. *Design with Nature*; Natural History Press: New York, NY, USA, 1969.
141. Reumer, J. *Wildpark Rotterdam. De stad als Natuurgebied*; Historische Uitgeverij: Groningen, The Netherlands, 2014.
142. Roggema, R. Landscape First! Nature-Based Design for Sydney's Third City. In *Nature-Driven Urbanism. Contemporary Urban Design Thinking*; Springer: Berlin/Heidelberg, Germany, 2020; Volume 2, pp. 81–110.
143. Girardet, H. *Creating Regenerative Cities*; Routledge: Boca Raton, FL, USA, 2014.
144. Girardet, H. Regenerative Cities. In *Studies in Ecological Economics*; Springer International Publishing: Berlin/Heidelberg, Germany, 2017; pp. 183–204.
145. Du Plessis, C. Towards a regenerative paradigm for the built environment. *Build. Res. Inf.* **2012**, *40*, 7–22. [CrossRef]

146. Thompson, G.; Newman, P. Urban fabrics and urban metabolism–from sustainable to regenerative cities. *Resour. Conserv. Recycl.* **2018**, *132*, 218–229. [CrossRef]
147. Roggema, R. *ReciproCity, Giving Instead of Taking. Inaugural Lecture*; Hanze University of Applied Sciences: Groningen, The Netherlands, 2019.
148. Sijmons, D. Contrast, Contact & Contract, Pathways to Pacify Urbanization and Nature. In *Nature-Driven Urbanism. Contemporary Urban Design Thinking*; Roggema, R., Ed.; Springer: Berlin/Heidelberg, Germany, 2020; Volume 2, pp. 9–42.
149. Rouse, W.B. *Design for Success: A Human-Centered Approach to Designing Successful Products and Systems*; Wiley: New York, NY, USA, 1991.
150. Harland, R.G.; Santos, M.C. From greed to need: Notes on human-centred design. In Proceedings of the AHRC Postgraduate Conference 2009, Loughborough, UK, 1–2 July 2009; pp. 141–158.
151. Brennan, A.; Yeuk-Sze, L. Environmental Ethics. In *The Stanford Encyclopedia of Philosophy (Fall Edition)*; Zalta, E.N., Ed.; Stanford University: Stanford CA, USA, 2011. Available online: http://plato.stanford.edu/archives/fall2011/entries/ethics-environmental (accessed on 21 December 2020).
152. Davoudi, S. Climate Change, Securitisation of Nature, and Resilient Urbanism. *Environ. Plan. C Gov. Policy* **2014**, *32*, 360–375. [CrossRef]
153. Stoltz, J. Perceived Sensory Dimensions: A Human-Centred Approach to Environmental Planning and Design. Ph.D. Thesis, Stockholm University, Stockholm, Sweden, 2019.
154. Mccormick, K. *Cities, Nature and Innovation: New Directions*; Lund University: Lund, Sweden, 2020.
155. European Commission. *Towards an EU Research and Innovation Policy Agenda for Nature-Based Solutions & Re-Naturing Cities*; Final Report of the Horizon 2020 Expert Group on 'Nature-Based Solutions and Re-Naturing Cities'; Publications Office of the European Union: Luxemburg, 2015.
156. Simpson, S.; Marshall, P.; Crumley, C.L. Nature's Web: Rethinking Our Place on Earth. *Environ. Hist.* **1996**, *1*, 106–108. [CrossRef]
157. Capra, F. *The Web of Life: A New Synthesis of Mind and Matter*; HarperCollins: New York, NY, USA, 1996.
158. Roggema, R.; Tillie, N.; Keeffe, G. Landscape first: Thriving nature-based urbanism in Western Sydney. *Land.* forthcoming.
159. Roggema, R. (Ed.) *Nature-Driven Urbanism*; Contemporary Urban Design Thinking, Volume 2; Springer: Dordrecht, Switzerland; Berlin/Heidelberg, Germany; London, UK, 2020; 359p.
160. Willett, W.; Rockström, J.; Loken, B.; Springmann, M.; Lang, T.; Vermeulen, S.; Garnett, T.; Tilman, D.; DeClerck, F.; Wood, A.; et al. Food in the Anthropocene: The EAT–Lancet Commission on healthy diets from sustainable food systems. *Lancet* **2019**, *393*, 447–492. [CrossRef]
161. Schuetze, T.; Chelleri, L. Integrating Decentralized Rainwater Management in Urban Planning and Design: Flood Resilient and Sustainable Water Management Using the Example of Coastal Cities in The Netherlands and Taiwan. *Water* **2013**, *5*, 593–616. [CrossRef]

sustainability

MDPI

Article

Stakeholder Participation in the Planning and Design of Nature-Based Solutions. Insights from CLEVER Cities Project in Hamburg

Alessandro Arlati [1,*], Anne Rödl [2], Sopho Kanjaria-Christian [3] and Jörg Knieling [1]

[1] Department of Urban Planning and Regional Development, HafenCity University Hamburg, Henning-Voscherau-Platz 1, 20457 Hamburg, Germany; joerg.knieling@hcu-hamburg.de
[2] Institute of Environmental Technology and Energy Economics, Hamburg University of Technology, Eissendorfer Straße 40, 21073 Hamburg, Germany; anne.roedl@tuhh.de
[3] Free and Hanseatic City of Hamburg, District of Harburg, Harburger Rathausplatz 4, 21073 Hamburg, Germany; sophio.konjaria-christian@harburg.hamburg.de
* Correspondence: Alessandro.arlati@hcu-hamburg.de

Abstract: Cities are essential players in responding to the present complex environmental and social challenges, such as climate change. The nature-based solution (NbS) concept is identified in the scientific discourse and further recognized by the European Commission as a part of the solution to address such challenges. Deploying NbS in urban contexts requires the cooperation of different public and private stakeholders to manage those processes. In this paper, the experiences of establishing and managing NbS-related processes following a co-creation approach in the city of Hamburg within the framework of an EU-funded research project (CLEVER Cities) are described and analyzed. The paper identifies and discusses the main emerging factors and challenges from (1) a procedural and methodological perspective and (2) concerning the different roles of the diverse stakeholders involved. This discussion is grounded in the context of existing regulations and novel concepts for citizens' participation in urban decision-making processes. As research results, the article defines the leading players involved in the process and their roles and interrelationships, along with recommendations for future policy agendas in cities when dealing with NbS planning.

Keywords: stakeholder participation; nature-based solutions; multi-level governance; co-creation; urban living lab; sustainable urban development; urban planning

Citation: Arlati, A.; Rödl, A.; Kanjaria-Christian, S.; Knieling, J. Stakeholder Participation in the Planning and Design of Nature-Based Solutions. Insights from CLEVER Cities Project in Hamburg. *Sustainability* **2021**, *13*, 2572. https://doi.org/10.3390/su13052572

Academic Editor: Israa H. Mahmoud

Received: 4 December 2020
Accepted: 23 February 2021
Published: 27 February 2021

Publisher's Note: MDPI stays neutral with regard to jurisdictional claims in published maps and institutional affiliations.

1. Introduction

Climate change poses cities complex environmental and social challenges. After an era of mainly favoring economic growth to the detriment of natural capital, the dual objective of addressing both elements entered European cities' political agendas [1] (p. 121). The inherent complexity of dealing with environmental and social demands requires a paradigm shift in policy-making [2].

In the 2010s, the concept of nature-based solutions (NbS) emerged in the political agendas of cities that are committed to becoming more "resilient, invest into green infrastructure and integrate nature-based solutions to improve microclimate, limit urban heat island phenomenon and improve air quality" [3] (p. 93). Given the fact that NbS are "designed to address various societal challenges in a resource-efficient and adaptable manner and to provide simultaneously economic, social and environmental benefits" [1] (p. 121), it appears that the simultaneity of addressing challenges related to the three pillars of sustainability is one of the main objectives that can be reached through NbS. Furthermore, Frantzeskaki et al. [4] argue that NbS can be potent tools to mitigate the effects of extreme weather events and provide additional adaptation strategies for urban settlements. The European Commission has also largely adopted the NbS concept [5], such as in the Horizon 2020 Funding Programme [6]. IUCN [7] has recently published criteria for verification, design, and scaling up of NbS to support national governments, local governments, planners, businesses, or organizations. Among the IUCN-defined criteria, the fifth states that, NbS should be based on inclusive, transparent, and empowering governance processes [7] (p. 14). This implies using the existing regulatory framework concerning participatory processes and eventually stimulating the finding of novel tools towards conducting a transparent and open process of co-creation. In this context, co-creation means allowing stakeholders to collaborate in the process of solution design, implementation, and monitoring [3,8]. In this sense, the co-creation of NbS is understood as a combination of various expertise from different scientific fields and the local knowledge of civil society representatives [9].

Kemp and Loorbach [10] argue that working towards sustainable development requires simultaneous communication between different governance levels. As also Frantzeskaki et al. [11] (p. 23) state, it is necessary to involve a wide range of stakeholders in decision-making processes at every level to create collective action for a more sustainable approach to shaping cities. Hence, decision-making processes within the field of sustainable development occur by participative momenta of exchange among composite governmental and non-governmental stakeholder constellations. This is reflected in the need to establish an everyday discourse based on the broad participation that includes both practitioners and laypersons. In this context, cities' governance structures may contain elements that can hinder or encourage participation depending on their hierarchies/political structures and processes, and they might require modifications.

This article was developed in the framework of the European-funded H2020 project CLEVER Cities, which deploys NbS to address urban challenges and social inclusion in cities [12]. CLEVER Cities' activities focus on the impacts of NbS on social cohesion, citizen security, environmental justice, and human health. Accordingly, the development of NbS happens through the active participation of local stakeholders following a co-creation logic called the Co-Creation Pathway [3]. This pathway is described in more detail in Section 2. The idea behind the Co-Creation Pathway is the broader concept of Urban Living Labs (ULL), which are conceived here as forums of innovation where resources and agencies are moving towards governed sustainable development [13] with the long-term objective of achieving resilient and climate-responsive cities [3].

By discussing how the co-creation process of NbS—including planning, design, and implementation phases—happened for the case study of Hamburg, the paper aims to answer the following question: which stakeholders should be involved in the co-creative process of the planning and design of NbS and which roles do they play in the different phases? The article explores which types of stakeholders contributed to the definition of the NbS and discusses their roles in the three Urban Living Labs (ULL) that were part of the CLEVER Cities project. Insights are provided into the tools and methods that supported the co-creation process's goals and facilitated stakeholders' inclusion. As an outcome, the article defines recommendations for future policy agendas in cities when dealing with NbS.

2. Materials and Methods

This section illustrates the methodology delineated for answering the research question and a brief introduction to the CLEVER Cities project area.

2.1. Methodology

The Co-Creation Pathway elaborated within the CLEVER Cities project by Morello et al. [14] describes a five-phase concept of co-creation to be applied in the development of NbS—namely, (i) urban innovation partnership (UIP) establishment, (ii) co-design, (iii) co-implementation, (iv) co-monitoring, and (v) co-development. Within the local project area, stakeholders are engaged to form partnerships (i) to go through the entire process from (ii) to (v). The first phase considers the establishment of a UIP. Morello et al. [14] (p. 90) describe the UIP as a "city-wide or district-focused informal alliance" between various local authorities and community groups, businesses, and academics to promote NbS to foster urban regeneration. Ideally, this alliance formation follows the quadruple helix concept [15], which denotes the neo-institutional networks between the government, business sector, academia, and civil society that have the task of steering and facilitating the co-creation process in the project area.

During the co-design phase (ii), the UIP members organize workshops to jointly design nature-based interventions that help to solve local, social, environmental, and economic challenges. To guide the co-design process effectively, the methodological approach Theory of Change (ToC) [16] represents the primary reference for the definition of the NbS. The method consists of a systematic process that brings the attendees to address local challenges through the conception of a long-term vision. It is then necessary to work backward by setting out the overall, intermediate, and short-term outcomes and outputs to achieve the defined vision [17] (p. 12; adapted from [16]).

The second phase's solutions are operationalized in the third phase (co-implementation phase) by involving and working closely with citizens and other relevant stakeholders. The fourth phase of the pathway comprises the co-monitoring process, in which the interventions' impact, durability, and quality are evaluated. The involvement of citizens is expected in all four phases.

The final phase, co-development, describes the UIP members' and citizens' joint efforts to maintain the interventions and eventually replicate them in other parts of the city (upscaling). The presented Co-Creation Pathway results in introducing shared governance arrangements [18] that facilitate and guide the transition process with multi-level [19,20] and multi-stakeholders approaches [21].

At the time this paper was written (November 2020), the CLEVER Cities project is between phase two (ii), co-design, and phase three (iii), co-implementation. Hence, only the first two phases (namely, UIP establishment and co-design) are discussed here, including the descriptions of tools, methods, and procedures. Additionally, stakeholders' participation in NbS planning, design, and implementation is analyzed for the Hamburg case study's practical example. Therefore, a stakeholder analysis was conducted to depict the stakeholders' constellation and their characteristics in the NbS planning process—namely, providing resources and goals and taking on decision-making power and roles. The analysis was performed based on stakeholder categorization adapted from Dente [22]. Furthermore, their relations were investigated and represented on a power-interest matrix [23]. The complete analysis can be seen in Konjaria-Christian et al. [24]. The analysis provided insights into how stakeholders were interrelated in the context of the co-creative design of NbS projects and allowed identifying elements of success and failure in stakeholder participation. Based on these experiences, the paper identifies and discusses the main positive factors and challenges from (1) a procedural and methodological point of view and (2) concerning stakeholders' experiences. This discussion is grounded in the context of existing regulations and novel concepts for citizens' participation in urban decision-making processes.

2.2. General Description of the Project Area

The Free Hanseatic City of Hamburg (FHH) is one of the three city-states in the Federal Republic of Germany, with almost 1.9 million people, and has recently experienced rapid population growth [25]. The pilot area of CLEVER Cities in Hamburg is located in the district council of Hamburg–Harburg in the urban district of Neugraben–Fischbek (NF), located in the south-west of Hamburg, close to the border to the Federal State of Lower Saxony. It is the largest urban district among the 17 urban districts of Harburg in terms of surface and inhabitants [26]. The project area stretches from the center of the neighborhood of Neugraben to the new development area of Vogelkamp in the east and from the Fischbek–Falkenberg district school and old village structure to the Sandbek residential area in the west. The project area includes both existing settlements and new development areas. Additionally, the project area is surrounded by two nature reserves: Fischbeker Heide and Fischbeker Moor (Figure 1).

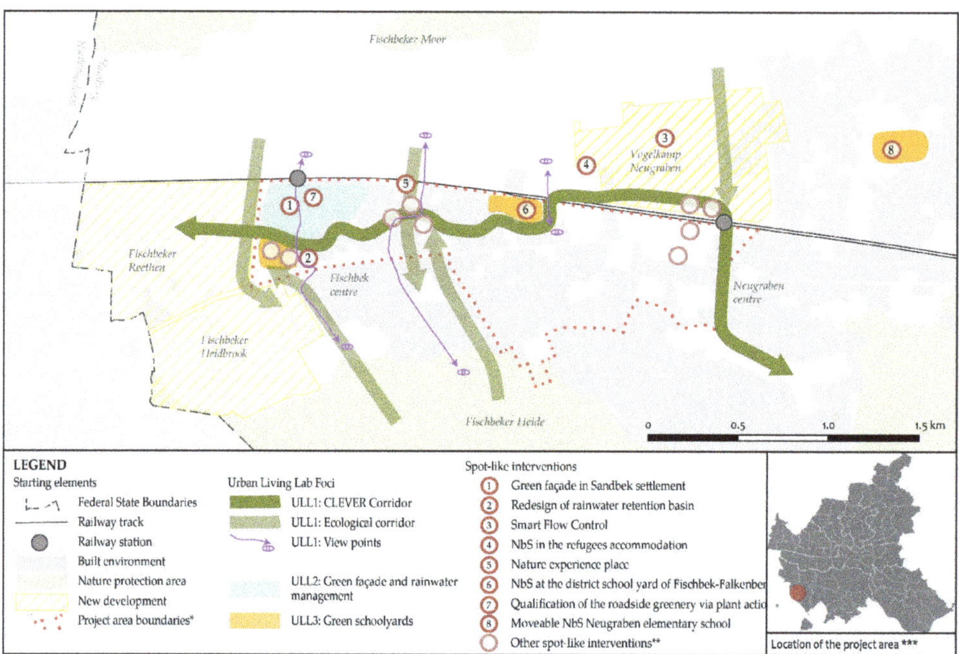

Figure 1. The CLEVER Cities project area in Hamburg. * Project area boundaries as defined in the CLEVER Cities Grant Agreement. ** Other spot-like interventions that are part of the CLEVER Cities Hamburg strategy but are not mentioned in this paper. *** Location of the project area [27]. (Own elaboration).

The project area is connected to the city center and the federal transport infrastructure through two local and federal train stations. Social housing developments are mainly located in the western part of the project area. At the time of the CLEVER activities (November 2020), three new large construction developments were under construction at the existing built area's fringe. Due to these new developments, the population is expected to increase in the district by about 40% [28]. According to the Social Monitoring Plan of 2019 [29], NF is considered to have a low or very low value in terms of social conditions and is therefore eligible for receiving special Hamburg funding for its requalification and further development (Integrated City Development Programme—RISE). Concerning the social structure, it is essential to mention that the refugees' accommodation facility, located in the neighborhood of Vogelkamp–Neugraben, has been included in the CLEVER Cities project activities (point 4 in Figure 1).

Notably, NF presents a distinct social and economic situation by hosting a varied social and spatial mix. In this context, CLEVER Cities decided to implement a range of NbS initiatives as an experimental pilot to explore the social co-benefits and environmental and economic improvements generated by the implementation of NbS. The simultaneous and reciprocal strengthening of the local community and natural resources constitutes an opportunity to address four urban regeneration challenges: human health and well-being, sustainable economic prosperity, social cohesion and environmental justice, and citizen security.

As Hamburg is a city-state, it is crucial to define the three governance levels involved in the process that will appear in the text. The term "federal level" refers to Hamburg as a federal city-state; with the term "district level," the Harburg district is meant; lastly, the term "local level" implies the urban district of NF.

3. Results

According to the framework illustrated in Section 2, the project team was set up in Hamburg before starting the co-creation process. The project team includes the District

Office of Hamburg-Harburg (DHH), three governmental institutions of the state Hamburg (Senate Chancellery; the Ministry of Environment, Energy, Climate, and Agriculture—BUKEA; and the State Agency for Geoinformation and Surveying—LGV), the urban development agency (steg), and three scientific partners (HafenCity University—HCU; Hamburg University of Technology—TUHH; and Hamburg Institute of International Economics—HWWI). An overview is provided in Table 1.

Table 1. Project team members' categorization (Own elaboration).

Institution Name	Level	Type	Resources [1]
DHH	District	Public	Political, economic, legal
Senate Chancellery	Federal	Public	Political, legal
BUKEA	Federal	Public	Political, legal, cognitive
LGV	Federal	Public	Cognitive
steg	local	Private	Political, cognitive, relational
HCU	Federal	University	Cognitive, relational
TUHH	Federal	University	Cognitive, relational
HWWI	Federal	Research	Cognitive

[1] Type of resources according to Dente [22].

The District Office of Hamburg-Harburg (DHH) is the institution responsible for coordinating all project partners and processes and implementing the local interventions. The tasks of the DHH include coordination of the Hamburg interventions and evaluation and further concretization of the project ideas together with partners and UIPs. Moreover, DHH is responsible for planning processes, contracting third parties to commission project implementation, and keeping a constant dialogue between the parties involved to ensure innovation and co-creation in the design process. The DHH acts as an intermediary for Hamburg interventions both within the district office's administrative departments and for the project's local, district, federal, and international partners. Most importantly, the DHH is the primary contact concerning issues around the CLEVER Cities project in Hamburg. Presenting and raising awareness on the project at different scales (within the administration and civil society) is one of the DHH work's cornerstones.

The other three governmental institutions represent the federal level of the city-state of Hamburg. Senate Chancellery is the leading international contact point and the coordinator for the entire CLEVER Cities project and is in charge of communicating at the state level. BUKEA is the ministry at Hamburg-level in charge of policies regarding the environment, energy, and climate. Within the CLEVER Cities project, BUKEA is engaged in developing and upscale the environmental strategy learnt from the CLEVER Cities experience at the federal level. LGV holds the georeferenced database and the cadastre land register for the Federal State of Hamburg and provides technical measurements. In the CLEVER Cities project, LGV has the task of developing, implementing, and integrating the urban data platform with new information gathered during the project's lifetime. Though not physically involved with the interventions at the local level, they contribute substantially with their specific expertise on particular aspects of the NbS interventions, and they represent the direct link to the federal level.

The local development agency, steg, runs an on-site district office in the project area to improve visibility through various activities promoted at the local level in recent years. The local presence of steg is of significant importance, especially regarding co-creation processes and citizen participation for the various project activities.

Lastly, the three research institutions involved support the local activities with their scientific competencies in urban planning, policies, and landscape architecture (HCU), environmental technology, energy, and water management (TUHH), and socio-economic studies (HWWI).

3.1. Initiating the Co-Creation Process

In 2018, to inaugurate the CLEVER Cities project's activities and ensure visibility among local stakeholders, a large-scale kick-off event was organized by DHH, BUKEA, and steg. The event's main intention was to raise awareness about the project's objectives and provide a factual basis for further co-creation steps. The event brought together around 130 people, including citizens and representatives from private and public sectors. An innovative tool for digital participation (DIPAS) has facilitated the process of gathering valuable insights and ideas from the participants, which laid the groundwork for specifying particular interventions within three main Urban Living Labs (ULL) (Figure 2a). The content of each ULL will be briefly presented in Section 3.2.

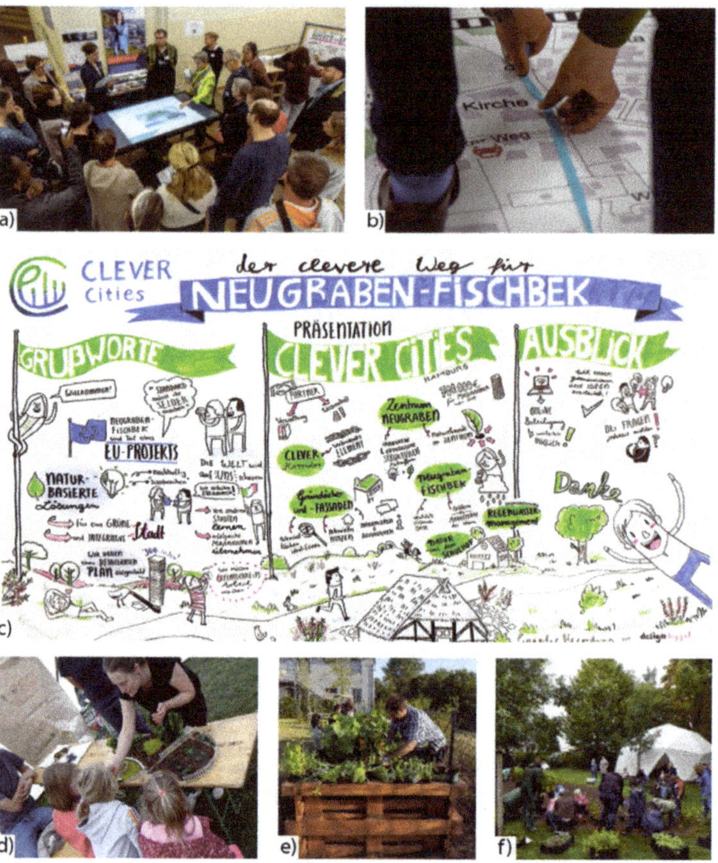

Figure 2. (**a**) Digital participation system (DIPAS) used in the kick-off meeting (DHH, 2018); (**b**) carpet with the orthophoto of Neugraben–Fischbek (NF) in the kick-off meeting (DHH, 2018); (**c**) graphic recording during the kick-off meeting (DHH, 2018); (**d**) model of the garden in the school Fischbek–Falkenberg built with pupils and teachers (steg, 2019); (**e**) high bed realized by refugees (steg, 2019); (**f**) planting action in the Sandbek settlement (steg, 2019).

The DIPAS tool has been used not only during the event day: it was also available online a few weeks after, allowing for further contributions from the local population to the co-creative process. Additionally, a huge carpet representing the Neugraben-Fischbek urban district's orthophoto was used as a basis for further discussion and commenting during the inaugural event (Figure 2b) and other events (district festival Neugraben). Another addition to the event was the graphic recording of the event and its results (Figure 2c).

After the kick-off event, the work in each ULL started with defining the local UIPs. The definition of the local UIP is denoted as phase 1 (i) in the Co-Creation Pathway of

Morello et al. [30]. Depending on the intervention types, the UIPs were organized into various formats, such as jours fixes, workshops, and multilateral or bilateral strategic planning meetings. UIP meetings have been carried out in face-to-face format, and online meetings due to Covid-19 restrictions and dynamically adapted depending on the specific steps and goals of each ULL. Generally, the DHH initiates the UIP formation process and accompanies it throughout its development, supported by the other project team members.

During the initiation phase of the local UIPs, the project team applied the Theory of Change (ToC) method. The ToC meeting brought together the main interested local stakeholders in defining visions and incremental outcomes needed to achieve the visions set for each ULL under the moderation of the project team.

In September 2019, the project team organized its first annual UIP event, which registered around 140 participants. The meeting took place simultaneously at three different locations throughout the project area, strategically selected to implement NbS using a hands-on approach. As an example of the activities conducted, pupils were involved in building a physical model of their ideal school garden (see Figure 2d); in another location, young and old representatives were brought together for a planting action project (Figure 2f).

Additionally, the co-creation process within CLEVER Cities was not limited to the activities conducted in the three ULL. The project team brought the CLEVER Cities project's experience to other external activities and events, aiming at broader participation. Of particular importance, during the urban district festival "Neugraben Erleben," the "Sensafety-App" was presented by LGV [31]. It is a mobile application that allows users to evaluate specific locations in the project area based on their subjective impressions and perceptions of safety. The citizens' participation via the "Sensafety-App" defines one of the integral elements in the co-monitoring phase (iv).

3.2. Co-Creation Processes in Hamburg

The following paragraphs describe the implemented projects in Hamburg for each of the three ULL, with a particular focus on the co-creation process, which included the previously mentioned co-creation phases "UIP establishment*"' (i) and "co-design"(ii) with their respective stakeholders, procedures, and tools. Three different focus topics have been defined for the project area: (1) a green corridor (ULL 1), (2) green roofs and façade, and rainwater management (ULL 2), and (3) green schoolyards (ULL 3).

Funding opportunities from the Horizon 2020 Program are covering interventions and activities within the project timeframe, demanding for taking decisions towards enabling mechanisms of ownership building for future maintenance for both technical solutions, such as aquaponics (Section 3.2.2) and social solutions, such as the high beds constructed together with refugees (Section 3.2.1). Notably, because of the COVID-19 situation, alternative participation tools had to be considered to continue the co-creation processes.

3.2.1. Focus—"CLEVER Corridor" (ULL 1)

The "CLEVER Corridor" aims to establish a connection among several NbS interventions spatially spread across the CLEVER Cities project area (Figure 1). The interventions have a broad objective of connecting the two surrounding nature conservation areas with a potential bridging function. The connection effort was translated into a set of small interventions developed organically under the corridor's frame. A guiding system that will be co-created with diverse stakeholders and inclusive formats, such as workshops, will function as a recognizable sign for the corridor. Private and public entities and individual citizens were strongly involved. The CLEVER Corridor will reciprocally link all these diverse spot-like interventions and will emphasize and highlight the existing path connections between the two nature conservation areas. This ULL consists of two levels: the individual spot-like interventions and the guiding system that creates the umbrella for all projects. Therefore, co-creation in this ULL is organized in multiple UIPs established in phase 1 (i) of the co-creation process. The UIPs within the focus topic "CLEVER Corridor" are practicing and representing diverse forms of collaboration between local stakeholders, private entities, the public administration, and universities and signal the active participation of different social and age groups in many actions along the corridor. Another intervention within the corridor's scope is the "nature experience place" project coordinated by the public management office department of DHH and facilitated by a landscaping architecture company. In this case, face-to-face participation during the co-design phase (ii) was complemented by online participation due to COVID-19 prevention regulations. For this purpose, the project team used the DIPAS tool supported by LGV and local NGOs, where citizens for three weeks had the opportunity to directly participate in the planning by choosing their favorite options of natural elements.

In the context of the "CLEVER Corridor" ULL, it is worth mentioning that the project also realized a collaboration with the refugee accommodation facility close to the new development area Vogelkamp–Neugraben (Figure 1). With the facilitation of steg and the help of translators, in the co-design phase (ii), ideas for designing mobile elements for the common exterior area of the accommodation were collected from the refugees via workshops. Successively, under the guidance of steg and the facilitation of the manager of the refugees' accommodation, refugees were involved in the co-designing (ii) of multi-purpose islands with seats and planting areas constructed out of timber and destined as places for gathering and entertainment. Additionally, this participatory process involved refugees in the third phase of co-implementation (iii) (construction and planting) of the high beds to transform the area into a place that "invites them to stay" (Figure 2e). A local NGO and a carpenter were supporting the construction works. Figure 3 shows the stakeholders' constellation in ULL 1.

ULL 1: CLEVER Corridor

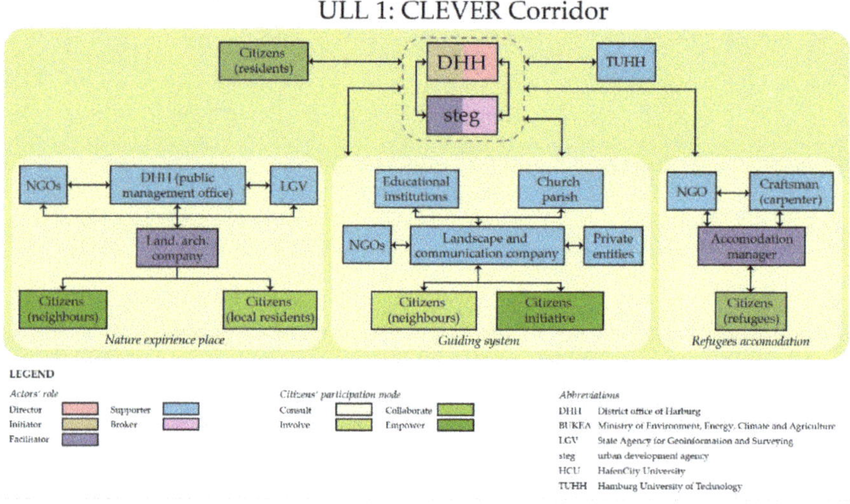

Figure 3. Representation of the stakeholders' constellation for Urban Living Lab (ULL) 1. Roles were adapted from Dente [22] (see Supplementary Material A for definitions). Roles of citizens refer to the degrees of participation according to Morello and Mahmoud [32] based on Arnstein [33]. This graph has to be understood as a simplified representation of a more comprehensive and complex stakeholder constellation. The stakeholders represented here are the project team members and the main stakeholders involved in the local urban innovation partnerships (UIPs). The constellation represented here is the one that can activate and bring on board other stakeholders for achieving the vision defined in the Theory of Change (ToC) process.

3.2.2. Focus—"Green Roof and Façade, and Rainwater Management" (ULL 2)

The focus of ULL 2 consists of two main pillars: (a) implementation of green roofs and façades, and (b) rainwater management. Interventions include the greening of a noise barrier at the train station Neugraben and installing a green façade in the Sandbek residential area. The co-creation process covers various activities, such as the ToC workshop with the federal housing company's participation and continuous monthly meetings with project partners and Integrated City Development Programme (RISE) representatives. Throughout the period marked by strict COVID-19 prevention regulations, informational letters were sent to the residents via mail. However, on-site events (e.g., the planting action of the façade) needed to be postponed to the future.

Furthermore, awareness-raising measures regarding green roofs and façades funding possibilities have been carried out. During the annual UIP event, information about funding opportunities has been disseminated thanks to the direct involvement of BUKEA. Furthermore, press releases were issued, and additional informational material was disseminated during the urban district festival (see Section 3.1).

Building greenery is also closely related to the second pillar within ULL 2—an analysis of the Neugraben–Fischbek area's vulnerability against heavy rainfall events. The concept that will be developed based on this analysis is innovative per se, as it will be applied at the entire urban district Neugraben–Fischbek, a new scale for conducting such analysis in Hamburg. The co-creation framework has brought together many stakeholders, from the public sector to academic representatives. The local UIP members are meeting regularly through online sessions wherein the progress and next steps are discussed.

Concerning the rainwater management topic, two more projects need to be mentioned. The collaboration with the Hamburg Water management company (Hamburg Wasser) has succeeded in a pilot project using innovative Smart Flow Control (SFC) technology [34]. The public-private collaboration project is planned to be tested on a ca. 10 sq.m green roof to further optimize the retention capacity and application-controlled release and discharge water into the sewer system. Another part of the ULL 2 envisions redesigning a rainwater retention basin by building a retention soil filter. The co-creative process will see the involvement of Hamburg Wasser and landscape architecture studios under the guidance of steg and HCU. Figure 4 shows the stakeholders' constellation in ULL 2.

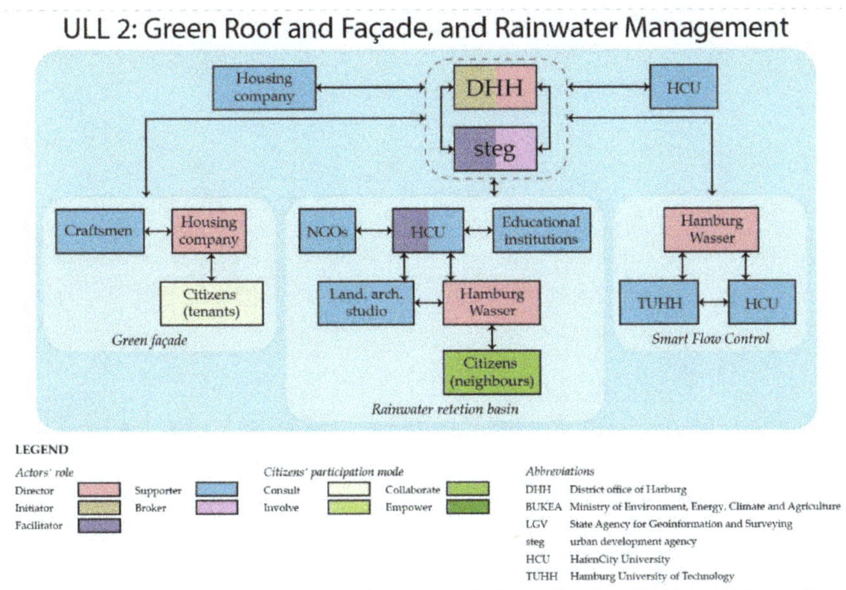

Figure 4. Representation of the stakeholders' constellation for the ULL 2. Roles were adapted from Dente [22] (see Supplementary Material A for definitions). Roles of citizens refer to the degrees of participation according to Morello and Mahmoud [32] based on Arnstein [33]. The stakeholders represented here are the project team members and the main stakeholders involved in the local UIPs. The constellation represented here is the one that can activate and bring on board other stakeholders for achieving the vision defined in the ToC process.

3.2.3. Focus—"Green Schoolyards" (ULL 3)

The third ULL is fostering the redesign of schoolyards located in the project area. At the beginning of the co-creation process, a workshop was held in the district school of Fischbek–Falkenberg to gather innovative ideas from pupils, parents, and teachers. The school staff is taking the lead in the process, which is planned to be further replicated by at least one other school in the project area. Planned interventions include the so-called researchers' garden that combines the curriculum with gardening and outdoor activities and the realization of aquaponics.

One of the participating schools (Neugrabe elementary school—point 8 in Figure 1) introduced a specific challenge for NbS implementation into the project because the school and its yard will be completely renovated in the coming years. New elements included in the schoolyard will be therefore positioned only temporarily. Hence, participants of the co-creative process were engaged to think about mobile and temporary NbS. The local UIP is composed mainly of the schools' administrations, the school building company, a teacher, pupils, CLEVER project partners (steg, DHH, and HCU), and students of the architectural faculty HCU. The solutions developed in the actual school shall be an excellent example for other schools in the urban district to undertake NbS projects with a dedicated focus on sustainability. Plans developed during the co-design process will be realized together with teachers, pupils, and parents, and with the help of steg in 2021. The ULL 3 activities are supported by an increase in teaching hours linked to environmental aspects, including sustainability topics in the school curriculum. Figure 5 shows the stakeholders' constellation in ULL 3.

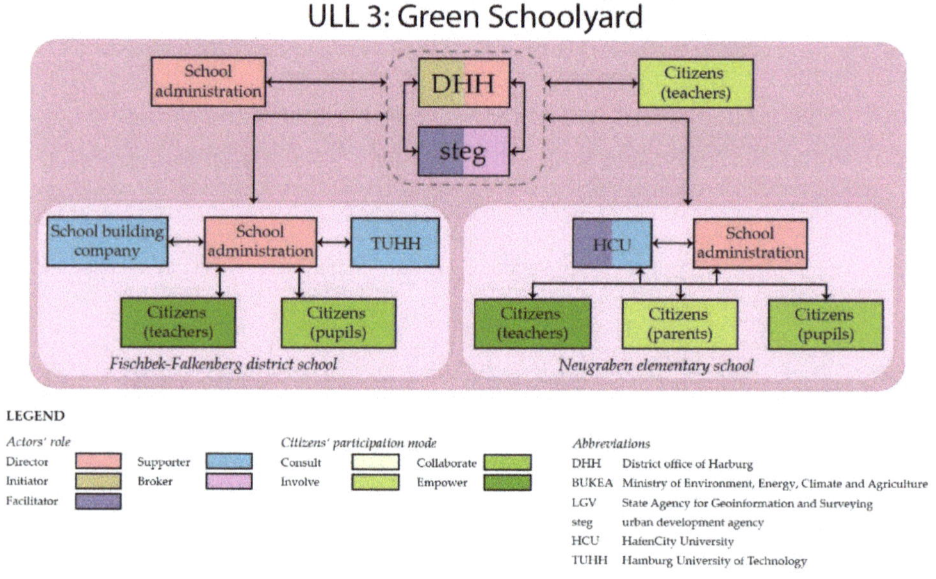

Figure 5. Representation of the stakeholders' constellation for the ULL 3. Roles were adapted from Dente [22] (see Supplementary Material A for definitions). Roles of citizens refer to the degrees of participation according to Morello and Mahmoud [32] based on Arnstein [33]. The stakeholders represented here are the project team members and the main stakeholders involved in the local UIPs. The constellation represented here is the one that can activate and bring on board other stakeholders for achieving the vision defined in the ToC process.

4. Discussion

This section identifies and discusses the main results of the co-creation experience in the Hamburg case study and their challenges regarding stakeholders' participation.

As explained in Section 3.2, various procedures and methods have been used to manage the different phases of the co-creation process. Since only the first two steps (i) and (ii) of the Co-Creation Pathway have been discussed, procedures, methods, tools, and stakeholder experiences deployed during these phases are analyzed. The first part of the discussion will concentrate on procedural and methodological issues in the first two co-creation phases, and the second part will analyze the stakeholder constellation and its implications during these phases. Lastly, recommendations are considered. Figures 3–5 represent the stakeholders and their constellation within the three ULL synthetically.

It is essential to mention that the commitment towards the EU commission and the strict allocation of funds as an incentive to apply the process is not to underestimate and brought the adoption of individual specifications and principles, such as the use of the NbS concept as a strict requirement. However, the network built during the process is expected to continue for the project's duration and even beyond. This is also obtained by establishing new and strengthening existing stakeholders' networks and applying formal and informal instruments.

4.1. Procedural and Methodological Aspects of NbS Co-Creation in the Hamburg Case-Study

Regarding the co-creation process of NbS, the systematic application of the Theory of Change (ToC) [17] guaranteed a particular path dependency of the co-creation process, which translated in a rigid but structured logic. As mentioned in Section 3, the application of ToC guided the project team and the stakeholders involved from a common understanding of the problems towards a shared definition of the objectives. Especially the inherent common sense of the ToC approach helped to structure the process. Additionally, it favored an easy understanding of the process itself also from the participants' perspective. In fact, the passage between steps is based on understanding how and why certain activities produce effects on the local context [35] and drives towards the resolutions of eventual challenges encountered. In Hamburg, the ToC process resulted in the visions and outcomes that currently guide local stakeholders in their activities on site. Further, the ToC workshop results have been fed into the co-monitoring phase (iv) of the NbS implemented in the project area.

For the co-creation process, the local UIPs played a core role as experimental interventions. They have been established for different interventions, and each case was composed of a varying set of stakeholders. Additionally, the focus area of the corridor of ULL 1 requires many different stakeholders and combinations from ULL 2 and ULL 3. Mostly classical formats like workshops and meetings to bring together the stakeholders were chosen and adapted to co-creation principles. With this, it is essential to mention that Hamburg's stakeholders had already experienced co-creation approaches through the RISE program. Hence, previous experiences have facilitated the implementation of the project framework.

Concerning the co-creation formats, the participation formats deployed in the schools (ULL 3) were suitable for catching pupils' attention for the project activities and objectives to raise their awareness for NbS specifically and sustainability issues in a broader sense. Hence, hands-on workshops to grasp people's attention and integrate them into the development processes revealed their potentials in Hamburg's case.

Nevertheless, the concept of NbS is mostly founded on capacity building and bottom-up participation (e.g., [7] p. 14) in the conviction that only participation-based processes can raise awareness on complex topics such as sustainable development (see [36]). Accordingly, a shared definition of common objectives plays a relevant role in the next steps of the NbS development process. Lastly, the ULL approach was addressed as a challenge in terms of time and resource deployed: especially, bringing together local, district, and federal stakeholders required specific facilitation and coordination competencies. Nonetheless, the organizational process has resulted in more vital and more solid networking among the key stakeholders.

The integration of the Living Lab approach with a large event format, on the one hand, provided updates cyclically on the project and allowed citizens to get informed on the current initiatives and to contribute with different ideas and suggestions (as in the kick-off event). On the other hand, organizing and steering such a process required a considerable investment of time and human resources: activities must flow across scales in a continuous effort of communication and decisional steps, investigating and deepening each element and their interconnection [35]. Nevertheless, the two big events organized in Hamburg were useful for two reasons: increasing the project's visibility to the broader public (130 and 140 participants respectively) and synthesizing the work done so far.

The several moderation tools and methods deployed along the co-creation process were advantageous, particularly the online-tool DIPAS, the orthophoto-carpet, and the graphic recording. Additionally, the Hamburg CLEVER Team has deployed a combination of traditional and innovative tools to facilitate knowledge transfer and support the discussion along the process. The DIPAS tool, with its participatory data-table, was used in the kick-off event to collect stakeholders' opinions and visualizing issues within the project area. The advantage of using such digital participation tools is that all identified issues are immediately linked to geo-data coordinates. After the event, the conversion of stakeholder comments into a digital format was more straightforward and less time-consuming. Therefore, this tool's use led to reduced operational costs and time within the stakeholder participation processes. DIPAS was further used on a second occasion to support citizen participation in the nature experience park along the corridor in ULL 1.

As a further tool, the carpet realized with the orthophoto of Neugraben–Fischbeck was applied successfully during the kick-off event in autumn 2018 and at the district festival "Neugraben Erleben" in 2019. It attracted people to express their interests and issues in the project area: in fact, a carpet of such size catches people's eye and animates them to participate. Therefore, it was regarded as a useful tool for stakeholder activation during phase (i) of the co-creation process.

Lastly, the graphic recording of the discussion was useful to depict the outputs of the meeting visually and, at the same time, create a recognizable design for the participants.

4.2. Stakeholder-Related Aspects of NbS Co-Creation in the Hamburg Case Study

Co-creation requires managing the challenge of engaging local stakeholders to listen to all opinions and empower them to participate in policy-making actively [37]. Concerning policy-making, individual stakeholders' role is discussed in the next subsection concerning their constellations and experiences within the project activities.

Firstly, in a co-creation process, the local stakeholders' network has to be created and, secondly, to be maintained. This includes the management of the various activities conducted. While many scholars claim that local public authorities do not have power or interest in this sense (see [21]), the situation for Hamburg was somewhat different due to the central role of co-creation for the CLEVER Cities project around which the project team organized all other activities in the first two analyzed phases (i) and (ii).

Additionally, the kick-off event format with the large-scale participation created a favorable environment for the citizens to play a central role in defining the topics to debate upon [8]. Concerning the refugee accommodation facilities, the activities developed together with the residents resulted already in physical interventions (see Section 3.2.1). As a result, engagement is one of the critical elements of the NbS design process, as previously discussed.

The district public administration thus functions as initiator and coordinator of the entire process, and it was one of the main stakeholders having interests in the success of the initiatives put in place by the CLEVER Cities project (Figures 3–5). It has the authority to initiate and foster co-creation activities within its project partner capacities and by subcontracting third parties or directly commissioning works entailing the planning, realization, and implementation of different project co-creation activities. With various stakeholders' involvement, the coordination, supervision, and approval of the commissioned work rested with the district public administration, which remained *de facto* always indirectly involved. In this way, the respect of the co-creation principles can be guaranteed if the administration is backing them. Meetings within the project team were held regularly to update on the status of the various interventions.

Additionally, the district public administration participated in several political committee meetings at the district and city level, where the CLEVER Cities project and its activities were documented. However, multiple spot-like interventions, tailored co-creation approaches for these interventions, diverse stakeholders with different institutional settings and experiences, and multilateral agreement rounds for fine-tuning the processes in some cases lead to lengthy procedures and a high workload [38]. Furthermore, since the budgetary authority rested with the district public administration, together with the coordination role and the communication activities with the Hamburg partners as well as internationally in the overall project consortium (cf. [39] p. 14), its decisional power resulted in being very relevant, compared to the other partners. However, the political constellation's multi-level character in Hamburg puts the district public administration in a constant dialogue with the Hamburg ministries level.

The BUKEA contributed with scientific expertise on the natural environment and provided support from and within institutional levels. The Ministry shares a political interest at a city-wide level, being in charge of green roofs strategy while being engaged through CLEVER Cities at the local level (see ULL 2). One of the benefits of having the different city authority levels on board is the potential for upscaling the districts' results to the city-level (vertical integration) and, similarly, its transfer to other districts (horizontal integration).

Having an urban development agency being physically present in the area allowed the project activities to be adequately spread among the residents. Organizing events and attracting people to participate were not the only advantages; also, the profound knowledge of the local (social) situation and their agency's daily contact with citizens were crucial elements for the success of the initiatives (see the role of "Broker" in Busetti and Vecchi [40]). Indeed, the presence of an intermediary organization active in citizen involvement at the local level and knowing some of the most important local stakeholders in person has proved to be fundamental in establishing a stakeholder network in the urban district. This helped the project team in several situations to get into contact with key players and to solve conflicts. Nevertheless, the local development agency's inner knowledge and moderation skills were considered substantially useful for raising awareness on NbS among local stakeholders and citizens and contributing to capacity-building processes.

Furthermore, the co-creation process was largely supported by scientific partners' involvement, both federal and international. While the structure of the framework was provided by the international project partners (UIP, ToC, Co-Creation Pathway), the federal research institutions were in charge of adapting it to the local context and translating the general framework instructions to the specific implementation level. Steering activities, data collection, and analysis of the ToC workshops' results were carried out with the help of the federal scientific partners. The practicality degree was also challenging throughout the project while providing sparks for reflections based on real case implementation. Nevertheless, the involvement of scientific partners facilitated the further elaboration of the main results to be coupled with a broader context and to be able to respond with novel and sound scientific background to the local challenges encountered.

4.3. Recommendations

Based on the aspects discussed in the previous subsections, some critical issues for NbS co-creation organization can be derived, comprising the horizontal and vertical dialogue, the essential characteristics of the stakeholder constellation, and the presence of an overarching guiding framework.

According to the contract signed with the European Commission, the project team initiated the complex participation and implementation process. The public administration played a steering role, among others, because it holds an intermediate position, guiding and supporting the initiatives on the ground and participating at the strategic level. The guiding role often signifies that the public administration plays a central role in the co-creation process. On the local side, the local development agency has been working in many directions, e.g., as a coordinator of the local activities in the urban district, as a mediator between local stakeholders and the district public administration, and as a facilitator for enhancing social inclusion. Since the administration was involved in all processes, it was perceived as close to the citizens' challenges and wishes. The district public administration's leading role in the initiation and coordination of co-creative processes has proven to be beneficial. However, it was only through close cooperation with the urban development agency that residents could be reached and co-creative processes carried out. Therefore, it is highly recommended to combine one planning (public administration) and one implementing body (local development agency) to implement co-creative processes.

The establishment of such cooperation opened up new involvement opportunities and enriched the process with additional ideas. In this sense, communication was particularly relevant in Hamburg, with the scope of informing and involving the broader public and requiring a continuous adaptation to search for a common and understandable language. Additionally, to achieve the desired outcomes defined through the co-creative process, it was necessary to establish dialogues and cooperation not only across administrative levels [41] (p. 26) but also within the institutions themselves [42]. This cross-sectoral dynamic is revealed to be of enormous importance to reaching the project's objectives concerning the inherent characteristics of NbS, which requires a certain degree of interdisciplinary, cross-sectoral cooperation and a broader view of the local challenges. Barriers could be overcome by establishing contacts, building relationships, subscribing to formal and informal cooperation formats. To some extent, changes to correct the process trajectory were envisioned and enabled by facilitation tools.

Concerning the role of civil society and academia, the entire process should be conceived to let citizens and, most importantly, disadvantaged social groups play a direct role in the implementation. This direct involvement that will be translated into co-implementation activities in the next phases of the project fosters the sense of ownership of the various co-implemented interventions and further maintains their interest to continue to take care of these interventions after their realization. Direct involvement also contributes to a certain degree of empowerment. The process's learning effects are fundamental to continuity in applying co-creation activities at the local level. Thus, thanks to the first-hand experience in dealing with NbS through hands-on workshops, it can help enhance the understanding of such complex topics to the general public. In this sense, the workshops serve as a knowledge transfer tool.

Generally, combining the several local aggregation and meeting formats (local UIPs) with the more comprehensive and outreaching annual UIP events can be considered a reasonable practice. This integration was useful for connecting the various local interventions under a broader and shared vision and informing and mobilizing a broader and more diverse group of people, thus enhancing the potential for creativity in the intervention design. Furthermore, it helps gather and synthesize the work conducted locally in a presentable way for dissemination and visibility purposes.

Working with citizens usually entails a more significant effort to prepare the various steps within the co-creation process. The timing and content of communication with stakeholders affect their willingness to participate. Keeping their interest high and showing the results coming from the discussions regularly increases motivation. Further, laypersons' involvement requires finding suitable communication formats, instruments, and wording where all can meet and agree upon, which should avoid reaching only certain groups and excluding others. The risk of excluding specific participant sets is amplified when working with vulnerable social groups, such as in ULL 1. Starting the dialogue with the refugees requires more extended and more careful preparation. Besides the necessity of hiring translators for various languages, the main challenge is to awaken their interest in a place where they presumably intend to live only temporarily (e.g., opened in 2016, the refugee camp in Vogelkamp– Neugraben is planned to be closed by 2026).

In the current scientific discourse, co-creation processes are discussed as a potent tool to sustain the development of NbS through the involvement of all social groups [7,42] when addressing sustainability and resiliency in cities. The Co-Creation Pathway provides a guide for consistent implementation of this process yet allowing for adjustments when applied at the local level as presented in [14,30]. In Hamburg, the adaptation of the Co-Creation Pathway to the local context was facilitated by the scientific institutions by a broader involvement of civil society and other social groups in addition to the usual suspects [21]. As previously mentioned, communication and conflicts were fundamental for achieving the project's objectives because they helped reach new levels of knowledge and understanding of sustainable urban development and the role of NbS. The continuous integration of local knowledge, the support from academia, and the business sector's involvement create a potent base for developing further the initial ideas of a project. Improved organization and participation strategies include providing a foundation for a discourse, collecting examples to implement possible ideas, and proposing alternative and ad hoc approaches [4,37]. This is achieved due to a mutual and constant dialogue with the local partners and agreement on project continuation.

5. Conclusions

This paper provided insight into the current situation (November 2020) of the advancement of the NbS interventions achieved in the CLEVER Cities project in Hamburg and the interplay of the several stakeholders that contributed to the process. Hence, the analyses conducted to draft this paper refer to the interim project results. Nevertheless, some relevant outcomes can be derived to continue the work within the CLEVER Cities project timeframe and for the scientific discussion on stakeholder involvement in planning and designing NbS and some thoughts for their implementation.

Given the intrinsic multi-dimensionality of NbS, they can be identified to address complex issues characterized by uncertainty and interdependence. NbS are claimed to answer several current societal challenges [43], and foster local economies and allow inclusion simultaneously [7]. According to a co-creative model, the CLEVER Cities project's answer to this complexity is to bring various views, knowledge, and areas of expertise under the same roof.

As discussed, the project team gathered representatives from a wide range of backgrounds (Section 3.1), complemented by the sectorial expertise of the different stakeholders involved that suited the best specific interventions. It is possible to notice that the project team is frequently present as a core stakeholder within the ULL (Figures 3–5). These are the district public administration and the local development agency: for legal, political, and economic resources, the former; for proximity with the local population and experience with co-creation processes, the latter. These two stakeholders had to overview all activities being conducted and punctually activate the necessary stakeholders to address specific challenges of the selected areas of intervention.

The co-creation processes [18] (p. 273) might be a difficult and tortuous path, implying a considerable amount of time and resources to dedicate to its sustainment. Instruments and cooperation modes are critical and should not be underestimated. Concerning the Hamburg experience, it can be stated that the co-creation process benefited from the support of an overarching strategy. As an additional benefit, co-creating the NbS contributed to generating a learning effect among the participants. From the point of view of the co-creation approach's resilience in the case-site, it became apparent that formal and informal cooperation mechanisms have to be considered early in the process and should outlast the research project's duration reach a self-sustaining state.

The case study experiences showed that the experimental approach of research-based interventions could lead to new insights that will transform the existing governance settings. Hence, the co-creation principles that sustain the NbS development effort are demanding a restructuring of the decision-making processes by learning from the approaches mentioned above and becoming common practice (see [44,45]). The enlargement of participatory design solutions includes foreseeing a certain degree of flexibility, which allows reacting to problems, offering alternative solutions, and deploying different mechanisms to connect, commit, and share decision-making power with ad hoc governance models.

The described co-creation activities, the chosen pathway, and the involved stakeholders can be taken as examples of how NbS co-creation can be steered, supported, and facilitated. It was recognized that the NbS topic is of great interest among the different social groups and contributes to uniting people while achieving beneficial results for their neighborhoods and cities [8]. Additionally, district public administrations fostering the NbS idea can profit from enhanced visibility in the district and resulting benefits from new networks within and beyond the authorities' boundaries.

All this said, by addressing governance and decision-making structures, bringing together different expertise in the joint effort to address significant societal challenges, NbS are claimed to unlock potential for building resilient cities and fostering more shared sustainable development.

Supplementary Materials: The following are available online at https://www.mdpi.com/2071-105 0/13/5/2572/s1.

Author Contributions: Conceptualization, A.A., A.R., S.K.-C., and J.K.; methodology, A.A. and A.R.; writing—original draft preparation, A.A., A.R., and S.K.-C.; writing—review and editing, A.R., S.K.-C., and J.K.; funding acquisition, J.K. All authors have read and agreed to the published version of the manuscript.

Funding: This research was funded by the European Union's Horizon 2020 innovation action program under grant agreement number 776604.

Institutional Review Board Statement: Not applicable.

Informed Consent Statement: Not applicable.

Data Availability Statement: The data presented in this study are available on request from the corresponding author. Data are taken from Konjaria-Christian et al. [24]. The data are not publicly available due to the internal character of the deliverable.

Acknowledgments: This work has been presented at the Greening Cities Shaping Cities Symposium in October 2020 (https://www.greeningcities-shapingcities.polimi.it/ accessed on 18 November 2020). The work was prepared in the framework of the European Horizon 2020 project Clever Cities. Herewith, the authors would like to thank Jan Pastoors (DHH) and steg Hamburg mbH for their comments and suggestions for the draft and the provision of information related to local stakeholders and activities; and Fabrizio M. Amoruso from the Sustainable Architecture Integrated Design Lab of SKKU (Republic of Korea) for his general feedback on the text. Lastly, the authors would like to thank Israa Mahmoud and Eugenio Morello (Politecnico di Milano) as editors of this Special Issue.

Conflicts of Interest: The authors declare no conflict of interest.

References

1. Maes, J.; Jacobs, S. Nature-Based Solutions for Europe's Sustainable Development. *Conserv. Lett.* **2017**, *10*, 121–124. [CrossRef]
2. Biermann, F. The future of 'environmental' policy in the Anthropocene: Time for a paradigm shift. *Environ. Politics* **2020**, *21*, 1–20. [CrossRef]
3. Mahmoud, I.; Morello, E. Co-Creation Pathways as a catalyst for implementing Nature-based Solution in Urban Regeneration Strategies: Learning from CLEVER Cities framework and Milano as test-bed. *Urban Inf.* **2018**, *278*, 204–210.
4. Frantzeskaki, N.; McPhearson, T.; Collier, M.J.; Kendal, D.; Bulkeley, H.; Dumitru, A.; Walsh, C.; Noble, K.; van Wyk, E.; Ordóñez, C.; et al. Nature-Based Solutions for Urban Climate Change Adaptation: Linking Science, Policy, and Practice Communities for Evidence-Based Decision-Making. *BioScience* **2019**, *69*, 455–466. [CrossRef]

5. European Commission (EC). Nature-Based Solutions: Nature-Based Solutions and How the Commission Defines Them, Funding, Collaboration and Jobs, Projects, Results and Publications. Available online: https://ec.europa.eu/info/research-and-innovation/research-area/environment/nature-based-solutions_en (accessed on 15 February 2021).

6. European Commission (EC). Nature-Based Solutions Research Policy: EU Research Policy, What Nature-Based Solutions Are, Background, News and Documents. Available online: https://ec.europa.eu/info/research-and-innovation/research-area/environment/nature-based-solutions/research-policy_en (accessed on 15 February 2015).

7. IUCN. Global Standard for Nature-based Solutions. In *A User-Friendly Framework for the Verification, Design and Scaling up of NbS*, 1st ed.; IUCN: Gland, Switzerland, 2020; ISBN 978-2-8317-2058-6.

8. Ferreira, V.; Barreira, A.; Loures, L.; Antunes, D.; Panagopoulos, T. Stakeholders' Engagement on Nature-Based Solutions: A Systematic Literature Review. *Sustainability* **2020**, *12*, 640. [CrossRef]

9. Frantzeskaki, N.; Kabisch, N. Designing a knowledge co-production operating space for urban environmental governance—Lessons from Rotterdam, Netherlands and Berlin, Germany. *Environ. Sci. Policy* **2016**, *62*, 90–98. [CrossRef]

10. Kemp, R.; Loorbach, D. Governance for Sustainability through Transition Management. In Proceedings of the Open Meeting of Human Dimensions of Global Environmental Change Research Community, Montreal, QC, Canada, 16 October 2003; (accessed on 26 February 2021).

11. Frantzeskaki, N.; Loorbach, D.; Meadowcroft, J. Governing societal transitions to sustainability. *Int. J. Sustain. Dev.* **2012**, *15*, 19–36. [CrossRef]

12. CLEVER Cities. About the Project. Available online: https://clevercities.eu/the-project/ (accessed on 13 September 2020).

13. Bulkeley, H.; Coenen, L.; Frantzeskaki, N.; Hartmann, C.; Kronsell, A.; Mai, L.; Marvin, S.; McCormick, K.; van Steenbergen, F.; Voytenko Palgan, Y. Urban living labs: Governing urban sustainability transitions. *Curr. Opin. Environ. Sust.* **2017**, *22*, 13–17. [CrossRef]

14. Morello, E.; Mahmoud, I.; Gulyurtlu, S. CLEVER Cities Guidance on Co-Creating Nature-Based Solutions: PART II-Running CLEVER Action Labs in 16 Steps. 2018. Available online: https://clevercities.eu/resources/deliverables/ (accessed on 3 September 2020).

15. Carayannis, E.G.; Campbell, D.F. Triple Helix, Quadruple Helix and Quintuple Helix and How Do Knowledge, Innovation and the Environment Relate To Each Other?: A Proposed Framework for a Trans-disciplinary analysis of Sustainable development and Social Ecology. *Int. J. Soc. Ecol. Sustain. Dev.* **2010**, *1*, 41–69. [CrossRef]

16. Weiss, C.H. Nothing as practical as a good theory: Exploring theory-based evaluation for comprehensive community initiatives for children and families. In *New Approaches to Evaluating Community Initiatives: Concepts, Methods and Contexts*; Connell, J., Kubisch, A.C., Schorr, L.B., Weiss, C.H., Eds.; The Aspen Institute for Humanistic Studies: New York, NY, USA, 1995; ISBN 0-89843-167-0.

17. Perez, I.G.; Cantergiani, C.; Boelman, V.; Davies, H.; Murphy-Evans, N. CLEVER Monitoring and Evaluation Framework. Available online: https://clevercities.eu/resources/deliverables/ (accessed on 14 January 2021).

18. Mahmoud, I.; Morello, E. Co-creation Pathway for Urban Nature-Based Solutions: Testing a Shared-Governance Approach in Three Cities and Nine Action Labs. In *Smart and Sustainable Planning for Cities and Regions: Results of SSPCR 2019—Open Access Contributions*; Bisello, A., Vettorato, D., Ludlow, D., Baranzelli, C., Eds.; Springer International Publishing: Cham, Switzerland, 2021; pp. 259–276. ISBN 978-3-03-057764-3.

19. Hooghe, L.; Marks, G. Types of Multi-Level Governance. *J. Eur. Integr.* **2002**, *5*. [CrossRef]

20. Betsill, M.M.; Bulkeley, H. Cities and the Multi-level Governance of Global Climate Change. *Glob. Gov.* **2006**, *12*, 141–159. [CrossRef]

21. Frantzeskaki, N.; Rok, A. Co-producing urban sustainability transitions knowledge with community, policy and science. *Environ. Innov. Soc. Transit.* **2018**, *29*, 47–51. [CrossRef]

22. Dente, B. *Understanding Policy Decisions*; Springer: Cham, Switzerland, 2014; ISBN 978-3-319-02520-9.

23. Bryson, J.M. What to do when stakeholders matter: Stakeholder identification and analysis techniques. *Public Manag. Rev.* **2014**, *6*, 21–53. [CrossRef]

24. Konjaria-Christian, S.; Pastoors, J.; Arlati, A.; Rödl, A.; Berghausen, M.; Quanz, J.; Robert, J.; Carlini, G. CAL Specific co-Implementation Plan. 2020. Available online: https://clevercities.eu/resources/deliverables/ (accessed on 3 September 2020).

25. Statistisches Amt für Hamburg und Schleswig-Holstein. Bevölkerung in Hamburg am 31.12.2019: Auszählung aus dem Melderegister. Available online: https://www.statistik-nord.de/fileadmin/Dokumente/Statistische_Berichte/bevoelkerung/A_I_S_1_j_H/A_I_S1_j19.pdf (accessed on 15 October 2020).

26. Statistisches Amt für Hamburg und Schleswig-Holstein. Hamburger Stadtteil-Profile Berichtsjahr 2018. Available online: https://www.statistik-nord.de/fileadmin/Dokumente/NORD.regional/NR21_Statistik-Profile_HH-2018.pdf (accessed on 15 October 2020).

27. Hamburg-stadtteile.de. Hamburg Stadtteile. Available online: http://www.hamburgs-stadtteile.de/hamburg/allnsbin/ (accessed on 4 January 2021).

28. Bezirk Harburg. *Neugraben-Fischbek: Aktives Fördergebiet*; Bezirk Harburg: Hamburg, Germany, 2017. Available online: https://www.hamburg.de/harburg/rise-neugraben-fischbek/ (accessed on 3 September 2020).

29. Frei und Hansestadt Hamburg (FHH). Sozialmonitoring Integrierte Stadtteilentwicklung: Ergebnisbericht 2019. Available online: https://www.hamburg.de/contentblob/13278936/8e978b2127057b0e459f30d81ef9f00c/data/d-sozialmonitoring-bericht-2019.pdf (accessed on 3 September 2020).

30. Morello, E.; Mahmoud, I.; Gulyurtlu, S.; Boelman, V.; Davis, H. CLEVER Cities Guidance on Co-Creating Nature-Based Solutions: PART I-Defining the Co-Creation Framework and Stakeholder Engagement. 2018. Available online: https://clevercities.eu/resources/deliverables/ (accessed on 3 September 2020).
31. Bezirk Harburg. *Das Projekt CLEVER Cities Nahm am Stadtteilfest 'Neugraben Erleben' 2019 Teil*; Bezirk Harburg: Hamburg, Germany, 2019. Available online: https://www.hamburg.de/harburg/horizon-2020-clever-cities/13703868/clever-cities-bei-neugraben-erleben/ (accessed on 3 September 2020).
32. Morello, E.; Mahmoud, I. CLEVER Cities-Co-Design Planning: Steps 07/08-Co-Creation Pathway. Available online: https://clevercitiesguidance.files.wordpress.com/2019/10/steps-7-and-8-instructions.pdf (accessed on 18 February 2021).
33. Arnstein, S.R. A Ladder Of Citizen Participation. *J. Am. Inst. Plan.* **1969**, *35*, 216–224. [CrossRef]
34. Optigrün Dachbegrünung. Smart Flow Control SFC. Available online: https://www.optigruen.de/produkte/ablaufdrosseln/smart-flow-control-sfc/ (accessed on 30 September 2020).
35. Connell, J.; Kubisch, A.C. Applying a Theory of Change Approach to the Evaluation of Comprehensive Community Initiatives: Progress, Prospects, and Problems. Available online: http://dmeforpeace.org/sites/default/files/080713%20Applying+Theory+of+Change+Approach.pdf (accessed on 30 September 2020).
36. Mauser, W.; Klepper, G.; Rice, M.; Schmalzbauer, B.S.; Hackmann, H.; Leemans, R.; Moore, H. Transdisciplinary global change research: The co-creation of knowledge for sustainability. *Curr. Opin. Environ. Sustain.* **2013**, *5*, 420–431. [CrossRef]
37. Vignola, R.; Locatelli, B.; Martinez, C.; Imbach, P. Ecosystem-based adaptation to climate change: What role for policy-makers, society and scientists? *Mitig. Adapt. Strateg. Glob. Chang.* **2009**, *14*, 691–696. [CrossRef]
38. Frantzeskaki, N.; Borgström, S.; Gorissen, L.; Egermann, M.; Ehnert, F. Nature-Based Solutions Accelerating Urban Sustainability Transitions in Cities: Lessons from Dresden, Genk and Stockholm Cities. In *Nature-based Solutions to Climate Change Adaptation in Urban Areas: Linkages between Science, Policy and Practice*; Kabisch, N., Korn, H., Stadler, J., Bonn, A., Eds.; Springer Imprint: Cham, Switzerland, 2017; pp. 65–88. ISBN 978-3-319-53750-4.
39. Dushkova, D.; Haase, D. Not Simply Green: Nature-Based Solutions as a Concept and Practical Approach for Sustainability Studies and Planning Agendas in Cities. *Land* **2020**, *9*, 19. [CrossRef]
40. Busetti, S.; Vecchi, G. Process tracing change management: The reform of the Italian judiciary. *Int. J. Public Sect. Manag.* **2018**, *31*, 566–582. [CrossRef]
41. Cantergiani, C.; Perez, I.G.; Menny, M.; Murphy-Evans, N.; Casagrande, S. UIP Launch: Urban Innovation Partnership (UIP) Launching. Available online: https://clevercities.eu/resources/deliverables/ (accessed on 11 January 2021).
42. Mahmoud, I.; Morello, E. Are Nature-based solutions the answer to urban sustainability dilemma? The case of CLEVER Cities CALs within the Milanese urban context. In *Proceedings of the Atti della XXII Conferenza Nazionale SIU., L'Urbanistica italiana di fronte all'Agenda 2030. L'Urbanistica italiana di fronte all'Agenda 2030. Portare territori e comunità sulla strada della sostenibilità e della resilienza, Matera-Bari, 5–7 June*; Planum Publisher: Roma Milano, Italy, 2019; ISBN 9788899237219.
43. IUCN. *Nature-Based Solutions to Address Global Societal Challenges*; Cohen-Shacham, E., Walters, G., Janzen, C., Maginnis, S., Eds.; IUCN: Gland, Switzerland, 2016; ISBN 978-2-8317-1812-5.
44. Bulkeley, H.; Kern, K. Local Government and the Governing of Climate Change in Germany and the UK. *Urban Stud.* **2006**, *43*, 2237–2259. [CrossRef]
45. Kemp, R.; Loorbach, D.; Rotmans, J. Transition management as a model for managing processes of co-evolution towards sustainable development. *Int. J. Sustain. Dev. World Ecol.* **2009**, *14*, 78–91. [CrossRef]

sustainability

MDPI

Article

Valuing the Invaluable(?)—A Framework to Facilitate Stakeholder Engagement in the Planning of Nature-Based Solutions

Sophie Mok [1,*], Ernesta Mačiulytė [2], Pieter Hein Bult [1] and Tom Hawxwell [1,3]

1   Fraunhofer Institute for Industrial Engineering IAO, Fraunhofer IAO, 70569 Stuttgart, Germany; Pieter-Hein.Bult@iao.fraunhofer.de (P.H.B.); tom.hawxwell@hcu-hamburg.de (T.H.)
2   Institute of Human Factors and Technology Management (IAT), University of Stuttgart, 70741 Stuttgart, Germany; ernestam5@gmail.com
3   Faculty of Urban Planning and Regional Development, HafenCity University, 20457 Hamburg, Germany
*   Correspondence: sophie.mok@iao.fraunhofer.de

Abstract: Nature-based solutions (NBS) have emerged as an important concept to build climate resilience in cities whilst providing a wide range of ecological, economic, and social co-benefits. With the ambition of increasing NBS uptake, diverse actors have been developing means to demonstrate and prove these benefits. However, the multifunctionality, the different types of benefits provided, and the context-specificity make it difficult to capture and communicate their overall value. In this paper, a value-based framework is presented that allows for structured navigation through these issues with the goal of identifying key values and engaging beneficiaries from the public, private, and civil society sector in the development of NBS. Applied methods such as focus groups, interviews, and surveys were used to assess different framework components and their interlinkages, as well as to test its applicability in urban planning. Results suggest that more specialized "hard facts" might be needed to actually attract larger investments of specific actors. However, the softer and more holistic approach could inspire and support the forming of alliances amongst a wider range of urban stakeholders and the prioritization of specific benefits for further assessment. Consequently, it is argued that both hard and soft approaches to nature valuation will be necessary to further promote and drive the uptake of NBS in cities.

Keywords: nature-based solutions; greening cities; urban governance; urban planning

Citation: Mok, S.; Mačiulytė, E.; Bult, P.H.; Hawxwell, T. Valuing the Invaluable(?)—A Framework to Facilitate Stakeholder Engagement in the Planning of Nature-Based Solutions. *Sustainability* **2021**, *13*, 2657. https://doi.org/10.3390/su13052657

Academic Editor: Israa H. Mahmoud

Received: 12 December 2020
Accepted: 23 February 2021
Published: 2 March 2021

Publisher's Note: MDPI stays neutral with regard to jurisdictional claims in published maps and institutional affiliations.

1. Introduction

Nature-based solutions (NBS) have been widely promoted as an important means for cities to combat the pressing challenges of climate change and ongoing urbanization. Beyond their direct contributions to lowering risks associated with climate-related pressures such as flooding and heat stress, NBS can also deliver diverse indirect co-benefits related to aspects such as health and well-being, economic opportunities, or addressing social issues [1]. According to the European Commission they are defined as "actions inspired by, supported by or copied from nature and which aim to help societies address a variety of environmental, social and economic challenges in sustainable ways" [2]. As cities seek to increase their resilience against climate change impacts whilst dealing with budgetary constraints, the effective valuation and communication of these multiple functions of NBS have become increasingly important [3,4]. However, the strong context-dependency, the diffuse nature of many ecosystem services, as well as the multifunctionality of NBS make it difficult to paint a clear picture of the specific value NBS create for different urban stakeholders. It is at least partly for these reasons that the development of alternative business, governance, and financing models, taking into account diverse public and private actors, has been identified as an important challenge for mainstreaming NBS in urban planning and design [1].

This paper presents and discusses the logic and individual components of a framework, which aims to enable targeted and user-friendly navigation through this complex issue of NBS valuation. It has been developed within the SCC2 project Urban Nature Labs (UNaLab), which is funded by the European Commission under the Horizon 2020 research and innovation program and seeks to contribute to "the development of smarter, more inclusive, more resilient and more sustainable cities through the implementation of nature-based solutions" [5]. The framework aims to support urban planners and practitioners with identifying NBS values in a structured way to inform decision-making and stakeholder engagement around NBS implementation [6]. The first section discusses the challenges and trade-offs associated with nature valuation and makes a case for "softer" value assessment tools to encourage awareness-raising, stakeholder engagement, and mobilize local actors around NBS to complement "harder" valuation mechanisms. The second section provides an overview of the methods applied to develop the framework. The third section describes the different framework components and their linkages. In the final section, some reflections are made on the utility of such an approach and its relevance for future (applied) research.

1.1. Challenges and Trade-Offs in Approaches to NBS Valuation

The term "value" is rooted in philosophy and only from there entered economics and financial theory [7]. Against the background of nature valuation, Costanza 2003 states that "value ultimately originates in the set of individual and social goals to which a society aspires" [8] (p. 24). Whereas conventional economic value is mostly focused on the maximization of individual utility and its expression in monetary terms, NBS also strongly relate to other goals (and resulting values), such as sustainability or social wellbeing [7–9]. As a result, an assessment of value should be based on the contribution to achieving these multiple goals [8].

Many initiatives and approaches have been developed which seek to better describe and capture the values of NBS and green infrastructure components. It is assumed that through a better understanding and being able to effectively quantify the benefits, the evidence base will pave the way for the development of new financing and business models to facilitate planning and uptake of NBS in cities. In "The Economics of Valuing Ecosystem Services and Biodiversity," Pascual et al. give an overview of different approaches to nature and ecosystem valuation. Most importantly, they distinguish between biophysical valuation which derives value by measuring the physical costs involved (such as labour, energy, or material input), and preference-based methods that build on the assumption that value derives from the subjective preferences of humans [9]. Whereas such approaches may contribute to better value assessment, capture, and communication, controversies remain with regards to expressing the variety of different values and benefit types in a single unit of measure, such as money [10,11]. Out of this discussion, multi-criteria analyses have emerged as an alternative to formally integrate multiple values of different units in decision-making [12].

Against this background, various methods and instruments exist, ranging from calculative approaches, such as established market valuation techniques, to more holistic assessments incorporating wider sustainability impacts [4,9]. These methods of valuation serve different purposes and should collectively play an important role in mainstreaming NBS. Whilst hard quantification and monetization techniques are typically highly technical and NBS specific (i.e., developed to assess impacts of specific NBS types), softer valuation approaches are emphasizing advocacy and awareness-raising [9]. However, there are certain trade-offs involved in the development of tools to assess the diffuse and complex nature of the benefits associated with NBS and to take context-specific preconditions into account. Along with the benefits, the associated beneficiary structure and the individual potential and willingness to invest in such an NBS differ greatly. For example, an intensive green roof with public access in a cold northern European city will likely create very different benefits for different beneficiaries than it would on a private building in a warm southern European city. This is because the various functions of NBS translate differently into benefits depending on the demands and preconditions of the surrounding context. This strong context specificity, the diffuse nature of many ecosystem services, as well as the multifunctionality of NBS imply that the establishment of the generic business models for NBS that would attract sustainable sources of finance is rather challenging. This uncertainty factor significantly increases when trying to anticipate the impacts of a planned future intervention (ex-ante). Thus, the creation of tools to better understand value creation through NBS will inevitably create trade-offs between their effectiveness in quantifying the benefits and their role in communication, advocacy, and mobilizing support. While both are of significant importance, the former will allow NBS to be more effectively integrated into prevailing value calculation systems, while the latter can function as a means to expand comprehensions of value creation, provide insights into other points of view and integrate different types of benefits and meanings that stakeholders associate with them.

This framework was developed within a large European Project involving eight European cities and two observer cities, all with varying cultural, institutional, and climatic contexts. It should also function as a mechanism to support future interventions (ex-ante). Thus, the imperative of interoperability of the framework between diverse contexts further limits the potential for hard valuation approaches. Therefore, the ambition of the described approach is not to quantify or monetize specific benefits of NBS, but rather to serve as a means for awareness-raising and mobilizing local support in diverse urban contexts.

1.2. A Case for "Softer" Approaches to NBS Valuation

One of the central strengths of the NBS concept is its broader more holistic consideration of links between social and ecological systems, as well as the potential broader societal impacts of interventions in complex systems. Due to its integrative, systemic approach, NBS has the potential to overcome a prevailing bias for developments that focus on short-term economic gains and effectiveness [13]. Additionally, associated benefits can be more public or private in nature (or often a combination of both), making them susceptible to collective action problems [14]. The importance of mobilizing local support behind NBS is reflected in their proponents' emphasis on open stakeholder engagement processes with diverse actors that are hoped to support with the bridging of social and economic interests [1,3]. This is reflected in recent rounds of European funding associated with NBS, in which NBS are typically combined with prescriptions of integrating alternative governance models and a high degree of stakeholder engagement and co-creation [2]. As a rather recent, rapidly evolving, and applied concept co-creation thereby describes the co-design process in a group of stakeholders [15]. Effective partnerships amongst local actors have been identified as a central enabling factor for effective NBS rollout, while awareness-raising activities and mobilizing amongst local actors can play an important role in addressing some of the central barriers to NBS uptake including countering path dependencies, institutional fragmentation, and the uncertainty associated with NBS implementation processes and benefits [16]. Due to the abstract benefits associated with a given NBS, effective discussions and decision making amongst stakeholders are challenging when its values are not represented in a visible way. By presenting soft benefits in a systematic and adaptable way, informed discussion can be possible, and arguments may be steered through visible examples. This could help overcome the routine lock-in by increasing awareness amongst urban planners (and other urban actors) of the impacts of NBS from their own and other actor perspectives, facilitating communication between silos. Beyond building intersectoral bridges, such approaches can support improved integration of knowledge between academics and planners, which has been identified as an important factor for supporting NBS uptake [17].

Others have articulated the importance of effectively communicating nonmaterial benefits of NBS in a persuasive manner in ways that benefits can be accounted for and traded off in common framings [18–20]. It is in this context that the described framework has been developed. Taking into account the trade-offs associated with nature valuation and attempting to develop an approach that is applicable in multiple contexts ex-ante, it aims to support urban planners and practitioners with identifying NBS values in a structured way to inform decision-making and stakeholder engagement around NBS implementation in early stages of NBS development. This approach does not aim to discount "harder" approaches to nature valuation, but rather tries to integrate these through directing users towards potential valuation methods, based on their own selections. Nevertheless, the primary function of the framework is communication and awareness-raising, and it should be applied at the early stages of collaboration processes around NBS development.

2. Methodological Review

The development of the framework followed a rather applied research approach. Broad desktop research was performed to develop an overview of existing evidence related to NBS functions, benefits, and beneficiaries, as well as existing valuation and quantification tools and financing options. After organizing the findings, diverse focus group sessions, expert interviews, and surveys were conducted to fill existing knowledge gaps and link the identified components of the resulting framework. Focus group sessions consisted of interdisciplinary working groups of researchers (e.g., from economics, environmental sciences, social sciences, urban planning, or biology background) and city experts from the UNaLab consortia. All focus groups were conducted in the frame of UNaLab project activities and, depending on the topic, different experts were involved. Finally, the logic of the framework was operationalized and applied as part of a series of NBS roadmapping workshops in the five cities Stavanger (NO), Cannes (FR), Castellón (ES), Prague (CZ), and Başakşehir (TU). Furthermore, feedback on its applicability and usefulness in current planning processes was gathered through targeted and semi-structured interviews with 12 experienced urban planners across Western Europe (Eindhoven (NL), Arnhem (NL), Tilburg (NL), Apeldoorn (NL), Zwolle (NL), London (UK), Gent (BE), Freiburg (DE), Hamburg (DE), and Ludwigsburg (DE)). The gathered reactions helped to finetune and improve the framework, as well as to better define potential application areas. Table 1 summarizes the different methods used to develop the different framework components and highlights the links to other research activities within the H2020 UNaLab project.

Table 1. Data sources and methods used to develop the different framework components.

Framework Component	Method	Link to UNaLab Research
Challenges, nature-based solutions (NBS), and functions	Desktop research on NBS functions and existing valuation and performance measurement approaches (2018–2019)	NBS Technical Handbook [21]
Linking functions to beneficiaries	Expert survey (December 2019 to January 2020) involving 26 participants from 15 different countries, all involved in NBS-related projects (65% answering from research and 35% from city perspective)	
Beneficiaries, benefit types, and value capture potential	Desktop research and expert focus groups on beneficiaries, benefit types, and value capture potential (2018–2019)	
Financing Options	Desktop research and expert focus groups on financing options and strategies (2018–2019)	Business Models and Financing Strategies for NBS [22]
Testing of the framework	Semi-structured interviews with 12 urban planners (March 2019 to June 2019) to test the usability of the concept Application and discussion of the framework logic in Roadmapping Workshops in five European cities	Use of the framework in strategy development within the UNaLab follower cities [23]

3. Results

3.1. A Value-Based Framework for Stakeholder Engagement in NBS Development

The presented framework should enable planners and users to take different types of benefits into account (even if not quantified) and identify key beneficiaries and potential financing options that could be involved and applied in a given NBS project. It builds on three major pillars:

1. The different functions of NBS and their relation to specific urban challenges
2. The key beneficiaries of NBS functions and their individual benefits
3. Different financing options for NBS

Figure 1 shows the different components, linkages, and the general logic of the framework, all of which will be further discussed in the following sections.

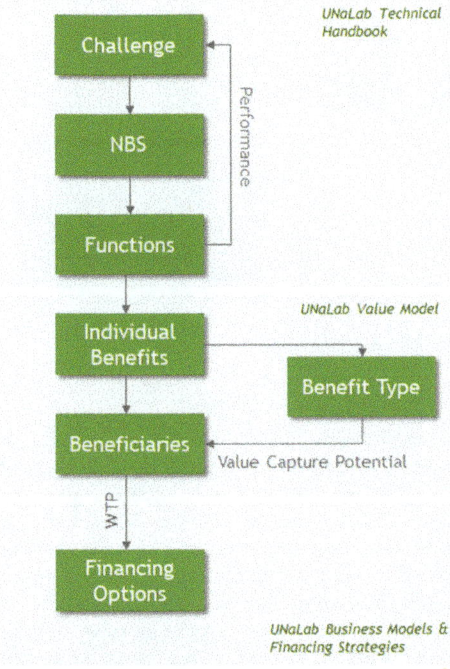

Figure 1. Components and logic of a value-based framework for stakeholder engagement around NBS (further developed from Mok et al., 2019 [6]).

3.2. NBS to Solve Urban Challenges

The concept of NBS has only recently evolved and covers different existing approaches, such as green infrastructure or ecosystem-based adaptation. It can thus be best understood as an overarching umbrella concept, which unites those various approaches from different research domains or policy contexts. However, a common feature of NBS is the focus on addressing societal and urban challenges through a range of different functions and ecosystem services [24,25]. In the frame of the UNaLab project, a technical handbook is being developed, which builds on these urban challenges, describes different NBS, and indicates their multiple functions alongside an estimate of the individual performance. Functions are thereby of a technical nature and can for instance relate to the ability of NBS to provide shade, retain surface water runoff, or filter the air. As context-specificity plays a big role in estimating the individual performance, approximate values in terms of no, low, medium, or high performance are indicated for each NBS and function [21]. Whilst the handbook will be further developed until the end of the project, Figure 2 summarizes the different categories and functions, which were used in the presented framework. The focus was thereby laid on functions that address climate and water-related urban challenges, as well as such that cover social services and biodiversity aspects.

Figure 2. NBS function categories (in bold) and functions (in bullets) based on the UNaLab Technical Handbook [21], which were used to build the value-based engagement framework.

3.3. From NBS Functions to Beneficiaries

In this step, the technical NBS functions were translated into individual benefits of different beneficiary groups. Public authorities are frequently grouped as a single homogeneous actor, which can oversimplify the different prevailing logics and priorities between municipal departments and public agencies. Thus, potential beneficiaries in the framework differentiate between municipal departments or public agencies and private entities in the urban realm, as well as citizens and other stakeholders from civil society. Figure 3 shows the results of an expert survey in which the extent of benefit from the given technical functions of NBS was estimated for different types of urban stakeholders. Knowing both, the technical performance of different NBS functions, as well as an estimation of the extent to which urban stakeholders might benefit from these, may help in understanding the potential beneficiary constellation of a selected solution. It is assumed that the identified stakeholders with high benefits across the different functions will have a deeper material vested interest in implementing such NBS and could thus be valuable partners to engage in the planning, design, implementation, and financing stages of related projects. In this framework, the material vested interest could be expressed in terms of Willingness to Invest (WTI), which is introduced in the subsequent sections. If such an interest appears to be absent, this might suggest an information gap or lack of awareness of such benefits. This perspective does not take aspects related to power and power imbalances into account, which must be considered by the planner on a case-by-case basis due to their typical context specificity.

Several trends could be observed when analyzing the survey results. Overall, it was indicated that socio-economic functions, biodiversity, and water purification are the easiest functions to rate, with 89%, 82%, and 81% of participants describing the ease of linking beneficiaries to functions as "rather easy" or "very easy." On the other hand, provisioning services, cooling, and climate regulation were perceived as the most difficult ones with 50%,

37%, and 36% of participants indicating the rating to be "rather difficult" or "very difficult." Generally, research representatives tend to rate benefits higher than city representatives—a tendency that was particularly noticeable with regards to the "role" of the urban planning department and what it would benefit from (up to 30% higher benefit rating for individual functions by the research representatives). Varying perceptions and some uncertainty could also be perceived as to how much the health sector (health agencies and insurance companies) would benefit from NBS functions. Lastly, the overarching category of "citizens" received some criticism and, due to its broadness and the nature of NBS services, was rated as "top beneficiary" for almost all of the considered functions.

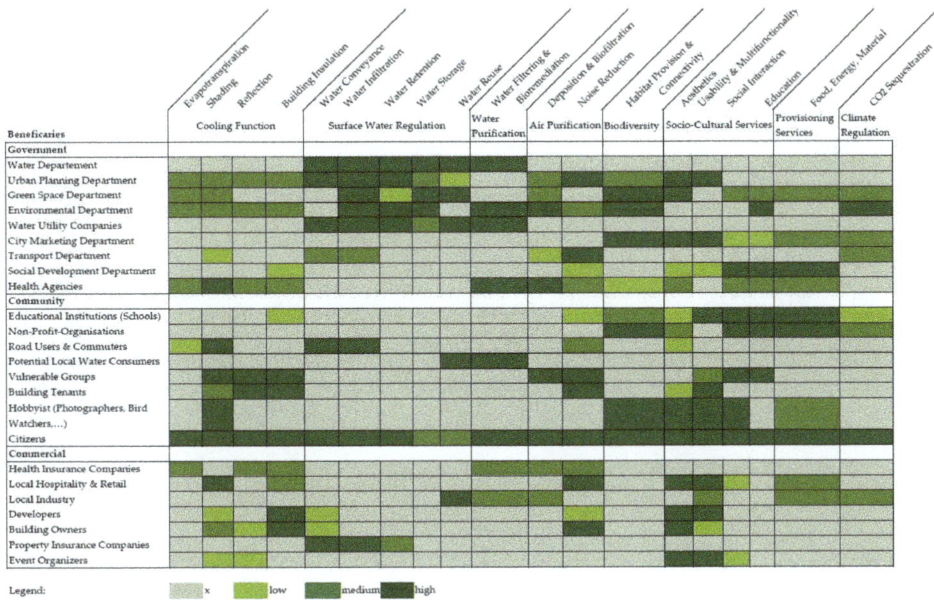

Figure 3. Survey results ranking the extent (low, medium, high) to which different urban stakeholders from public, private, and civil society sector would benefit from different NBS functions. Fields indicated with an "X" were not assessed.

3.4. Benefit Types and Value Capture Potential

Next to the beneficiary constellation, the value capture potential of a given benefit will have a big effect on the individual interest and willingness to invest. In an attempt to better integrate the benefits provided by nature into conventional accounting structures, monetary value capture has been a popular and much-discussed approach (see Figure 4 for more specific methods concerning monetary value capture). It has also been applied to various ecosystem services [9,11,26]. However, most of these approaches are highly context-specific, based on many assumptions, and thus come with a high degree of abstraction and uncertainty. Against this background, we argue that depending on the nature of an individual benefit it might be easier or harder to materialize. In cases where such an approach would lead to a high abstraction level or require in-depth expert knowledge, it might be more tangible to use alternative or complementary assessment methods to capture and communicate the value at hand. To better depict this issue, the potential benefits from NBS were categorized in six different benefit types, which provide a basis for value allocation that includes both monetary as well as non-monetary value of NBS (Figure 4).

Benefit Type	Description	Examples	Assessment Method Examples		
Revenue & Income	The beneficiary directly increases his/her income through the intervention	Increased property values, improved sales through increased foot traffic in business areas	Hedonic pricing, cash flow analysis, business model canvas, NPV analysis	Higher	
Cost Savings	The beneficiary saves money due to the intervention	Better insulation and reduced energy costs, flood risk mitigation	Contingent valuation, cash flow analysis, business model canvas, NPV analysis		Monetary Value Capture Potential
Compliance	The intervention helps the beneficiary to fulfill a mandate or comply with regulations	Fulfilling environmental standards, achieving city goals, risk reduction	Audit, stakeholder consultations, risk and policy mapping, performance indicators		
Active Use	The beneficiary can make direct use of the intervention	Opportunities for recreation and sports	People and visitor counters, qualitative surveys and interviews		
Local Identity & Image	The beneficiary gains recognition and visibility or identifies better with the place	Improved city marketing, CSR, sense of place	Qualitative surveys and interviews, media analysis		
General Wellbeing	The beneficiary's quality of life/health/wellbeing is improved through the intervention	Better air quality, increased contact with blue green spaces, improved mental and physical health	Physical indicators (temperature, air quality, stress levels,…), qualitative surveys and interviews	Lower	

Figure 4. Benefit types and related value capture (exemplary compilation by the authors, further developed from Mok et al., 2019 [6]. Icons by flaticon.com).

3.5. From Beneficiaries to Financing Options

Building on the identified NBS functions and beneficiaries pillars, the framework concludes with different financing options. The objective of this pillar is to inspire the urban stakeholders to consider several potential NBS financing options based on the stakeholder constellation relevant in the local context. In this framework, the connection between the beneficiaries and the financing options relies on the concept of the Willingness to Invest (WTI). It is an adaptation of the Willingness to Pay (WTP) method, where the users of the ecosystem are asked to assign monetary values to the services upon which, following a contingent valuation approach, an aggregate WTP of the population for the specified ecosystem service can be derived. This framework is based on the assumption that beneficiaries have differing interests and capabilities to invest in NBS. It encourages to explore the potential WTI of selected beneficiaries in a given context when planning NBS. The development of the financing options attempts to move from existing case studies whose governance constellations are tied to the site-specific context toward more generic options that can be used to inform the development of new NBS financing constellations.

3.6. NBS Financing Options

Whether a good or service is more of a private or a public nature has direct implications on how it will be governed, financed, and managed [14]. For instance, private sector finance is more likely to be available for NBS that provide marketable products with private good characteristics (e.g., property price premium or agricultural produce). Similarly, NBS, which generate mostly public services (e.g., enhanced water retention due to public green infrastructure, reduced urban heat island effect, etc.) would more heavily rely on public investments [27]. From a private sector perspective, investing in NBS with predominantly public benefits is often not very attractive, due to a weak business model. This is a result of the spatially and temporally diffuse benefits implying a high risk of investments with a questionable availability of high return to compensate for the risk [28–30]. Overcoming this challenge of the limited ability to pinpoint a suitable and rather traditional business model for financing NBS calls for identifying the viable combinations of public-private financing and partnership models, which would allow for sharing of the risks and benefits over time [19]. The potential governance and subsequent financing constellations around NBS interventions should be documented to provide an evidence base for generating further knowledge and inspiration on the available sources of investment.

The present framework builds on 11 financing options that have been compiled as part of the UNaLab project as stated in Table 1 [6,22]. It organizes the identified options following the common government-market-community trichotomy. In its traditional form, this trichotomy implies that any productive activity or resource is owned/executed by the government, market, or community [31]. However, the traditional model has been expanded to account for the hybrid solutions that can occur between the three extremes of the model, such as user fees. Additionally, comprehensive research on the potential external NBS financing schemes and sources was performed. The EU, as well as other international organizations and financial institutions, can, in some cases, be a major contributor to NBS implementation financing. Subsequently, an overview of the funds, financing facilities, and platforms has been compiled to draw the cities' attention to such financing possibilities, as well as their major eligibility criteria that often call for national and/or local political and financial support.

New governance constellations (such as grassroots initiatives or community-managed public space) around (nature-based) urban interventions have been included as well. These stakeholder constellations could sometimes be omitted from "financing models" of NBS as such, since the parties involved might not undertake capital investments in NBS per se. However, they can reduce the financial burden on the city through contributing labor, providing maintenance, or even supporting the construction of NBS. Hence, the framework includes these community-led initiatives under the term "invest" and encourages the user of the model to consider them among other financing options. Additionally, partnerships between private actors and the city for building and maintaining NBS often need to be complemented with a governance model to ensure that the benefits of the public good are realized and that long-term contractual arrangements are upheld (e.g., in Public-Private Partnerships). For these reasons, the framework attempts to integrate both financing, business, and governance models to facilitate a broader understanding of these aspects. Figure 5 illustrates the identified financing options and their allocation in the public-private-community domain.

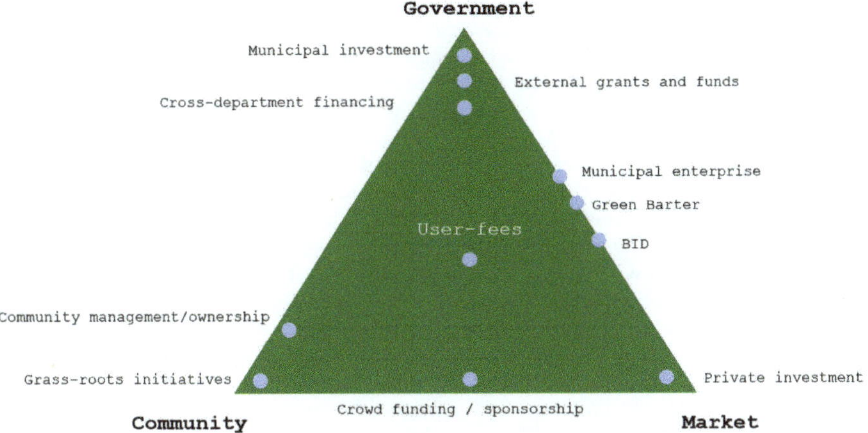

Figure 5. Financing options from UNaLab research, which were used to build the value-based engagement framework (taken from Mok et al., 2019 [6]).

3.7. Reactions to the Framework

Whilst applying and discussing the concept and logic of the presented framework, it became apparent that respondents of the interviews were still very much exploring the field of planning instruments that promote NBS and targeted tools are not structurally used. A much-heard argument against higher uptake was that these tools are very complicated and need expert information, which can become very costly for smaller public administrations. Further, tools were mainly used to monitor whether an intervention works or has the expected impact, and therefore not in a proactive planning manner.

Reacting to the logic and sequence of the presented framework, respondents were positive about how different steps of the framework take the user through a process that is often taken for granted by urban officials with a "green" background. Many indicated that the "soft" benefits are sometimes used as arguments for more NBS inclusion in urban planning and development, but not in such a systematic way, as most benefits are assumed to be known by everyone. It was agreed that by showing the benefits and beneficiaries of NBS more systematically, all actors could reach a similar understanding as to why an intervention is an improvement or necessary.

Reflecting the focus on hard facts and financial aspects in current planning practices, the main points of critique for this approach involved the lack of quantified values in monetary terms, as promoted by other natural capital valuation frameworks such as "The Economics of Ecosystems and Biodiversity" (TEEB) or iTree [9,32]. For instance, some respondents doubted that the soft benefits will be convincing enough to change existing processes and decision-making. However, it was acknowledged that the context specificity does not allow for generalized statements regarding financial outputs. Tools that try to account for different contexts were found to often end up with a large range of figures, which need to be interpreted and are difficult to apply in accounting processes as well. Furthermore, the issue was raised that care should be taken to also account for conflicting interests and potential disservices of NBS, as assuming that everyone will only benefit and want to build NBS could lead to unforeseen drawbacks and conflicts.

More practical feedback mainly focused on the importance of visible examples. There was broad agreement among interview partners that integrating case studies and showing NBS with pictures that describe the atmosphere and impact convinces more actors to include or support NBS in a planning process. Most respondents were not aware of co-investment strategies to include external financing outside of government sources of NBS. This shows that including components such as the different financing options may have a lot of potential to inform and make more people aware of alternative approaches. However, including and linking to many different factors and new insights inevitably raises the complexity of an issue, therefore good and careful moderation, as well as practical examples, were deemed essential to ensure the usefulness of the approach.

All in all, respondents indicated that the framework could be used in various situations. Firstly, it could be used to generally raise the awareness of urban planners regarding the impacts of NBS and inspire them to include more nature-based elements in their planning. Such a framework makes it easier for urban planners to include climate adaptation measures as it uses a discourse focusing on the NBS functionality. Secondly, it could be used for a better network management. It was mentioned that climate adaptation is a very abstract subject and therefore it is important to show the challenge to the partners and show what you want to do or be done about it. By showing benefits not only from a climate adaptation perspective but also including other themes and co-benefits such as aesthetics, health, leisure or education, it can connect targets from various stakeholders and help in forming alliances. It also proposes a standardized language on climate adaptation, functions, and benefits, which makes it easier for different departments to communicate. Thirdly, the framework was perceived as helpful for stakeholders to reach a consensus on what benefits should be prioritized. Lastly, because of the relative simplicity, it could also play a role in participation processes to show the importance of climate adaptation, but also the other functions of NBS.

3.8. Applying the Framework in a Workshop Setting

As a specific use case, the logic and components of the framework were also applied in individual NBS roadmap development processes in five UNaLab follower cities: Stavanger, Cannes, Castellón, Prague, and Başakşehir [23]. Different sets of inspiration cards with information on the three pillars (NBS functions and performance, value creation and beneficiaries, and financing options) were used to inform NBS project development sessions in interdisciplinary groups (see Figure 6). In accordance with the statements above, it could be observed that the link between NBS functions and urban challenges was rather intuitive to most, whereas the steps around identifying beneficiaries and alternative financing options needed more moderation (also deriving from the fact that most of the workshop participants had a more technical or planning-related professional background). Discussions around stakeholder engagement and financing strategies became more thorough the more concrete and advanced a specific NBS project idea was. However, many participants mentioned that dealing with these issues from the very beginning of a project development stage helped in building a more solid case and considering new aspects and alternative ways for realization from early on. It also led to interdisciplinary and cross-departmental discussions around which actors are (or should be) responsible for and involved in NBS implementation in public space. Unclear responsibilities are an important barrier to NBS uptake which could be tackled by more collaborative and joint approaches in project development and implementation [16,33]. It was mentioned by the participants that based on the identified beneficiary structures and financing options, a more in-depth analysis of the local stakeholders, as well as more detailed technical studies on the performance and extent of certain benefits would be necessary next steps to leverage on the learnings and activate the beneficiaries around the defined project. Overall, the approach helped to navigate the process and discover various available solutions and alternative approaches to realization, sparked discussions around different ways to capture identified functions and benefits, and proved to be a good tool for developing a joint and solid "storyline" and strategy for project implementation at an early stage. However, as most of the projects developed in these workshops were still rather vague or fictional (with no immediate intention of implementation), further operationalization and testing of the framework in real-life use cases will be necessary to fully evaluate its potential and impacts.

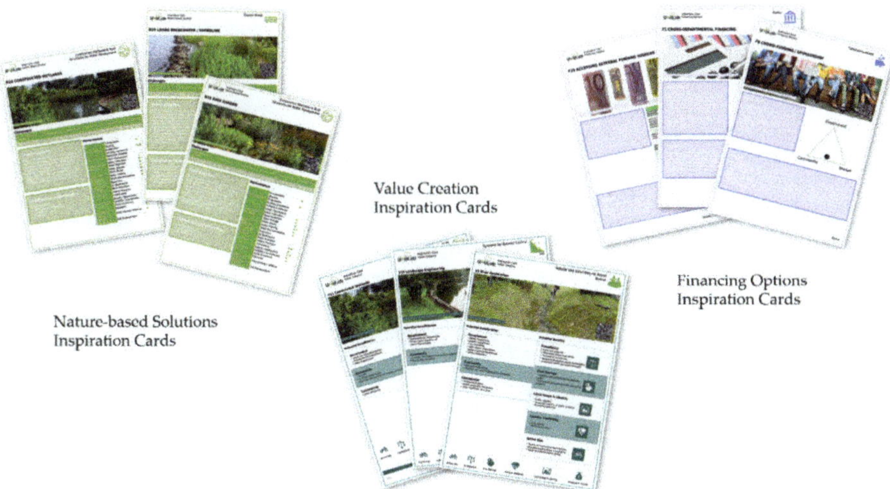

Figure 6. Examples of inspiration cards used to operationalize the framework in a workshop setting (compiled from Den Ouden et al., 2020 [23]).

4. Discussion

The presented framework can be applied to different types of NBS and highlights their multi-functional nature by suggesting a structured approach in which different functions, benefits, and beneficiaries can be factored in. Rather than opting for specialized quantification, the framework offers targeted integration of user knowledge and allows for flexibility of application, taking into account the importance of context-specificity regarding NBS value creation. These features, alongside the educative and communicative character, make it an interesting tool for urban planners and decision-makers in the early stages of NBS project development. Based on the findings in this research, it appears that integrating and operationalizing the framework in urban planning processes can widen the view on potential benefits and beneficiaries and help in forming alliances and joint NBS projects between different urban stakeholders. Forming partnerships with different local actors and understanding their perceptions and preferences has shown high potential for increasing success in the planning and implementation of NBS projects, e.g., by encouraging trust, ecosystem stewardship, and social learning [34]. Still, the fear of conflicting interests and a lack of consensus which would slow down urban planning and decision-making processes represents an important barrier to multi-stakeholder involvement [1]. With this regard, several cases have shown that an early involvement of key stakeholders in a dialogue to identify common goals and communicate concerns, as well as the building of a "common language" can help minimizing conflicts of interest and issues with green space management [35,36]. However, to avoid scientific biases and the influence of power asymmetries among the stakeholders, transparent processes and good moderation are necessary—aspects which often require additional resources [37].

Another potentially positive feature of the framework is the established link to different financing options. It allows for the exploration of alternative ways of financing and encourages non-expert audiences to think about the investment capacity and interest of different beneficiaries. While the framework does not provide definite conclusions on how a certain NBS could or should be financed, it aims to demonstrate the array of possibilities, especially among urban planners. Adding to the discussions on public and private actor activation for NBS, it also includes local community actors among the potential financiers of NBS. A common problem in public and especially citizen engagement is the fact that NBS stewardship is often perceived by residents as sole responsibility of the government [34,38]. Here, the argumentation with multiple and individual benefits on the local and community level could improve social mobilization. Studies on energy cooperatives have shown that while large, less spatially-bound communities are dominated by the return on investment as the key driver, smaller and spatially-close communities tend to put greater emphasis on the social and environmental aspects of their investment [39]. Perhaps a similar approach could hold for NBS investments as well, especially seeing that citizens are often perceived as the main beneficiary across many of the discussed NBS functions. As NBS tend to enhance the sense of place and provide a range of local social and environmental benefits to beneficiaries, local community initiatives for setting up or maintaining NBS have been emerging (e.g., "Adopt a Place" initiative in London, numerous adopt-a-tree initiatives). While such initiatives might remain on a relatively small scale, they could hold cost-saving potential for local municipalities by mobilizing small-scale private investment and in-kind contribution from the local beneficiaries. Hence, the value-added of this framework also lies in its ability to depict a range of potential constellations and promote a holistic dialogue between the different entities.

It is hoped that raising awareness and shedding light on these various aspects and their interlinkages can inspire, mobilize, and involve more stakeholders from the private, public, and civil society sector in future NBS development, allowing for new financing and governance constellations, and thus helping to tackle some of the key challenges in mainstreaming NBS implementation in cities [3,16,19]. The application and testing of the framework in more concrete real-life NBS case studies and a close monitoring of the resulting project outcomes would be suggested next steps. These could help to evaluate, whether the use of such an approach will actually increase the priority of NBS in urban planning, achieve a higher involvement of different urban stakeholders, or lead to choice of alternative financing options. It would also give the opportunity to see the influence of information flow and qualitative valuation on the empowerment of different stakeholders, which we acknowledge as a very important aspect in the planning process.

Contrarily, our research has highlighted the tradeoffs between different ecological valuation methods. Valuation of any kind always involves a certain degree of simplification and it is important to be aware of the potential pitfalls involved with reductionist approaches to understanding interventions in complex socio-ecological systems [40,41]. It is not contended here that more narrow (reductionist) mechanisms of value quantification do not have an important role in supporting the establishment of interventions with broad (often abstract) value creation. The intention here is rather to underline the importance of establishing a multiplicity of tools to assess value creation and to demonstrate the utility of softer, more subjective, forms of value assessment. Thus, responding to the call to analyze and describe NBS from multiple perspectives which take the complex interlinkages between social and ecological systems into account [4,42]. As quantitative evaluation techniques might provide a stronger evidence base at the cost of a broader conception of value creation, softer approaches, such as the one discussed here, maintain a more holistic scope but at the cost of the "hard quantitative facts" that are considered imperative for justification of investment according to prevailing logics [43].

In accordance with this finding, the lack of the quantified outputs might be regarded as a fundamental limitation of the presented framework, especially seeing that urban infrastructure investments need to be deemed economically feasible to be undertaken in the first place. Whereas the built narratives may help stimulate informed discussions and convince decision-makers why NBS are worth the investment, the framework will not be able to provide any certainty about the final performance and return on investment. Feedback from different urban stakeholders have indicated that such information, as a result of more expert and data-intensive tools such as TEEB, Natural Capital Accounting, or other forms of quantified impact modeling will be needed to add more financial certainty and attract larger investor attention. However, setting out the full spectrum of potential benefits and beneficiaries associated with a given NBS can help considering NBS as a crucial urban infrastructure asset worth to be further investigated in the first place, and prioritize specific values and KPIs to be further studied.

However, in terms of further value assessment methods, we argue that capturing value in economic terms will be easier for certain types of benefits (such as increased revenue and income or cost savings), whereas others might be more efficiently captured by other means of impact assessment (e.g., compliance, active use, local image and identity, or general health and wellbeing). Under the title "Ecosystem services: The economics debate, "Farley discusses the implications of treating ecosystem services as market commodities, e.g., through assigning monetary values. He points out that the use of (monetary) exchange values inevitably implies a certain degree of substitutability and non-essentiality of natural functions, which has to be carefully applied and continuously re-evaluated, especially when ecological thresholds are approached [11]. Thus, keeping a certain degree of diversification in the approaches to valuation may lessen the extent of abstraction and allow for differentiated and targeted communication of different types of benefits, thereby highlighting crucial functions and dependencies. It is hoped that by presenting soft benefits in a more structured and holistic way, their relevance for the overall urban system and the indirect influence on different budgets can be shown. Moreover, targeted and differentiated communication may increase the individual perception of benefits and the meanings that different stakeholders assign to NBS, which can fundamentally shape the final allocation of value and influence the demand and individual support for NBS [20].

5. Conclusions

The present paper describes a value-based framework that was developed in the frame of the UNaLab project to facilitate stakeholder engagement in NBS implementation. It is shown that the multifunctionality and the range of (co-)benefits that NBS provide hold great potential to engage different stakeholders from the public, private, and civil society sector around one given project. It may also help in linking climate adaptation to other important goals of cities. Providing a structured approach which can be used in multiple contexts can facilitate navigation through the complexity of the issue. It may further help in building a common understanding and language between actors from different backgrounds, and thus support the formation of new alliances for NBS planning and implementation. Whilst the link between given values and alternative financing options is not so well-established in current NBS planning processes, taking this issue into account from early on may help unlocking new potentials for final project realization. Additionally, the paper highlights different ways to value NBS and argues that hard and soft approaches serve different purposes, but should complement each other in building a case for NBS.

Author Contributions: Conceptualization, S.M., E.M. and T.H.; data curation, S.M., E.M. and P.H.B.; formal analysis, S.M., E.M. and P.H.B.; methodology, S.M., E.M., P.H.B. and T.H.; project administration, S.M. and E.M.; supervision, S.M. and E.M.; validation, S.M. and P.H.B.; visualization, S.M. and E.M.; writing—original draft, S.M., E.M., P.H.B. and T.H.; writing—review and editing, S.M., E.M., P.H.B. and T.H. All authors have read and agreed to the published version of the manuscript.

Funding: This paper is based on work undertaken in the context of the UNaLab project funded by the European Commission under the SCC2 Horizon 2020 programme (grant number 730052).

Institutional Review Board Statement: Not applicable.

Informed Consent Statement: Not applicable.

Data Availability Statement: Referred deliverables and supporting project activities, can be found on the UNaLab website (www.unalab.eu (accessed on 15 February 2021)).

Acknowledgments: The authors would like to thank all project partners for their valuable feedback and support, as well as the external experts from the SCC community and beyond, who participated and shared their experience and expertise in the surveys and interviews. We would also like to thank the reviewers for their constructive comments. This work has been presented at the Greening Cities Shaping Cities Symposium in October 2020 (https://www.greeningcities-shapingcities.polimi.it/ (accessed on 15 February 2021)).

Conflicts of Interest: The authors declare no conflict of interest.

References

1. Raymond, C.M.; Frantzeskaki, N.; Kabisch, N.; Berry, P.; Breil, M.; Nita, M.R.; Geneletti, D.; Calfapietra, C. A framework for assessing and implementing the co-benefits of nature-based solutions in urban areas. *Environ. Sci. Policy* **2017**, *77*, 15–24. [CrossRef]
2. European Commission. Towards an EU Research and Innovation Policy Agenda for Nature-Based Solutions & Re-Naturing Cities: Final Report of the Horizon 2020 Expert Group on "Nature Based Solutions and Re-Naturing Cities". Publications Office of the European Union. 2015. Available online: https://data.europa.eu/doi/10.2777/765301 (accessed on 25 January 2021).
3. Kabisch, N.; Korn, H.; Stadler, J.; Bonn, A. *Nature-Based Solutions to Climate Change Adaptation in Urban Areas*; Springer International Publishing: Berlin/Heidelberg, Germany, 2017.
4. Wild, T.; Henneberry, J.; Gill, L. Comprehending the multiple 'values' of green infrastructure—Valuing nature-based solutions for urban water management from multiple perspectives. *Environ. Res.* **2017**, *158*, 179–187. [CrossRef] [PubMed]
5. UNaLab. Main Website I Home. Available online: www.unalab.eu (accessed on 15 February 2021).
6. Mok, S.; Hawxwell, T.; Kramer, M.; Maciulyte, E. *NBS Value Model*; UNaLab Public Deliverable D6.4; UNaLab: Brussels, Belgium, 2019.
7. Brosch, T.; Sander, D. *Handbook of Value: Perspectives from Economics, Neuroscience, Philosophy, Psychology and Sociology*; OUP Oxford: Oxford, UK, 2015; 700p.
8. Costanza, R. Social goals and the valuation of natural capital. *Environ. Monit. Assess.* **2003**, *86*, 19–28. [CrossRef] [PubMed]
9. Pascual, U.; Muradian, R. *The Economics of Valuing Ecosystem Services and Biodiversity*; Taylor and Francis: Abingdon, UK, 2010; 133p.
10. Georgescu-Roegen, N. Energy Analysis and Economic Valuation. *South. Econ. J.* **1979**, *45*, 1023. [CrossRef]

11. Farley, J. Ecosystem services: The economics debate. *Ecosyst. Serv.* **2012**, *1*, 40–49. [CrossRef]

12. Munda, G. Social multi-criteria evaluation: Methodological foundations and operational consequences. *Eur. J. Oper. Res.* **2004**, *158*, 662–677. [CrossRef]

13. Nesshöver, C.; Assmuth, T.; Irvine, K.N.; Rusch, G.M.; Waylen, K.A.; Delbaere, B.; Haase, D.; Jones-Walters, L.; Keune, H.; Kovacs, E.; et al. The science, policy and practice of nature-based solutions: An interdisciplinary perspective. *Sci. Total Environ.* **2017**, *579*, 1215–1227. [CrossRef] [PubMed]

14. Ostrom, E. Beyond Markets and States: Polycentric Governance of Complex Economic Systems. *Am. Econ. Rev.* **2010**, *100*, 641–672. [CrossRef]

15. Gioia, S. A Brief History of Co-Creation—The XPLANE Collection—Medium. 2015. Available online: https://medium.com/the-xplane-collection/a-brief-history-of-co-creation-2e4d615189e8 (accessed on 18 February 2021).

16. Sarabi, S.E.; Han, Q.; Romme, A.G.L.; De Vries, B.; Wendling, L. Key Enablers of and Barriers to the Uptake and Implementation of Nature-Based Solutions in Urban Settings: A Review. *Resources* **2019**, *8*, 121. [CrossRef]

17. Thompson, M.A.; Owen, S.; Lindsay, J.M.; Leonard, G.S.; Cronin, S.J. Scientist and stakeholder perspectives of transdisciplinary research: Early attitudes, expectations, and tensions. *Environ. Sci. Policy* **2017**, *74*, 30–39. [CrossRef]

18. Díaz, S.; Demissew, S.; Carabias, J.; Joly, C.; Lonsdale, M.; Ash, N.; Larigauderie, A.; Adhikari, J.R.; Arico, S.; Báldi, A.; et al. The IPBES Conceptual Framework—connecting nature and people. *Curr. Opin. Environ. Sustain.* **2015**, *14*, 1–16. [CrossRef]

19. Frantzeskaki, N.; McPhearson, T.; Collier, M.J.; Kendal, D.; Bulkeley, H.; Dumitru, A.; Walsh, C.; Noble, K.; Van Wyk, E.; Ordóñez, C.; et al. Nature-Based Solutions for Urban Climate Change Adaptation: Linking Science, Policy, and Practice Communities for Evidence-Based Decision-Making. *Bioscience* **2019**, *69*, 455–466. [CrossRef]

20. Van Wyk, E.; Breen, C.; Freimund, W.A. Meanings and robustness: Propositions for enhancing benefit sharing in social-ecological systems. *Int. J. Commons* **2014**, *8*, 576. [CrossRef]

21. Eisenberg, B.; Polcher, V. *Nature-Based Solutions—Technical Handbook Part II*; UNaLab Public Deliverable D5.1; UNaLab: Brussels, Belgium, 2019.

22. Mačiulytė, E.; Cioffi, M.; Zappia, F.; Duce, E.; Ferrari, A.; Kelson Batinga de Mendoca, M.F.; Loriga, G.; Suška, P.; Vaccari Paz, B.L.; Zangani, D.; et al. *Business Models & Financing Strategies*; UNaLab Public Deliverable D6.3; UNaLab: Brussels, Belgium, 2018.

23. Den Ouden, E.; Valkenburg, R.; Mok, S.; Hawxwell, T. *Replication Roadmaps for UNaLab Follower Cities*; UNaLab Public Deliverable D6.7; UNaLab: Brussels, Belgium, 2020.

24. Cohen-Shacham, E.; Walters, G.; Janzen, C.; Maginnis, S. *Nature-Based Solutions to Address Global Societal Challenges*; IUCN: Gland, Switzerland, 2016.

25. Raymond, C.M.; Berry, P.; Nita, M.R.; Kabisch, N.; de Bel, M.; Enzi, V.; Frantzeskaki, N.; Geneletti, D.; Cardinaletti, M.; Lovinger, L.; et al. *An Impact Evaluation Framework to Support Planning and Evaluation of Nature-Based Solutions Projects: Prepared by the EKLIPSE Expert Working Group on Nature-Based Solutions to Promote Climate Resilience in Urban Areas*; Centre for Ecology and Hydrology: Lancaster, UK, 2017.

26. Costanza, R.; d'Arge, R.; de Groot, R.; Farber, S.; Grasso, M.; Hannon, B.; Limburg, K.; Naeem, S.; O'Neill, R.V.; Paruelo, J.; et al. The value of the world's ecosystem services and natural capital. *Nature* **1997**, *387*, 253–260. [CrossRef]

27. Toxopeus, H.; Polzin, F. *Characterizing Nature-Based Solutions from a Business Model and Financing Perspective*; Naturvation: Berlin, Germany, 2017.

28. Faber, A.; Frenken, K.K. Models in evolutionary economics and environmental policy: Towards an evolutionary environmental economics. *Technol. Forecast. Soc. Chang.* **2009**, *76*, 462–470. [CrossRef]

29. Polzin, F. Mobilizing private finance for low-carbon innovation—A systematic review of barriers and solutions. *Renew. Sustain. Energy Rev.* **2017**, *77*, 525–535. [CrossRef]

30. Tompkins, E.L.; Eakin, H. Managing private and public adaptation to climate change. *Glob. Environ. Chang.* **2012**, *22*, 3–11. [CrossRef]

31. Kolbjørnsrud, V. Collaborative organizational forms: On communities, crowds, and new hybrids. *J. Organ. Des.* **2018**, *7*, 11. [CrossRef]

32. i-Tree. Tools. 2021. Available online: https://www.itreetools.org/tools (accessed on 15 February 2021).

33. Kabisch, N.; Frantzeskaki, N.; Pauleit, S.; Naumann, S.; Davis, M.; Artmann, M.; Haase, D.; Knapp, S.; Korn, H.; Stadler, J.; et al. Nature-based solutions to climate change mitigation and adaptation in urban areas: Perspectives on indicators, knowledge gaps, barriers, and opportunities for action. *Ecol. Soc.* **2016**, *21*, 39. [CrossRef]

34. Ferreira, V.; Barreira, A.P.; Loures, L.; Antunes, D.; Panagopoulos, T. Stakeholders' Engagement on Nature-Based Solutions: A Systematic Literature Review. *Sustainability* **2020**, *12*, 640. [CrossRef]

35. Ugolini, F.; Sanesi, G.; Steidle, A.; Pearlmutter, D. Speaking "Green": A Worldwide Survey on Collaboration among Stakeholders in Urban Park Design and Management. *Forests* **2018**, *9*, 458. [CrossRef]

36. Khoshkar, S.; Balfors, B.; Wärnbäck, A. Planning for green qualities in the densification of suburban Stockholm—opportunities and challenges. *J. Environ. Plan. Manag.* **2018**, *61*, 2613–2635. [CrossRef]

37. Gómez-Baggethun, E.; de Groot, R. Natural capital and ecosystem functions: Exploring the ecological grounds of the economy. *Ecosistemas* **2007**, *16*, 4–14.

38. Moskell, C.; Allred, S.B. Residents' beliefs about responsibility for the stewardship of park trees and street trees in New York City. *Landsc. Urban Plan.* **2013**, *120*, 85–95. [CrossRef]

39. Bauwens, T. Analyzing the determinants of the size of investments by community renewable energy members: Findings and policy implications from Flanders. *Energy Policy* **2019**, *129*, 841–852. [CrossRef]
40. Kallis, G.; Gómez-Baggethun, E.; Zografos, C. To value or not to value? That is not the question. *Ecol. Econ.* **2013**, *94*, 97–105. [CrossRef]
41. Mazzocchi, F. Complexity in biology. *EMBO Rep.* **2008**, *9*, 10–14. [CrossRef] [PubMed]
42. Angelstam, P.; Andersson, K.; Annerstedt, M.; Axelsson, R.; Elbakidze, M.; Garrido, P.; Grahn, P.; Jönsson, K.I.; Pedersen, S.; Schlyter, P.; et al. Solving Problems in Social–Ecological Systems: Definition, Practice and Barriers of Transdisciplinary Research. *Ambio* **2013**, *42*, 254–265. [CrossRef]
43. Gasparatos, A. Embedded value systems in sustainability assessment tools and their implications. *J. Environ. Manag.* **2010**, *91*, 1613–1622. [CrossRef]

 *sustainability*

MDPI

Article

Exploring Challenges and Opportunities of Biophilic Urban Design: Evidence from Research and Experimentation

Maria Beatrice Andreucci [1,*], **Angela Loder** [2], **Martin Brown** [3] **and Jelena Brajković** [4]

[1] Department of Planning, Design, Technology of Architecture, Sapienza University of Rome, 00196 Rome, Italy
[2] International WELL Building Institute, New York, NY 10001, USA; angela.loder@wellcertified.com
[3] Fairsnape, Inglewhite, Lancashire PR3 2LE, UK; fairsnape@gmail.com
[4] Faculty of Architecture, University of Belgrade, 11120 Belgrade, Serbia; jelena.brajkovic@arh.bg.ac.rs
* Correspondence: mbeatrice.andreucci@uniroma1.it

Abstract: Global health emergencies such as Covid-19 have highlighted the importance of access to nature and open spaces in our cities for social, physical, and mental health. However, there continues to be a disconnect between our need for nature and our daily lived experience. Recent research indicates that our connectedness and relationship with nature, and in particular biophilic design, may be key for improving both health and quality of life. Rather than relying on abstract universal ideas of "nature", using evidence-based biophilic design and policy at a building, neighborhood, and city scale, to link our daily lives with biodiversity, may encourage sense of place and make environmental action more meaningful. Then, improving our natural capital in the urban built environment might help address the current climate and disease crisis, as well as improving our physical and mental health. Drawing from emerging research and innovative practice, the paper describes key research and design paradigms that influence the way we understand the benefits of nature for different environments, including the workplace, neighborhood, and city, and explains where biophilic design theory sits in this field. Examples from recent research carried out in London and Chicago are provided, aiming at demonstrating what kind of research can be functional to what context, followed by a detailed analysis of its application supporting both human and ecological health. The study concludes indicating key policy and design lessons learned around regenerative design and biophilia as well as new directions for action, particularly with regard to climate change, sense of place, and well-being.

Keywords: biophilia; greening cities; health and well-being; nature-based solutions; urban design; urban green infrastructure

Citation: Andreucci, M.B.; Loder, A.; Brown, M.; Brajković, J. Exploring Challenges and Opportunities of Biophilic Urban Design: Evidence from Research and Experimentation. *Sustainability* **2021**, *13*, 4323. https://doi.org/10.3390/su13084323

Academic Editors: Israa H. Mahmoud, Eugenio Morello, Giuseppe Salvia and Emma Puerari

Received: 27 February 2021
Accepted: 10 April 2021
Published: 13 April 2021

Publisher's Note: MDPI stays neutral with regard to jurisdictional claims in published maps and institutional affiliations.

1. Introduction

Improved environmental and human health outcomes have long been associated with the integration of nature into our urban form [1–3]. Pandemics such as Covid-19 have highlighted again the importance of access to nature and open spaces in our cities for our social, physical, and mental health [4]. People living in neighborhoods with worse air pollution—which also often lacks greenspace—have been shown to have a higher death rate from Covid-19 [5]. Access to urban nature has also been shown to be influential in stress reduction and socialization [6,7], with urban parks receiving attention on the benefits of nature as urbanites seek out safer outdoor space in which to work, socialize, and play [8]. This renewed attention is supported by a trend in urban planning and design that is trying to provide opportunities to connect urbanites with nature through community-based ecosystem services projects, regenerative and biophilic design interventions, and residential greenspace, all of which have been linked to increased well-being, concentration, socialization, sense of place, and a connection with nature [9].

However, there continues to be a disconnect between our need for nature, our daily lived experience, and sustainable behavior. This is a missed opportunity given that a recent systematic research [6,7] has suggested that our connectedness and relationship with nature, and in particular our experience of biophilic design, may be key for improving both sustainability and our quality of life. However, though there is over forty years of research on the benefit of access to nature for human and climate health, there is still confusion in the sustainability and design fields on exactly what types of nature can lead to which types of benefit, and for whom. This confusion is partly rooted in a failure to understand how to interpret and apply research on nature and health to different design and policy interventions at different scales [10]. Specifically, issues arise from a disconnection between biophilic design principles, urban planning interventions, and specific health and well-being outcomes, as well as from a lack of integration between different disciplines. This confusion has real implications as buildings, cities, and regions attempt to align regenerative design goals with human health ones but often lack the tools and knowledge to do so, which can result in a lack of evidence to support the effectiveness of these interventions.

The identification of these issues has led to the research objectives of this paper. Specifically, this paper aims to (a) give researchers, designers, and urban planners a better understanding of the types of research on the benefits of nature, particularly studies following an adaptive or utilitarian paradigm, (b) compare this research to the most well-known application of these principles, i.e., biophilic design; (c) evaluate how real-world case studies in London and Chicago have used (or not used) this research and design foundation for positive human and ecological outcomes, and (d) provide detailed analysis of where biophilic design is working well and highlight new directions and opportunities that can help to address current shortfalls. Drawing from established and emerging theories and innovative practice, this contribution evaluates key research and design paradigms that influence the way we understand the benefits of nature, and then uses this foundation to assess the effectiveness of three applied case studies according to different pathways, and at different scales: the workplace, the neighborhood, and the city. The paper finally reflects on key policy and design lessons learned about regenerative design and biophilia and how these can be leveraged for a better connection with nature and a sense of place, which may make environmental action more meaningful. The study is structured as follows: Section 2 explains the methodology; Section 3 presents the conceptual framework, in which the theoretical and practical interrelations between regenerative design and biophilia are highlighted; Section 4 introduces and develops the London and Chicago case studies; Section 5 elaborates results and their discussion; and Section 6 presents the conclusions.

2. Methods

In order to achieve the objectives mentioned, the work has adopted a mixed-qualitative methodology that has been structured developing a combination of critical literature review and field research. A critical, in-depth review of the theoretical paradigms, underlying the most influential scientific programs on nature and health, was undertaken with the goal to understand how the paradigms influenced the kind of study that comes out of these research programs, the goals of this investigation, as well as how and why this research has been influential in policy circles, highlighting limitations and new directions. A more extensive analysis, from which this review is based, can be found in [11], as well as in [6,7], two systematic reviews (Cochrane style) on green and blue open spaces and mental health, developed by a multidisciplinary expert working group, led by one of the authors, under the Horizon 2020-funded programme EKLIPSE.

In the second phase, the research designed the protocol for the development of the case studies [12,13] and applied it to two different cities. The case study was selected as the method to undertake this part of the work as it allows investigating the phenomenon under study, in relation with its urban context, using different sources of evidence. Field research was conducted focusing on the analysis on cities that have already demonstrated good capacity to integrate biophilic design at multiple scales, i.e., cities with good potential to innovate and with more financial, technical, and institutional capacity and experience in running regenerative architecture and urban projects. The objective was to understand the level of integration of biophilic design, the theoretical foundation, and the policy, and implementation process for this, as well as drivers and limitations. The selection of cities was based on the following criteria: (1) focusing on two cities for different biophilic design scales, i.e., workplace, neighborhood, city; (2) sufficient secondary sources to develop the analysis; (3) availability to conduct interviews to designers, public servants, and/or other stakeholders. The cities selected for the development of the case studies were finally London and Chicago.

For London, field research was developed by the authors also within the wider scope of working group activities of the COST (Cooperation in Science and technology) Action "RESTORE Rethinking Sustainability Towards a Regenerative Economy", in the period 2017–2020.

For Chicago, key stakeholder interviews and media and policy analysis were conducted, in the period 2016–2019, as part of a larger project on Chicago's urban greening, climate change, and resilience initiatives. The Resilient Corridors project emerged as a pilot in 2019 from the City of Chicago.

3. Theoretical Frameworks on Health and Nature

3.1. Adaptive and Utility Paradigms

The link between access to nature and human health benefits is supported evidence accumulated over the last 40 years [14–18]. This has been of interest to designers who include access to nature for its diverse benefits, such as in the workplace [19], and city planners who are interested in the socio-cultural benefits of green infrastructure for human health and well-being [20,21]. Although the evidence points to clear benefits between access to nature and human health outcomes, there remains a lack of alignment between this large body of research and the type of evidence that convinces stakeholders that adding nature will reap tangible and trackable benefits for their unique project [11]. This misalignment is partly due to the types of research—and the paradigms that support them—that undergird the vast majority of findings that have gotten the attention of policy makers and building owners. Furthermore, these types of research tend not to align with the more holistic approach of designers using a biophilic framework [11,22].

Comparing research on nature is complicated by the wide variety of types and measures used, which can complicate the establishment of robust results between them [23]. The most influential research programs in the last forty years have been based on adaptive or utility paradigms. The adaptive paradigm is based on the assumption that evolution, or biological survival, motivates physiological and psychological responses to the experienced environment, and that some environments are better suited to human health and well-being than others. There are two research programs that have emerged out of an adaptive paradigm that have garnered the most attention and subsequent research. The first focuses on restorative environments that help with the restoration of attention or to improve cognition, notably Stephen and Rachel Kaplan's Attention Restoration Theory (ART) [2,24]. The second focuses on the ability of restorative environments to support stress recovery and positive mood, notably Roger Ulrich's Psychophysiological Stress Reduction Theory (PSR) [25].

The original ART research argued that nature possesses four attributes necessary to hold our attention involuntarily and be experienced as restorative: fascination, mystery, coherence, and the feeling of being away, and this research has been heavily tested in subsequent studies [24,26,27]. A key component of research testing ART has looked at aesthetic preferences for different types of nature. These studies argue that some types of nature are more favorable to restoration than other types of nature, and that nature overall is more restorative than urban environments [10,17,24,28]. Research testing the PSR theory also uses an evolutionary biology theory but tends to focus on the affective or emotional aspects of this relationship. At its core, evolutionary biology argues that because we evolved in nature, we tend to feel connected with things that remind us of nature; this attitude is called biophilia (translated as a love of nature) [25,29]. This love of nature has begun to be studied for its potential to link to our connectedness to nature, which has been shown to improve health and well-being outcomes as well as sustainability behaviors and belief in climate change [30–35]. While the utility paradigm also draws on the idea that our natural environment is connected with our well-being, it focuses on the role that nature plays as a quality of an environment to satisfy current personal or interpersonal needs. These are often measured by known benefits of access to nature, such as increased levels of physical activity, restorative experiences, or social cohesion, interaction, and safety [36–38].

3.2. Understanding Nature-Health Research through the Adaptive and Utility Paradigms

The adaptive and utility theories underlie the vast majority of research linking access to nature and improved physiological and mental health and well-being. Some researchers have continued to develop these theories and have proposed that these relationships can be viewed as a series of pathways that have formed the basis of multiple research streams: (1) stress reduction, (2) physical activity, (3) social cohesion, and (4) air quality [15]. Understanding the key types of research on the benefits of nature and the aim of these research streams can help designers and planners determine which research is relevant to their project goals.

Stress reduction has traditionally received the most empirical and theoretical attention. Research looking at stress reduction has tended to follow the ART and PSR restoration theories outlined above. These two theories rely mostly on the visual and aesthetic qualities of nature, and they link to the assumed characteristics of nature seen in evolutionary and related biophilia (or biophobia—fear of nature) theory [25,39]. While the variety of contexts for this research supports the strength of the research, it has been harder to evaluate their application at a building scale given the high number of variables involved.

Physical activity has been gaining attention and follows the utility paradigm. As opposed to sedentary behavior, outdoor physical activity has been shown to have positive effects on mental health, showing for example better outcomes in green areas than indoor, or non-green urban areas [40]. However, the results have been unclear in cross-sectional and/or epidemiological studies at the neighborhood scale [41], showing the difficulty of applying lab-based studies to real-world situations. Real-world situations have other explanatory variables that may influence health outcomes. Furthermore, lab-based studies do not always take into account other factors such as green space characteristics, location, and other influences, or mediators, on behavior or preferences. Studies have found that multiple factors over and above the amount of greenspace—including quality and accessibility—determine urban greenspace use and physical activity [42–44].

The third pathway looks at how access to nature is linked to improvements in social interactions (at the individual level) and social cohesion (at the neighborhood level) and varies in its research paradigms—ranging from utilitarian, which focuses on characteristics of parks that influence desired uses, to the design of parks, which influences social cohesion [45]. Although the link between social interaction and mental health has been firmly established [46], the link between social interactions, social cohesion, and green space has received less research attention than the first two pathways.

The research linking air pollution, nature, and health has equally received less attention. While the link between air pollution and negative effects on physical health and mortality has been long established [47], newer studies have also linked air pollution with negative impacts on mental health [48], and cognitive performance [49]. Some researchers have gone further and proposed that air pollution, together with traffic-related sounds, can put a constraint on the restorative potential of an environment as a whole [50]. This holistic approach is important for understanding negative environmental influences or ecosystem disservices. This last pathway can be one of the most easily integrated into regional-level planning and regenerative policies and can be a good way to balance synergies and trade-offs at this scale.

Lastly, the concept of Topophilia [51] has received renewed interest recently among planners, designers, and academics in Europe, who see the focus on personal identity and meaningful attachment with place and landscapes as a powerful design tool for reconnecting urbanites with local nature and thus inspiring sustainable behavior. While in theory, place attachment can be used to inform a regenerative approach to urban and regional planning, it has not been used much in application to date due to its more theoretical and qualitative approach and the lack of alignment with design and planning practice.

While there has been some qualitative research conducted in the adaptive and utility paradigms, the vast majority of this research follows a psychometric research approach, which aims to generalize relationships through quantifiable measures [52,53]. The psychometric approach aligns well with the kind of data promoted by urban planning and green building researchers and has created a vast amount of data on the benefits of access to nature (outlined below). It has also been very influential in public policy [54,55] and has provided much of the support for adding nature into buildings, neighborhoods, and cities to date. However, the type of linear and somewhat mechanistic approach to nature and health in psychometric research does not always align well with the more holistic, design-thinking approach seen in biophilic design and green infrastructure work to support human health. There has also been some criticism from social scientists that research based in the adaptive paradigm tend to not address the larger context of place and that the underlying evolutionary paradigm—i.e., that love of nature is innate—can hide cultural, socio-economic, and power differences that can influence the success of urban nature interventions and the equitable access to nature. The utilitarian paradigm has also been criticized for its limited understanding of the socio-economic and socio-cultural factors influencing access to nature, the reduction of environmental values to utility, and the general lack of acknowledgement of the symbolic aspect of nature [56]. In short, while research following the adaptive and utility paradigms have provided strong evidence to support the health goals of biophilic design, biophilia's focus on sense of place, lived experience, and holistic design-thinking may be more aligned with some of the relational and sense of place work on the human relationship to nature that rarely gets cited [57–59] outside of academia.

3.3. Research to Practice: Design Theory, Research, and Application

One of the most commonly understood "popular" urban greening and design approaches is biophilia. Popularized by the biologist E. O. Wilson's biophilia hypothesis, which prompted the modern biophilic design movement, biophilia is defined as the "[. . .] innate emotional affiliation of human beings to other living organisms. 'Innate' means hereditary and hence part of ultimate human nature" [22] (p. 31). Kellert and Wilson operationalized this concept to the built environment [39], and it was further developed in Kellert's proposed attributes for biophilic design [60], where he introduced key dimensions, elements, and attributes of biophilic design. As two main dimensions, the author identified organic/naturalistic and place-based/vernacular. Organic dimension refers to "shapes and forms in the built environment that directly, indirectly, or symbolically reflect the inherent human affinity for nature" [60] (p. 5). Vernacular dimension refers to "buildings and landscapes that connect to the culture and ecology of a locality or geographic area" [60] (p. 6). Further classifications refer to six main elements, which then break out into more than 70 biophilic design attributes. These attributes can be as simple and straightforward as the presence of water, air, sunlight, plants, animals, as well as more articulated, such as sensory variability, information richness, exploration and discovery, or geographic, historic, ecological, and cultural connection to place. Importantly, biophilic designers need to understand that the environment can be an atmosphere, a process, an experience.

There have been some further revisions to Kellert's work, an example of which is Terrapin Bright Green's 14 Patterns of Biophilic Design—Improving Health and Well-Being in the Built Environment. This report [61] defines 14 patterns of biophilic design organized into Nature in the Space, Natural Analogues, and Nature of the Space Patterns.

Another is the Biophilic Interior Design Matrix [62], which adopts and adapts Kellert's work to operationalize it for interior environments, in order to provide tangible and clearer guidance for designers.

3.4. Experimental Biophilic Design Approaches

In addition to more traditional biophilic design, alternative approaches strive to explore biophilic principles, ideas, and attributes in more experimental, even esoteric ways, trying to grasp the essence of human experience of space and model it in line with biophilic principles. Design, and architecture in particular, has been called a hybrid discipline, relying and building upon different elements within science, technology, and art [63]. This is in direct conflict with many of the quantitative and linear approaches favored by many researchers. Not all qualities of architectural space can be quantifiable and not all qualities of our experience of space can be translated into rational language. The process of developing space for designers and architects is that of an artist providing experience and hopefully emotional attachment.

One of the foremost thinkers in experimental biophilic design is Juhani Pallasmaa, who explores the art of building, elements of architectural experience, and meaningful spaces that stimulate people and provide existential encounters. Key components of his work include the experience of architecture through mental and physical frameworks which shape our identity, attachment, and sense of place. He argues that the mental component of experience has been widely neglected "in the field of architecture . . . where scientific criteria or methods have mainly been applied in its technical, physical and material aspects, whereas the mental realm has been left to individual artistic intuition" [63] (p. 4). He hopes for neuroscience to provide a deeper understanding of the mental implications and impacts of "the art of building". Pallasmaa also argues that our architectural experience is multi-sensory, and we experience architecture with our physical, emotional, mental, and social bodies, and that environments have the potential to stimulate our imagination and identity. Pallasmaa argues that the architectural attributes of hierarchies, information richness, order and complexity, affection and attachment, attraction and beauty, reverence, and spirituality [60], are all attributes that are also in biophilic design, and that they should be studied following an artistic and scientific approach.

Some of the limitations of the adaptive and utilitarian approaches to nature–human research may be addressed by Pallasmaa's suggestion of a biological historicity approach. This approach blends sense of place and biological and historical aspects of the place. For example, sense of place as an attribute does not rely only on biological, geographical, or natural features of the place but also on its historical layers, site-specific social developments, and cultural layers embedded in its core. These include historic and cultural connections to place, the integration of culture and ecology, age, change, and the patina of time [60]. Biophilic design addresses some of the gaps in adaptive and utilitarian research by acknowledging these social and cultural dimensions of places. Pallasmaa calls this a bio-psychological heritage, which—he argues—particularly influences the qualities of refuge and prospect, which are key factors in the evolutionary approach to the benefits of access to nature for humans [63]. Pallasmaa also connects biophilic attributes such as fear and awe [60] to the pleasure principle, which understands our experience of space through the dichotomy of pleasures and displeasures that drive our behavior and perception of space. Combined, these experiential approaches to the experience of space and design have the potential to create a more embodied and place-based understanding of the impact of biophilic design and access to nature on our health well-being and sense of place, which may help foster better nature–human connections, attachment, and a sustainable ethos.

4. Mainstreaming Biophilic Design: Research, Design, and Practice

How cities can build resilience has become a major undertaking and priority. It requires cities to address a variety of pressing global and local challenges through multi-functional strategies, including climate change, community health, economic downturns, and political uncertainty.

The integration of evidence-based research and design on nature and health has already proved to be successful toward these long-term goals, but it requires a genuine acknowledgement and a deep understanding of how it can be applied at different scales. This is particularly true when attempting to align, in policy and practice, building-level, neighborhood-level, or city-level initiatives with community resiliency or climate change measures.

As a reminder, a taxonomy of biophilic elements can be identified at three main levels (Table 1).

Table 1. Taxonomy of biophilic elements. Adapted from [64].

Scale of Biophilic Design	Forms of Biophilic Elements	Taxonomy of Biophilic Elements
Building scale	Green roofs, green walls, shade trees, vegetation, and natural elements inside and around the building	Green roofs, green walls, shade trees, vegetation, and natural elements inside and around the building
District and neighborhood scales	Street trees, pocket parks, orchards and community gardens, business parks	Many installations, small-medium in size, restoration possible, high technical and technological requirements, public and private properties
City scale	City parks, urban forests, urban agriculture, waterfronts	Few installations, large in size, restoration possible, high technical and technological requirements, public land

The selected case studies demonstrate the application of research and design practices on the benefits of nature in cities and will be followed by a discussion of limitations and suggested next steps. The first two use biophilic design, while the third uses a more socio-ecological approach to the benefits of nature. In order to support the relevance of a multiscale design investigation and related knowledge transfer from research to practice and policy, the implementation of "informed" biophilic design is illustrated in the following sections describing a research study conducted in the City of London, which is focused on biophilic implementation at different scales. The emphasis is on the value of biophilic design principles for people and the lived environment in application at multiple scales for regenerative design and community resilience.

By 2041, the population of London is forecasted to reach 10.3 million people, which is an increase of 1.2 million people when compared with 2019 [65]. London is also one of the greenest cities in the world [66]. All across London, a network of Royal Parks, pocket gardens, planted roofs, rain gardens, living walls, urban forests, community gardens, and street trees are greening the city, making its public spaces accessible, colorful, and vibrant places to visit, live, and work.

This nature is a vital part of the complex organism of the city bringing benefits right into the places where people work and live. Moreover, as London's population grows, and its neighborhoods experience more development, that will be more important than ever.

4.1. Building-Scale Applications: Living Lab at the Shard, London

With the growing research and interest focused on biophilic design, it is interesting to look at buildings specifically designed and constructed as a model to highlight the biophilic indoor attributes.

DaeWha Kang Design has created an experimental work environment on the 12th floor of the Shard, in London, that has the express purpose of measuring the impact of biophilic design on worker wellness and productivity.

Working in collaboration with Mitie (the client) and Dr. Marcella Ucci (head of the MSc in Health, Wellbeing and Sustainable Buildings at the University College of London), the designers have designed a pilot study to measure the impact on employees through a detailed post-occupancy evaluation.

Biophilia, as said, refers to human beings' innate need for a connection with nature. Human physiology is wired to seek qualities of light, view, material, and other factors common in the natural world. This project comprises two spaces designed according to those principles: a "Living Lab" that functions as an immersive work environment, and two "Regeneration Pods" that provide short-term rest and meditation functions for the Mitie employees.

The Living Lab is fully immersive, with rich and intricate patterns, natural materials, and interactive dynamic lighting. The room gains privacy through bamboo screens that wrap onto the ceiling above. The floor, desks, and task lights are also formed from different shades and textures of bamboo, providing an organic language for the entire space. The lighting in the room is circadian and linked to an astronomical clock—cool blue in the morning, brilliant white in the afternoon, and fire-like orange as the day winds down. The light softly breathes, very subtly shifting intensity in an almost imperceptible way, giving additional dynamism to the experience.

In the study, Mitie employees worked at these desks for four weeks at a time, answering daily surveys about their comfort, satisfaction, and emotional response (Figure 1).

Figure 1. Post-occupancy evaluation at the Shard Living Lab in London. Photo by Kyungsub Shin, with graphics by DaeWha Kang Design. Courtesy DaeWha Kang Design.

Then, they spent four weeks working in a control area on the same floor with similar environmental conditions but without biophilic design, and their responses were compared between the two spaces.

While studies have established the positive impact of daylight, natural materials, and a direct visual connection with nature, aesthetic design also has a strong impact. The bamboo screens strike a balance between the regular rhythm of structural ribs and the variation and playfulness of discrete leaves that maintain a sense of transparency and intricacy in the space. The leaves catch natural light but also diffuse embedded lighting within the screen itself.

While the Living Lab creates a sense of enveloping enclosure toward the rest of the office, it opens up toward the façade, providing long vistas and a strong connection to the sky. The Shard has a high-tech aesthetic of glass and metal, and the warm bamboo palette of the Living Lab establishes a strong counterpoint to that material language.

Mitie is one of the leading outsourcing and facilities management companies in the UK, and they have created a new "Connected Workspace" initiative that incorporates sensor technology, big data, and machine learning to revolutionize the way their portfolio of buildings is managed and maintained. The Living Lab was commissioned as part of the health, wellness, and user-experience aspect of Connected Workspace.

Following biophilic principles, the desks are originally crafted from natural bamboo and incorporate living plants directly into their workspace, and not only relevant technology. From a scientific point of view, achieving a meaningful experimental study on the users requires adapting for confounding environmental factors between the lab space and the control space, while on-desk sensors detect air quality, light levels, temperature, and humidity. An access card reader identifies the users and allows them to activate the task lights and charging strips, while an under-desk sensor records when they are actively working at the desk. All of these data are collected in Mitie's data lake and can be correlated with the survey results.

Direct access to living nature is also shown to have a host of benefits, and planters are organically integrated directly into the desks together with the task lights [67].

In the second section, the "Regeneration Pods" are once again constructed from bamboo, following Mitie's mental health and wellness initiative, providing a tech-free space for meditative moments within the workday. Similar to the "Living Lab", the bamboo construction provides a sense of shelter, while workers access the views outside. The routed featherlike panels slot into the seventeen identical spines, with minimal cross support. Upholstered seating is fitted within the spines, also with circadian LED lighting. Environmental sensors—monitoring light, movement, humidity, and temperature—were also integrated into the structures, making this an ambitious technical build for the architects and team and a good example of research-based design (Figure 2).

Figure 2. The "Regeneration Pods" provide a sense of shelter while workers access the views outside. Photo by Tom Donald for Aldworth James & Bond. Courtesy DaeWha Kang Design.

4.2. Scale Jumping: District- and City-Level Applications of Integrated Design and Research

Looking next to the city scale, the translation of biophilic design interventions for human health and well-being, inspired and informed by research paradigms, is also found in the larger City of London.

Working together, the Mayor, Natural England, major landowners, and the wider business community, represented by Business Improvement Districts (BIDs), have recognized the increasing importance of biophilic planning and design principles (Table 2) for future-proofing the capital.

Table 2. Attributes of biophilic design. Adapted from [68].

Direct Experience of Nature	Indirect Experience of Nature	Experience of Space and Place
Light	Images of Nature	
Water	Natural Materials	Prospect and
Vegetation	Natural Colors	Refuge
Animals	Simulated Natural	Organized Complexity
Weather Conditions	Light and Air	Integration of Parts to Wholes
Natural Landscape	Naturalistic Shapes	Transitional Spaces
and Ecosystems	and Forms	Mobility and Wayfinding
Fire	Evoking Nature	Cultural and Ecological
	Information Richness	Attachment to place
	Age, Change, and	
	Patina of Time	
	Natural Geometries Biomimicry	

4.3. Urban Park-Scale Applications: Queen Elizabeth Olympic Park

The Queen Elizabeth Olympic Park is one of the largest urban parks (102 ha) created in Western Europe for more than 150 years, which was designed by LDA Design in conjunction with Hargreaves Associates (2012), to enrich and preserve the local environment by restoring wetland habitats and planting native species of plants.

Its environmental features include the restoration of the River Lea, in the northern section of the park, the habitat-creation strategy, and the park's connection with its hinterland ecosystem. The landscape is dominated by native trees and flowering meadows of designed plant communities (Figure 3). "Flowing schemes are not arbitrary but have carefully thought-out shapes running through them: S-curves, lines of grasses, successive waves of plants, rising up through the season, anchor plants with satellites and fuzzy edges between one habitat and another" [69] (p. 24). Sensory and spatial variability, information richness, and natural shapes and forms encourage exploration, fostering sense of place and the human–nature relationship. Other elements and attributes of biophilic design include a lighting scheme designed by Speirs + Major, integrating natural light and shadows with filtered, diffused, or reflected light, all emphasizing spatial variability and harmony.

Figure 3. The European Garden at Queen Elizabeth Olympic Park, a distillation of the "meadow aesthetic": a visually dramatic, highly designed, and enhanced evocation of a wildflower meadow (Nigel Dunnett and Sarah Price). Photo by Maria Beatrice Andreucci.

4.4. District-Scale Applications: Greenwich Millennium Village

Greenwich Millennium Village (GMV) is a mixed-tenure modern development on an urban village model, which is located on the Greenwich Peninsula, in Greenwich, in southeast London, and it is part of the Millennium Communities Programme, under English Partnerships. GMV was originally designed by visionary architect Ralph Erskine as part of the regeneration of the whole brownfield site of East Greenwich Gas Works. The whole district landscape considers wildlife in the design of the soft estate around the built forms, through choice of species and inclusion of artificial refuges, in appropriate locations and numbers.

In particular, the Ecology Park is an exemplar of biophilic and biodiverse design, providing a significant boost to the value of the GMV in terms of exploration/discovery, affection/attachment, security/protection, and attraction/beauty. An Ecology Park Centre manages biodiverse areas of Southern Park as well as new habitats associated with future developments (Figure 4).

Figure 4. The biophilic features of the Greenwich Peninsula Ecology Park has been playing an important role in the area's regeneration and community life since its creation in 2002. Photo by Maria Beatrice Andreucci.

Greenwich Peninsula and GMV offer to residents and visitors alike multiple connections to place—i.e., historic (Maritime Greenwich heritage site), geographical (the Prime Meridian of the world, Greenwich Mean Time, and the Observatory), cultural (colleges and universities, artworks, museums, etc.), and ecological (Ecology Park)—fostering place-based relationships.

4.5. Neighborhood-Scale Applications: The Barbican

The Barbican is Europe's largest arts and conference complex, and it also includes a significant residential community. It is a noted example of uncompromising modernist architecture, built mostly in the 1970s. The original design aimed to create a self-contained "urban village", with the residential and public spaces separated completely from vehicle traffic. Most of the landscape elements, including the water bodies, are "podium landscapes" or "landscapes above structure": roof gardens and green roofs, with car parks, the arts complex, and recreational facilities beneath [69]. In 2013, following re-waterproofing of the roof gardens, the opportunity arose for completely new plantings to be installed. The new design takes a radically different approach. The concept for The Barbican plantings is to create continuous and successive waves of color over long periods of time through orchestrating a series of dramatic color washes over the entire site, from spring through to late autumn, and then to finish off the year with a textural array of seeds heads, plant structures, and foliage. Although the plantings are very diverse, at any one time, there are only two or three plant species that create the main flowering display. However, these species are repeated over the whole area, creating maximum impact. Planting in layers allows for one set of plants growing up and through the preceding set of plants, leading to a continuous succession. Naturalistic swathes of perennials and grasses are framed and contained within clumps, groupings, and scatterings of multi-stemmed trees and shrubs to give solidity and a three-dimensional framework throughout the year [69].

There is no precise planting plan for most of the species, but the proportion of each species in a mixture is carefully considered, and the plants are placed within the planting areas according to a set of rules and instructions aiming at replicating natural patterns and processes. Plants that are adapted to extreme dry conditions often have gray or silvery leaves (Figure 5), and there is a natural unity to plantings that comes from bringing plants together from similar habitats [69].

Figure 5. The "shrub steppe" plantings at the Barbican combine mixes of perennials and grasses to the steppe plantings, with additional low-density shrubs and multi-stemmed trees, to create multi-layered plantings with year-round structure and interest. Photo by Maria Beatrice Andreucci.

4.6. Community-Scale Applications: Mudchute Park and Farm at the Isle of Dogs

The Mudchute Park and Farm was established by the local island community. Originally, it was a piece of derelict land created during the last century from the spoil of construction from dredging Millwall Dock. For decades, Mudchute environmental features, natural patterns, and processes remained untouched. However, in 1974, the site was earmarked by the Greater London Council for the construction of a high-rise estate. The resulting public campaign against these plans reflected the affection that local people, and those working on the island, felt for the Mudchute. Their success secured it as the "People's Park" for the area. In 1977, the Mudchute Association was formed to preserve and develop the area. Farm animals and horses were introduced, trees and plants were planted by generous volunteers and corporate teams, and the educational benefits of the area were also recognized. Local schools are encouraged to use the project to study the natural world on their doorsteps (Figure 6). Since the establishment of the association, the Mudchute has steadily built a reputation for providing place-based relationship and direct nature experience through a variety of educational and leisure activities, on a London-wide basis.

Figure 6. School children are encouraged to experience the "biophilia effect" at Mudchute Garden and Farm. Photo by Maria Beatrice Andreucci.

4.7. Exploring Multi-Sensory Experiences through Experimental Biophilic Design: Olafur Eliasson at Tate Modern

Attempts to humanize architecture through the exploration of issues such as multi-sensory experiences and human perception, physical and psychological boundaries, the role of imagination and empathy in space, and the pleasure principle can provide very effective experiences of biophilic design in space.

One example is the practice and work of Olafur Eliasson, a Dutch–Islandic artist who is fusing many disciplines into his explorations of the human–nature–built environment nexus. Similar to many researchers who explore sense of place [70], he is concerned with phenomenological experiences. Eliasson is an artist, but he could also be called an architect, as many of his works are immersive environments with ephemeral spatial qualities that question perception, trigger the senses, and create a feeling of temporary community between people experiencing the environment (Figure 7).

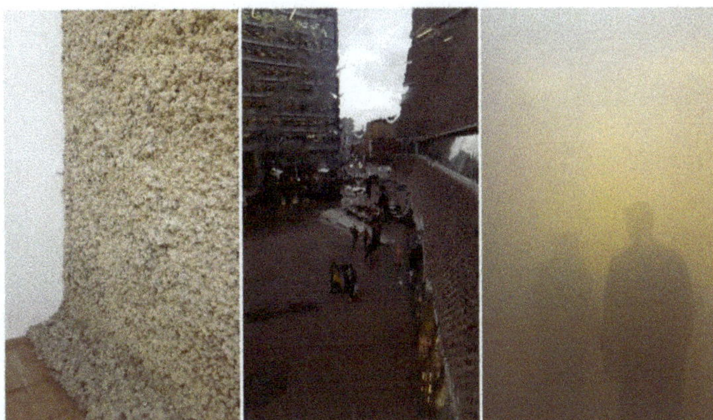

Figure 7. (**left**) "Moss Wall" (1994), (**middle**) "Regenfenster" (1999), (**right**) "Your Blind Passenger" (2010), artworks by Olafur Eliasson, exhibited at "In Real Life" exhibition, Tate Modern, 2019. Photos by Jelena Brajković.

Of relevance for biophilic design, his work contains many biophilic principles and attributes, such as affection and attachment, attraction and beauty, and reverence and spirituality. His spaces also include transitional spaces, a dynamic balance and tension, and generally, almost all attributes outlined by Kellert in his principles of biophilic design [60].

4.8. Linking Resilience with Social Justice and Economic Revitalization: Learning from Chicago

Similar to many cities, Chicago faces environmental challenges linked to climate change, such as increasingly hot summers and flooding from heavy rain and older stormwater systems [71]. After the heatwave of 1995, in which over 700 people died, many of them low-income and people of color [72], Chicago has undertaken a series of initiatives to increase the environmental and ecological resilience of the city. While some of these initiatives—such as the beautification of key boulevards with seasonal flowers—were more focused on economic neighborhood revitalization than ecological goals, many of the initiatives combined urban greening with ecological resilience. Key initiatives include their Building Green Matrix (now called Sustainability Development Matrix), which required nature-based design choices for projects in select neighborhoods, extensive use of TIF (tax incremental financing) at the district scale to incentivize sustainability, and urban revitalization projects in both high-profile (Figure 8) and disadvantaged neighborhoods, greening alleyways (Figure 9) that replaced pavement with permeable pavement, and their signature green roof program, supported by their Sustainability Matrix, which led them to be North American leaders in green roof implementation for over a decade [71,73,74].

These initiatives have been supported by larger policy plans, such as their 2015 Climate Change Action Agenda, their stormwater management plans [75,76], and their nomination as one of Rockefeller's 100 Resilient Cities, and subsequent resilience plan [11,73,77,78]. These policy plans regularly cite evidence of benefits of nature from research programs, which tend to use an adaptive and utilitarian paradigm. However, concerns about equity have meant that they have needed to also address social and economic aspects of urban nature. Chicago has also implemented a vacant lot revitalization and neighborhood stabilization plan, as well as a creative re-use of an abandoned elevated railway into a linear park, thus supporting active transportation that connects lower-income neighborhoods, in the west of the city, with wealthier neighborhoods, closer to the lake, in the east [79–82].

Figure 8. Crown Fountain, an interactive work of public art and video sculpture featured in Chicago's Millennium Park, in the Loop community area. Designed by Catalan artist Jaume Plensa, it features themes of dualism, light, and water. Photo by Maria Beatrice Andreucci.

Figure 9. Resilient Corridor stormwater street-level feature, Chicago. Photo by Michael Berkshire.

Despite the leadership of a neighborhood association, the involvement of a non-profit who did extensive stakeholder engagement, the inclusion of equity goals and artist's work, and the provision of a safe bikeway for active transportation, along a busy corridor, there have still been complaints that the project has spurred gentrification and is potentially displacing some of the more vulnerable residents in the eastern end of the 606 trail [83,84]. While this has been challenged by some groups involved in the project, who have claimed that such a large investment is an easy target for larger-scale gentrification forces, it is still a good example of the challenges of implementation for cities wishing to balance ecological, social, and economic goals in urban greening projects. It is also a good example of the need to include social and economic values into any discussion about ecological or regenerative urban initiatives.

The City of Chicago is a good example of a new hybrid approach to urban greening and is well aware of these challenges. For some of these projects, city administrators have deliberately framed them as urban stabilization projects, instead of environmental projects in economically distressed neighborhoods and have worked hard to ensure that their work on resilience, and the resulting Resilient Chicago plan, define resilience as inclusive and incorporate economic and social resilience into any environmental agenda [73,85]. One of the most innovative urban greening projects to come out of Chicago recently balances ecological, social, and biophilic goals. It is an instructive case study on how to use stakeholder engagement and collaboration to fill some of the gaps outlined above, which are typical in nature–health research approaches, from adaptive and utilitarian paradigms.

In 2015, after Hurricane Sandy, there was a significant amount of funding available to municipalities to address resilience and adaptation for extreme weather events. The City of Chicago began to examine which areas in the city had a combination of the most vulnerable populations and extreme weather, looking at sociodemographic data, health data, street and basement flooding, and urban heat island areas, finding that many disadvantaged neighborhoods, in the south and west of the city, suffered from extreme weather events as well as health and economic disadvantages. Learning from their experience in previous urban greening and environmental projects, they held a series of meeting organized by a local non-governmental organization (NGO) to discuss where the biggest issues were with the communities. They applied for funding for using green infrastructure, such as stormwater management, bioswales, green roofs, etc. as a strategy to address both vulnerability and extreme weather issues with the funding. While their application in that round was unsuccessful, in 2017, there was another found of disaster relief money called the Community Development Block Grant (CDBG), which they repurposed the project for, and they were successful. The City argued for using city-owned vacant land and turning them into storm water management landscapes, similar to what the City of Philadelphia had been doing [86,87].

The project builds on a 2008 stormwater ordinance that requires projects to keep half an inch of rainwater on site or increase the permeability of the site by 15%—thereby reducing both volume and rate of stormwater flow, which aligns with their Sustainability Matrix, which awards points for exceeding the stormwater ordinance requirements [87]. The project uses traditional green infrastructure strategies, such as permeable pavement, bioswales, and rain gardens, in combination with large underground storage and filtering strategies to drain stormwater from surrounding streets and alleys into these new landscapes, thus getting water out of people's basements. There are multiple linked parcels of land in the project, including three corridors, ten distinct projects, and 23 formerly city-owned vacant parcels, but the one with the strongest biophilic attributes is the parcel on 16th street (Figure 9). Of particular interest for the implementation of ecological, biophilic, and equity goals is the collaborative and ecological approach taken by the City. The City worked very closely with community groups, whom the City had previous experience working with, and who had the ability to help manage the projects. The City provided the community groups with a list of possible plants but let the community groups chose the plants and trees, and they collaborated extensively on the goals and design of the projects. The final project combines a nursery, green roof on an affordable housing project, and a runnel between the street and sidewalk, where the runoff will drain. The runnel blends big rock outcroppings in a serpentine pattern that crosses over the runnel so that children going to school can walk on the outcroppings and cross the runnel, mimicking a forest creek. The combination of interacting water features, community engagement, and native plantings is a good example of blending biophilic design with ecological and community-benefit goals.

The City collaborated with local neighborhood groups on a maintenance and stewardship plan, which is often a weak point with urban greening projects, and even hired locally for the installation [11]. They estimate that the entire Resilient Corridors project will provide over half a million gallons of storage of rainwater, lowering the level of water in the combined sewer area by 0.2 to 8.2 inches, and reducing the risk of basement flooding of almost 600 buildings in the area [87]. The project won an American Society of Landscape Architects (ASLA) award in 2018 [88]. While research on the outcomes of the project is ongoing, initial responses from the community have been very positive, with one resident commenting that she "couldn't believe they were doing this for them, that they listened to them, that they are getting exactly what they wanted, and that it is beautiful" [87] (p. 87).

5. Discussion

5.1. Understanding the Application of Research to Practice

The review of the two key paradigms underlying most research programs on nature and human health highlighted the strengths and limitations of these initiatives, emphasizing their easy transfer to policy, due to their psychometric methods, but also their tendency to miss socio-cultural and power dynamics of place. The conducted study also pointed to the disconnect between the design and lived experience of place goals of biophilic design and research programs used in policy as well as new directions in somatic experience of place that can be used to connect urbanites to biophilic design.

The case studies exemplified the translation of research to practice, and the use of a diversity of evidence in real-world contexts. The City of London case study represents a good example of what kind of research can be applied to which context, supported by several applications at different scales. The translation of research to practice at a building scale could benefit from a critical analysis of which studies can be applied to the workplace and why, combined with an attention to biophilic design principles and a sense of place. At a larger scale, urban parks, wetlands, and community gardens in London's initiatives can help achieve ambitious goals to green and re-wild the city for people and nature. These implementations represent "an acknowledgement to how vitally urban lives are bound up with and enriched by nature" [89]. The City of Chicago Resilient Cities project is equally an innovative example of bridging research and practice while envisioning a more resilient and just neighborhood through green infrastructure and biophilic design principles. It blends known research on the benefits of access to nature and lived experience of place with active and adaptive collaboration with community partners, so that the new "place" is both ecologically and socially important to the community while addressing real climate change and economic vulnerabilities.

The discussion below draws on the insights of the different research paradigms, design practices, and case studies, aiming to provide key lessons learned that designers and planners can apply to their practice.

5.1.1. Benefits of Nature in the Workplace

One of the outcomes of great interest to business and industry is the potential improvement in human performance from access to nature in the workplace. However, biophilic design, which translates research to practice, has been criticized for not linking specific studies to specific design outcomes. One way to do this is to examine which studies address the desired outcomes and then analyze if they can be applied to the context of the design intervention. Multiple studies have shown improved task performance from access to nature, which is measured often through cognitive tests and proxies for productivity. These studies have been criticized, in turn, for not replicating the actual day-to-day tasks of office workers, and there has been limited research done in situ for office workers.

However, the benefits of improved task performance from better concentration are supported by multiple studies in nature [90,91]. These studies should not be used alone to prove increased performance from biophilic design interventions, due to limitations in how performance is measured, its applicability to different types of work and workplaces, and a lack of research specifically looking at nature, performance, and the workplace. However, if studies showing improved concentration from access to nature at multiple scales are combined with other performance measures, at an individual and organizational level, such as absenteeism, or commercial output—such as at the Living Lab at the London Shard— they can provide a reasonable indication that design interventions that increase access to nature for workers will likely lead to improved cognitive function and performance in the workplace.

5.1.2. Biophilic Design and Mental Health

Getting out of buildings, into natural green space, walking, or forest bathing, has long been recognized as beneficial and a prescribed option for general practitioners. Even observing the ordered complexity of fractals, which are self-similar scales found within nature, can reduce stress [92]. This is a key relief that is especially needed during Covid-19.

Covid-19 has highlighted the role of nature in mental health and socialization [6,7]. We have been forced to slow down and pay attention to nearby nature and value the role it can play in our mental health and well-being. Urban parks, or the lack thereof, are making headlines for their role in nurturing quarantine people's mental and physical health [8]. Throughout the lockdown, governments, regional and city officials have recognized the importance of space, from country parks, to city parks, urban green spaces, as vital for physical and mental well-being.

The (re)discovery of the joy and refuge of nature, specifically local nature on doorsteps and in backyard gardens, has led to newfound delight in fractal minutiae around us and a slowing down of the pace of urban life. This slowed pace—at the core of neighborhood projects such as The Barbican and GMV, in London—may be key to mainstreaming the restorative benefits of nature.

5.1.3. Biophilic Design and Connectedness to Nature: A Tool for Environmental Behavior Change

This review proposes biophilic design as a possible framework or pathway to connecting humans with nature through design that encourages sensory contact, emotion, meaning, beauty, and compassion, and which builds on the biophilic elements from Kellert. This is aligned with calls to improve human–nature relationships as a way to address our climate crisis and ecological separation [93], as well as research that has shown that connectedness with nature is linked to pro-environmental behavior [32]. However, research has also found that some elements of the human–nature relationship are not covered by biophilic design and follow a more dominion-utility framework [94], or values [95]. The research of Lumber, Richardson, and Sheffield [96] found that four of Kellert's [39] values of biophilia were unrelated to nature connectedness. These were fear of nature [97], dominion over nature [98], the utilitarian use of nature [99], and a purely scientific relationship [31]. These types of relationship are often emphasized within capitalistic societies and can be seen as essential pathways for human survival and progress that, unchecked, have led to nature's decline [100–103]. For transformative change, there is clearly an urgent need for a new relationship with nature, yet these negative types of relationship with nature still dominate [103]. Addressing these underlying values and perceptions will be essential to creating effective biophilic design interventions as well as fostering a connection with nature [104].

5.1.4. Connecting Biophilic Design with Environmental Justice, Health, and Climate Change

At a global scale, climate change has been described as: "[…] the most serious threat to global economic, social, and environmental stability in recorded history […] with many […] prevalent human diseases linked to climate fluctuations" [105].

Authors [106] have argued that it is our destruction of natural habitats that helped the current Covid-19 pandemic and that we can expect more zoonotic-originated diseases in the future: "There is a single species that is responsible for the Covid-19 pandemic—us. As with the climate and biodiversity crises, recent pandemics are a direct consequence of human activity." [106].

In figuring out how to address future global emergencies, such as climate change and Covid-19, our relationship with nature, and in particular biophilic design, may be key for improving sustainable behavior and, ultimately, our well-being. Rather than relying on abstract universal ideas of nature to encourage sustainable behavior, using design and policy at a building, neighborhood, and city scale to connect our daily lives with nature may encourage connection, improve our health and well-being, and make action feel more meaningful. Then, improving sustainable behavior might help address the current climate and disease crisis. While inaction and business as usual has plagued climate change policies, Covid-19 has exposed the connection between climate change and infectious disease, with those who have been exposed to air pollution dying at a higher rate [5]. This direct and personal connection between climate change and health may prove to be more effective in shifting policy and practice around climate change.

From a health perspective, this may require a shift from risk reduction and the treatment of illnesses to biophilic research and practice that embraces salutogenic thinking, i.e., the medical concept [107] that encourages a focus on factors that improve and support human health and well-being, rather than on factors that reduce illness [108]. With the health and well-being of humans intrinsically linked to the health and well-being of the planetary ecosystems, the combination of biophilic and salutogenic design approaches may provide a more holistic framework to link ecosystem, human, and non-human dimensions. Considering that, at a building scale, research attention has tended to focus on threats to health, a more holistic way of thinking would also be useful to foster health-promoting environments [109].

5.1.5. Looking Forward: Engaging with Nature and Fostering a Systems-Thinking Approach

Engaging with nature necessitates a mindset focused on developing the capacity and capability for systems evolution. It is not about a sustainability that maintains what it is—or is attempting to restore something to what it was by only reducing impacts. Rather, it is about creating systems (places, buildings, communities, organizations) that have the capacity to evolve and regenerate toward states of health that thrive over time. The understanding of our position on the planet has a crucial role in building the awareness for regenerative sustainability.

On a larger scale, an emerging trend is the Bio-Leadership, i.e., a concept of an ecosystem made of people and projects transforming leadership by working with nature [110]. Within the design and policy world, the concept switches from a mechanistic perspective (where the world is seen to function as a machine), to a natural fluid approach. This framework has been used to describe the hoped-for next era of our relationship with the environment. This new way of envisioning the nature–human relationship in design and policy aims to nurture a co-evolving mutuality [111] and may provide hope for both a more equitable and regenerative future. If combined with work on equitable access to nature, along with evidence on the benefits of access to nature at multiple scales, this large-scale application of biophilic principles can play a part in restoring both human and ecological health.

6. Conclusions

Humans' disconnection with nature has already negatively impacted mental and physical health. Buildings today are often designed, constructed, and operated apart from nature, rather than as a part of nature. Over the last thirty years (since Brundtlandt, 1987) [112], sustainability in design and construction has been a core element in the built environment, and yet climate and biodiversity indicators have worsened, while the impact of building design and practice on health conditions is increasingly researched but still remains opaque. Evidence from the last forty years has shown that contact with nature in general can improve human health, but there are gaps in the application at different scales and a lack of understanding of which research to apply to which situation.

Conversely, biophilic design is growing in popularity, but it still suffers from a lack of specificity on research outcomes and variables. There is a tendency for it is to be dismissed from many design circles as "nice to have but dispensable" versus an effective intervention to improve health and performance. The research on nature and health to date supports many of the biophilic design attributes outlined above; however, in practice, biophilic design is often limited to a few variables, which limits its application in design practice. Furthermore, there is still much that is not known about the potential benefits of biophilic design interventions, individually and as a whole. This gap has not been overcome by the confusion of green design interventions in green buildings and green infrastructure over the last few decades, which may or may not have had any link to evidence-based or biophilic design. It is also complicated by the differing underlying paradigms in nature and health research and design: research that examines nature as a linear input with an expected outcome does not align well with the more philosophical sense of place and lived experience goals of biophilic design. Drawing on some experiences developed in experimental biophilic design, it may help to bridge some of the gaps in traditional nature–health research and address the nuances and complexities of the holistic lived experience, as connected to nature or biophilic design projects. Connecting to sense of place, historicity, and embodied experience in biophilic design may soften some of the criticisms of the adaptive and utilitarian approaches to nature–health research while creating design solutions that work for real people in real contexts.

Lastly, there is still a need to provide a synthesis with respect to the available knowledge about the relationship between nature design and policy interventions, natural systems, and health. This seems to be confirmed by the growing demand from policy makers.

For instance, in the "Urban green spaces: brief for action", which was published recently, the World Health Organization [113] emphasized the need for a change in urban health initiatives with a strong focus on the creation, promotion, and maintenance of green spaces, with an explicit call for expert advice. How this expertise is developed is a current gap in both education and practice.

The discussion above argues that understanding the strengths and limitations of the most influential research on health and nature can help it support and align with biophilic design at multiple scales. This knowledge can result in a more effective and holistic understanding of how nature can be incorporated into our buildings, neighborhoods, and cities. Critically combining research on health and nature with biophilic design principles may also provide a more holistic and just approach to connecting us with nature and encouraging sustainable behavior. This can further support regenerative policy and action. As we look to life with and after Covid-19, the shape of the future built environment remains unknown, but it provides an opportunity for re-evaluation and new insights about our human, natural, and built environment relationships.

Author Contributions: Conceptualization, M.B.A. and A.L.; methodology, M.B.A. and A.L.; writing—original draft preparation, M.B.A. and A.L.; writing—review and editing, M.B.A., A.L., M.B., J.B.; visualization, M.B.A.; supervision, M.B.A., A.L.; funding acquisition, M.B.A., M.B., J.B. All authors have read and agreed to the published version of the manuscript.

Funding: This article is based upon work from COST Action (www.cost.eu) CA16114 'RESTORE' Rethinking Sustainability Toward a Regenerative Economy, supported by COST (European Cooperation in Science and Technology).

Institutional Review Board Statement: Not applicable.

Informed Consent Statement: Not applicable.

Data Availability Statement: No new data were created or analyzed in this study. Data sharing is not applicable to this article.

Acknowledgments: This article is based upon work from COST Action (www.cost.eu) CA16114 'RESTORE' Rethinking Sustainability Toward a Regenerative Economy, supported by COST (European Cooperation in Science and Technology). COST (European Cooperation in Science and Technology) is a funding agency for research and innovation networks. COST Actions help connect research initiatives across Europe and enable scientists to grow their ideas by sharing them with their peers. This boosts their research, career, and innovation. Authors wish to thank DaeWha Kang Design for authorizing (March 29th, 2021) the use in this article of two images (Figures 1 and 2) of their project "Shard Living Lab", in London.

Conflicts of Interest: The authors declare no conflict of interest.

Declaration: An initial version of this paper was presented at the Greening Cities Shaping Cities Symposium in October 2020. https://www.greeningcities-shapingcities.polimi.it/.

Abbreviations

The following abbreviations have been used in this manuscript:

ART	Attention Restoration Theory
PSR	Psychophysiological Stress Reduction
COST	Cooperation in Science and Technology
RESTORE	Rethinking Sustainability Toward a Regenerative Economy

References

1. Hadavi, S.; Kaplan, R.; Hunter, M.C.R. Environmental affordances: A practical approach for design of nearby outdoor settings in urban residential areas. *Landsc. Urban Plan.* **2015**, *134*, 19–32. [CrossRef]
2. Kaplan, S. The restorative benefits of nature: Toward an integrative framework. *J. Environ. Psychol.* **1995**, *15*, 169–182. [CrossRef]
3. Thompson, C.W.; Roe, J.; Aspinall, P.; Mitchell, R.; Clow, A.; Miller, D. More green space is linked to less stress in deprived communities: Evidence from salivary cortisol patterns. *Landsc. Urban Plan.* **2012**, *105*, 221–229. [CrossRef]
4. Acuto, M. COVID-19: Lessons for an Urban(izing) World. *One Earth* **2020**, *2*, 317–319. [CrossRef]

5. Wu, X.; Nethery, R.C.; Sabath, B.M.; Braun, D.; Dominici, F. Exposure to air pollution and COVID-19 mortality in the United States. Available online: https://projects.iq.harvard.edu/covid-pm?gsBNFDNDN=undefined&utm_campaign=wp_the_energy_202&utm_medium=email&utm_source=newsletter&wpisrc=nl_energy202 (accessed on 12 May 2020).
6. Beute, F.; Andreucci, M.B.; Lammel, A.; Davies, Z.; Glanville, J.; Keune, H.; Marselle, M.; O'Brien, L.A.; Olszewska-Guizzo, A.; Remmen, R.; et al. *Types and Characteristics of Urban and Peri-Urban Green Spaces Having an Impact on Human Mental Health and Wellbeing. Report Prepared by an EKLIPSE Expert Working Group*; Centre for Ecology & Hydrology: Wallingford, UK, 2020.
7. Beute, F.; Davies, Z.; de Vries, S.; Glanville, J.; Keune, H.; Lammel, A.; Marselle, M.; O'Brien, L.; Olszewska-Guizzo, A.; Remmen, R.; et al. *Types and Characteristics of Urban and Peri-Urban Blue Spaces Having an Impact on Human Mental Health and Wellbeing. Report Prepared by an EKLIPSE Expert Working Group*; Centre for Ecology & Hydrology: Wallingford, UK, 2020.
8. Surico, J. The Power of Parks in a Pandemic. Available online: https://www.bloomberg.com/news/articles/2020-04-09/in-a-pandemic-the-parks-are-keeping-us-alive (accessed on 23 June 2020).
9. Beatley, T. *Handbook of Biophilic City Planning and Design*; Island Press: Washington, DC, USA, 2016.
10. Frumkin, H.; Bratman, G.N.; Breslow, S.J.; Cochran, B.; Kahn Jr, P.H.; Lawler, J.J.; Wolf, K.L. Nature contact and human health: A research agenda. *Environ. Health Perspect.* **2017**, *125*, 075001. [CrossRef] [PubMed]
11. Loder, A. *Small-Scale Urban Greening: Creating Places of Health, Creativity, and Ecological Sustainability*; Routledge: Abingdon, UK, 2020.
12. Yin, R.K. *Application of Case Study Research*; Sage: London, UK; New Delhi, India, 1993.
13. Yin, R.K. *Case Study Research: Design and Methods*; Sage: London, UK; New Delhi, India, 1994.
14. Keniger, L.E.; Gaston, K.J.; Irvine, K.N.; Fuller, R.A. What are the Benefits of Interacting with Nature? *Int. J. Environ. Res. Public Heal.* **2013**, *10*, 913–935. [CrossRef] [PubMed]
15. Hartig, T.; Mitchell, R.; De Vries, S.; Frumkin, H. Nature and Health. *Annu. Rev. Public Heal.* **2014**, *35*, 207–228. [CrossRef]
16. Dzhambov, A.M.; Markevych, I.; Hartig, T.; Tilov, B.; Arabadzhiev, Z.; Stoyanov, D.; Gatseva, P.; Dimitrova, D.D. Multiple pathways link urban green- and bluespace to mental health in young adults. *Environ. Res.* **2018**, *166*, 223–233. [CrossRef]
17. Bratman, G.N.; Anderson, C.B.; Berman, M.G.; Cochran, B.; De Vries, S.; Flanders, J.; Folke, C.; Frumkin, H.; Gross, J.J.; Hartig, T.; et al. Nature and mental health: An ecosystem service perspective. *Sci. Adv.* **2019**, *5*, eaax0903. [CrossRef]
18. Labib, S.; Lindley, S.; Huck, J.J. Spatial dimensions of the influence of urban green-blue spaces on human health: A systematic review. *Environ. Res.* **2020**, *180*, 108869. [CrossRef]
19. Ballard, B. Biophilic office designs drive productivity and creativity. European CEO. Available online: https://www.europeanceo.com/business-and-management/biophilic-office-designs-drive-productivity-and-creativity/ (accessed on 16 June 2019).
20. Millennium Ecosystem Assessment. *Ecosystems and Human Well-being: Synthesis*; Island Press: Washington, DC, USA, 2005.
21. Gómez-Baggethun, E.; Barton, D.N. Classifying and valuing ecosystem services for urban planning. *Ecol. Econ.* **2013**, *86*, 235–245. [CrossRef]
22. Wilson, E.O. *Biophilia: The Human Bond with Other Species*; Harvard University Press: Cambridge, MA, USA, 1984.
23. Somarakis, G.; Stagakis, S.; Chrysoulakis, N. (Eds.) ThinkNature Nature-Based Solutions Handbook. ThinkNature project funded by the EU Horizon 2020 research and innovation programme under grant agreement No. 730338. 2019. Available online: https://platform.think-nature.eu/system/files/thinknature_handbook_final_print_0.pdf (accessed on 15 March 2021). [CrossRef]
24. Kaplan, R.; Kaplan, S. Preference, Restoration, and Meaningful Action in the Context of Nearby Nature. In *Urban Place: Reconnecting with the Natural World*; Barlett, P., Ed.; MIT Press: Cambridge, UK, 2005; p. 330.
25. Ulrich, R.S. Biophilia, Biophobia, and Natural Landscapes. In *The Biophilia Hypothesis*; Kellert, S.E.O.W., Ed.; Island Press: Washington, DC, USA, 1993; pp. 74–137.
26. Kaplan, R.; Kaplan, S. *The Experience of Nature: A Psychological Perspective*; Cambridge University Press: Cambridge, UK, 1989.
27. Hartig, T.; Mang, M.; Evans, G.W. Restorative Effects of Natural Environment Experiences. *Environ. Behav.* **1991**, *23*, 3–26. [CrossRef]
28. Korpela, K.M.; Ylén, M.; Tyrväinen, L.; Silvennoinen, H. Determinants of restorative experiences in everyday favorite places. *Heal. Place* **2008**, *14*, 636–652. [CrossRef]
29. Fromm, E. *The Anatomy of Human Destructiveness*; Holt, Rinehart and Winston: New York, NY, USA, 1973.
30. Church, S.P. Exploring Green Streets and rain gardens as instances of small scale nature and environmental learning tools. *Landsc. Urban Plan.* **2015**, *134*, 229–240. [CrossRef]
31. Davison, A. The trouble with nature: Ambivalence in the lives of urban Australian environmentalists. *Geoforum* **2008**, *39*, 1284–1295. [CrossRef]
32. Liuna, G.; Jingke, X.; Lijuan, Y.; Wenjun, Z.; Kexin, Z. Connections with Nature and Environmental Behaviors. *PLoS ONE* **2015**, *10*. [CrossRef]
33. Perrin, J.L.; Benassi, V.A. The connectedness to nature scale: A measure of emotional connection to nature? *J. Environ. Psychol.* **2009**, *29*, 434–440. [CrossRef]
34. Wang, J.; Geng, L.; Schultz, P.W.; Zhou, K. Mindfulness Increases the Belief in Climate Change: The Mediating Role of Connectedness With Nature. *Environ. Behav.* **2019**, *51*, 3–23. [CrossRef]

35. Wyles, K.J.; White, M.P.; Hattam, C.; Pahl, S.; King, H.; Austen, M. Are Some Natural Environments More Psycho-logically Beneficial Than Others? The Importance of Type and Quality on Connectedness to Nature and Psychological Restoration. *Environ. Behav.* **2019**, *51*, 111–143. [CrossRef]

36. Braubach, M.; Egorov, A.; Mudu, P.; Wolf, T.; Thompson, C.W.; Martuzzi, M. Effects of Urban Green Space on Environmental Health, Equity and Resilience. In *Nature-Based Solutions to Climate Change Adaptation in Urban Areas: Linkages between Science, Policy and Practice*; Springer International Publishing: Cham, Switzerland; Berlin/Heidelberg, Germany, 2017; pp. 187–205.

37. Kim, D.; Jin, J. Does happiness data say urban parks are worth it? *Landsc. Urban Plan.* **2018**, *178*, 1–11. [CrossRef]

38. Han, B.; Cohen, D.A.; Derose, K.P.; Marsh, T.; Williamson, S.; Raaen, L. How much neighborhood parks contribute to local residents' physical activity in the City of Los Angeles: A meta-analysis. *Prev. Med.* **2014**, *69*, S106–S110. [CrossRef]

39. Kellert, S.R.; Wilson, E.O. *The Biophilia Hypothesis*; Island Press: Washington, DC, USA, 1993.

40. Barton, J.; Griffin, M.; Pretty, J. Exercise, nature and socially interactive-based initiatives improve mood and self-esteem in the clinical population. *Perspect. Public Heal.* **2011**, *132*, 89–96. [CrossRef]

41. Berg, M.M.V.D.; Van Poppel, M.; Van Kamp, I.; Ruijsbroek, A.; Triguero-Mas, M.; Gidlow, C.; Nieuwenhuijsen, M.J.; Gražule-vičiene, R.; Van Mechelen, W.; Kruize, H.; et al. Do Physical Activity, Social Cohesion, and Loneliness Mediate the Association Between Time Spent Visiting Green Space and Mental Health? *Environ. Behav.* **2019**, *51*, 144–166. [CrossRef]

42. Schipperijn, J.; Bentsen, P.; Troelsen, J.; Toftager, M.; Stigsdotter, U.K. Associations between physical activity and characteristics of urban green space. *Urban For. Urban Green.* **2013**, *12*, 109–116. [CrossRef]

43. PennPraxis. Green2015: An Action Plan for the First 500 Acres. Philadelphia: City of Philadelphia. Available online: http://planphilly.com/green2015 (accessed on 30 August 2020).

44. City of Philadelphia. *Greenworks: A vision for a sustainable Philadelphia*; Office of Sustainability: Philadelphia, PA, USA, 2016.

45. Peters, K.; Elands, B.; Buijs, A. Social interactions in urban parks: Stimulating social cohesion? *Urban For. Urban Green.* **2010**, *9*, 93–100. [CrossRef]

46. Holt-Lunstad, J.; Smith, T.B.; Layton, J.B. Social Relationships and Mortality Risk: A Meta-analytic Review. *PLoS Med.* **2010**, *7*, e1000316. [CrossRef]

47. Sun, Z.; Zhu, D. Exposure to outdoor air pollution and its human health outcomes: A scoping review. *PLoS ONE* **2019**, *14*, e0216550. [CrossRef]

48. Klompmaker, J.O.; Hoek, G.; Bloemsma, L.D.; Wijga, A.H.; van den Brink, C.; Brunekreef, B.; Janssen, N.A. As-sociations of combined exposures to surrounding green, air pollution and traffic noise on mental health. *Environ. Int.* **2019**, *129*, 525–537. [CrossRef]

49. Calderón-Garcidueñas, L.; Torres-Jardón, R.; Kulesza, R.J.; Park, S.B.; D'Angiulli, A. Air pollution and detrimental effects on children's brain. The need for a multidisciplinary approach to the issue complexity and challenges. *Front. Hum. Neurosci.* **2014**, *8*, 613.

50. von Lindern, E.; Hartig, T.; Lercher, P. Traffic-related exposures, constrained restoration, and health in the residential context. *Health Place* **2016**, *39*, 92–100. [CrossRef]

51. Tuan, Y.F. Rootedness versus sense of place. *Landscape* **1980**, *24*, 3–8.

52. Williams, D.R. Making sense of 'place': Reflections on pluralism and positionality in place research. *Landsc. Urban Plan.* **2014**, *131*, 74–82. [CrossRef]

53. Zufferey, C.; King, S. Social work learning spaces: The Social Work Studio. *High. Educ. Res. Dev.* **2016**, *35*, 395–408. [CrossRef]

54. Cornell Health. Nature Rx. Cornell University. Available online: https://health.cornell.edu/resources/health-topics/nature-rx (accessed on 16 June 2019).

55. Kallen, C. NaturePHL Bringing 'Nature Prescriptions' to Local Doctors' Offices. Family Focus Media, Philadelphia. Available online: http://familyfocus.org/nature-phl-nature-prescriptions-philadelphia (accessed on 16 June 2019).

56. Lachowycz, K.; Jones, A.P. Towards a better understanding of the relationship between greenspace and health: Development of a theoretical framework. *Landsc. Urban Plan.* **2013**, *118*, 62–69. [CrossRef]

57. Trentelman, C.K. Place Attachment and Community Attachment: A Primer Grounded in the Lived Experience of a Community Sociologist. *Soc. Nat. Resour.* **2009**, *22*, 191–210. [CrossRef]

58. Stedman, R.C.; Beckley, T.M. If we knew what it was we were doing, it would not be called research, would it? *Soc. Nat. Resour.* **2007**, *20*, 939–943. [CrossRef]

59. Williams, D.R.; Patterson, M.E. Snapshots of What, Exactly? A Comment on Methodological Experimentation and Conceptual Foundations in Place Research. *Soc. Nat. Resour.* **2007**, *20*, 931–937. [CrossRef]

60. Kellert, S.R. Dimensions, Elements, and Attributes of Biophilic Design. In *Biophilic Design: The Theory, Science, and Practice of Bringing Buildings to Life*; Kellert, S.R., Heerwagen, J., Mador, M., Eds.; John Wiley & Sons: Hoboken, NJ, USA, 2018.

61. Browning, W.D.; Ryan, C.O.; Clancy, J.O. *14 Patterns of Biophilic Design*; Terrapin Bright Green LLC: New York, NY, USA, 2014.

62. McGee, B.; Park, N.; Portillo, M.; Bosch, S.; Swisher, M. Diy Biophilia: Development of the Biophilic Interior Design Matrix as a Design Tool. *J. Inter. Des.* **2019**, *44*, 201–221. [CrossRef]

63. Pallasmaa, J.; Amundsen, M. Q&A with Juhani Pallasmaa on Architecture, Aesthetics of Atmospheres and the Passage of Time Questions-réponses avec Juhani Pallasmaa sur l'architecture, l'esthétique des ambiances et les effets du temps. Ambiances. Environnement sensible, architecture et espace urbain Comptes-rendus. Available online: http://journals.openedition.org/ambiances/1257 (accessed on 30 August 2019).

64. Reeve, A.; Desha, C.; Hargroves, K.; Newman, P. Informing healthy building design with biophilic urbanism design principles: A review and synthesis of current knowledge and research. In Proceedings of the 10th International Conference on Healthy Buildings, Brisbane, Australia., 8–12 July 2012.

65. Statista. Forecasted population in London (UK) from 2019 to 2041. Available online: https://www.statista.com/statistics/379035/london-population-forecast/ (accessed on 16 March 2021).

66. Greenspace Information for Greater London CIC (GiGL). Key London Figures. Available online: https://www.gigl.org.uk/keyfigures/ (accessed on 16 March 2021).

67. DaeWha Kang Design and Aldworth James & Bond. Using cutting-edge fabrication technology to construct unique working spaces in the UK's tallest building. Kellert, S.R., and Calabrese, E.F. 2015. The practice of Biophilic design. Available online: www.biophilic-design.com (accessed on 10 February 2021).

68. Kellert, S.R.; Calabrese, E.F. The practice of Biophilic design. Available online: www.biophilic-design.com (accessed on 10 February 2021).

69. Dunnett, N. *Naturalistic Planting Design: The Essential Guide*; Filbert Press: London, UK, 2019.

70. Smith, C.J.; Relph, E. Place and Placelessness. *Geogr. Rev.* **1978**, *68*, 116. [CrossRef]

71. Loder, A. 'There's a meadow outside my workplace': A phenomenological exploration of aesthetics and green roofs in Chicago and Toronto. *Landsc. Urban Plan.* **2014**, *126*, 94–106. [CrossRef]

72. Klinenberg, E. *Heat Wave: A Social Autopsy of Disaster in Chicago*; University of Chicago Press: Chicago, IL, USA, 2002.

73. City of Chicago, Department of Transportation. The Chicago Green Alley Handbook: An Action Guide to Create a Greener, Environmentally Sustainable Chicago. Available online: https://www.chicago.gov/dam/city/depts/cdot/GreenAlleyHandbook.pdf (accessed on 12 June 2019).

74. Berkshire, M. Chicago Green Projects. In *Small-Scale Urban Greening: Creating Places of Health, Creativity, And Ecological Sustainability*; Loder, A., Ed.; Routledge: Abingdon, UK, 2020; p. 94.

75. City of Chicago, Department of Water Management. Green Stormwater Infrastructure Strategy. Available online: https://www.chicago.gov/content/dam/city/progs/env/ChicagoGreenStormwaterInfrastructureStrategy.pdf (accessed on 10 July 2015).

76. City of Chicago, Department of the Environment. New Stormwater Management Ordinance. Available online: https://www.chicago.gov/content/dam/city/depts/water/general/Engineering/MS4/MS4_Stormwater_Plan.pdf (accessed on 4 September 2018).

77. Andreucci, M.B. *Progettare L'involucro Urbano. Casi Studio di Progettazione Tecnologica Ambientale*; Wolters Kluwer: Milano, Italy, 2019.

78. 100 Resilient Cities. Available online: https://100resilientcities.org/ (accessed on 10 May 2019).

79. Gobster, P.H.; Stewart, W.P.; van Riper, C.J.; Williams, A.R. Vacant Lot Stewardship and the Creation of New Natures in Chicago. In Proceedings of the International Association for Landscape Ecology Conference, Chicago, IL, USA, 8–12 April 2018.

80. WLS-TV Chicago. City of Chicago selling more than 4K vacant lots for 1.ABC7EyewitnessNews. Available online: https://abc7chicago.com/news/city$-$of$-$chicago$-$selling$-$more$-$than$-$4k$-$vacant$-$lots$-$for$-$1.ABC7EyewitnessNews.November292019.https://abc7chicago.com/news/city$-$of$-$chicago$-$selling$-$more$-$than$-$4k$-$vacant$-$lots$-$for$-$$1-/1630671/ (accessed on 29 November 2019).

81. Lindsey, G.; Qi, Y.; Gobster, P.H.; Sachdeva, S. The 606 at Three: Trends in Use of Chicago's Elevated Rail- Trail, Proceedings of the Fábos Conference on Landscape and Greenway Planning, 6, 37. Available online: https://scholarworks.umass.edu/fabos/vol6/iss1/37 (accessed on 10 June 2020).

82. The Trust for Public Land. Our Story. *The 606*. Available online: https://www.the606.org/about/story/ (accessed on 26 May 2019).

83. Smith, G.; Duda, S.; Lee, J.M.; Thompson, M. Measuring the Impact of the 606: Understanding How a Large Public In-vestment Impacted the Surrounding Housing Market. Chicago: Institute for Housing Studies at DePaul University. Available online: https://www.housingstudies.org/media/filer_public/2016/10/31/ihs_measuring_the_impact_of_the_606.pdf (accessed on 12 June 2019).

84. Rodkin, D. Was gentrification around the 606 inevitable? Crain's Chicago Business. Available online: https://www.chicagobusiness.com/residential-real-estate/was-gentrification-around-606-inevitable (accessed on 13 December 2019).

85. Wessel, M. Chicago's Resiliency Plan Aims for Equity. Next City. Available online: https://nextcity.org/daily/entry/chicagos-resiliency-plan-aims-for-equity (accessed on 24 April 2019).

86. Green Stormwater Infrastructure Partners. Sustainable Business Network of Greater Philadelphia. The Economic Impact of Green City, Clean Waters: The First Five Years. 2016. Available online: https://gsipartners.sbnphiladelphia.org/wp-content/uploads/2014/07/Local-Economic-Impact-Report_First-Five-Years-GCCW_full-downloadable-web2.pdf (accessed on 12 June 2019).

87. Berkshire, M. Resilient Corridors. In Rockefeller Foundation 100 Resilient Cities, Chicago: A plan for Inclusive Growth and a Connected City, City of Chicago. Available online: https://resilient.chicago.gov/download/Resilient%20Chicago.pdf (accessed on 5 June 2019).

88. American Society of Landscape Architects. Chicago Resilient Corridors. Available online: https://il-asla.org/award/chicago-resilient-corridors/ (accessed on 2 October 2020).

89. Macfarlane, R. London Becomes the World's First National Park City. London National Park City. Available online: https://www.nationalparkcity.london/press/24-media/130-london-becomes-the-world-s-first-national-park-city (accessed on 14 June 2019).
90. Choudhry, K.Z.; Coles, R.; Qureshi, S.; Ashford, R.; Khan, S.; Mir, R.R. A review of methodologies used in studies investigating human behaviour as determinant of outcome for exposure to 'naturalistic and urban environments'. *Urban For. Urban Green.* **2015**, *14*, 527–537. [CrossRef]
91. Li, D.; Deal, B.; Zhou, X.; Slavenas, M.; Sullivan, W.C. Moving beyond the neighborhood: Daily exposure to nature and adolescents' mood. *Landsc. Urban Plan.* **2018**, *173*, 33–43. [CrossRef]
92. Taylor, R.P. Reduction of Physiological Stress Using Fractal Art and Architecture. *Leon* **2006**, *39*, 245–251. [CrossRef]
93. Richardson, M.; Dobson, J.; Abson, D.J.; Lumber, R.; Hunt, A.; Young, R.; Moorhouse, B. Applying the pathways to nature connectedness at a societal scale: A leverage points perspective. *Ecosyst. People* **2020**, *16*, 387–401. [CrossRef]
94. Schultz, W.P.; Zelezny, L.C. Values as predictors of environmental attitudes: Evidence for consistency across cultures. *J. Environ. Psychol.* **1999**, *19*, 255–265. [CrossRef]
95. Stern, P.C.; Dietz, T.; Kalof, L.; Guagnano, G.A. The new environmental paradigm in social psychological perspective. *Environ. Behav.* **1995**, *27*, 723–745. [CrossRef]
96. Lumber, R.; Richardson, M.; Sheffield, D. Beyond knowing nature: Contact, emotion, compassion, meaning, and beauty are pathways to nature connection. *PLoS ONE* **2017**, *12*, e0177186. [CrossRef]
97. Nash, R. Wilderness and the American Mind. Yale University Press: New Haven, CT, USA, 2001.
98. Merchant, C. Reinventing Nature: Western Culture as a Recovery Narrative. In *Uncommon Ground*; Cronon, W., Ed.; W. W. Norton & Company: New York, NY, USA, 1995; pp. 132–159.
99. Smith, N. The Production of Nature. In *Future Natural: Nature, Science, Culture*; Robertson, G.M.M., Tichner, L., Curtis, B., Putnam, T., Eds.; Routledge: London, UK, 1996; pp. 35–54.
100. Baskin, J. Paradigm dressed as epoch: The ideology of the antropocene. *Environ. Values* **2015**, *24*, 9. [CrossRef]
101. Catton, W.R.; Dunlap, R.E. Environmental Sociology: A New paradigm. *Am. Sociol.* **1978**, *13*, 41–49.
102. IPBES. *Summary for Policymakers of the Global Assessment Report on Biodiversity and Ecosystem Services of the Intergovernmental Science-Policy Platform on Biodiversity and Ecosystem Services*; Díaz, S., Settele, J., Brondízio, E.S., Ngo, H.T., Guèze, M., Agard, J., Arneth, A., Balvanera, P., Brauman, K.A., Butchart, S.H.M., et al., Eds.; IPBES: Bonn, Germany, 2019.
103. Ison, R.; Straw, E. *The Hidden Power of Systems Thinking: Governance in a Climate Emergency*; Routledge: London, UK, 2020.
104. Loder, A. Regeneration. Between Ecological and Human Systems. In *Progettare l'involucro Urbano: Casi Studio di Progettazione Tecnologica Ambientale*; Andreucci, M.B., Ed.; Wolters Kluwer: Milano, Italy, 2019; p. 179.
105. Africa, J.; Heerwagen, J.; Loftness, V.; Ryan Balagtas, C. Biophilic Design and Climate Change: Performance Parameters for Health. *Front. Built Environ.* **2019**, *5*, 28. [CrossRef]
106. Settele, J.; Díaz, S.; Brondizio, D.; Daszak, P. COVID-19 Stimulus Measures Must Save Lives, Protect Livelihoods, and Safeguard Nature to Reduce the Risk of Future Pandemics. Intergovernmental Science-Policy Platform on Biodiversity and Ecosystem Services (IPBES Report). Available online: https://ipbes.net/covid19stimulus (accessed on 10 June 2020).
107. Antonovsky, A. *Unravelling the Mystery of Health*; Jossey-Bass Inc: San Francisco, CA, USA, 1987.
108. Brown, M. *FutuREstorative: Working Towards a New Sustainability*; RIBA Publishing: London, UK, 2016.
109. Loder, A.; Gray, W.A.; Timm, S. The international WELL Building Institute's Global Research Agenda. Available online: https://marketing.wellcertified.com/global-research-agenda (accessed on 10 February 2021).
110. Roberts, A. How would nature Change Leadership? Available online: https://www.ted.com/talks/andres_roberts_how_would_nature_change_leadership (accessed on 7 May 2020).
111. Mang, P.; Haggard, B. *Regenerative Development and Design: A framework for Evolving Sustainability*; Wiley: Hoboken, NJ, USA, 2016.
112. World Commission on Environment and Development. *Our Common Future*; Oxford University Press: Oxford, UK, 1987.
113. WHO Regional Office for Europe. Urban Green Space Interventions and Health: A Review of Impacts and Effectiveness. Copenhagen: WHO Regional Office for Europe. Available online: http://www.euro.who.int/__data/assets/ (accessed on 12 March 2018).

 *sustainability*

Article

Evaluating the Relationship between Park Features and Ecotherapeutic Environment: A Comparative Study of Two Parks in Istanbul, Beylikdüzü

Didem Kara [1,*] **and Gülden Demet Oruç** [2]

1 Urban Design Master Program, School of Engineering, Science and Technology,
 Istanbul Technical University, Istanbul 34469, Turkey
2 Urban and Regional Planning Department, Faculty of Architecture, Istanbul Technical University,
 Istanbul 34437, Turkey; orucd@itu.edu.tr
* Correspondence: karadi@itu.edu.tr; Tel.: +90-5534-921-003

Abstract: The impacts of problems related to dense, unplanned, and irregular urbanization on the natural environment, urban areas, and humankind have been discussed in many disciplines for decades. Because of the circular relationship between humans and their environment, human health and psychology have become both agents and patients in interactions with nature. The field of ecopsychology investigates within this reciprocal context the relationship between human psychology and ecological issues and the roles of human psychology and society in environmental problems based on deteriorated nature–human relationships in urbanized areas. This approach has given rise to ecotherapy, which takes a systemic approach to repairing this disturbed nature–human relationship. This study aims to uncover the relationship between the physical attributes of urban green areas and their potential for providing ecotherapy service to users, first by determining the characteristics of ecotherapeutic urban space and urban green areas given in studies in the ecopsychology and ecotherapy literature, and then by conducting a case study in two urban parks from the Beylikdüzü District of the Istanbul Metropolitan Area. The impacts of these parks' changing physical characteristics on user experiences are determined through a comparison of their physical attributes and the user experiences related to their ecotherapy services.

Keywords: greening cities; urban design; ecopsychology; ecotherapy

Citation: Kara, D.; Oruç, G.D. Evaluating the Relationship between Park Features and Ecotherapeutic Environment: A Comparative Study of Two Parks in Istanbul, Beylikdüzü. *Sustainability* **2021**, *13*, 4600. https://doi.org/10.3390/su13094600

Academic Editor: Israa H. Mahmoud

Received: 24 February 2021
Accepted: 12 April 2021
Published: 21 April 2021

Publisher's Note: MDPI stays neutral with regard to jurisdictional claims in published maps and institutional affiliations.

1. Introduction

There has always been a bidirectional relationship between humankind and its environment. While humanity changes the environment based on its needs, the environment has in turn played an essential role in human evolution and development. Urban areas are one of the best examples of anthropogenic impacts on the environment. Such places are structured based on human needs and lifestyles under the influence of other anthropogenic factors such as industrialization, population growth, migration, development levels, and national policies. The phenomena born of these factors, such as rapid and distorted urbanization, have negatively affected natural areas and resources, leading to the creation of problematic and substandard urban areas. Moreover, the establishment of unplanned urban areas has resulted in both direct and indirect harm upon their inhabitants [1].

The indirect impacts of these areas are felt mostly in the natural environments that provide vital services for human life, resulting in shortages of environmental resources, the destruction of necessary ecosystems, the loss of biodiversity, and rises in global warming and pollution [2]. The direct impacts involve the damage caused by these urban areas to people's physical and mental health, e.g., diseases that can spread quickly in dense urban areas with poor physical conditions and a lack of infrastructure [3–8], lifestyle-related illnesses [3,4,6,9,10] resulting from the lack of physical activity and unhealthy dietary habits and food provision in some urban areas, afflictions related to exposure to urban pollution [3–6,8,9], and, lastly, mental issues caused by urban features such as the lack of social infrastructure [11], poor physical conditions, pollution [3,4,8,11–13], high population densities, and overcrowding [3,4,8,12,13].

The role of environmental sciences in solving the problems above has become more prominent; however, world ecosystems and human populations are still facing constant ecological issues. Considering the importance of individual behavior and awareness, there is a need for systemic (and comprehensive) approaches and solutions in terms of the reconstruction of individuals' relationship with nature and the environment. In this way the outputs of the social sciences examining human–nature relationships may offer valuable inputs for the urban planning and design disciplines. Environmental psychology stands out for this purpose, as it has been examining since the 1960s the bidirectional relationship between humans and the environment, its focus ranging from the physical and social effects of urban space to the impacts of natural areas on human psychology. Moreover, discussions on sustainability have included the claim that environmental psychology has evolved as a "psychology of sustainability" [14].

First coined by Theodore Roszak in 1992, ecopsychology has helped to develop environmental awareness and change behavior toward ecological problems through examination of the relationship between the environmental issues and human spiritual or psychological ones. Roszak argued that human activities and economic systems have changed, detailing the harmful effects of this changing activity and economic order on the ecosystem. Roszak noted that disconnection from nature and other people due to urbanization both increases negative impacts on the environment and deepens psychological problems [15]. To this end, the field examines the roles of human psychology and society in environmental issues within the framework of the deteriorated nature–human relationship [16].

In his treatment of the relationship between people and the environment, Scull posited a more experiential role for ecopsychology in theory and practice, asserting that many things can be learned through contact with nature [16]. This approach is speculative, philosophical, and theoretical, preparing a basis for the reconstruction of the nature–human relationship with a new language and model; it may also have a role to play in environmental protection and in solving human psychological problems through the adoption of practices such as environmental activism and ecotherapy.

At this point, it is clear that ecopsychology offers a solution to the problems of urbanization and urban areas based on the individual's perspective of and connectedness to nature (CNS). In addition to Scull's approach, through which strong ties are established to fields such as deep ecology and environmental activism, ecotherapy studies have introduced a systemic therapy method for repairing the disturbed nature–human relationship. Clinebell defined ecotherapy as "recovery and growth with a healthy relationship with the world" [17], using it as an inclusive term in the context of nature-based physical and psychological recovery methods. This approach to ecotherapy deals with psychotherapy and psychiatry in the context of nature and nature–human relationships. Clinebell labeled ecological deterioration the most profound health issue of all time owing to the vital role of ecosystems in the continuity of our kind and offered as a solution to this problem the raising of awareness about lifestyles through ecotherapy and early childhood eco-education [17].

Ecotherapy thus may offer a help to solve environmental problems and the psychological issues caused by disconnection from nature. Ecotherapeutic studies are based on a three-phased process: (1) acknowledgment of the healing presence of nature, (2) recognition of more-than-human experiences and self-relocation in the natural world, and (3) the sharing of this experience with other people and involvement in activities that care for the planet [17]. Ecotherapy is the name given to a wide range of programs aiming to improve mental and physical health through activities in natural areas and connection to nature. These activities include working in or experiencing nature [18]. However, the fulfillment of this reconnection is conditional on spending time and being active in natural areas and making these activities a part of daily life.

It is thus essential to be in nature and understand that being a part of the ecosystem is vital to solving both physical and mental health problems as well as to help the environmental crisis. Accordingly, ecotherapy helps people to recognize nature, appreciate it more, and be respectful to the earth. The necessity of addressing this approach through spatial studies has arisen from an emphasis on the importance of natural areas and spending time in nature, as issues related to disconnection from nature occur most frequently in urban spaces where natural areas and elements are scarce. Natural areas and urban greeneries have been subjects for examination in environmental disciplines for many decades because of their services to humans, their recreational functions, and their importance to urban quality and ecosystems [19,20]. Studies have investigated the benefits of these qualities for mental and physical health, in particular, their role in encouraging people to do physical exercise [21,22]. However, apart from their impacts on overall health, spatial studies have focused on the role of ecopsychology and ecotherapy to help people to be aware of environmental problems. For this purpose, it is helpful to understand their therapy functions for citizens in addition to their impacts on the quality of urban areas. The design of urban green spaces should be reviewed based on the features of therapeutic environments that create environmentally conscious individuals who can address the source of their health problems and environmental problems.

This study aims to reveal the relationship between the physical attributes of the urban green areas and their potential for providing ecotherapy service to citizens. The first section contains a brief explanation of the aspects of ecotherapeutic environments, determining the characteristics of ecotherapeutic urban spaces and urban green areas through an examination of the benefits obtained from green or natural places, their effects on human psychology, the attributes of therapeutic areas, and the types of therapeutic activities. These have been classified by discourse analysis in accordance with their contribution to the urban design process. In determining the attributions above, literature research was conducted in the Scopus' database in August 2019. A total of 249 papers were found in the database with the "ecopsychology" keyword and 57 with the "ecotherapy" keyword. Out of these articles, those related to psychology, social sciences, and environmental sciences were filtered, and 37 articles remained to be examined. The findings of this literature review were presented in detail at the 28th Symposium of Urban Design and Implementations and published as an article in the *Design+Theory Journal* in Turkish [1].

The second part of the study examines two parks from the Beylikdüzü District of the Istanbul Metropolitan Area in order to compare the impressions of the results obtained from a literature review of space and user experience. This comparison is twofold: (1) physical characteristics and (2) user experience. The physical characteristics of the parks are analyzed and presented via several maps, satellite images, diagrams and pictures. Data concerning user experience were obtained through a survey conducted with the users of these parks. This study adds to the ecopsychology literature by evaluating the ecotherapeutic benefits of green spaces and how these differ according to the urban design principles adopted when designing the spaces. In addition to highlighting the ecological and recreational benefits of urban green spaces, this study provides guidance for planning and designing green areas with improved ecotherapeutic features that may further enhance the psychological health and environmental awareness of city residents.

2. Characteristics of Ecotherapeutic Environment

The characteristics of ecotherapeutic environments and their effects on human psychology can be evaluated according to ecotherapeutic activities, type, benefits and features of ecotherapeutic environments. Ecotherapeutic activities are examined within two categories: working in nature and experiencing nature. Working in nature includes various athletic activities defined as the green and blue gym [23–29], the most significant of which is walking [23–26,30–35]. Apart from athletics, this group comprises activities such as meditation/therapy [23,25,28,33,34,36,37], art [28,38,39], and production in/with nature (frequently gardening and horticulture) [36,38]. Experiencing nature involves spending time observing and listening in nature [23,26,28–30,32,36,38]. Activities from both groups can be conducted in natural areas to obtain ecotherapy services, and an understanding of these activities allows designers to provide proper facilities or places to citizens in these areas.

The types of ecotherapeutic environments are grouped according to their location in inner, peripheral, and outer urban areas. Ecotherapeutic areas located on the outer and peripheral parts of an urban area include various natural areas and landscapes, of which forests [23,24,30,31,40–43] and wilderness areas [23,25,32,36,44,45] are the most prominent types. Ecotherapeutic areas located in inner urban areas include many public and private green areas; urban parks [24–26,30–33,36,38,41–43,46,47] and private gardens [23,26,36,43,48–50] are the most prominent examples of this type. These results reveal the need for natural spaces and urban greeneries in the urban texture because of their ecotherapy benefits. Moreover, they underline the importance of providing and protecting these areas both within and outside of the urban texture. Knowledge of the types of ecotherapeutic areas can help planners and designers consider these areas in their spatial decisions.

The benefits of therapeutic environments on human psychology comprise two categories: (1) mental and emotional benefits and (2) advancement in self-placement and perception. The most prominent mental and emotional benefits are relaxation [24,26,27,36,51], improvement in attention [24,28,30,34,39,41,48,52], concentration [26,31,34,53], and mood [23,26,29,34,48,51], and declines in stress [23,24,26,28–31,33,34,36,43,48,51,52], anxiety, depression [25,28,30,33,39,48], and anger [39,41]; better self-esteem is the most prominent manifestation of advancement in self-placement and perception [25–28,34,39,48,54]. These results demonstrate that spending time in natural areas helps people to cope with mental problems and gains importance in tandem with the growing negative impacts of urban areas on human mental health. Ecotherapy services increase the quality of life of citizens. Recognizing these benefits offers a new perspective for urban studies and design practices, especially in terms of designing cities and their green areas in a way that will provide ecotherapy services.

The features of ecotherapeutic areas, which can serve as the most directing outputs to environmental designers, are grouped into the categories of accessibility and size, design features, the fauna of therapeutic areas, and the sensations the areas create. First, as mentioned above, spending time in nature daily is essential in the provision of ecotherapy services. So the accessibility [34,36], inner circulation [39], and size [36,55] of these areas should be suitable for the daily use of citizens in their activities. The second group, design features, includes subgroups such as vegetation and natural elements, facilities and furniture, physical environmental control (daylight, wind, etc.), inner view and perception, and relationship with surrounding urban space. Vegetation and natural elements consists of the existence of landscapes and green areas with trees [29,30,32,36,45,55], bushes [26,30,55], grass [55], and flowers [24], and their type [30], density [31,42,56], and diversity [42,50]. Natural and artificial water elements [24,26,30,32,36,41,45] are evaluated under this subgroup. The facilities and furniture subgroup involves, rather than specific facility or furniture types, the compatibility of the furniture materials [26,55] with the natural characteristics of the area. It also includes certain exercise equipment [26] that encourages people to be more active. The inner view and perception subgroup comprises the necessity of structuring depth, complexity, enclosure, and vegetation density, each in a well-balanced manner, allowing for open views and remote landscapes [42,56]. Moreover, the visual relationship between ecotherapeutic areas and urban texture is a critical part of providing pristine and more natural perception [36,50] in an area. Consequently, it is better to obscure visibility of the urban pattern from ecotherapeutic areas [56] and increase the visibility of these areas from urban spaces [45,50] through regulations such as those that limit the number of floors in new buildings, lower urban density around green spaces [32,57], and create mild transitions from parks to urban areas [56]. Additionally, the presence of fauna enhances the natural image of the area, and encounters with wild animals and hearing animal sounds (bird sound, etc.) increase therapy service [26,29,31,41,45,53]. Lastly, ecotherapeutic areas create sensations helpful in obtaining therapy services such as peacefulness [41,58], quiet [26], solitude, distance [55], aesthetic pleasure [26], beauty [26,41], and fascination [35]. In order to obtain ecotherapy services as they are defined, people require the presence of sensations that oppose those endemic to dense urban areas such as overcrowding, noise pollution, etc. Natural elements and characteristics have thus become prominent in the design of therapeutic areas.

3. Method

The methodology of the study was twofold: examining the spatial characteristics of selected urban parks and examining the change of user experience according to the features of the parks. For the spatial examination of selected parks, the characteristics of the surrounding urban fabric and demographic structure of the population they serve were kept constant for the purpose of comparing their internal characteristics and the relationships they established with the surrounding urban fabric. Accordingly, two parks located close to each other were selected for the examination. The study also compared the different features of these two parks, such as type, size, form, design, and vegetation. A detailed examination of vegetation was conducted for this study with the help of site observation, 28 videos and 691 photographs that have geo-positioning data.

In order to evaluate user experience, a survey was conducted in the selected urban parks (Table S1). The first section of the survey contained a scale measurement of "connectedness to nature" to gauge individuals' effective and experiential connection to nature [59]. The scale was developed for the empirical studies on the basis of Leopold's claim that the environmental awareness depends on the feeling of belonging to the wider natural world [60]. Dependently, CNS included 14 questions about one's perspective of being a member of the natural world, feeling a sense of kinship with it, seeing themselves as belonging to the natural world as much as it belongs to them, and considering that their welfare depends on the welfare of natural world [59]. It was developed as a 5-point Likert scale, and scores were calculated as a mean value of the answers. CNS was selected, first, to seek out a relationship between the frequency of time spent in selected urban parks and consciousness about the value of the natural environment, and, second, to determine a relationship between the user profile regarding connectedness to nature and ecotherapy service. An understanding of user profile relation to environmental issues and connectedness to nature was essential in revealing whether or not the ecotherapy service provided by the city parks was available regardless of the user profile and ecological consciousness.

The second section of the survey consisted of 5-point Likert scales and open-ended questions about the features, activities, and feelings highlighted in ecopsychology and ecotherapy literature. The section made inquiries concerning types of activities, the adequateness of the parks for users, the impact of park characteristics on park preference, satisfaction with park characteristics, the influence of interior and exterior features or factors on the natural image of the parks, the relationship with the surrounding urban area, and the emotions/mental states that participants experienced during park use.

The survey was conducted in two selected parks at the same time, on four days from 12–15 September 2020 (two days during the week and two days on the weekend) from 8 a.m. to 8 p.m. Participants were chosen randomly within the two parks. The researchers first introduced themselves, informed the participant about the study, and the participant's consent was obtained for conducting the survey. A total of 90 subjects (49 male, 41 female) participated in the survey, 45 from each park. As the data on the total daily users of the park were unavailable, the decision on minimum sample size was based on the Central Limit Theorem, which defines the accurate sample size as more than 30. Besides, according to the calculations made on the population of the neighborhoods surrounding the parks, the ideal sample size was found to be 96 people, yet the sample was limited to 90 people in total because the proposed park users did not volunteer to participate in the survey during the pandemic.

4. Case Study

The history of the Istanbul Metropolitan Area goes back to the ancient settlements of 7–8 thousand years ago. It is a city that later became the capital city of important empires such as the Byzantine and Ottoman. The most important period of time that changed the face of the city took place in the Republic period. With the industrialization process, migration from rural to urban areas and rapid urbanization have taken place since the 1950s; the city has begun to sprawl and lose the important natural and green areas [61,62]. The development direction of the city shifted from the east–west direction to the north where the forests and other natural areas are rich, after the construction of the bridges over the Bosporus. In addition, due to the increased accessibility and uncontrolled urbanization, the historical core has become denser in time [62]. Today, with its diverse cultural and historical layers and over 15 million inhabitants, Istanbul is the biggest metropolitan area of Turkey [63]. Beylikdüzü District, where selected urban parks are located, is a newly urbanized settlement in comparison to the history of the city. Urban development of the district was pioneered by the housing cooperatives in 1990s [64]. This district was selected for the case study due to the presence of green areas of various sizes and shapes in the similar urban pattern, for an accurate comparison.

In the case study, two parks were selected to help gauge the relationship between spatial features and user experience. The first of these parks is a linear park system consisting of the Fatih Sultan Mehmet (FSM) and Mehmet Akif Ersoy (MAE) Woods, and the other is the Municipality Park (Figure 1).

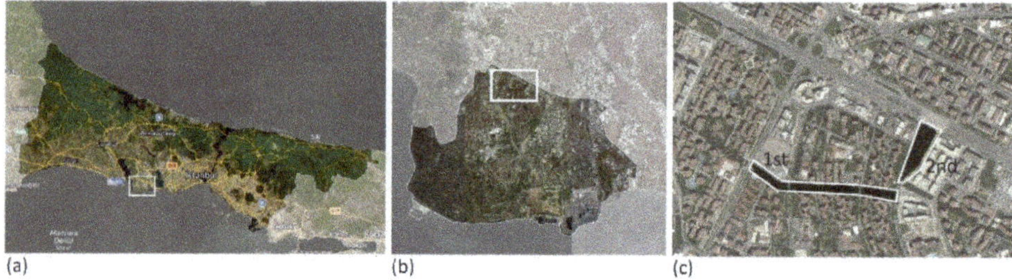

Figure 1. (**a**) Location of the Beylikdüzü District in Istanbul; (**b**) Location of selected parks in Beylikdüzü District; (**c**) Location of parks (Figure is produces by researchers. Source: Yandex Maps Satellite Image, date: 15 May 2018, accessed on 15 December 2020 [65]).

Spatial Analysis/Characteristics of Fatih Sultan Mehmet (FSM) and Mehmet Akif Ersoy (MAE) Woods and Municipality Park

The FSM and MAE Woods are located between a street and a residential dwelling unit. The total length of the park system is 702 m, and its width varies from 16 to 25 m (see Figure 2). It has a surface area of 15,000 square meters. On the other hand, Municipality Park, located at the eastern end of the MAE Woods, is a vaguely triangular-shaped park 285 m in length and with a 14,107 square meter surface area. The park is adjacent to an urban square designed for pedestrian passage above the E-5 highway (Figure 2).

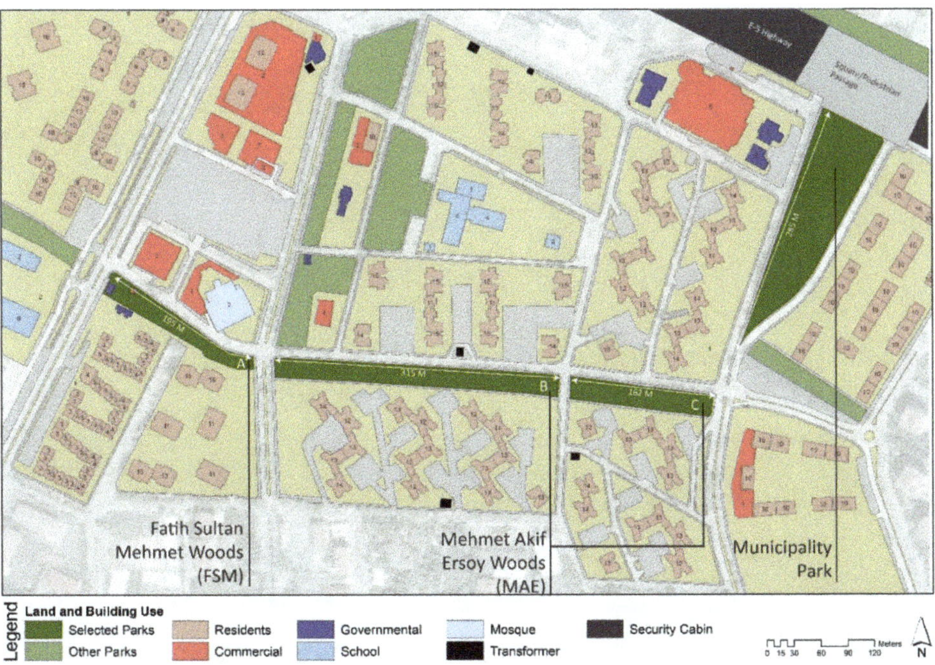

Figure 2. Land uses and no. of building floors in the surrounding urban area (reproduced by researchers from the Municipality Base Map [66]).

These parks and the surrounding urban area have a nearly flat topography, which provides easy access and mobility for pedestrians. The land uses of the surrounding urban area consist mostly of highly populated and gated residential dwelling units and a few large facilities such as mosques, schools, and malls. While of the number of floors in facilities like schools and mosques varies from one to six, nearby residential buildings generally have a higher number, ranging from four to 16 (Figure 2).

While the vegetation of both parks consists mainly of evergreen trees such as the Lawson cypress, palm tree, and nut, black, and Tenasserim pine, there are some deciduous trees such as the common ash, Norway maple, horse chestnut, acacia and plum. As shown in Figure 3, crown closure of the canopy is very high due to the density of the trees in both parks. Because of the prominence of evergreens, these parks have a very closed and forest-like atmosphere in every season (Figure 3).

The distribution of trees and bushes shows that all sections have different characteristics. The bushes and shrubs were evaluated based on density and length (Figure 4). Sparse and short in other sections, bushes and shrubs are dense in the FSM Wood (Figure 5). In the Municipality Park, there is a high plant variety in both trees and bushes, with three vegetation layers consisting of the tallest pine trees at the top, various deciduous trees of relatively shorter height in the middle, and various bushes, shrubs and annual wild plants on the floor. Along the eastern border, the park is separated from the road by a wall of Lawson cypresses. On the western edge, various other tall and medium-height bushes and shrubs act as separators. The density of bushes and shrubs becomes higher and more irregular in the inner part of the park and decreases in the southern region (Figure 6). There are no planted flowers, grass-covered surfaces, or wide-open spaces in either of the parks.

Figure 3. Satellite image of the parks showing the crown closure (Source: Google Earth Satellite Image, date: 08 December 2020, accessed on 08 April 2021 [67]).

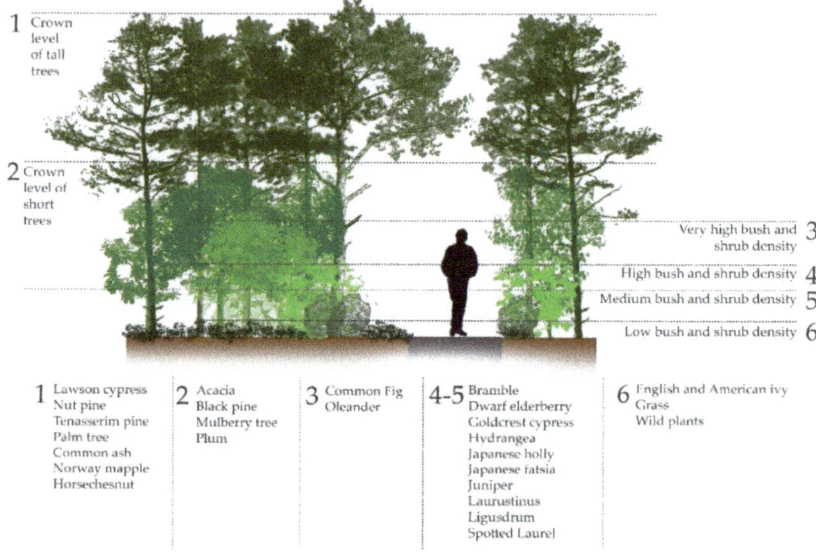

Figure 4. Height and density of trees, bushes, and shrubs.

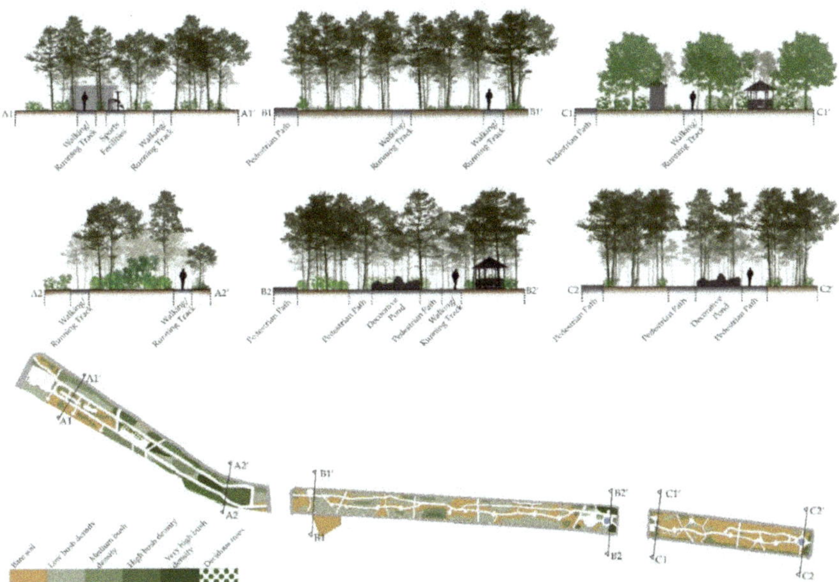

Figure 5. Height and density of trees, bushes, and shrubs in the FSM and MAE Woods (reproduced by researchers based on the Municipality Base Map [66]).

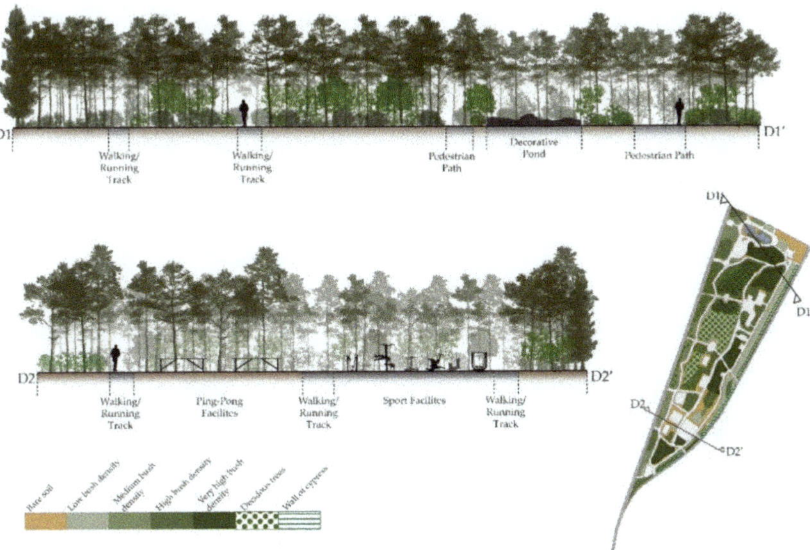

Figure 6. Height and density of trees, bushes, and shrubs in Municipality Park (reproduced by researchers based on the Municipality Base Map [66]).

In regard to the facilities and fixed furniture, the parks house various common urban furniture and sports facilities. Most of the sports facilities are located in Municipality Park, with one such area in the FSM Wood. The walking/running tracks, paved with rubber, mainly run along the main circulation routes of the parks.

By and large, the parks' sports facilities and areas where seating elements are clustered function as focal points (Figures 7–9). Decorative ponds also create an attraction point with their surrounding seating elements (Figure 9). Most of the circulation lines serve as landscape vistas that consist of tunnel-like views of plantation, especially in Municipality Park due to the vegetation density (Figure 10). Examination of the visibility of the surrounding urban pattern determined that due to the high density of trees, building visibilities are similar in both parks. However, the impacts of nearby roads are higher in the FSM and MAE Woods, as their sparse bushes and shrubs, especially between sidewalks and woods, do not create a strong barrier between the park and the surrounding urban area, and their width does not allow for any great distance from adjacent roads (Figure 11).

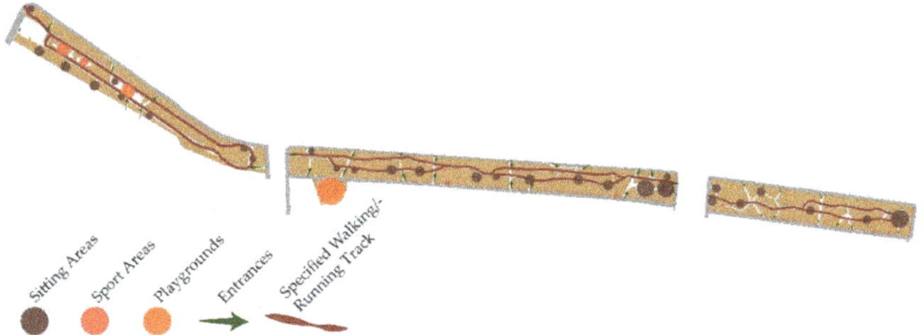

Figure 7. Activity points and axes in the FSM and MAE Woods (reproduced by researchers based on the Municipality Base Map [66]).

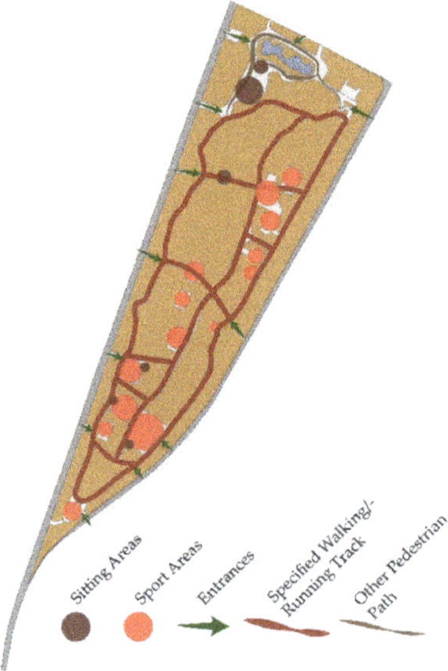

Figure 8. Activity points and axes in Municipality Park (reproduced by researchers based on the Municipality Base Map [66]).

Figure 9. Sports facilities in Municipality Park (**left**) and pond in the MAE Wood (**right**).

Figure 10. Landscape vistas from Municipality Park (**left**) and Woods (**right**).

Figure 11. Road visibility from Woods (**left**) and Municipality Park (**right**).

Overall, because of its size and shape, Municipality Park has a different structure and characteristics from FSM and MAE Woods, which allows for more plantation and facilitations. On one hand, with its sports facilities and long walking tracks, Municipality Park attracts people who want to be (physically) active. On the other hand, in the Woods, people are mostly passing through or resting, which is expected when its length and shape as a linear park are considered. Moreover, the varying vegetation types and densities of these differently shaped parks affect their exposure to the urban view and the impacts of the adjacent road. These data comprised the base for the investigation of changes in user experience, especially in terms of ecotherapy service.

5. Findings and Discussion

The participants from the FSM and MAE Woods consisted of 21 males and 24 females ranging in age from 18 to 75; the participants from Municipality Park consisted of 28 males and 17 females ranging in age from 18 to 76. The average participant age was 42.5 in the Woods and 33.7 in Municipality Park. There were 11 students and 11 retired people in the FSM and MAE Woods' sample, with other participants occupied as medical technician, salesman, beautician, cashier, architect, accountant, teacher and so on. In Municipality Park there were 19 students, with other participants occupying varying professions such as homemaker, retired, machine engineer, biologist, and accountant.

5.1. Correlations of Connectedness to Nature and Other Scales

The average CNS scores of each park were almost the same, with 4.1 out of 5 points in the FSM and MAE Woods and 4.06 out of 5 in Municipality Park. User connectedness to nature was thus similar in both parks. Such close CNS values for the two parks demonstrated a constant user profile essential for understanding the relationship between ecotherapy service and park features.

A Spearman's rho correlation analysis for each park was conducted in SPSS among all scales of the survey study, such as the time (A) and frequency (B) of park usage, the number of activities conducted in these parks (C), impacts of design characteristics on park preference (closeness to home, size, physical environment, facilities and furniture and vegetation) (D), satisfaction with design characteristics (E), the impacts of natural elements (F) and urban texture on the park's natural appearance (G), and emotions/mental states (H) related to ecotherapy service. The results of the survey demonstrated a moderate correlation between CNS and the variable frequency of park usage (B) ($r_s = 0.470$, $p < 0.05$), number of activities conducted in the parks (C) ($r_s = 0.473$, $p < 0.01$), impacts of design characteristics on park preference (D) ($r_s = 0.419$, $p < 0.01$), impacts of natural elements on the park's natural appearance (F) ($r_s = 0.404$, $p < 0.01$) and emotions/mental state (H) ($r_s = 0.550$, $p < 0.01$). As seen above, it was clear that the CNS score of the participant was moderately related to certain features of the park. In Municipality Park, only satisfaction with design characteristics and emotions/mental states were significant; however, they displayed shallow correlation values ($r_s = 0.376$, $p < 0.05$ and $r_s = 0.386$, $p < 0.01$). Because of the bidirectional relationship between emotional/mental states and CNS score, it was unclear whether those more connected to nature receiveed slightly higher ecotherapy services, or those receiving greater ecotherapy services had an increased connection to nature. However, the data proved a clear relationship between the ecotherapy service and connectedness to nature, which the ecopsychology approach has put forward as a solution to the problem of separation from nature in the urban space.

Besides CNS, another correlation analysis was conducted to reveal the relationship between other scales of the survey. For the FSM and MAE Woods, the frequency of use increased with the age of the participants ($r_s = 0.452$, $p < 0.01$), the years of service ($r_s = 0.506$, $p < 0.01$), and the effect of park features such as closeness to home, size, physical environment, facilities and furniture and vegetation, on park choice ($r_s = 0.449$, $p < 0.01$). However, it also detected a negative correlation ($r_s = -0.530$, $p < 0.01$) between the number of activities the participants performed and their satisfaction with the suitability of the park for these activities, when the number of participant activities increased, their satisfaction decreased. This result, however, was to be expected upon consideration of the limited facilities in the FSM and MAE Woods. Lastly, emotional services were correlated with participant years of use ($r_s = 0.309$, $p < 0.05$), their age ($r_s = 0.409$, $p < 0.01$) and satisfaction with design characteristics ($r_s = 0.450$, $p < 0.01$). For Municipality Park, frequency of use correlated solely with the impact of park features on park choice ($r_s = 0.396$, $p < 0.01$). Moreover, the emotional experience related to ecotherapy service within the park correlated with the CNS score ($r_s = 0.386$, $p < 001$) and satisfaction with the interior characteristics of the park ($r_s = 0.539$, $p < 0.01$).

These results revealed that users who preferred either park due to factors such as proximity to home, size, and adequacy and compatibility of equipment were using them more frequently. Therefore, in cases where frequent use is intended, parks should be designed in such a way that they are accessible and suitable in size for users, with appropriate facilities, ideally tuned physical environmental conditions, and adequate planting. Moreover, because the negative correlation between number of activities and park satisfaction in the FSM and MAE Woods was not observed in Municipality Park, the former can be assumed to provide more limited opportunities for therapeutic activities than the latter.

5.2. Park Usage and Ecotherapeutic Activities

The survey put out more descriptive results in addition to the correlation analysis of the scales. Participants were first asked if they used another park anywhere in the city and, if so, what the purpose of that use was. The survey also inquired whether there were other places in İstanbul that made them feel more connected to and integrated with nature and, if so, the reasons for these feelings. A total of 79% of the participants preferred the parks in the Beylikdüzü District, while 21% preferred parks located mostly along the Bosporus coasts for their social activities and spaces, sports activities, and walking pathways; others preferred these parks for their natural appearance, vegetation density, available grass for sitting, and closeness to home or work. The concentration of the selected parks in the Beylikdüzü District showed the importance of proximity in park preference. Moreover, social and sports activities and natural appearance were essential criteria in park preference.

Affirmative answers to the second question about the places where participants feel truly in nature and integrated with nature demonstrated an expected preference for the natural areas of Istanbul over the inner city parks; the reasons most often given included natural characteristics of these areas (58%), such as natural appearance, tree and vegetation density, natural landscapes such as sea and rural views, as well as emotional responses (43%) such as satisfaction with the quiet, peacefulness, and being away from the city. While the presence of people, social activities, and sports were essential criteria for park usage; the presence of people and the visibility of urban patterns were negative factors in the feeling of connectedness to nature, and green areas were not enough to provide this feeling in their current state. Responses about the types of preferred areas and the reasons for such preferences corresponded to findings in ecotherapy literature, which indicated that these features should be evaluated in design decisions in order to increase the natural appearance and therapy service of urban parks.

Participants were also asked whether they would engage in 15 given activities (see Figure 12) in these parks, nine of which fell under the working in/with nature group, and five of which were taken from the experiencing nature group. Additionally, the survey inquired, on a 5-point Likert scale, about the level of sufficiency of the park for these activities. In the FSM and MAE Woods, the most frequent activities were sitting/resting for a short time, sitting, walking, and passing through. The park's highest suitability rating (mean value) was for passing through, sitting, and walking (Figure 12). In Municipality Park, the most frequent activities were sports, walking, sitting, sitting/resting for a short time, and passing through. The highest sufficiency ratings belonged to passing through and running (Figure 12). The average number of activities was close in both parks; however, the parks differed in terms of activity groups. While the number of participants choosing activities in the experiencing nature group was similar in both parks, 27 more participants chose activities in the working in/with nature group in the Municipality Park because of the sports facilities located on the site. Consequently, users of Municipality Park spent their time more actively than those of the FSM and MAE Woods. These results indicated that the provision of such sports spaces and equipment can encourage people to be active.

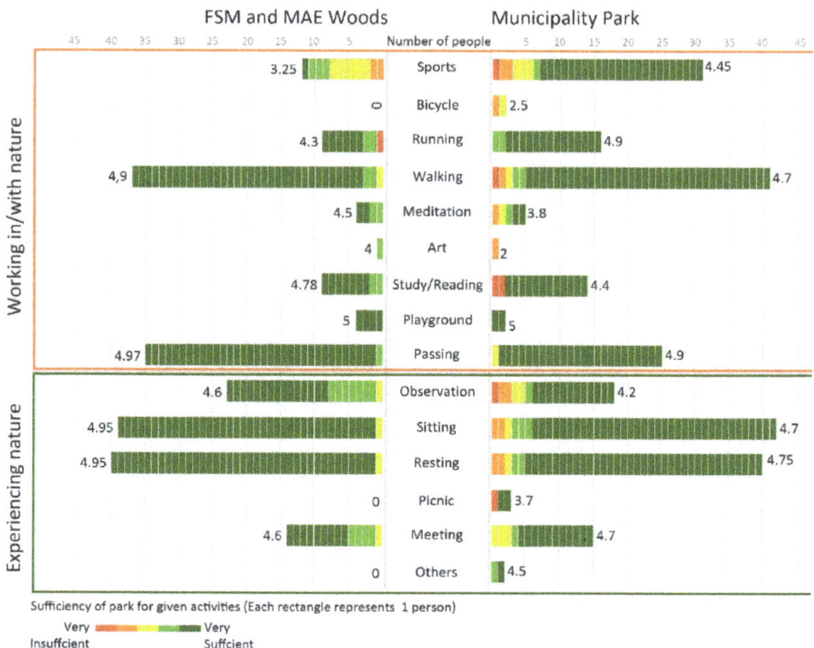

Figure 12. Comparison of ecotherapeutic activities in the FSM and MAE Woods and Municipality Park (number of people and mean values of satisfaction).

5.3. Park Characteristics, Image of Naturalness and Ecotherapeutic Experiences

Analysis of the impacts of park features on the park preferences of participants indicated that "closeness to home" had the lowest value for each park, with a mean score of 4.1 points. However, while the scores of Municipality Park were higher than those of the Woods in the categories of sufficient size for use, adequate physical environment (sunlight, fresh air, etc.), and sufficient vegetation, they were lower in the categories of sufficient facilities and furniture value (Figure 13). The findings indicated that both parks were similar in terms of preference due to their closeness to users' homes. However, size, physical environment, and vegetation had a greater impact on a preference for Municipality Park. Still, its sports equipment and furniture were evaluated as insufficient compared to that of the FSM and MAE Woods. It was expected that users of FSM and MAE Woods would be satisfied by the numerous seating elements, but the number of facilities and furniture, consisting of mostly sports equipment and a few seating elements, were not sufficient for users in Municipality Park.

Inquiries concerning the interior characteristics and appearance of the parks attempted to gauge participants' ease of mobility and finding their bearings, ability to be alone and in nature and feel distant from the urban center, and the sufficiency and suitability of facilities, furniture, pavements, and water elements. The results indicated that FSM and MAE Woods had higher scores than Municipality Park for all statements, with the exception of "I feel away from urban area" (Figure 14). Moreover, the results of the "I feel away from urban area" statement demonstrated that both parks were affected by the urban pattern. However, this impact was higher in the FSM and MAE Woods.

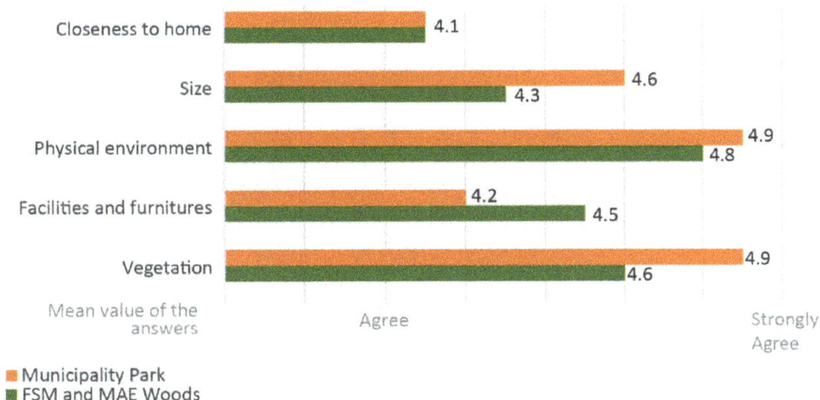

Figure 13. Impact of park features on user preference scores (mean values).

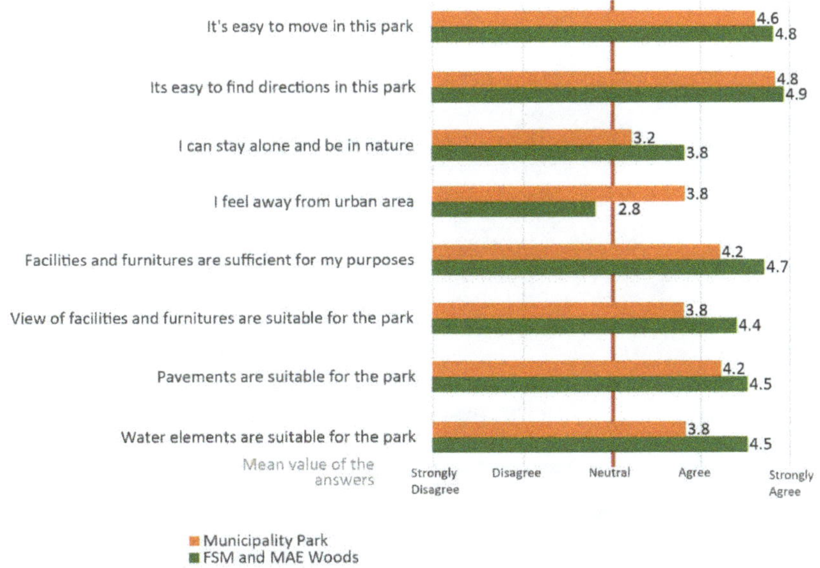

Figure 14. Satisfaction with interior characteristics and appearance (the red line indicates the neutral score of the Likert scale, the threshold of satisfaction and dissatisfaction).

In addition to these statements, participants were asked two open-ended questions, the first of which concerned the elements that might interrupt the natural appearance of the parks (Figure 15). The answers revealed that 80% of the park users thought that there was nothing interrupting the natural appearance of both parks. However, others responded that, in both parks, certain facilities and furniture were not compatible with the parks' natural appearance.

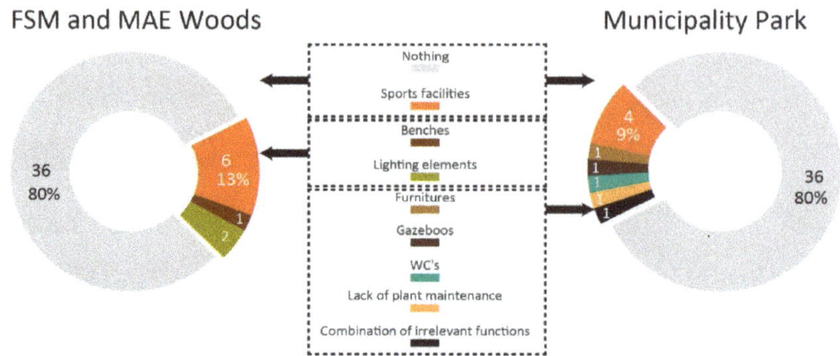

Figure 15. Answers to the question "What are the inner features that interrupt the natural appearance of the park?" (number of people and percentage).

The second open-ended question concerned the exterior factors that interrupted the parks' natural appearance. In Municipality Park, 21 (47%) participants responded "nothing". However, others mentioned factors such as crowds, building visibility, noise from the roads, garbage, and dust. In the FSM and MAE Woods, 32 (71%) people mentioned noise from the roads. A total of 20% of participants also gave answers such as building visibility, garbage, crowds, car visibility, and lack of maintenance (Figure 16). These answers indicated that most of the negative factors were similar in both parks; however, noise from the roads was a severe problem in the FSM and MAE Woods. Additionally, a juxtaposition of the third and fourth statements of the previous question (Figure 14), which referred to the feeling of being "in nature", with the answers of the two open-ended questions aided in understanding the relationship between the feeling of being "in nature" and the interior and exterior features that respondents felt interrupted the parks' natural appearance. Of the 32 people, 25 who gave three or fewer points to the third (I can stay alone and be in nature) and/or fourth (I feel away from urban area) statements also mentioned an interior element and/or exterior factor that broke the natural appearance of Municipality Park (Figure 17); 30 of 34 respondents in the FSM and MAE Woods did the same. These findings demonstrated that the parks' interior elements and exterior factors affected the feeling of being in nature and the natural appearance of the parks and should be evaluated in the design process.

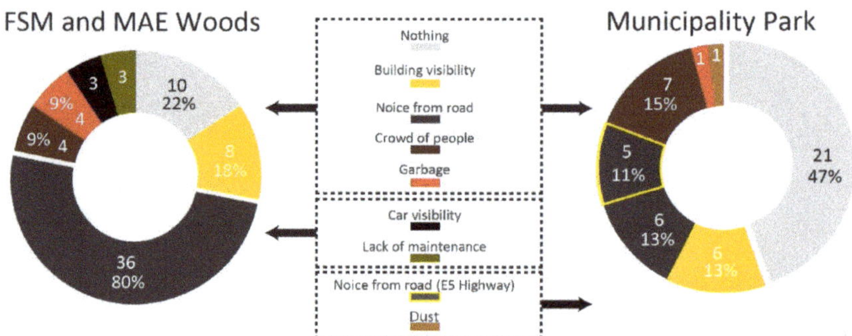

Figure 16. Answers to the question "What are the exterior factors that interrupt the natural appearance of the park?" (number of people and percentage).

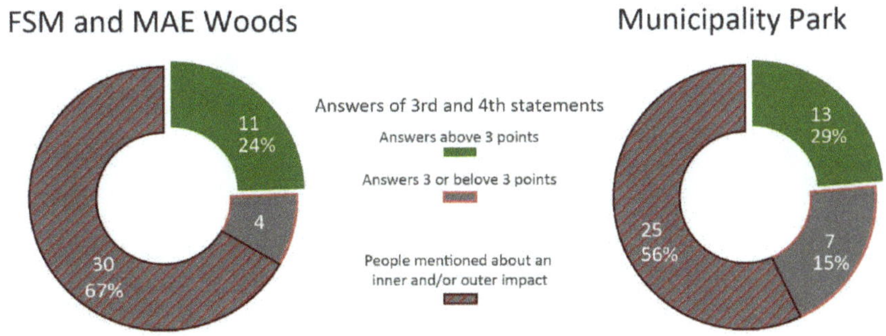

FSM and MAE Woods

Municipality Park

Figure 17. Ratio of people who gave a 3 or below score to the third (I can stay alone and be in nature) and/or fourth (I feel away from urban area) statements and also mentioned an inner and/or outer negative impact on the natural appearance of the park.

The other part of the survey study included seven statements about the impacts of vegetation, one statement on the effect of the amount of paved surfaces in the parks, and three statements about the effects of the urban pattern on the parks' natural appearance. Both parks had almost equal mean scores in the first four questions (Figure 18). However, while participants in the FSM and MAE Woods recorded higher positive views on the effects of the presence of bushes and shrubs on the park's natural appearance, those in Municipality Park responded that the impact of the volume and order (wild-like) of bushes and shrubs had a positive effect on the park's association with nature. Based on similarities in the parks' vegetation types and order and the number of trees shown in the spatial analyses, the two parks had similar values for the impact of these elements on their natural appearance. However, considering the difference in bush and shrub density in the parks, high values were expected for the impact of bushes and shrubs on the natural appearance of Municipality Park, which has a higher density of this type of vegetation.

The questions about the impact of the urban pattern concerned the effects of building visibility, the number of building floors, and road noise and car visibility. The impacts of building visibility and number of floors had slightly higher values in Municipality Park, while the impacts of road noise and car visibility had higher values in the FSM and MAE Woods (Figure 18). Both parks were similar in the perceived impacts of building visibility; however, the FSM and MAE Woods were more affected by the surrounding roads. Considering the similarities in the surrounding urban pattern for both parks, the reason for this difference may lie in the parks' vegetation. Both parks have a similar tree pattern, consisting of pine trees with bare stems and high crowns that function to block the view of surrounding buildings; nevertheless, they do not diminish park goers' views of adjacent roads. The two parks, however, vary significantly in shrub density. Municipality Park, which is adjacent to a highway, received a lower score than the FSM and MAE Woods because of its dense bush and shrub vegetation, which served to better block the view of the road. This difference indicated that the presence of vegetation contributed to the perception of naturalness by acting as a visual barrier separating the park from the city.

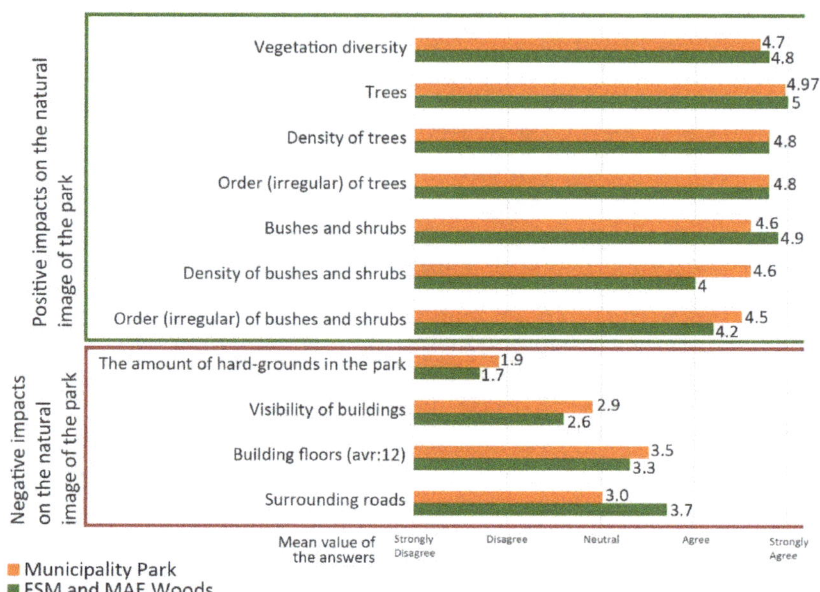

Figure 18. Impacts of the vegetation, paved surfaces, and urban pattern on nature-like appearance of parks (mean values).

After this evaluation of urban impact, participants were asked to point out which parts of the park seemed more natural than others and the reasons for their opinion. In the FSM and MAE Woods 72% made their selection for reasons related to vegetation and natural characteristics, 13% for atmosphere (quiet, silent, isolated), and a few for reasons related to water elements and design. In Municipality Park, 29 (64%) people selected the inner part of the park and the middle walking track due to the density and variety of vegetation, forest-like and wild-like views, quiet environment, and low building visibility. A total of 79% of the answers were related to vegetation and natural characteristics, while others were based on water elements, emotions, and design. These answers were in line with the literature concerning the features of ecotherapeutic environments, including vegetation density, water elements, and feelings such as quiet, isolation, silence and the urban visibility for both parks. Based on this concordance, it was clear that both parks have elements and features that provide ecotherapy service.

Participants were then asked to point out the part of the parks they liked the most. In FSM and MAE Woods, 15 (33%) people selected the MAE Wood due to its dense vegetation, size, and the breezy, isolated, and quiet environment. Moreover, the ponds (due to the dense vegetation and sound of water), sports facilities in FSM Woods and the whole park system (due to the length of the park, which allows long walking) were indicated by some other participants. In Municipality Park, 13 people (29%) selected the sports facilities and the park's quiet, breezy and open environment. Eight selected the northern inner part due to its vegetation density and quiet atmosphere, and five chose the northern sports facilities for their quiet, breezy, and open atmosphere.

Analysis of the answers to these two questions indicated no severe shift between the responses for the first and second questions in the FSM and MAE Woods (Figure 19). Most of the selected parts remained the same in both questions for similar reasons. However, in Municipality Park, there was a contrast between the parts that people saw as more natural and the parts they liked, the principal cause of which was the park's sports facilities and activities (Figure 19). These results indicated that the FSM and MAE Woods were favored for their natural characteristics and comfortable environment. However, Municipality Park has different features that provide more than just natural views. Moreover, because of the importance of exercise for both a healthy lifestyle and ecotherapy service, this park has the potential to offer more varied ecotherapeutic activities. All these results were in line with the previous ones regarding park preference, which indicated that Municipality Park stood out because of its available sports activities, illustrating the critical role of facilities that provide opportunities for these activities in park preference and enjoyment.

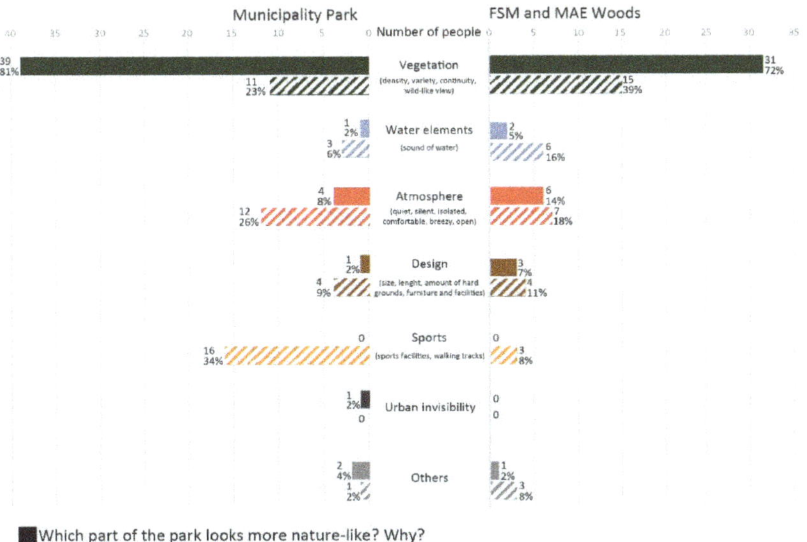

Figure 19. Themes for why respondents preferred different parts of the parks and which ones they regarded as more nature-like.

Finally, participants were asked about the shift in their emotions and mental states when they spent time in the parks. All of the participants confirmed that spending time in these parks improved their mental status. While the differences were small, the values of Municipality Park were found to be slightly higher (Figure 20). These results can be interpreted as both parks being capable of providing an environment where users could experience emotions related to ecotherapy.

In addition to the evaluations, several participants commented on their wishes or mental situations without being asked. These comments were valuable due to the presence of such sentiments in the ecopsychology and ecotherapy studies about the effects of ecotherapy on mental states and emotions.

- My self-esteem increases when I spend time here (FSM).
- This park is the place where I can ask questions and find answers (MAE).
- I'm discharging (MAE).
- Whenever I feel suffocated, I come here (MAE).
- This park is a therapy area (MAE).
- Parks need to be accessible and ubiquitous (MAE).
- There should be areas like this all over the city (MP).
- I'd love to see animals like squirrels (MP).
- It should look pristine (MP).
- This park is like my home (MP).
- It is much more pleasant and beautiful to exercise in the greenery here than in the indoor fitness hall (MP).

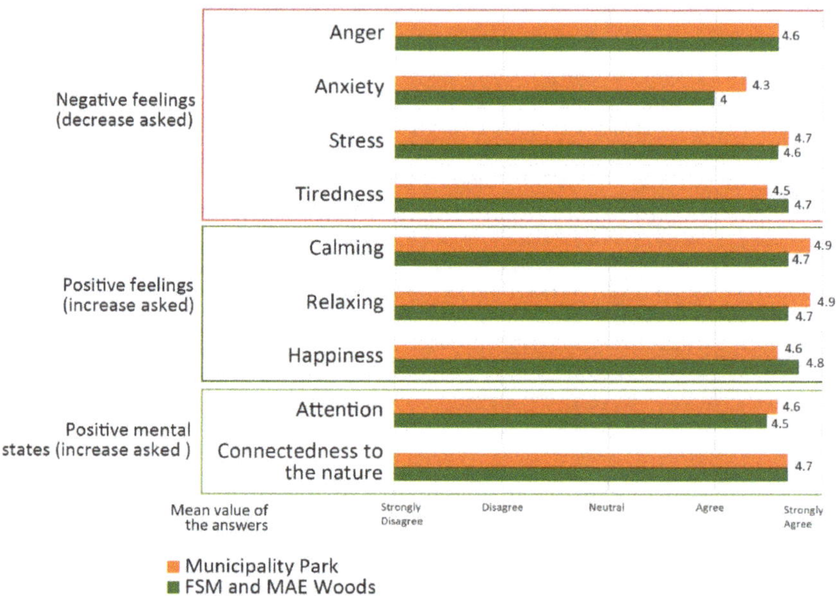

Figure 20. Experience scores of emotions/mental states related to ecotherapy service.

These statements demonstrated that people were experiencing the benefits of ecotherapy service; they also showed the role of the parks' spatial features in providing this service.

6. Conclusions

In today's world, many problems have arisen, dependent on the form of urbanization and the nature of urban areas; these problems not only cause environmental problems by affecting natural areas but also negatively affect the health and psychology of citizens in urban areas. The ecopsychology approach, which seeks a solution to these issues in the re-establishment of a relationship with nature, notes the role of the ecotherapy service obtained through spending time in natural areas and defines the main features of therapeutic spaces. As mentioned before, Istanbul lost its valuable natural lands and urban green areas as a result of rapid urban growth and densification that started in 1950s. Besides the loss of natural areas, ecotherapy services cannot be provided in these unplanned and dense urban areas due to the lack of green spaces. In this manner, protecting both natural and urban green areas from uncontrolled and rapid urban growth should be considered due to their immanent values, the ecosystem services they provide and ecotherapy potentials for citizens.

Within the scope of this study, the spatial and functional characteristics of ecotherapy service were evaluated through two parks selected from the Istanbul Metropolitan area. The evaluation was made primarily by comparing the spatial characteristics and user experiences of these two parks with the aim of measuring the effects of different spatial features on experiences and activities related to ecotherapy service, as well as discussing the contributions of the characteristics that can increase the ecotherapy service potential of urban green areas in the urban design processes. The findings of the study parallel the descriptions of therapeutic spaces made by ecopsychology and ecotherapy literature in terms of the effects on the users' connectedness to nature, obtaining therapy services, and the natural perception of parks. The importance of creating more natural landscapes underlined by the ecotherapy literature was demonstrated by the high scores on the experience of emotions/mental states associated with ecotherapy service in both parks. Hence, the ecotherapeutic effect of the space can be increased by creating a dense and wild appearance in the urban green areas or by preserving and enriching the existing vegetation as it is. At the same time, both natural features and criteria such as calmness and silence contribute to naturalness and influence the user preferences of the parks. Similarly, findings concerning the effect of different spatial features, the type and appearance of facilities, and the visual and auditory relationship established with the city based on the perception of naturalness show that the therapeutic potential of a park can be increased when these features and relationships are adequately designed. It is possible to obtain more natural, calm, quiet and isolated spaces by using vegetation both to cut off the relationship of the park with the city and to reduce the urban effect, and to separate different functions (sports areas, seating areas, etc.) in the park. In addition to spatial features, the provision of space and equipment for therapeutic activities is influential in the choice and enjoyment of the park while at the same time encouraging people to engage in an active lifestyle.

To conclude, in accordance with the contributions of ecopsychology and ecotherapy studies to design processes, both the physical and psychological health of urban citizens can be increased, and environmental problems may be solved through the transformation of individuals. For further research in this direction, more spatial studies and design examples taken up from this perspective are needed to assess the contributions of the ecopsychology discipline.

Supplementary Materials: The following are available online at https://www.mdpi.com/article/10.3390/su13094600/s1, Table S1: Survey Form.

Author Contributions: Conceptualization, D.K. and G.D.O.; methodology, D.K. and G.D.O.; software, D.K. and G.D.O.; validation, D.K. and G.D.O.; investigation, D.K. and G.D.O.; resources, D.K. and G.D.O.; data curation, D.K. and G.D.O.; writing—original draft preparation, D.K.; writing—review and editing, G.D.O.; visualization, D.K.; supervision, G.D.O. All authors have read and agreed to the published version of the manuscript.

Funding: This research received no external funding.

Institutional Review Board Statement: Not applicable.

Informed Consent Statement: Informed consent was obtained from all subjects involved in the study.

Data Availability Statement: The data presented in this study are available on request from the corresponding author. The data are not publicly available due to privacy.

Conflicts of Interest: The authors declare no conflict of interest.

Declaration: This work has been presented on Greening cities shaping cities symposium in October 2020, www.greeningcities-shapingcities.polimi.it.

References

1. Kara, D.; Oruç, G.D. Birey-Doğa İlişkisinin Yeniden Kurgulanması Bağlamında Ekoterapötik Mekânlar. *Tasar. Kuram* **2020**, *16*, 257–277. [CrossRef]
2. Singh, R.L.; Singh, P.K. *Principles and Applications of Environmental Biotechnology for a Sustainable Future*; Singh, R.L., Ed.; Springer: Singapore, 2017; ISBN 978-981-10-1865-7.
3. Flies, E.J.; Mavoa, S.; Zosky, G.R.; Mantzioris, E.; Williams, C.; Eri, R.; Brook, B.W.; Buettel, J.C. Urban-associated diseases: Candidate diseases, environmental risk factors, and a path forward. *Environ. Int.* **2019**, *133*, 105187. [CrossRef] [PubMed]
4. Jackson, L.E. The relationship of urban design to human health and condition. *Landsc. Urban Plan.* **2003**, *64*, 191–200. [CrossRef]
5. Moore, M.; Gould, P.; Keary, B.S. Global urbanization and impact on health. *Int. J. Hyg. Environ. Health* **2003**, *206*, 269–278. [CrossRef] [PubMed]
6. Mutatkar, R. Public health problems of urbanization. *Soc. Sci. Med.* **1995**, *41*, 977–981. [CrossRef]
7. Phillips, D. Urbanization and human health. *Parasitology* **1993**, *106*, S93–S107. [CrossRef] [PubMed]
8. Weimann, A.; Oni, T. A Systematised Review of the Health Impact of Urban Informal Settlements and Implications for Upgrading Interventions in South Africa, a Rapidly Urbanising Middle-Income Country. *Int. J. Environ. Res. Public Health* **2019**, *16*, 3608. [CrossRef] [PubMed]
9. Kumar, P.; Druckman, A.; Gallagher, J.; Gatersleben, B.; Allison, S.; Eisenman, T.S.; Hoang, U.; Hama, S.; Tiwari, A.; Sharma, A.; et al. The nexus between air pollution, green infrastructure and human health. *Environ. Int.* **2019**, *133*, 105181. [CrossRef] [PubMed]

10. Restivo, V.; Cernigliaro, A.; Casuccio, A. Urban Sprawl and Health Outcome Associations in Sicily. *Int. J. Environ. Res. Public Health* **2019**, *16*, 1350. [CrossRef] [PubMed]
11. Qiu, Y.; Liu, Y.; Liu, Y.; Li, Z. Exploring the Linkage between the Neighborhood Environment and Mental Health in Guangzhou, China. *Int. J. Environ. Res. Public Health* **2019**, *16*, 3206. [CrossRef] [PubMed]
12. Bhugra, D.; Castaldelli-Maia, J.M.; Torales, J.; Ventriglio, A. Megacities, migration, and mental health. *Lancet Psychiatry* **2019**, *6*, 884–885. [CrossRef]
13. Reichert, M.; Braun, U.; Lautenbach, S.; Zipf, A.; Ebner-Priemer, U.; Tost, H.; Meyer-Lindenberg, A. Studying the impact of built environments on human mental health in everyday life: Methodological developments, state-of-the-art and technological frontiers. *Curr. Opin. Psychol.* **2020**, *32*, 158–164. [CrossRef]
14. Steg, L.; Van den Berg, A.E.; de Groot, J.I.M. *Environmental Psychology: An Introduction*, 2nd ed.; Steg, L., de Groot, J.I.M., Eds.; BPS textbooks in psychology; Wiley-Blackwell: Hoboken, NJ, USA, 2018; ISBN 978-1-119-24111-9.
15. Rozsak, T. *Voice of the Earth: An Exploration of Ecopsychology*; Simon & Schuster: New York, NY, USA, 1992; ISBN 0-671-72968-3.
16. Scull, J. Ecopsychology: Where Does It Fit in Psychology in 2009? *Trumpeter J. Ecosophy.* **2008**, *24*, 18.
17. Clinebell, H. *Ecotherapy: Healing Ourselves, Healing the Earth*, 1st ed.; Routledge: London, UK, 2013; ISBN 978-1-315-79977-3.
18. Wheeling, S.D. Making Sense. *West Va. Med. J.* **1993**, *89*, 113.
19. Haase, D.; Larondelle, N.; Andersson, E.; Artmann, M.; Borgström, S.; Breuste, J.; Gomez-Baggethun, E.; Gren, Å.; Hamstead, Z.; Hansen, R.; et al. A Quantitative Review of Urban Ecosystem Service Assessments: Concepts, Models, and Implementation. *Ambio* **2014**, *43*, 413–433. [CrossRef]
20. Parker, J.; De Baro, M.E.Z. Green Infrastructure in the Urban Environment: A Systematic Quantitative Review. *Sustainability* **2019**, *11*, 3182. [CrossRef]
21. Qin, B.; Zhu, W.; Wang, J.; Peng, Y. Understanding the relationship between neighbourhood green space and mental wellbeing: A case study of Beijing, China. *Cities* **2021**, *109*, 103039. [CrossRef]
22. Triguero-Mas, M.; Dadvand, P.; Cirach, M.; Martínez, D.; Medina, A.; Mompart, A.; Basagaña, X.; Gražulevičienė, R.; Nieuwenhuijsen, M.J. Natural outdoor environments and mental and physical health: Relationships and mechanisms. *Environ. Int.* **2015**, *77*, 35–41. [CrossRef]
23. Summers, J.K.; Vivian, D.N. Ecotherapy—A Forgotten Ecosystem Service: A Review. *Front. Psychol.* **2018**, *9*, 1389. [CrossRef] [PubMed]
24. Wilson, N.; Ross, M.; Lafferty, K.; Jones, R. A review of ecotherapy as an adjunct form of treatment for those who use mental health services. *J. Public Ment. Health* **2009**, *7*, 23–35. [CrossRef]
25. Sackett, C.R. Ecotherapy: A Counter to Society's Unhealthy Trend? *J. Creat. Ment. Health* **2010**, *5*, 134–141. [CrossRef]
26. Schebella, M.F.; Weber, D.; Lindsey, K.; Daniels, C.B. For the Love of Nature: Exploring the Importance of Species Diversity and Micro-Variables Associated with Favorite Outdoor Places. *Front. Psychol.* **2017**, *8*, 2094. [CrossRef] [PubMed]
27. Stevens, P. Embedment in the environment: A new paradigm for well-being? *Perspect. Public Health* **2010**, *130*, 265–269. [CrossRef] [PubMed]
28. Kamitsis, I.; Simmonds, J.G. Using Resources of Nature in the Counselling Room: Qualitative Research into Ecotherapy Practice. *Int. J. Adv. Couns.* **2017**, *39*, 229–248. [CrossRef]
29. Abdelaal, M.S.; Soebarto, V. Biophilia and Salutogenesis as restorative design approaches in healthcare architecture. *Arch. Sci. Rev.* **2019**, *62*, 195–205. [CrossRef]
30. Pasanen, T.; Johnson, K.; Lee, K.; Korpela, K. Can Nature Walks With Psychological Tasks Improve Mood, Self-Reported Restoration, and Sustained Attention? Results from Two Experimental Field Studies. *Front. Psychol.* **2018**, *9*, 2057. [CrossRef] [PubMed]
31. Reese, R.F.; Lewis, T.F. Greening Counseling: Examining Multivariate Relationships between Ecowellness and Holistic Wellness. *J. Humanist. Couns.* **2019**, *58*, 53–67. [CrossRef]
32. Barnes, M.R.; Donahue, M.L.; Keeler, B.L.; Shorb, C.M.; Mohtadi, T.Z.; Shelby, L.J. Characterizing Nature and Participant Experience in Studies of Nature Exposure for Positive Mental Health: An Integrative Review. *Front. Psychol.* **2019**, *9*, 2617. [CrossRef] [PubMed]
33. Wolsko, C.; Hoyt, K. Employing the Restorative Capacity of Nature: Pathways to Practicing Ecotherapy among Mental Health Professionals. *Ecopsychology* **2012**, *4*, 10–24. [CrossRef]
34. Greenleaf, A.T.; Bryant, R.M.; Pollock, J.B. Nature-Based Counseling: Integrating the Healing Benefits of Nature into Practice. *Int. J. Adv. Couns.* **2013**, *36*, 162–174. [CrossRef]
35. Bornioli, A.; Parkhurst, G.; Morgan, P.L. The psychological wellbeing benefits of place engagement during walking in urban environments: A qualitative photo-elicitation study. *Health Place* **2018**, *53*, 228–236. [CrossRef]
36. Burls, A. People and green spaces: Promoting public health and mental well-being through ecotherapy. *J. Public Ment. Health* **2007**, *6*, 24–39. [CrossRef]
37. Chatalos, P.A. Sustainability: Ecopsychological insights and person-centered contributions. *Pers. Exp. Psychother.* **2013**, *12*, 355–367. [CrossRef]
38. Brazier, C. *Ecotherapy in Practice: A Buddhist Model*; Routledge: London, UK; Taylor & Francis Group: New York, NY, USA, 2018; ISBN 978-0-415-78595-2.

39. Wilson, N.; Fleming, S.; Jones, R.; Lafferty, K.; Cathrine, K.; Seaman, P.; Knifton, L. Green shoots of recovery: The impact of a mental health ecotherapy programme. *Ment. Health Rev. J.* **2010**, *15*, 4–14. [CrossRef]
40. Davis, K.M.; Atkins, S.S. Ecotherapy: Tribalism in the Mountains and Forest. *J. Creat. Ment. Health* **2009**, *4*, 272–282. [CrossRef]
41. Ibes, D.; Hirama, I.; Schuyler, C. Greenspace Ecotherapy Interventions: The Stress-Reduction Potential of Green Micro-Breaks Integrating Nature Connection and Mind-Body Skills. *Ecopsychology* **2018**, *10*, 137–150. [CrossRef]
42. Juan, C.S.; Subiza-Pérez, M.; Vozmediano, L. Restoration and the City: The Role of Public Urban Squares. *Front. Psychol.* **2017**, *8*, 2093. [CrossRef]
43. Pálsdóttir, A.; Wissler, S.; Nilsson, K.; Petersson, I.F.; Grahn, P. Nature-Based Rehabilitation in Peri-Urban Areas for People with Stress-Related Illnesses—A Controlled Prospective Study. *Acta Hortic.* **2015**, 31–35. [CrossRef]
44. Cole, D.N.; Hall, T.E. Experiencing the Restorative Components of Wilderness Environments: Does Congestion Interfere and Does Length of Exposure Matter? *Environ. Behav.* **2010**, *42*, 806–823. [CrossRef]
45. Hartig, T.; Staats, H. Guest Editors' introduction: Restorative environments. *J. Environ. Psychol.* **2003**, *23*, 103–107. [CrossRef]
46. Jordan, M.; Marshall, H. Taking counselling and psychotherapy outside: Destruction or enrichment of the therapeutic frame? *Eur. J. Psychother. Couns.* **2010**, *12*, 345–359. [CrossRef]
47. Pedersen, E.; Weisner, S.E.; Johansson, M. Wetland areas' direct contributions to residents' well-being entitle them to high cultural ecosystem values. *Sci. Total Environ.* **2019**, *646*, 1315–1326. [CrossRef]
48. Clatworthy, J.; Hinds, J.; Camic, P.M. Gardening as a mental health intervention: A review. *Ment. Health Rev. J.* **2013**, *18*, 214–225. [CrossRef]
49. Kusmane, A.S.; Ile, U.; Ziemelniece, A. Importance of Trees with Low-growing Branches and Shrubs in Perception of Urban Spaces. *IOP Conf. Ser. Mater. Sci. Eng.* **2019**, *471*, 092061. [CrossRef]
50. Stoltz, J.; Schaffer, C. Salutogenic Affordances and Sustainability: Multiple Benefits with Edible Forest Gardens in Urban Green Spaces. *Front. Psychol.* **2018**, *9*, 2344. [CrossRef]
51. Grassini, S.; Revonsuo, A.; Castellotti, S.; Petrizzo, I.; Benedetti, V.; Koivisto, M. Processing of natural scenery is associated with lower attentional and cognitive load compared with urban ones. *J. Environ. Psychol.* **2019**, *62*, 1–11. [CrossRef]
52. Phelps, C.; Butler, C.; Cousins, A.; Hughes, C. Sowing the seeds or failing to blossom? A feasibility study of a simple ecotherapy-based intervention in women affected by breast cancer. *Ecancermedicalscience* **2015**, *9*. [CrossRef] [PubMed]
53. Pedretti-Burls, A. Ecotherapy: A Therapeutic and Educative Model. *J. Mediterr. Ecol.* **2007**, *8*, 19–25.
54. Wang, D.; Macmillan, T. The Benefits of Gardening for Older Adults: A Systematic Review of the Literature. *Act. Adapt. Aging* **2013**, *37*, 153–181. [CrossRef]
55. Bagot, K.L.; Allen, F.C.L.; Toukhsati, S. Perceived restorativeness of children's school playground environments: Nature, playground features and play period experiences. *J. Environ. Psychol.* **2015**, *41*, 1–9. [CrossRef]
56. Hauru, K.; Lehvävirta, S.; Korpela, K.; Kotze, D.J. Closure of view to the urban matrix has positive effects on perceived restorativeness in urban forests in Helsinki, Finland. *Landsc. Urban Plan.* **2012**, *107*, 361–369. [CrossRef]
57. Lindal, P.J.; Hartig, T. Architectural variation, building height, and the restorative quality of urban residential streetscapes. *J. Environ. Psychol.* **2013**, *33*, 26–36. [CrossRef]
58. Gill, C.; Packer, J.; Ballantyne, R. Spiritual retreats as a restorative destination: Design factors facilitating restorative outcomes. *Ann. Tour. Res.* **2019**, *79*, 102761. [CrossRef]
59. Mayer, F.S.; Frantz, C.M. The connectedness to nature scale: A measure of individuals' feeling in community with nature. *J. Environ. Psychol.* **2004**, *24*, 503–515. [CrossRef]
60. Austin, E.S.; Leopold, A. A Sand County Almanac with Other Essays on Conservation from Round River. *Bird-Banding* **1967**, *38*, 252. [CrossRef]
61. Karakuyu, M.; Tezer, S.T.; Balik, H. İstanbul'un Tarihsel Topoğrafyası ve Literatür Değerlendirmesi. 2010. Available online: https://dergipark.org.tr/tr/download/article-file/652793 (accessed on 15 March 2021).
62. Tekeli, E.; Kuşuluoğlu, D.; Ersoy, M. Kentleşme ve Yeşil Alan Değişiminde İstanbul Boğaz Köprülerinin Rolü. *Anadolu Doğa Bilim. Derg.* **2015**, *6*, 211–219.
63. 2020 Adrese Dayalı Nüfus Kayıt Sistemi Sonuçları. 2021. Available online: https://data.tuik.gov.tr/Bulten/Index?p=Adrese-Dayali-Nufus-Kayit-Sistemi-Sonuclari-2020-37210 (accessed on 15 April 2021).
64. Keçeli, A.; Sariusta, F.; Karakuyu, M. Kamu Hizmetlerinin Kentsel Yaşanabilirlik Üzerine Etkisi: Beylikdüzü Örneği. *Marmara Cograf. Derg.* **2014**, *29*. [CrossRef]
65. Yandex Satellite Image 2018. Available online: https://yandex.com.tr/harita/107757/beylikduzu/?ll=28.648786%2C41.007837&z=16.09 (accessed on 15 December 2020).
66. Beylikdüzü Municipality Municipality Basemap. Available online: https://ebys.beylikduzu.bel.tr/ebelediye (accessed on 15 April 2021).
67. Google Earth Sattelite Image 2020. Available online: https://earth.google.com/web/@41.00961869,28.64648813,169.68696015a,1356.4063303d,35y,0h,0t,0r (accessed on 8 April 2021).

 *sustainability*

Article

Parque Augusta (São Paulo/Brazil): From the Struggles of a Social Movement to Its Appropriation in the Real Estate Market and the Right to Nature in the City

Wendel Henrique Baumgartner

Department of Geography, Federal University of Bahia, Salvador 40170-110, Brazil;
wendel_henrique@hotmail.com

Abstract: Through a dialectical approach, building a thesis, an antithesis and a synthesis, our goal in this article is to discuss the implementation of the Parque Augusta, in the center of São Paulo, Brazil. For years, an organized social movement struggled with the municipality and real estate developers for the protection of the park and its green area. The demanded and desired park, collectively designed and managed, physically structured on the principles of the nature-based solutions (NBS), should represent a victory. However, in a capitalist urban space, the future park has already been appropriated in the real estate market to enhance development values and to increase the density of its environs with the construction of new skyscrapers. In a city tagging its climate actions using NBS concepts, the project in implementation by the municipality has fewer NBS elements than the co-designed with citizens participation. Here we present the narratives of the park creation and some indicators about its appropriation, based on land use and real estate market prices. The theoretical critical perspective was fundamental to reveal the contradictions within the park construction, called attention to the consideration of the surrounding area in greening projects and promoted a synthesis towards the universalization of the right to nature in the city.

Keywords: Parque Augusta; nature-based solutions; social movements; appropriation of nature; green gentrification; right to nature

Citation: Baumgartner, W.H. Parque Augusta (São Paulo/Brazil): From the Struggles of a Social Movement to Its Appropriation in the Real Estate Market and the Right to Nature in the City. *Sustainability* **2021**, *13*, 5150. https://doi.org/10.3390/su13095150

Academic Editor: Israa H. Mahmoud, Eugenio Morello, Giuseppe Salvia and Emma Puerari

Received: 6 April 2021
Accepted: 30 April 2021
Published: 5 May 2021

1. Introduction

The city, one of the greatest human achievements, an artifact par excellence, and an apparent denial of nature, becomes the main place for observing a new relationship between society and nature. Nature, as ideas, metaphors, metonymies, objects, amenities, realm, body, organism, among several possible interpretations, already reified and incorporated into social life throughout history, is now being (re)produced, (re)incorporated or (re)introduced, in scale, in the cities, through greening or renaturing urban projects, as green-blue infrastructures [1], ecosystem services [2], or nature-based solutions [3–6]. The prefix 're' is used here to stress the dialectical and the historical content in these new forms of (re)connection between city and nature [7–9]. This (re)valuation of nature in the city can be read, in a critical and dialectical way, as a possibility for a radical transformation of the urban space, towards the production of resilient towns [10], adapted to climate change and sustainable [11] in its existence and at the ordinary level of the everyday life. Nevertheless, the spaces of nature in the cities are, as well, an open arena for contradictions and appropriations of collective resources, demands and desires [12].

Since our main goal is to produce a right to nature in the city [13], founded upon social equity, spatial/environmental justice and just sustainability [14–16], through the construction of spaces of nature in our cities, related with nature-based solutions, it is still important to reveal some contradictions presented in the production of urban space [17].

If in the commonsense nature is opposed to culture and history [18,19], and consequently to the city [20], on the other side, there is a long tradition, especially within the dialectic's approaches [21–29] to understand nature also as a product of a long human history due to its incorporation into the social life. Fulfilled with ideologies, sentiments and expectations, the evolution of the ideas and concepts of nature [30,31], reveled and development by everyday life practices and techniques, shows possibilities to understand the role of nature the material-immaterial human production [32–34].

In this article, we investigated the history of the conception, creation and implementation of Parque Augusta, in the center of São Paulo/Brazil, after an intense dispute between real estate developers and a social movement, with the municipality trying to mediate the conflict [35]. The idea of this park, nourished by the residents and the civil society, demonstrates the importance of popular participation in the discussion and production of spaces of nature in the city, since the space is now protected, due to its green area. The vegetation was the key natural element to safeguard the park and include it in the green area system of the city, as a green space and public green infrastructure. During years of collaboration inside the movement, organized in 2013, a citizen's plan, co-created, co-designed and self-managed was produced after consulting residents, users and scholars [36]. In this plan, the park has a clear connection with the concept of the NBS, even if it was not defined as a nature-based solution (NBS). The ideas, design, concepts and infrastructures presented in this space, that remains as close as possible to nature, guaranteeing aspects such as permeability of urban soil to retain and purify water, indicate that the lack of the NBS concept does not invalidate the interpretation and use of the park and its content as a NBS.

However, there were several conflicts in the history of the Parque Augusta, not produced by the social movement or by the citizen's park proposal, but by the other agents in the production of urban space, such as the real estate developers and the municipal government [37]. The contradictions resulting from the clashes between possibilities of economic growth, environmental security and social equity are materialized in the urban landscape [38]. The commodification of the green spaces [39] and the appropriation of the successful results of the social movement fight and efforts [40] show that the economic sphere is still unbalancing the idea of sustainability.

It is important to mention that we do not intend to make an academic appropriation of the social movement struggles or to present a superficial and negative critique about the population participation in movements demanding greening projects and protecting spaces of nature in the cities. Our proposal is to build a critical reading showing how the real estate agents appropriate the achievements of these movements, to increase the values of estate developments. In the case of Parque Augusta, the area of the park is now safeguarded, but not its surroundings, which have no protective measures to avoid speculation and densification around the park. Unfortunately, contradictorily to the vision and desires of the collective movement, the park is being used in the real estate market, increasing housing and rental prices, creating an economic-based selection of who can live around the park. Understanding and revealing these processes are, in our perspective, fundamental to building a socially equal, greener and sustainable urban space [41].

2. Notes about the Method

The use of a dialectical approach could serve as an alert to future movements and contributes to raising awareness about the importance to include in the debate the surroundings of parks, green areas, NBS, and other definitions and concepts for environmental amenities and produced urban natural spaces, in an attempt to avoid speculation, displacement and spatial selectivity. In this way, considering the contradictions and a wider spatial idea about the natural spaces (beyond the objects in themselves) can contribute to uncover possible negative aspects of greening projects, and improve the universalization of the right to nature in the city.

The epistemological construction of the idea to support the conceptualization about the production of nature was built within the domain of dialectical knowledge [25]. Recently, the idea of a produced nature expands from the territory of dialectical studies and is reaching a broader audience [42–44]. Additionally, it has been gaining form, shape and application with the conception and acceptance of the nature-based solutions (NBS) [45,46], which are infrastructures built by society to mimic and reproduce the functioning nature, to offer ecosystem services and solve urban environmental problems.

The incorporation of the NBS in urban and environmental plans overcomes the idea that, in the city, nature is no longer an antithesis of socially produced space. On the contrary, it is an ally and a 'Work' [31] incorporated into the very existence and survival of the city. The Lefebvrian conceptualization of 'Work' and 'Product' can provide an important contribution in the necessity of upscaling the NBS. While a 'Work' is fulfilled with uniqueness and it is genuine and authentic in its own existence and meaning, the 'Product' is the result of mass production, an object easily reproduced. In our idea, the NBS should be conceived and implemented as a 'Work', when the object (the infrastructure, the plan) tries to capture the uniqueness of nature and becomes an intrinsic and undissociated part of the city, materializing the universality of natural cycles with local particularities and singularities of nature. However, we need to be cautions, because the NBS can also become a 'Product', when incorporated into the mass/industrial production, reproducing the same solution (infrastructure/plan) without taking consideration the singularities and particularities of places and local nature.

In our critical approach [47,48], we cannot fail to reflect on the fact that the social production of nature in the city happens within the capitalist logic of production of space. In this direction, it is central to produce a careful and strong critical analysis to expose commonplaces, contradictions, and examine the burdens and benefits (who pays and who receives) of policies, plans and projects, in the direction to overrule individualism and egoism, in the name of 'eternal prosperity and supreme happiness' [47] (p. 101). A critical way of thinking, that incorporates the dialectical sphere in the discussion about nature-based solutions, demands to consider possible appropriations of NBS by the real estate market or the uneven access to the benefits of greening and renaturing projects, creating an unfair distribution of natural spaces among high and low-income citizens. In this text, we will seek to bring, in a dialectical tradition:

-   A thesis, based in the analysis of the successful history of an urban park implementation, understood as a NBS, after several years of strong and combative actions of a social movement;
-   An antithesis, to confront the results of the social demand for a park with its appropriation by the real estate market;
-   The synthesis, where we can combine the two sides of the process (the production and the appropriation of the NBS), to transcend the contradictions, and, in our goal, to promote a universal right to nature in the city.

The association of the nature-based solutions and the promotion of the right to nature in the city [18] brings the urgency to consider the production and incorporation of nature in the urban space, as well as the relationships between its use and appropriation [49,50]. While some can effectively enjoy the benefits of living with nature, others (the poorest, more vulnerable, marginalized) end up being left out of the advantages, having to face nature as a problem (diseases, risks, dangers, etc.) [41]. The clashes between classes, communities, the winners and losers, the upper and lower-income residents, represent some conflicts that is fundamental to take into consideration when implementing a NBS. Revealing the contradictions and conflicts can bring light to the problems and make possible to solve and overcome the uneven access to nature and promote a real collective emancipation [26], inherent in the conception of the production and use of nature to safeguard and ensure urban life. Green spaces, the green-blue infrastructures, the NBS and other forms that natural spaces are taking shape cannot become a product or a commodity just for individual satisfaction [13,14,26], where the incorporation and appropriation of nature is other marketing strategy [39,51].

2.1. Presenting the Study Area

Our time presents a visible clash between a consumerist and a sustainable culture, where cities and nature are in the center of this disputed arena, and both cultures, contradictory, were created and coexist in the sphere of the capitalist society. This fundamental characteristic of the present is the path to follow and build our dialectical analysis about the process that resulted in the creation and implementation of the Parque (Park) Augusta, in the central district of São Paulo, Brazil (Figure 1).

Figure 1. Parque Augusta in the city center of São Paulo (December/2020). Credits: Baumgartner.

It was a long and strong battle for an environmentally important and economically valorized urban land, with several U-turns in an emotional rollercoaster story. With 24,000 m^2 the area contains one of the last reminiscences of Atlantic Rain Forest (Figure 2) in the heart of the most populous city in the Southern Hemisphere and of the World.

Figure 2. Landscape of the Parque Augusta (December/2020). Credits: Baumgartner.

The chronology of the disputed area shows an organized social movement in one side and real estate agents on the other, with the municipal government trying to intermediate the conflict (Table 1).

Several legal protections did not prevent the attempts of real estate developers to cut down the vegetation and build skyscrapers in the area. Just the intense participation the population throughout a self-organized movement succeeded to create the urban park, what was conceived, in their project with the principles of a nature-based solution in a densely built-up area of the city. Since 2020, for the municipality, the park is part of the system of green urban parks of the city, which is officially planed and managed has a nature-based solution [52].

Table 1. The chronology of the Parque Augusta. Credits: Baumgartner; [35].

Year	Event
1969	The schools which occupied the land were closed
1970	The area was declared of public interest by the municipality
1975	The municipality did not pay the compensation and the land was divided in two lots: 7000 m^2 with public interest and 17,000 m^2 for buildings projects
1986	Without any construction, the area was declared Environmental Monument
1996	Both lots were sold
2002	The first project for a park was designed
2004	The area is declared a Special Environmental Protection Zone
2012	The lots were sold again to two real estate developers to build 3 skyscrapers
2013	The social movement occupied the area and opened a park. The municipality approved a park. The owners had removed the occupants and closed the land
2015	The developers projected 3 towers. The area with public interest was reopened, occupied and closed again. A negotiation began.
2016	With the area closed, it was declared a Special Environmental Protected Zone
2017	Conciliation hearings: the owners sold the land to the municipality
2019	After 20 months, the deal was signed
2020	Beginning of the park construction
2021	The park remains closed

2.2. Methodology

The research design focused on the production of authorial examination of published documents and data; direct observation in the field; research about land and housing prices; and the dialectical construction of our discussion and results (Figure 3). The methodology is more qualitative than quantitative. For Brazil, since the last census with data at the individuals and house units' levels is from 2010, it is difficult to get an actual picture about, for example, the characteristics of the population living around the park (age, income or educational level, occupation, etc.) that will be important to support, statistically, the discussion about some of the themes such as exclusionary displacements [53,54], amongst others.

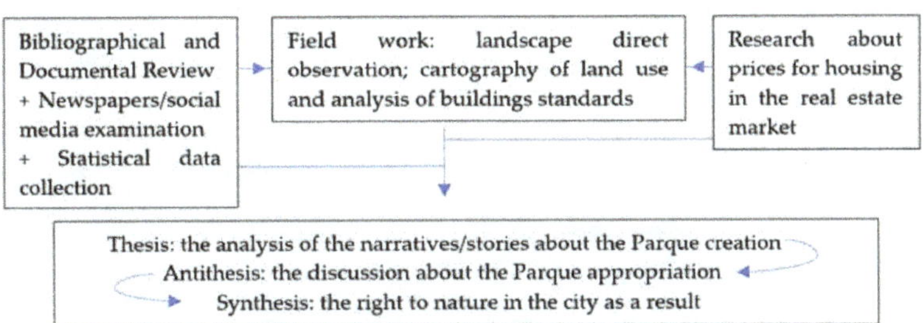

Figure 3. The research design. Credits: Baumgartner.

To overcome the lack of detailed data, we collected official general data to characterize the case study from the Brazilian Institute of Geography and Statistics [55] and Seade Foundation [56]; and legislation [57] and plans produced by the municipality about the park, such as the urban zoning map [58]. The data about the real estate market in the area around the park was important to visualize how the developers are appropriating the park in their strategies to produce a property value gain. The prices for housing purchase and the features of the estates were researched via real estate websites and the marketing material of the new developments [59–61]. The data we present about the price of housing unities available in the market is representing examples but are not statistically connected with all available unities for the city.

The qualitative research procedures, such as the topologic-morphologic analysis of the landscape, the analysis of marketing strategies for new developments in the area around the park and the data about the current prices in the real estate market produced some indicators about the profile of the desirable new residents and users for the park surroundings. The field trips also allowed the identification of possible spaces for new developments, increasing the already dense use of the area, visiting the new developments in the area. Due to the situation of the COVID-19 pandemic in Brazil and the sanitary protocols, we did not carry out interviews and inquiries with the local population, users, real estate customers and agents, government representatives, etc.

In May and December/2020, we carried out fieldworks to cartograph and identify the land use of lots and buildings in the immediate perimeter surrounding the Parque Augusta to confirm, or not, the appropriation of nature and the potentialities for a green gentrification process [12,38]. At that phase of the research, by direct observation, we cataloged 43 units with different typologies (individual and townhouses, multistory buildings, parking lots) and uses (residential, commercial, educational and institutional) around the park. Amongst these units, 17 are old, new (3) or under construction (1) residential buildings (some with commercial use on the ground floor), ranging from 10 to 33 stories; 2 are commercial buildings (from 8 to 29 stories); 1 is the campus of a private university; 1 hosts a police station; 2 are closed and unused parking lots; 16 are lower constructions (1 or 2 floors) with commercial or residential use (Figure 4). Amongst them, 6 are closed and without apparent use; 1 is for sale; 2 where demolished before December/2020 for the construction of a new residential building; 1 building in construction with 3 floors and mixes commercial and residential use.

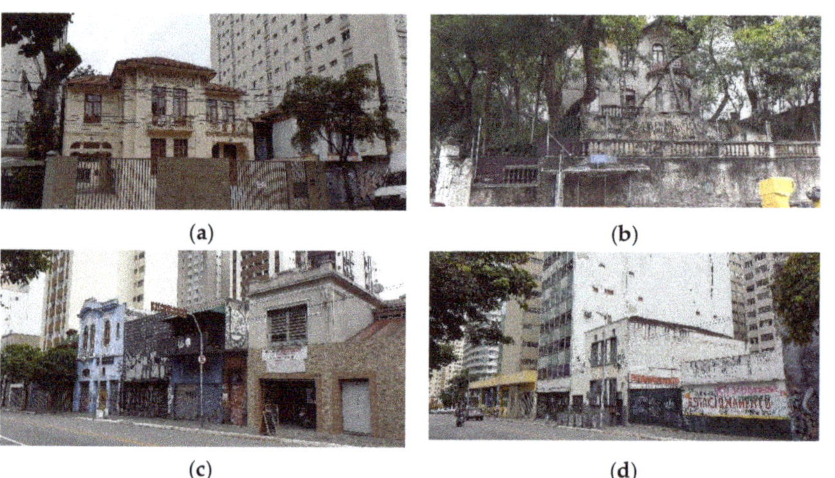

(a)

(b)

(c)

(d)

Figure 4. Aspects of the built environment around the park in December/2020. (**a**) A house for sale in front of the park (400 m^2, BRL 8,550,000 or BRL 21,375 per m^2); (**b**) the historical protected house was sold in an auction in 2018, but still has no use; (**c**) lower buildings at Rua Augusta. Just the first one is considered a historical monument and it is protected against demolition; (**d**) a closed parking lot, a potential area for new developments. Credits: Baumgartner.

Finally, before starting our explanation of this important victory of the social mobilization in favor of the Parque Augusta, it is important to clarify that this is an external study of the social movement (Movimento Parque Augusta), since we did not participate in the collective that demanded and proposed the park. Movimento Parque Augusta (Movement Augusta Park) is a self-organized social movement formed by three groups Aliados do Parque Augusta (Augusta Park Allies), Parque Augusta sem Prédios (Augusta Park without buildings) and Organismo Parque Augusta—OPA (Organism Augusta Park). OPA has the most extensive documented references about the history of the movement, including the production of two issues of a journal. Our analysis is based on the vast number of detailed documents, videos and social media texts that have been produced and organized, since 2013, by the movement [36,62–66]. We also checked for newspapers reports and news [37,40,67,68]; academic texts published in journals [69,70]; and social media [71,72].

3. Results and Discussions

3.1. The Thesis: The Action of The Social Movement for the Creation of Parque Augusta and its Institutionalization as a NBS

In 2013, emerged an organized social movement around the claim of the area located between Augusta, Caio Prado and Marquês de Paranaguá streets and the Catholic University property, for the creation of an urban park. In the past, the area was occupied by two primary and secondary education institutions, and part of a university campus. After the demolition of the internal buildings, the land was used as a parking lot, for circus tent and as a space for shows [69,73,74]. The land was divided in two lots, where one was of public interest declared by the municipality. This dual ownership was the cause of the intense dispute about the area. Even though the area was already protected, in 2012, two real estate developing corporations decided to build three high residential towers in the area, or four smaller ones, or even a shopping mall [75].

Although other social efforts demanded the implementation of the Parque Augusta prior to the Parque Augusta Movement, it was this group that successful accomplished the task and have a well-documented history [62–66]. The movement was organized through festivals, meetings and other events, where participants with different backgrounds (residents, activists, scholars, artists) and expertise (traditional, academic or non-academic knowledge) were congregated in support for the maintenance of the space and its vegetation, located in the gray and highly dense center of São Paulo. The goals of the movement were the preservation of local vegetation, maintaining the land as a permeable space, contributing to the infiltration of rainwater and control of flooding, and the creation of a citizen's designed, self-organized and managed public park. These goals show the awareness of the social movement about what would later be defined as a NBS. The group defined itself as 'self-managed, horizontal and heterogeneous', achieved a great impact in media coverage, with a large repercussion and mobilization of public opinion about the importance of the area and the vegetation of the Parque Augusta for the discussions about the production of the city and the right to the city [76].

In 2013, the first year of the published actions of the movement, the documents show an internal concern of the group in relation to the subject of gentrification [63]. A debate about the theme was promoted by the journalistic project Architecture of Gentrification [77]. Nevertheless, this first effort and the knowledge about gentrification did not prevent the park being used to promote green gentrification in the surroundings, as we will explore in the next section of this article.

The park conception created by the movement should be contemplate the citizens participation in all phases, from the project design to the construction stage, finishing with a form of self-governance or management shared among the residents, users and general population. The park, projected, built and administered by the community could become a living laboratory for urban environmental planning, according with the movement [62]. To promote these conceptions, the movement occupied the land and a opened their version of the park in July/2013. In this process of occupation, infrastructures were built; furniture produced and acquired by the movement itself through donations of materials (gardening tools, chairs, tables, dumpsters, etc.) and were placed throughout the park; money (for maintenance, rent chemical toilet, etc.) was collect and people collaborated in the construction of the park's signage.

With the park opened and self-managed by the citizens group, various educational, cultural and political activities took place until December/2013, attracting academic, political and cultural personalities. Strong social and political pressure ensured the approval, on 23 December 2013, of the law that institutionalized Parque Augusta [57].

Unfortunately, just six days later, on 29 December 2013, the park's occupants were removed, the provisional infrastructure dismantled and the real estate developers managed, through private security agents, to retake the private portion of the park's land, walled and guarded, making it impossible the access even to the public portion of the area.

Throughout 2014, the gates remained closed and heavily controlled. The Movimento Parque Augusta continued to meet and held, on a weekly basis, cultural, political and educational activities on the sidewalks around the area. They reinforced the concept of a citizen's collective movement active in the production of the city, going beyond the simple idea of the park as physical space only. In their notion, the land of the park is part of an urban ecosystem, including springs of a hidden urban basin that flows under asphalt; promotes better quality of life for humans, animals; and regulates and improves water and air quality [62].

In the beginning of 2015, the park is reopened and occupied again by the movement. However, soon after the reoccupation, the corporations managed to repossess the land [78,79]. At that time, judicially and politically favorable to the construction companies, there was a judicialization of the social movement, with the criminalization of activists who occupied the land and the surrounding streets [80]. Due to the negative repercussion, the municipality tried a first draft of a public-private agreement, where the City Hall would sell the public part of the land to the real estate companies to build their skyscrapers projects [78]. In exchange, the developers must create and manage an open park to the public, which was a reduced version of what the original Parque Augusta should be. Just a small portion of the land will be kept opened, the rest of the area could be used for the construction of residential buildings and a private park for the future residents, where most of the vegetation would be preserved.

With this setback, the movement reaffirms the strategies of struggle and mobilization of public opinion against the public-private agreement. Strengthening the idea of creating a park that occupies 100% of the land and the importance to protect the vegetation granted the public opinion support to the movement. With this repercussion, the municipality reversed the decision to sell part of the land, and due to U-turn, the companies reopened the public portion of the land.

Already in 2017, the impasse continued to block a solution for the park. Between April and August, conciliation hearings were held between the three agents involved in the demand for the park: the municipality, the real estate companies and the social movement [35]. Among some proposals, the first decision had been the exchange of the private portion of the land in the Parque Augusta for a public area in the other part of the city. But, since the proposed area had a higher value in the market, this proposal had been not approved by the City Council. The new attempt for a solution involved the exchange of the park land for Bonds of Potential Constructed Area—BPCA (Certificados de Potencial Adicional de Construção—CEPACs, in Portuguese), issued by the City Hall [67,68]. The BPCA are papers issued by the municipality to capitalize the created soil in the real estate market, allowing constructions between a basic and a maximum potential build area, according to the Brazilian urbanist laws [81]. With the agreement, the previous owners of the private parcel in Parque Augusta, received an amount of Bonds that they can use to build in other developments, or sell in the stock market. The final agreement was signed in August 2018.

With this decision, the judicial imbroglio was finally solved. In the social networks (Facebook and Instagram), practically, all the comments celebrate the park creation and the fundamental involvement of the social movement, emphasizing the importance of the park and its remaining green area in the center of São Paulo, its role in floods and urban heat island prevention and as a recreational area.

The Parque Augusta Movement accepted the decision [67] and reiterated the importance of building a space with effective community participation in the park's planning and co-creation process and a shared management and. In fact, during 2017, the movement organized several popular polls (approximately 300 contributions) to establish the main aspirations of the population for the park and promoted several co-creation workshops to design the park. The citizens/community proposal for the implementation of the Augusta Parque [36] was designed and have proposals constructed with the participation of organized groups and individuals (scholars, artists and residents). In this plan for the park (Figure 5), the area should be sustainable, greener, self-sufficient, integrate educational and recreational areas and have a blue infrastructure to increase the contribution of the park to the problems of floods and drought in a climate adaptation strategy. The main actions and structures in the citizens proposal for the park should contain, mainly outside the area covered by the original vegetation:

- Photovoltaic panels to produce energy;
- Infrastructure to treat and transform used waters and waste into biogas;
- A system to collect selective solid garbage;
- A composting area on site;
- Furniture and signage system produced with natural and sustainable materials;
- An increased green area, recovered with local fruit trees, contributing to the maintenance of the fauna and the regeneration of the ecosystem functions of the place;
- Basic infrastructures (toilets, dog training space and educational areas);
- A reservoir for rainwater harvesting;
- A seedling nursery, producing new trees and plants for the park and the community.
- A space for urban agriculture and communal garden.

Figure 5. Citizens park project co-created by the Movimento Parque Augusta. Credits: [36].

Comparing these proposals presented in the citizen's plan for the Parque Augusta with some catalogues [3,82–84], all these measures and infrastructures can define the whole park as nature-based solution, albeit the term is not appearing the document published by the social movement [36].

In June 2020, São Paulo published its territorial plan of Sustainable Development Goals implementation, based, at least in the mayor's speech [52] (p.23), on the conception of nature-based solutions. According to this document, the Parque Augusta, amongst others municipal parks, is considered by the municipality as a green space and a green infrastructure, joining a system of green areas that would be understood and configured as a strategy tackling climate change.

Within the 'Municipal plan for protected, green areas and open spaces' [52], the park and other green spaces are clearly defined as areas for provision of ecosystem or environmental services. In private areas, the plan projected a financial bonus program for the construction, restoration and maintenance of natural spaces that offer ecosystem or environmental services. In the city's conception, these services provided by the biodiversity present in the city have an aesthetic nature, but also are associated with the availability and control of water, soil and atmosphere quality, regulating the urban thermal sensation. Thus, the green and unpaved area would have some functions: filtering rainwater and regulate underground storage; promoting carbon sequestration and retention; improving air quality, reducing pollution rates; and regulating surface temperature, through evapotranspiration and direct shading. In addition, the park would be a space for sociability, recreation and a platform for educational activities, improving the indicators of habitability in the city.

Associating the assumptions of the plan and the potentialities observed in the Parque Augusta, considering the environmental sphere, the territorialization of following SDGs would be possible: 3 (health and well-being), 6 (drinking water and sanitation), 11 (sustainable cities), 13 (action against global climate change) and 15 (terrestrial life). The municipality also indicates that SDG 17 (partnerships and means of implementation) would be achievable in municipal parks.

However, the project of the park has a stronger partnership with the real estate developers than the social movements. The official project of the Parque Augusta (Figure 6) designed and approved by the City Hall will be managed by the former's private owners of the land, who are responsible for its implementation and maintenance for 2 years [85].

Figure 6. Official project for the park. Credits: [85].

Besides a dog training space (Cachorródromo), public toilets, composting area and seedling nursery, all the other 'attractions' (hammock lounge [redário], open gym for the elderly [academia da Terceira idade], slackline area, a grandstand [arquibancada], an elevated deck) were not in the scope of the citizens proposal for the park.

Several demands and strategies, including a much more sustainable and consistent with the NBS approach, presented in the community project, are not present in the official project produced by the municipality. Additionally, the project defines that 10% of the area can be impermeabilized, with some pavements constructed with concrete. This material could easily be replaced by some other type of material, increasing the permeability such as demonstrated in the NBS catalogues and standards [82,83].

In 2019, the official presentation of the project registered that it was showed to the social movement in 2017, but there is no information about any update or changing in the project with the movement [86]. In addition, the project is vague about the community participation in its designs, just informing that the suggestions from the population were received, but without further details about how many people participated and when/where the participation happened. The only sphere of the official project with clear citizen's involvement is the participation in a management committee. According to local regulations, the committee should control the planning and maintenance of the park and monitor the activities and events in the area. The local population (registered in the document as users) took part in the election, back in 2019, for four seats in a 2-year term, to form this group with members of the municipal government (presidency and other two seats), a member representing the park works, and another member from an organized social movement [87].

In March 2021, even with a simpler project in implementation, the bureaucracies solved and a local committee elected to follow the implementation and construction of the park infrastructures, the Parque Augusta remains closed. The works have not been completed and the inauguration date has been postponed since 2020. The expected date to open the gates for the population is July/2021, but there is already another conflict, involving the construction of a grandstand designed for art performances. Even if an amphitheater was present in the citizens proposal, it is now not desired by some local residents who does not want to have noise and higher number of visitors during cultural events in the park [88].

3.2. The Antithesis: The Appropriation of Nature by the Agents of the Real Estate Market

To dialectically confront the implementation process of the Parque Augusta, we take, as a premise, the fact that nature, materially and symbolically, is incorporated into the sphere of a capitalist world, where it is (re)produced as an instrumental rationality [89]. In the cites, from a set of intrinsic needs for humans and other forms of life, the values, the experiences and the spaces of nature are reified and commodified [90]. Converted into wishes and desires, transmuting its constitution as a value of use into value of exchange, nature is becoming another possibility for consumption, in a process that appear to be inherent to our society [91]. In the XIX century, Marx had already warned about the commodification of nature and its capitalist appropriation, indicating that it can be used as bait to capture, ironically, the essence of a person in the capitalistic society: its money [21].

In the contemporary city, highly artificialized places, nature is converted and perceived as a resource, a commodity or a service, and used to enhance premiums in the real estate market [39,51]. Since a price to live with nature is established, a difference in its accessibility and affordability, according to class or income, is also recognized. Even if all humans depend on nature to survive, and renaturing of cities is urgent and necessary, not all have the same capital to live with nature, or nearby. From luxurious condominiums to low-income housing projects, the spaces of nature are very different. While in high income urban or suburban neighborhoods' nature is well cared, produced for contemplation, relaxation and exhibition of prestige/privilege, in the low-income, vulnerable and marginalized areas, in the social or geographical peripheries of the capitalist city, the 'raw' and not controlled nature is a synonym of risks, dangers and threats, such as floods or diseases. The appropriation of nature by real estate companies denies collective benefits, hiding illusions of comfort, satisfaction and sustainability in the individual life.

Having in consideration these statements, we can try uncovering how the Parque Augusta, a necessary but also a desirable space, have been appropriated by the real estate developers. The commodification of the park, from its physical nature to its image, happens in the immediate surroundings of it, and had started even before its official implementation. As explained before, to have the full ownership of the land where the park was planned, the municipality exchange (paid) the lot for Bonds of Potential Construction Area (BPCA). The number of certificates issued represents a potential 3.3 million m^2 of constructed area, that can be used in several parts of São Paulo or even be sold in the regular stock market. Usually, the money received by the municipality in BPCA transactions is used to improve the basic urban infrastructure (sewage, water distribution, garbage collection, etc.) in areas of saturation or highly dense use. The strategy, at the surface, seems to be proper, since the municipality received the land without paying for it in advance, just had ceded future capital income. However, the Bonds issued overpriced the land in BRL 95 million (USD 24,1 million in the exchange rate of August/2018 or USD 16,5 million nowadays, due to intense devaluation of the Brazilian Real) [68]. It is important to remember, that even if the land had a contested private ownership, the area and the vegetation should be protected according to the municipal laws. On several occasions, documented in their webpage, the Movimento Parque Augusta complained about the lack of transparency in the financial agreements between the municipality and the developers [62].

But our biggest concern is the propagation of what some scholars are defining as green gentrification [12,38,92,93] and is already punctually identified around the Parque Augusta, in three recent real estate developments in front of the park, but with a potential growth. These new developments will be identified as Development 1 (with one tower), Development 2 (with two towers) and Development 3 (with one tower).

The green gentrification can be the result of exclusionary displacement [53] or exclusion of the poorest from the benefits of environmental amenities [94]. The green gentrification materializes the supremacy of the production and consumption of nature as an individual or personal satisfaction project in the access of nature in the urban space. Ecological gentrification [95] was the concept used to represent the denial of access to natural spaces in the cities, based in class, income or social hierarchy, including the perspectives of homeless people about the nature. Several homeless are living in the streets around the park, being naturalized and included in the urban landscape without the minimal basic rights, and quite far away from the right to nature in the city. Comments in news or posts about the park revels the need to remove/transfer these homeless people from the Parque Augusta area to avoid depreciation of real estate vale and the infeasibility for families to use the future park.

The region, known as Baixo Augusta (Lower Augusta) has a popular commerce profile, and it was famous for its dancing/sex clubs, that are now partially shut down [74]. Several other commercial properties are also closed, not only because a possible property price reevaluation, but also due to bankruptcy during the COVID-19 crisis.

The actual valid urban zoning, from 2016 (Figure 7) [58], includes 3 of the park perimeters (Augusta and Marques de Paranaguá streets, and the limit with a university campus) as Areas for Urban Structuring Transformation (ZEU) and 1 perimeter (Caio Prado street) as a Centrality Zone (ZC). The park itself is a Special Environmental Protection Zone (ZEPAM). The ZEU foresees measures to increase the density, allowing high construction potentials without story limits. The ZC is defined as medium density area, with incentives for commercial and residential uses, with heights limited up to 60 m.

Thus, according to this current zoning legislation, the areas of the two currently closed parking lots and 9 lower buildings (1 or 2 floors) are in the zone that allows high density. Four lower buildings are in the medium density zone. The height profile and the lower density of the park surroundings buildings, the current use, or its emptiness, indicate a great constructive potential in the area, with the replacement of individual houses, townhouses and parking lots by skyscrapers, fulfilling the projections of the urban zoning indicators. For example, in 2019, after the final decision about the park, a developer demolished a gas station and two townhouses to build a residential apartment tower, which is already conclude. This development has 24 floors and 250 units (from 67 to 125 m^2). In another space, a former parking lot gave space, also in 2019, to a 27-story residential building with 216 units (ranging from 24 to 58 m^2) [60].

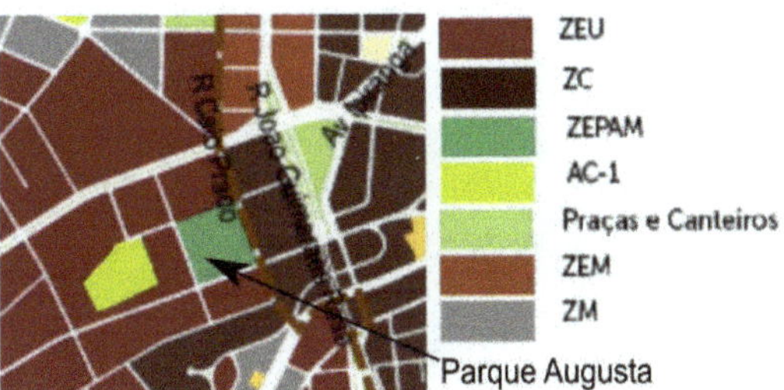

Figure 7. Extract from the zoning map of São Paulo center district. ZEU, Urban Structuring Transformation Zone; ZC, Centrality Zone; ZEPAM, Special Environmental Protection Zone; AC-1, recreational club; praças e canteiros, squares and flower beds; ZEM, Metropolitan Structuring Axis Zone; ZM, Mixed Zone. Credits: [74].

Besides the substitution of houses and others low-density used spaces for high residential developments, the prices of these new apartment's units are also an indicator of the valorization that the park brought to the area. Our survey on the prices of land, units and average value for constructed area (square meters—m^2), shows that the value of the constructed square meters, in the area surrounding the park increased from BRL 5522 (August/2018) to BRL 8817 (March/2021). The first value corresponds to the price on the date of the final agreement between the developers and the municipality [59,67] and the second one, from March/2021, covers the average price for m^2 (buying transactions) in the region where the park is located [68]. Besides the park, the only transformation in the area was the inauguration of a subway station in January/2018. Since the Brazilian Real suffered a large devaluation in the past years, we decide to present the values in the local currency, avoiding transferring the dollar devaluation to the prices that are practiced in 'reais' in the local real estate market. The conversion rate to American dollars was USD 1/BRL 3,93 (August/2018) and USD 1/BRL 5,76 (27 March 2021). In the same period, the accumulated inflation was 11.09%, meaning that the price per m^2 soared 143.73% above inflation.

Four new skyscrapers (in three different developments) are the first visual impact, recently built or nearing completion in the immediate perimeters of the park (Figure 8), are also priced for purchase with values above the neighborhood average values.

The Development 1 is most recent completed residential building has its marketing strategy focused on the direct relationship with the park, the urban renewal of the area and the gentrification of services and commerce [60]. In December/2020, we visited this development, which had gross (without upgrade options) values for m^2 ranging from BRL 11,558.44 (unit on the 9th floor with 77 m^2) to BRL 12,779.19 (unit on the 10th floor with 66 m^2).

The above-average value is also visible in other properties, which offer free views to the Parque Augusta, such as the two towers in the Development 2. In the residential one, we found a unit with 158 m^2 being sold for almost BRL 2 million, or BRL 11,958.81 per m^2. Apartments in older buildings, from the 70s and 80s, are also being offered in the market, even without a direct view of the park at average prices around BRL 9,000.00 per m^2.

The Development 3, in its last phase of conclusion [61], have units with 24 m^2 being sold for BRL 15,208.33; units with 37 m^2 for BRL 16,101.18 per m^2; and units with 57 m^2 for BRL 15,870.17 per m^2. These values represent almost a 100% valorization above the average prices in the region (BRL 8,817 m^2).

The neighborhood valorization and the attraction of new private investments to the area has been predicted by the municipality since the launch of the final and official project for the park [85].

Figure 8. New developments in the environs of the park (December/2020). (**a**) The two newest building, on the right Development 1 and Development 3, on the left side of the picture; (**b**) The view from an apartment for sale; (**c**) Development 2 towers. Credits: Baumgartner.

To understand how those values promote an exclusionary displacement, and a potentially green gentrification, it is important to have a clear picture about the economic situation of the majority of São Paulo's population. According to the Brazilian Institute of Geography and Statistics (IBGE) [55], 12,325,232 people are living in the city and 21,6 million in its Metropolitan Area (50% of the state population; approximately 10% of Brazil's population; and 33.3% of the Brazilian GDP). The average income (2018) of formal workers in the city is 4,3 times the minimum wage (BRL 954 in 2018 and BRL 1,102 in 2021), but just 45.8% of its population has a formal (regular/legalized) job; 31.6% of the inhabitants were living (2010, last available data from the census) with less than a half of the minimum wage. The most recent data about income is from 2019, and shows that the medium income per person is BRL 48.40 per day, or BRL 1,472, 16 monthly (in favelas/slums the income is BRL 13.47 per day, BRL 409.71 monthly) [56,81].

The center of São Paulo (2.67 million inhabitants), where the Parque Augusta is located, has an average income of BRL 2,366 (63% higher than the city), the highest amount of people with university or higher degree (28.3%), the lowest rate of unemployment, the highest rate of formal jobs and the lowest infantile and juvenile population [56].

The differences of the general revenue in the city, considering formal/informal jobs and their distribution along São Paulo, show the gap between the average income of the inhabitants and the prices found in the real estate market for housing around the park. The profile of the inhabitants, that is represented in the context of the people engaged in the social movement [72], indicates another debate about green gentrification: are the environmental amenities inducing gentrification or inhabitants of gentrified areas are demanding better green spaces?

In the case of the Parque Augusta, the documents, the data and the field trips show that both processes running together. First, the implementation of the park has a direct connection with the profile of the area and its population, which has better access to the legal and media channels, than the population of lower income area. The residents and the activists of the social movement, due to their professional background, have a clear picture about their strategies, the importance of this park and its green area to the neighborhood and to the whole city. Connecting the conceptualization of the park with the NBS principles, matching the climate action by the municipality, was an important strategy to receive support from the public opinion, resulting in the park creation. Second, the immediate vicinity of the park has several potential lots for high density new developments (indicated by the actual urban zoning of São Paulo), and the prices in the market (above local average prices and medium income of the population), are enhanced by the implementation of the park. These indicators visible in the urban landscape and support by the prices survey, demonstrate the potential exclusionary displacement is already taking place, promoting what could be characterized as green gentrification.

3.3. The Synthesis: The Right to Nature in the City and the NBS

If we seek to build a synthesis, in the direction to promote a right to nature in the city [13], a first step is to uncover the history and the ideology that produces the city [7,17] and the spaces of nature in the city [25,26], to understand conflicts, class/racial privileges or struggles, such as like in a green gentrification process [38].

The construction of our proposal for dissemination of the right to nature in the city has three basic principles:

- A new praxis in the relationship with nature. In an optimistic conception of human action with nature, the right to nature in the city is not a right to exploit, dilapidate or consume nature for satisfaction of few individuals, but it is a process of collective emancipation to change the future perspectives for humanity in times of climate change. It is necessary to rethink daily life society's action on nature, from the crops to the cities, passing through commerce, industries and finances [96–98].
- The idea of nature for all citizens and not for customers. The imposition of individual satisfaction also led nature to be understood as an object of consumption and people as the consumer. The installation of consumption's ideology in all spheres of social life is one of the greatest perversions of the current period, with the commodification of nature and its services. One way to overcome this process is the defetishization and demystification of nature, understanding it as an element integrated to social life through history. In addition, it is essential to combat class, age, social and racial discriminatory access to nature in the city [99–101].
- To rethink the place of nature in the city. First, it is necessary to rethink the city not as an object, as a machine or an organism, but as the place of people's life, a human ecosystem, restoring collectiveness and empathy as the purpose of urban life. It is necessary to inhabit the city, to conscious participate in collective social life. Then, we can build a collective place to nature and to the nature-based solutions in the center of urban life [102–107].

The Praxis: 10 Strategies to Promote the Right to Nature in the City

The optimistic or utopic view of the relationship between urban society and nature is based on the urgency of the physical presence of natural spaces throughout the city, opened and accessible for all inhabitants or communities, despite income, classes, race or age. To achieve this goal, the following propositions were elaborated, considering necessary:

1. To create a right to nature in the city, as a collective right, not dissolving history or the city to return to a primitive nature. Nature can also not be reduced to a view, a panorama or a landscape.
2. To destroy the ideology of consumption of nature and urban spaces, in the direction to renew an innovative and creative freedom to produce a new real sustainable urban humanism and urbanism, considering all aspects of sustainability.
3. To transform the access to natural spaces in an accessible and affordable essential right. Nature is necessary for everyone, regardless of income and address. Nature cannot be a luxury commodity neither an object of decoration.

4. To reaffirm the public aspect of nature, changing the current practice in which some high-income real estate developments have natural reserves in their areas, a space restricted to their residents, while in the poorest areas of cities the benefits of nature are denied, with, for example, in the absence of parks and green areas and the transformation of urban rivers in sewage. Preservation and production of nature and nature-based solutions in the city must be universal.

5. To consider the totality of nature in the city, overcoming its reduction as a synonym for green areas. Nature should be understood beyond trees, shrubs and grasses in the urban plans. The right of nature in the city must be expanded as the right to air, soil and water of good quality.

6. To understand the nature in the city as part of human work and history, overcoming the idea that nature in the city is a refuge for nature that no longer exists or an enclave in the urban fabric protected with a dome and separate from city life. Today nature is incorporated into human urban life. Therefore, the spaces of nature in the city must be used and lived, integrated in the everyday life. It is necessary to socialize its use, making nature a meeting place.

7. To promote ideas of nature-based solutions and blue-green urban infrastructures, radically transforming the cities, financing the incorporation of sustainable practices, especially in the poorest and vulnerable countries. Having green cities in some rich or high developed places will not create a resilient urban world.

8. To establish the connection with nature under the guidance of the concept of value of use and not exchange value. If nature is understood as a value of use it should not be commodified or commercialized, for example by real estate developers to enhance premiums in their properties. This will avoid green gentrification, spatial and economic segregation in the access to nature in our cities. There would be an appreciation of nature for its content, used by society, incorporated into the territory and benefiting all people.

9. To revel the perversion of neoliberal and public-private associations, where the public pays the burdens and the private sphere receives the bonus.

10. To understand the nature in the city dialectally, making it visible the history and the conflicts of the relations between society and nature, to overcome the traditional dichotomy between the city and nature.

4. Some Conclusions and Final Considerations

The project of the Parque Augusta produced by the social movement, with citizen's participation, had several principles and infrastructures which can define it as a nature-based solution, despite not using the term in the document. This collective design is much more integrated with nature, from its lake to the enhanced green area, but was not chosen by the São Paulo city's government to be implemented in the area. The official project in execution, designed by the municipality and constructed by the real estate developers who sold the land, although it is integrating a climate action, where the nature-based solutions are a key concept, has fewer physical elements of a NBS in its design. There is no space for rainwater harvesting, such as a lake or pond, the reforestation and other greening actions are vague and have more constructed and paved surfaces.

Considering how it was designed and co-created, the perspectives of a self-managed park and the NBS conceptualization, the project presented by the Movimento Parque Augusta was a much better choice for the area.

In our critical perspective, both projects are deficient in an integrated vision with the surroundings and the homeless people who live around the park. The integration and the participation of the homeless should be important to the construction of the park. Of course, the ideal solution will be the distribution of houses for them, but since this is not foreseen in the near horizon, due to deficiencies of public policies and social affordable housing projects, the homeless will be living around and inside the park. Regrettably, for some future users and inhabitants, they are and will be a problem, when using the park in their everyday life. Throughout the analysis of the content in some comments posted in social media (Facebook and Instagram) about the park creation, the problems of security and social selection is quite evident. Even without statistical accuracy, those comments provide some points for further discussion about the Parque Augusta, and other parks and open public green areas, since the fear and discomfort with diversity in public Brazilian spaces is a common felling [108]. The widespread and easiest solution have been fencing and gating the green spaces and public parks, controlling schedules and their access. As well, the integration of homeless people in the debate about the NBS, and other conceptual framework for projects designed to produce greener and resilient cities, is still an area for improvement.

Additionally, both projects did not incorporate instruments to prevent land speculation, fast values enhancing in housing prices, population displacement or green gentrification. Several instruments available in the Brazilian urban legislation, especially those in the 'Estatuto da Cidade' (City Statute) [81] were not taken into consideration to tackle, through an alternative to the current urban zoning, the increased construction potentials around the park, the liberation in the amount of floors and heights for new developments in three perimeters of the park, what can improve the densification of the in the immediate surroundings of the Parque Augusta.

In our conclusion, the focus just in the object (the park, the NBS, etc.), losing a wider spatial perspective, with the inclusion of the surroundings and the whole neighborhoods, can produce gaps where the real estate capital will prosper and appropriate the benefits of greening and renaturing projects. To promote social, environmental justice and the right to the nature in the city, the action of real estate developers must be controlled and regulated, otherwise the land and housing prices will rise. Exclusionary displacement, spatial selectiveness and green gentrification means less diverse urban environment. These side effects are contradictory with the ideas behind the NBS, the SDGs and other greening projects which reinforce the importance of diversity. However, in the urban space, the biodiversity must include the socio-diversity, as well.

A dialectical approach to understand the (re)production of urban space and the (re)production of nature in the city is fundamental to uncover the contradictions and conflicts, rooted in the urban fabric and visible in the landscape. As a theoretical perspective, it brings other layers of complexity to the debate about the scope of nature-based solutions and how them can reach and be beneficial for more people. The construction of a synthesis, more than a practical compromise between the production and the appropriation of nature, allowed us to thing further, establishing principles for a co-existence of nature and city in the same space and launch perspectives about the access to nature.

Climate change adaptation and life during/post the COVID-19 pandemic demonstrate how important and valuable nature in the city can be. The construction of a better, sustainable and collective world will be possible, dialectally, understanding the production of urban and natural spaces and the universalization of the right to nature in the city to all inhabitants, despite color, age, race, social-economic situation and geographical position on the city. The right to nature in the city is a right for humans to live and prosper with nature and for nature to exist and flourish with humans.

Funding: This research received no external funding.

Institutional Review Board Statement: Not applicable.

Informed Consent Statement: Not applicable.

Data Availability Statement: Statistical Data provided by the Brazilian Institute of Geography and Statistics (IBGE) with open access are available in www.ibge.gov.br (accessed on 16 March 2021).

Conflicts of Interest: The author declares no conflict of interest.

References

1. Kozak, D.; Henderson, H.; Mazarro, A.C.; Rotbart, D.; Aradas, R. Blue-Green Infrastructure (BGI) in Dense Urban Watersheds. The Case of the Medrano Stream Basin (MSB) in Buenos Aires. *Sustainability* **2020**, *12*, 2163. [CrossRef]
2. Birkhofer, K.; Diehl, E.; Andersson, J.; Ekroos, J.; Früh-Müller, A.; Machnikowski, F.; Mader, V.L.; Nilsson, L.; Sasaki, K.; Rundlöf, M.; et al. Ecosystem services—current challenges and opportunities for ecological research. *Front. Ecol. Evol.* **2015**, 1–12. [CrossRef]
3. EC (European Commission). *Towards an EU Research and Innovation Policy Agenda for Nature-Based Solutions and Re-Naturing Cities*; Publications Office of the European Union: Luxembourg, 2015. [CrossRef]
4. Colléony, A.; Shwartz, A. Beyond Assuming Co-Benefits in Nature-Based Solutions: A Human-Centered Approach to Optimize Social and Ecological Outcomes for Advancing Sustainable Urban Planning. *Sustainability* **2019**, *11*, 4924. [CrossRef]
5. Naumann, S.; Davis, M.; Iwaszuk, E.; Freundt, M.; Mederake, L. *Addressing Climate Change in Cities—Policy Instruments to Promote Urban Nature-Based Solutions*; Sendzimir Foundation: Warsaw, Poland, 2020.
6. Mahmoud, I.; Morello, E. Co-creation Pathway for Urban Nature-Based Solutions: Testing a Shared-Governance Approach in Three Cities and Nine Action Labs. In *Smart and Sustainable Planning for Cities and Regions, Green Energy and Technology*; Bisello, A., Ludlow, D., Vettorato, D., Baranzelli, C., Eds.; Springer: Cham, Switzerland, 2021; pp. 259–276. [CrossRef]
7. Mumford, L. *The City in History: Its origins, Its transformations, and Its prospects*; Harcourt, Brace and World: New York, NY, USA, 1961.
8. Glacken, C.J. *Huellas en la playa de Rodas. Naturaleza y cultura en el pensamiento occidental desde la Antiguedad hasta finales des siglo XVIII*; Ediciones del Serbal: Barcelona, Spain, 1996.
9. Férnandez-Armesto, F. *Civilizations. Culture, Ambition, and the Transformation of Nature*; The Free Press: New York, NY, USA, 2001.
10. Weichselgartner, J.; Kelman, I. Geographies of resilience: Challenges and opportunities of a descriptive concept. *Prog. Hum. Geogr.* **2015**, *39*, 249–267. [CrossRef]
11. Sachs, J. *The Age of Sustainable Development*; Columbia University Press: New York, NY, USA, 2015.
12. Anguelovski, I.; Connolly, J.J.T.; Pearsall, H.; Shokry, G.; Checker, M.; Maantay, J.; Gould, K.; Lewis, T.; Maroko, A.; Roberts, J. Why green "climate gentrification" threatens poor and vulnerable populations. *PNAS* **2019**, *116*, 26139–26143. [CrossRef]
13. Henrique, W. *O direito a natureza na cidade*; Edufba: Salvador, Brazil, 2009.
14. Gilbert, L. Social Justice and the "Green" City. *Rev. Bras. Gest. Urbana* **2014**, *6*, 1–18. [CrossRef]
15. Harvey, D. *Social Justice and the City*; University of Georgia Press: Athens, OH, USA, 1973.
16. Agyeman, J.; Bullard, R.D.; Evans, B. (Eds.) *Just Sustainabilities: Development in an Unequal World*; Taylor & Francis: London, UK, 2012.
17. Lefebvre, H. *The Production of Space*; Blackwell Publishing: Oxford, UK, 1991.
18. Henrique, W. O direito à natureza na cidade. Ideologias e práticas na história. Doctoral Thesis, Doctorate in Geography—State University of São Paulo, Rio Claro, Brazil, 24 September 2004.
19. Lenoble, R. *Histoire de l'idée de nature*; Éditions Albin Michel: Paris, France, 1969.
20. Kos, D. Nature in the city or the city in nature? *Urbani Izziv* **2008**, *19*, 129–132. [CrossRef]
21. Marx, K. *Capital—A Critique of Political Economy, Volume I, Book One: The Process of Production of Capital*; Penguin Classics: London, UK, 1992.
22. Engels, F. *Dialectics of Nature*; Lawrence and Wishart: London, UK, 1939.
23. Marsh, G.P. *Man and Nature: Or Physical Geography as Modified by Human Action*; Charles Scriber: New York, NY, USA, 1874.
24. Reclus, É. *Du sentiment de la nature dans le societés modernes et autres textes*; Édition Premières Pierres: Paris, France, 2002.
25. Smith, N. *Uneven Development: Nature, Capital, and the Production of Space*; University of Georgia Press: Athens, OH, USA, 2008.
26. Harvey, D. *Justice, Nature and the Geography of Difference*; Blackwell Publishers: Oxford, UK, 1996.
27. Leff, E. *Racionalidad ambiental. La apropiación social de la naturaleza*; Siglo XXI Editores: Ciudad de México, México, 2004.
28. Kovel, J.; Löwy, M. Ecosocialist Manifesto. Available online: http://green.left.sweb.cz/frame/Manifesto.html (accessed on 16 March 2021).
29. Löwy, M. Why Ecosocialism: For a Red-Green Future. Available online: https://greattransition.org/publication/why-ecosocialism-red-green-future (accessed on 16 March 2021).
30. Whitehead, A.N. *The Concept of Nature*; Cambridge University Press: Cambridge, UK, 1955.
31. Casini, P. *As Filosofias da Natureza*; Presença: Lisboa, Portugal, 1975.
32. Castree, N. Marxism and the Production of Nature. *Capital Class* **2000**, *24*, 5–36. [CrossRef]
33. Castree, N.; Braun, B. (Eds.) *Social Nature: Theory, Practice, and Politics*; Blackwell Publishers: Oxford, UK, 2001.
34. Loftus, A. *Everyday Environmentalism: Creating an Urban Political Ecology*; University of Minnesota Press: Minneapolis, MN, USA, 2012.
35. Marques Júnior, J.J. Parque Augusta no Município de São Paulo: Instituições estatais e não estatais envolvidas na consolidação desse equipamento urbano e as dinâmicas de negociação. *Cadernos Jurídicos* **2019**, *52*, 91–105.
36. Movimento Parque Augusta. Parque Augusta: Proposta de Implementação. Available online: https://drive.google.com/file/d/1mJzJvj1LOzPt9Zp4b_0ZGZ6v0nvzJQP4/view (accessed on 16 March 2021).
37. Carvalho, F. Parque Augusta, nova área verde de São Paulo e "negócio da China" para construtoras. Available online: https://brasil.elpais.com/brasil/2017/07/18/politica/1500413441_128758.html (accessed on 24 March 2021).
38. Gould, K.A.; Lewis, T.L. *Green Gentrification. Urban Sustainability and the Struggle for Environmental Justice*; Routledge: New York, NY, USA, 2017.

39. Henrique, W. A cidade e a natureza: A apropriação, a valorização e a sofisticação da natureza nos empreendimentos imobiliários de alto padrão em São Paulo. *GEOUSP Espaço e Tempo* **2006**, *10*, 65–77. [CrossRef]
40. Castilho, I. Parque Augusta, luta coletiva a um passo da vitória. Available online: https://outraspalavras.net/blog/parque-augusta-luta-coletiva-a-um-passo-da-vitoria/ (accessed on 24 March 2021).
41. United Nations Human Settlements Programme (UN-Habitat). *World Cities Report 2020*; United Nations Human Settlements Programme: Nairobi, Kenya, 2020.
42. Chen, W.Y.; Yong Hu, F.Z. Producing nature for public: Land-based urbanization and provision of public green spaces in China. *Applied Geography* **2015**, *58*, 32–40. [CrossRef]
43. Angelo, H. *How Green Became Good. Urbanized Nature and the Making of Cities and Citizens*; University of Chicago Press: Chicago, IL, USA, 2020.
44. Aalto, H.E. Projecting Urban Natures. Investigating Integrative Approaches to Urban Development and Nature Conservation. Doctoral Thesis, Architecture and the Built Environment—The Royal Institute of Technology in Stockholm, Stockholm, Sweden, 2017.
45. Bush, J.; Doyon, A. Building urban resilience with nature-based solutions: How can urban planning contribute? *Cities* **2019**, *95*, 1–8. [CrossRef]
46. Mendes, R.; Fidélis, T.; Roebeling, P.; Teles, F. The Institutionalization of Nature-Based Solutions—A Discourse Analysis of Emergent Literature. *Resources* **2020**, *9*, 6. [CrossRef]
47. Wacquant, L. Critical Thought as Solvent of Doxa. *Constellations* **2004**, *11*, 97–101. [CrossRef]
48. Morgan, T. Alienated Nature, Reified Culture: Understanding the Limits to Climate Change Responses under Existing Socioecological Formations. *Political Econ. Commun.* **2017**, *5*, 30–50.
49. Lefebvre, H. *Le droit à la ville*; Anthropos: Paris, France, 1968.
50. Lefebvre, H. *Espace et politique*; Anthropos: Paris, France, 1973.
51. Aroul, R.R.; Rodriguez, M. The Increasing Value of Green for Residential Real Estate. *J. Sustain. Real Estate* **2017**, *9*, 112–130. [CrossRef]
52. São Paulo (Municipality). *Relatório de localização dos Objetivos de Desenvolvimento Sustentável na cidade de São Paulo*; Agência Frutífera: São Paulo, Brazil, 2020.
53. Marcuse, P. Gentrification, Abandonment, and Displacement: Connections, Causes, and Policy Responses in New York City. *Wash. U.J. Urb. Contemp. L.* **1985**, *28*, 195–240.
54. Atkinson, R. Does Gentrification Help or Harm Urban Neighbourhoods? An Assessment of the Evidence-Base in the Context of the New Urban Agenda. *Urban Studies* **2002**, 1–26.
55. IBGE Cid@des. Available online: https://cidades.ibge.gov.br/ (accessed on 16 March 2021).
56. Fundação SEADE. Pesquisa SEADE. Available online: https://www.seade.gov.br/wp-content/uploads/2020/01/Pesquisa-SEADE_Aniversario-SP_23jan2020.pdf (accessed on 16 March 2021).
57. São Paulo (Municipality). Lei 15.941. 23/12/2013. Available online: http://legislacao.prefeitura.sp.gov.br/leis/lei-15941-de-23-de-dezembro-de-2013/detalhe (accessed on 16 March 2021).
58. São Paulo (Municipality). Lei 16.402/16. 22/03/2016. Available online: https://gestaourbana.prefeitura.sp.gov.br/marco-regulatorio/zoneamento/arquivos (accessed on 16 March 2021).
59. Proprietário Direto. Available online: www.proprietariodireto.com.br/preco-m2/consolacao-sao_paulo (accessed on 22 March 2021).
60. Olha Augusta. Available online: https://www.olharaugusta.com/ (accessed on 25 March 2021).
61. Faça tudo a pé em frente ao Parque Augusta—You, Central Park. Available online: https://www.youinc.com.br/empreendimento/you-central-park-venda-apartamento-studio-1-2-dormitorios-consolacao-rua-caio-prado-sao-paulo-sp/ (accessed on 25 March 2021).
62. Canteiro Vivo Parque Augusta. Available online: https://parqueaugusta.hotglue.me/ (accessed on 22 March 2021).
63. Organismo Parque Augusta. Available online: https://issuu.com/organismopa (accessed on 22 March 2021).
64. Parque Augusta. Available online: https://www.facebook.com/parqueaugustaja (accessed on 22 March 2021).
65. Aliados do Parque Augusta. Available online: https://www.facebook.com/parqueaugusta (accessed on 22 March 2021).
66. Parque Augusta Já. Available online: https://www.youtube.com/channel/UC9yKjcq4_CH-7pkBby_B6bQ/featured (accessed on 22 March 2021).
67. Amâncio, T. Construtoras e prefeitura de SP fecham novo trato sobre parque Augusta Administração municipal e empresas disputam área verde no centro da cidade. Available online: https://www1.folha.uol.com.br/fsp/fac-simile/2018/08/11 (accessed on 16 March 2021).
68. Zylberkan, Mariana. Empreiteiras podem lucrar até R$95 milhões com acordo do parque Augusta. Available online: https://www1.folha.uol.com.br/fsp/fac-simile/2018/11/28 (accessed on 16 March 2021).
69. Júnior, S.; Guimarães, A.P.M. Dialética do Parque Augusta. *Cidades Verdes* **2015**, *03*, 56–73. [CrossRef]
70. Maróstica, J.R.; Cortese, T.T.P.; Nascimento, A.P.B. Implantação do Parque Augusta: Critérios para unificar as diferentes demandas solicitadas pela população. In Anais do VII SINGEP. In Proceedings of the VII SINGEP, São Paulo, Brazil, 22–23 October 2018; Uninove: São Paulo, Brazil; pp. 1–17.
71. Rolnik, R. Parque Augusta 100% público. Available online: https://raquelrolnik.blogosfera.uol.com.br/2019/04/05/parque-augusta-100-publico/ (accessed on 16 March 2021).

72. Rolnik, R. Mitos e verdades sobre o Parque Augusta. Available online: https://raquelrolnik.wordpress.com/2015/01/29/mitos-e-verdades-sobre-o-parqueaugusta/ (accessed on 16 March 2021).

73. Antunha, A.; Entini, C.E. Parque na Augusta: Um impasse de 40 anos. Available online: https://acervo.estadao.com.br/noticias/acervo,parque-na-augusta-um-impasse-de-40-anos,9208,0.htm (accessed on 26 March 2021).

74. Pissardo, F.M. A rua Apropriada: Um Estudo Sobre as Transformações e usos Urbanos na Rua Augusta (São Paulo, 1891–2012). Master's Thesis, Architecture—USP, São Paulo, Brazil, 2013.

75. Pinho, M. Construtoras querem mudar área tombada para erguer prédios em SP. Empresários pretendem construir em terreno na Rua Augusta. Moradores, porém, pedem criação de um parque no local. Available online: http://g1.globo.com/sao-paulo/noticia/2012/03/construtoras-querem-mudar-area-tombada-para-erguer-predios-em-sp.html (accessed on 26 March 2021).

76. Brito, G.; Silva Junior, P. Luta pelo Parque Augusta dialoga com todas as pautas pelo direito à cidade. Available online: https://www.correiocidadania.com.br/index.php?option=com_content&view=article&id=10615:manchete230315&catid=63:brasil-nas-ruas&Itemid=200 (accessed on 26 March 2021).

77. Arquitetura da Gentrificação. Available online: https://www.catarse.me/ag#about (accessed on 22 March 2021).

78. Jardim, M.H. O Parque é na Rua: A (in)visibilidade da ocupação do Parque Augusta. Available online: https://research.uca.ac.uk/5254/1/1%20O_Parque_e_na_Rua_A_in_visibilidade_da_o.pdf (accessed on 26 March 2021).

79. Rossi, M. Parque Augusta levanta acampamento. Available online: https://brasil.elpais.com/brasil/2015/03/04/politica/1425486632_591073.html (accessed on 26 March 2021).

80. Scatolini, L.; Bampa, G. Um chamado à liberdade: Ocupe o Parque Augusta. Available online: http://vaidape.com.br/2015/03/um-chamado-liberdade-ocupe-o-parque-augusta/ (accessed on 26 March 2021).

81. Rolnik, R. *y Saule Júnior. N. Estatuto da cidade: Guia para implementação pelos municípios e cidadãos*; Caixa/Instituto Polis: Brasília, Brazil, 2002.

82. Iwaszuk, E.; Rudik, G.; Duin, L.; Mederake, L.; Davis, M.; Naumann, S.; Wagner, I. *Addressing Climate Change in Cities—Catalogue of Urban Nature-Based Solutions*; Sendzimir Foundation: Warsaw, Poland, 2019.

83. International Union for Conservation of Nature and Natural Resources (IUCN). *Global Standard for Nature-Based Solutions. A User-Friendly Framework for the Verification, Design and Scaling Up of NbS*; IUCN: Gland, Switzerland, 2020.

84. Somarakis, G.; Stagakis, S.; Chrysoulakis, N. (Eds.) *ThinkNature Nature-Based Solutions Handbook*; Hellas-FORTH: Heraklion, Greece, 2019. [CrossRef]

85. São Paulo (Municipality). Apresentação Projetos Parque Augusta. 04/08/2017. Available online: http://www.capital.sp.gov.br/arquivos/pdf/pmsp_smj_projetos_parque_augusta_apresentacao_20170804a.pdf/view (accessed on 16 March 2021).

86. São Paulo (Municipality). Prefeitura de São Paulo assina escritura do terreno do futuro Parque Augusta. Available online: http://www.capital.sp.gov.br/noticia/prefeitura-de-sao-paulo-assina-escritura-do-terreno-do-futuro-parque-augusta (accessed on 16 March 2021).

87. Atas Conselho Gestor Augusta. Available online: https://www.prefeitura.sp.gov.br/cidade/secretarias/meio_ambiente/parques/conselhos_gestores/index.php?p=302809 (accessed on 22 March 2021).

88. Quintela, S. Prometido para julho, Parque Augusta agora tem discórdia por arquibancada. Available online: https://vejasp.abril.com.br/cidades/capa-parque-augusta/ (accessed on 26 March 2021).

89. Habermas, J. *Toward a Rational Society: Student Protest, Science, and Politics*; Beacon Press: Boston, MA, USA, 1970.

90. Horkheimer, M.; Adorno T, W. *Dialectic of Enlightenment*; Stanford University Press: Stanford, CA, USA, 2002.

91. Marcuse, H. *One Dimensional Man: Studies in the Ideology of Advanced Industrial Society*; Beacon Press: Boston, MA, USA, 1991.

92. Anguelovski, I.; Connolly, J.J.T. Green gentrification in Barcelona. In *Renaturing Cities. Town Planning and Housing*; Ballester, E.J., Ed.; Barcelona Provincial Council's Press and Communication Office: Barcelona, Spain, 2019; pp. 100–112.

93. Ali, L.; Haase, A.; Heiland, S. Gentrification through Green Regeneration? Analyzing the Interaction between Inner-City Green Space Development and Neighborhood Change in the Context of Regrowth: The Case of Lene-Voigt-Park in Leipzig, Eastern Germany. *Land* **2020**, *9*, 24. [CrossRef]

94. Pearsall, H. From brown to green? Assessing social vulnerability to environmental gentrification in New York City. *Environ. Plan C Politics Space* **2010**, *28*, 872–886. [CrossRef]

95. Dooling, S. Ecological Gentrification: A Research Agenda Exploring Justice in the City. *IJURR* **2009**, *33*, 621–639. [CrossRef]

96. Hahn, T.; McDermott, D.; Ituarte-Lima, C.; Schultz, M.; Green, T.; Tuvendal, M. Purposes and degrees of commodification: Economic instruments for biodiversity and ecosystem services need not rely on markets or monetary valuation. *Ecosystem Services* **2015**, *16*, 74–82. [CrossRef]

97. Prieto, M. Practicing costumbres and the decommodification of nature: The Chilean water markets and the Atacameño people. *Geoforum* **2016**, *77*, 28–39. [CrossRef]

98. Bond, P. Water commodification and decommodification narratives: Pricing and policy debates from Johannesburg to Kyoto to Cancun and back. *Capitalism Nature Socialism* **2004**, *15*, 7–25. [CrossRef]

99. Rowland-Shea, J.; Doshi, S.; Edberg, S.; Fanger, R.; The Nature Gap. Confronting Racial and Economic Disparities in the Destruction and Protection of Nature in America. Available online: https://www.americanprogress.org/issues/green/reports/2020/07/21/487787/the-nature-gap/ (accessed on 28 March 2021).

100. Byrne, J. When green is White: The cultural politics of race, nature and social exclusion in a Los Angeles urban national park. *Geoforum* **2012**, *43*, 595–611. [CrossRef]

101. Strife, S.; Downey, L. Childhood Development and Access to Nature: A New Direction for Environmental Inequality Research. *Organ Environ.* **2009**, *22*, 99–122. [CrossRef]
102. Grimm, N.B.; Schindler, S. Nature of Cities and Nature in Cities: Prospects for Conservation and Design of Urban Nature in Human Habitat. In *Rethinking Environmentalism: Linking Justice, Sustainability, and Diversity*; Lele, S., Brondizio, E.S., Byrne, J., Mace, G.M., Martinez-Alier, J., Eds.; MIT Press: Cambridge, MA, USA, 2018; pp. 99–125.
103. Frantzeskaki, N. Seven lessons for planning nature-based solutions in cities. *Environ. Sci. Policy* **2019**, *93*, 101–111. [CrossRef]
104. Mok, S.; Mačiulytė, E.; Bult, P.H.; Hawxwell, T. Valuing the Invaluable(?) A Framework to Facilitate Stakeholder Engagement in the Planning of Nature-Based Solutions. *Sustainability* **2021**, *13*, 2657. [CrossRef]
105. Arlati, A.; Rödl, A.; Kanjaria-Christian, S.; Knieling, J. Stakeholder Participation in the Planning and Design of Nature-Based Solutions. Insights from CLEVER Cities Project in Hamburg. *Sustainability* **2021**, *13*, 2572. [CrossRef]
106. Longato, D.; Geneletti, D. Nature-based solutions: New challenges for urban planning. Book of Papers. In Proceedings of the AESOP 2019 Conference Planning for Transition, Venice, Italy, 9–13 July 2019; pp. 3785–3792.
107. Kabisch, N.; Frantzeskaki, N.; Pauleit, S.; Naumann, S.; Davis, M.; Artmann, M.; Haase, D.; Knapp, S.; Korn, H.; Stadler, J.; et al. Nature-based solutions to climate change mitigation and adaptation in urban areas: Perspectives on indicators, knowledge gaps, barriers, and opportunities for action. *Ecol. Soc.* **2016**, *21*, 1–15. [CrossRef]
108. Souza, M.L. *Fobópole: O medo generalizado e a militarização da questão urbana*; Bertrand Brasil: Rio de Janeiro, Brazil, 2008.

sustainability

Article

Nature-Based Solutions for Storm Water Management—Creation of a Green Infrastructure Suitability Map as a Tool for Land-Use Planning at the Municipal Level in the Province of Monza-Brianza (Italy)

Giulio Senes [1,*], Paolo Stefano Ferrario [1], Gianpaolo Cirone [1], Natalia Fumagalli [1], Paolo Frattini [2], Giovanna Sacchi [3] and Giorgio Valè [4]

[1] Department of Agricultural and Environmental Sciences, University of Milano, 20133 Milan, Italy; paolo.ferrario@unimi.it (P.S.F.); gianpaolo.cirone@unimi.it (G.C.); natalia.fumagalli@unimi.it (N.F.)

[2] Department of Earth and Environmental Sciences, University of Milano Bicocca, 20126 Milan, Italy; paolo.frattini@unimib.it

[3] Studio Geologia Sacchi, 24121 Bergamo, Italy; studio.giovannasacchi@gmail.com

[4] BrianzAcque s.r.l, 20900 Monza, Italy; giorgio.vale@brianzacque.it

* Correspondence: giulio.senes@unimi.it; Tel.: +39-02-503-16885

Citation: Senes, G.; Ferrario, P.S.; Cirone, G.; Fumagalli, N.; Frattini, P.; Sacchi, G.; Valè, G. Nature-Based Solutions for Storm Water Management—Creation of a Green Infrastructure Suitability Map as a Tool for Land-Use Planning at the Municipal Level in the Province of Monza-Brianza (Italy). *Sustainability* **2021**, *13*, 6124. https://doi.org/10.3390/su13116124

Academic Editors: Israa H. Mahmoud, Eugenio Morello, Giuseppe Salvia and Emma Puerari

Received: 15 February 2021
Accepted: 24 May 2021
Published: 28 May 2021

Abstract: Growing and uncontrolled urbanization and climate change (with an associated increase in the frequency of intense meteoric events) have led to a rising number of flooding events in urban areas due to the insufficient capacity of conventional drainage systems. Nature-Based Solutions represent a contribution to addressing these problems through the creation of a multifunctional green infrastructure, both in urban areas and in the countryside. The aim of this work was to develop a methodology to define Green Infrastructure for stormwater management at the municipal level. The methodology is defined on the basis of three phases: the definition of the territorial information needed, the production of base maps, and the production of a Suitability Map. In the first phase, we define the information needed for the identification of non-urbanized areas where rainwater can potentially infiltrate, as well as areas with soil characteristics that can exclude or limit rainwater infiltration. In the second phase, we constructed the following base maps: a "map of green areas", a "map of natural surface infiltration potential" and a "map of exclusion areas". In phase 3, starting from the base maps created in phase 2 and using Geographical Information Systems' (GIS) geoprocessing procedures, the "Green area compatibility map to realize Green Infrastructure", the "map of areas not suitable for infiltration" and the final "Green Infrastructure Suitability Map" are created. This methodology should help municipal authorities to set up Green Infrastructure Suitability Maps as a tool for land-use planning.

Keywords: spatial planning; nature-based solutions; green infrastructure; rainwater management

1. Introduction

Land take and soil sealing are the most evident and worrying consequences of growing and uncontrolled urbanization. Climate change, with an associated increase in the frequency of intense meteoric events, have led to an increasing number of flooding events in urban areas due to the insufficient capacity of conventional drainage systems.

Urban rainwater drainage systems are essential infrastructures for cities, which are needed to collect and convey rainwater away. Conventional stormwater management systems (so-called gray infrastructures) are systems primarily oriented towards a single objective: the control of water quantities. In Italy, most of the grey infrastructure is represented by the sewerage network [1]. It is a "mixed network" which collects both rainwater and wastewater, then transports it to the treatment plant. Due to this network characteristic, despite significant developments, it remains difficult to implement a fully efficient conventional urban drainage system [2].

Intense urbanization over the past few decades has significantly changed land-uses and greatly increased the proportion of impermeable surfaces around the world [3–5]. This rise in the percentage of sealed soils has changed urban hydrological systems, as demonstrated by the increase in surface water runoff and peaks of maximum flow, the decrease in the amount of rainwater infiltrated into the soil, alterations in the recharge cycle of aquifers and the deterioration of water quality [6–10].

These factors, along with the combined effect of climate change-induced intense meteoric events, have caused a higher frequency of flooding in urban areas [11].

At present, there exists a need to consider other important aspects of water management in urban environments: the quality of the water itself, the ecological and recreational value of locations and their visual quality, the aesthetic aspect and architectural form of drainage systems and the possibility of reusing the volumes of water conveyed [12].

Over the last two decades, the academic and professional worlds have increasingly investigated the effectiveness of using Nature-Based Solutions (NBS) for the creation of a multifunctional green infrastructure, both in urban areas and in the countryside. The European Commission defines NBS as "solutions that are inspired and supported by nature, which are cost-effective, simultaneously provide environmental, social and economic benefits and help build resilience. Such solutions bring more and more diverse, nature and natural features and processes into cities, landscapes and seascapes, through locally adapted, resource-efficient and systemic interventions. Nature-based solutions must benefit biodiversity and support the delivery of a range of ecosystem services" [13].

Nature-based solutions can contribute to stormwater management both by reducing the volume and flow rate of stormwater runoff and removing contaminants from stormwater. Nature-based solutions such as urban parks and open spaces, wetlands, green roofs, bioswales, rain gardens and detention and retention ponds promote water storage and infiltration, reducing stormwater runoff [14–16]. Cities with combined sewer infrastructure will see improvements from nature-based solutions arising from reductions in stormwater quantity and reduced sewage overflows [16].

Nature-based solutions are attractive in combination with grey infrastructure, not only for stormwater management but also for properly considering the full spectrum of co-benefits and their integration within wider GI networks [16,17].

The interdisciplinary concept of ecosystem services (the benefits in terms of goods and services to humans provided by nature) [18] helps to understand the suite of services that the different types of NBSs deliver to the human society, among which we can find stormwater management [16].

The scientific literature related to NBSs for stormwater management has presented a diversified terminology to describe their principles and practices, in relation to local situations and different contexts. Different terms have been used to define similar concepts in different parts of the world, sometimes generating contradictions and confusion [14].

The term Low Impact Development (LID), used mainly in North America, refers to small-scale water treatment works close to the point of origin of runoff [19], while the term Best Management Practice (BMP) describes interventions and practices designed to prevent pollution [14]. Water Sensitive Urban Design (WSUD) has been used since the early 1990s in Australia, with the main objective of managing the water balance, while the concept of Integrated Urban Water Management (IUWM) refers, more broadly, to the integrated management of all parts of the water cycle at the catchment level.

The term Sustainable Drainage Systems (SuDS) originated from the UK and it includes a range of techniques used to drain water by restoring drainage conditions existing prior to site development [20].

The concept of Green Infrastructure (GI) was developed in the USA in the 1990s [21] and represents a term referring not only to rainwater management. Its origins were derived from landscape architecture and landscape ecology.

Green Infrastructure can be defined as "a strategically planned network of high quality natural and semi-natural areas with other environmental features, which is designed and managed to deliver a wide range of ecosystem services and protect biodiversity in both rural and urban settings. [. . .] One of the key attractions of GI is its ability to perform several functions in the same spatial area, in contrast to most 'grey' infrastructures, which usually have only one single objective [. . .]. Green Infrastructure is made up of a wide range of different environmental features that can operate at different scales, from small linear features such as hedgerows or green roofs to entire functional ecosystems [. . .]. Each one of these elements can contribute to GI in urban, peri-urban and rural areas, inside and outside protected areas" [22].

Green Infrastructure plays an important role in stormwater management (in addition to the existing grey infrastructure), enhancing natural processes such as infiltration, evapotranspiration and filtering and reuse of water [23]. Green Infrastructure for stormwater management provides several benefits, such as rainwater detention, flood alleviation, fewer sewer overflow events and the reduction of management costs for grey infrastructure [24,25].

In Italy, several regional authorities have adopted laws and regulations aimed at planners and designers in order to satisfy the hydraulic-hydrologic invariance (HHI) principles in land-use plans and new developments design (i.e., the maximum outflow rate should be at greenfield runoff) [26]. These principles can be carried out by dimensioning appropriate grey infrastructure (e.g., water storage tanks) or nature-based solutions to balance the soil sealing effects [27].

In 2017, the Lombardy Region adopted a new regulation related to HHI which obliges municipalities to set up a hydraulic risk management plan that should identify, among other things, areas suitable for rainwater infiltration.

The aim of this work was to develop a methodology to identify areas where there was the potential to install Green Infrastructure for stormwater management at the municipal level, in particular infiltration SuDSs which reduce both the flow and the volume of runoff [28]. This methodology should help municipal authorities to set up Green Infrastructure Suitability Maps as a tool for land-use planning [29–33]. This work is part of a broader study involving several entities (universities and professionals), promoted by BrianzAcque SRL, a public water management company for municipalities in Monza-Brianza province (Italy). The paper presents the results of the first part of the study. In the second part (still ongoing) we are analyzing the existing drainage network (the sewer network) in order to define where and how to carry out specific interventions to solve the problems of insufficiency of the drainage network, through the modeling of the networks.

2. Materials and Methods

In a preliminary step, the factors to be taken into consideration for identifying the areas suitable for the realization of green infrastructure for stormwater management were analyzed through a review of the available literature.

Several authors [28,34–37] have agreed about the use of Geographical Information Systems (GIS) as support systems for the localization of NBS. GIS allows users to manage and consider many territorial characteristics, overlay geographic data layers, develop models based on raster and vector data, support choices of land-use planning, and define possible alternative scenarios.

From the review of the methodologies proposed by various authors, it emerged that the main factors considered are slope, soil type and land use.

The slope determines a considerable influence on the localization of NBS, limiting infiltration and increasing surface runoff. According to several authors [29,35,38], natural solutions for detention and infiltration should be realized in areas with slopes not exceeding 15%. In this sense, zones with inadequate slopes represent areas to be excluded in the definition of green infrastructure.

With regard to the type of soil, the most important parameter considered is the infiltration potential, expressed as the saturated hydraulic conductivity value (m/s) [35,39]. These data can be derived from pedological–lithological maps, if available, with a good level of detail, or through direct survey and infiltration testing.

In urban areas, the permeability characteristics of soils can vary significantly, due to compaction caused by buildings and other uses [40]; for these reasons, direct surveys are generally necessary to define the characteristics of the infiltration potential at a specific site. Tredway and Havlick [41] also underlined the usefulness of carrying out, where necessary, any work to improve the infiltration rate of the soil.

Land use is a fundamental topic as the possibility of natural infiltration depends on the presence of non-impervious surfaces. Land use and the proportion of green space available in a given area determine the behavior of surface waters and affect exposure to floods [28–30]. Land-use maps provide information about the green areas that are compatible with the possibility of infiltration and, at the same time, allow for the identification of different impervious surfaces and the estimation of possible surface runoff [28–31,42–44].

The proposed methodology was defined on the basis of three phases (see Figure 1): (I) Definition of the information needed; (II) Production of the base maps; and (III) Production of the Suitability Map. We further present, as a case study, an application of the methodology to the municipality of Caponago (in the province of Monza-Brianza).

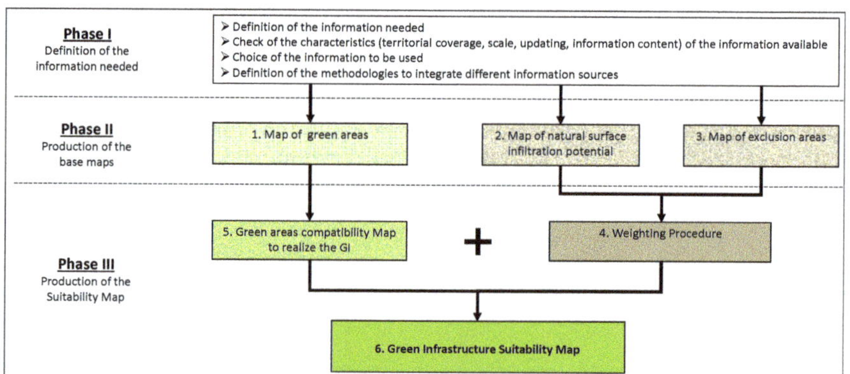

Figure 1. Phases of the methodology proposed for the definition of Green Infrastructure for stormwater management.

2.1. Definition of the Information Needed (Phase I)

During this phase, we defined the information needed for the identification, on the one hand, of non-urbanized areas where rainwater can potentially infiltrate and, on the other hand, areas with soil characteristics that can exclude or limit rainwater infiltration. Secondly, we selected the most suitable information based on their availability, territorial coverage, scale and updating.

Most of the data analyzed were either produced by the Lombardy Region and made available on their geoportal or produced by the municipalities involved.

2.2. Production of the Base Maps (Phase II)

In phase II, the following base maps were produced: a map of green areas, a map of natural surface infiltration potential and a map of exclusion areas.

2.2.1. Map of Green Areas

In relation to the green areas potentially available for infiltration (phase II, map 1 of Figure 1), we followed a specific procedure, which is shown in Figure 2.

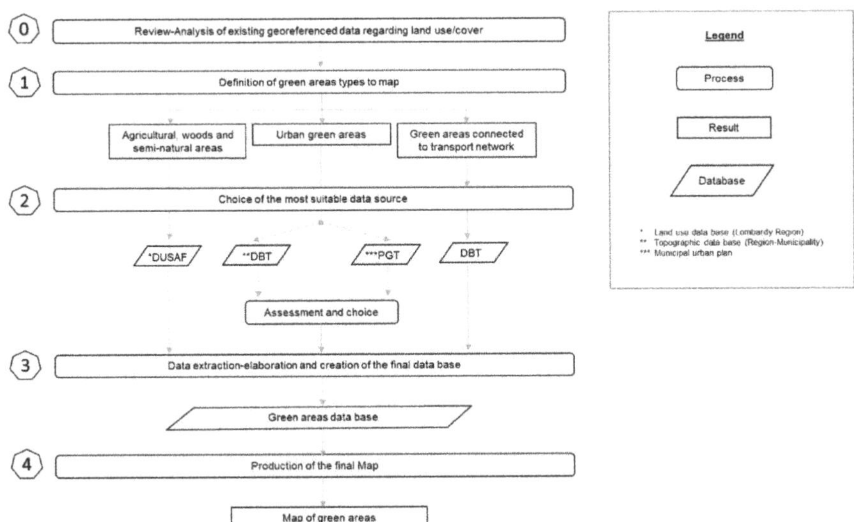

Figure 2. Procedure for mapping green areas potentially available for infiltration.

First of all, an analysis of the available geographical databases concerning the theme of land-use/land-cover was carried out. In particular, the following data sources were analyzed:

- The land-use database "DUSAF", produced by Lombardy Region;
- The topographic database "DBT", produced by Lombardy Region and Municipalities;
- Maps contained in Municipal urban plans, produced by Municipalities.

The Land-use database DUSAF is a geographical database produced for the last 20 years by the Lombardy Region. The first version was created by the photo-interpretation of digital orthophotos (IT2000 program, frames from 1998 to 1999); subsequent updates were made in 2005, 2006, 2007, 2009, 2012 and 2015, up to the latest update available with orthophotos in 2018 (DUSAF 6). The database has a level of detail compatible with the scale of 1:10,000 (minimum mapped area = 1600 m^2) and uses a legend structured in classes and subclasses. It is available, for the entire Region, in shapefile format.

The Topographic database DBT represents the reference base of the Regional Information System at the municipal level. Its creation was overseen by the regional law n.12/2005 and it has been produced by the municipalities at 1:1000–1:2000 scales. With reference to vegetation, the DBT includes agro-forest areas (agricultural crops, woods, pasture—uncultivated, areas temporarily not vegetated) and urban green areas (green areas, rows, trees). The update year differs from municipality to municipality and varies from 2007 to 2015; it is available as a shapefile for most of the regional territory.

The maps of the municipal urban plan are specific to each municipality; in the case of Caponago, information on green areas is contained in the maps "Destination of land use" and "Components of environmental systems" (2011, scale 1:5000), which identify agricultural areas, wooded areas and green areas of public interest, such as parks and gardens. They are only available in pdf format.

From a comparison of the above-mentioned databases, the following considerations emerged: The DUSAF database represents the official land-use map of the Lombardy Region, which is particularly suitable for obtaining information about green areas in extra-urban contexts (e.g., agricultural, wooded and semi-natural areas); however, the level of detail of the DUSAF may not be sufficient in the urban context. It has been updated to 2018.

The DBT contains information with a greater level of detail in the urban context and includes, in addition to public green spaces, private areas. It contains green areas connected to the transport network (e.g., traffic islands, roundabouts). The DBT of Caponago has been updated to 2015.

The maps of the Municipal urban plan (PGT) of Caponago contain information on green areas that are less detailed and less updated than the DUSAF and DBT. In the urban context, only public green areas are detected (not private).

Based on the information available, and the objectives of the project, we defined the classes of green areas to map as:

1. Agricultural, wooded and semi-natural areas;
2. Urban green areas;
3. Green areas connected to the transport network.

In order to have the most updated and accurate data, we decided to map the green areas by integrating information from different sources.

We used the DUSAF database as a data source for agricultural, wooded and semi-natural areas (1), mainly located in the extra-urban context and the DBT green areas connected to the transport network (3).

For urban green areas (2), the choice of data source (DBT or PGT) was made by evaluating the following conditions (in hierarchical order):

-   Availability of data in shapefile format, including not only public green areas but also private green areas (i.e., annexed to residential areas, industrial areas etc.);
-   Completeness of the data related to green cover;
-   Up-to-date data (we preferred to use the most recent data).

In this work, we used the DBT for urban green areas (2), (as shown in Table 1) as it satisfied all three conditions (unlike the PGT); it is available in shapefile format, it is complete with regard to green cover and it is more up-to-date.

Table 1. Urban green areas: availability, completeness and update for Caponago.

Urban Green Area Conditions	PGT	DBT
Availability	no	yes
Completeness	no	yes
Update	yes	yes

After selecting the databases to be used, we proceeded with the extraction of the data and the creation of the final database. With regard to agricultural, wooded and semi-natural areas mainly located in the extra-urban context, polygons were extracted from the DUSAF database. With regard to urban green areas, polygons were extracted from the DBT (only for the portion of the territory in the urban context). A check was carried out in order to identify any changes in land-use through visual analysis of the most recent satellite images available (Google, ESRI) and digital orthophotos available on the Lombardy geoportal.

With regard to green areas connected to the road network, polygons were extracted from the DBT. All the derived layers were then merged into a single shapefile, creating a field containing the final classification, as reported in Table 2.

As required by the Lombardy regional law, we analyzed the whole municipal territory. This is also coherent with the need to give land-use planners information related to all the not yet urbanized areas of the municipality, in order that they can take into account different factors for the decision of the new urbanizations to include in the land use plans.

The map of green areas (phase II, map 1 of Figure 1) was produced at 1:5000 scale (Figure 3).

Table 2. Map of green areas: final classes.

Class	Type
Urban green areas	garden
	flowerbed
	trees
	other green area
Green areas connected to transport network	traffic island/roundabout
	parking strip
Agricultural, wooded and semi-natural areas	crop
	crop with trees
	horticultural
	horticultural (greenhouse)
	floricultural and plant nursery
	vineyard
	grassland
	grassland with trees
	orchard
	vegetable garden
	deciduous wood
	reforestation
	riparian wood
	bushes
	bushes in abandoned agricultural lands

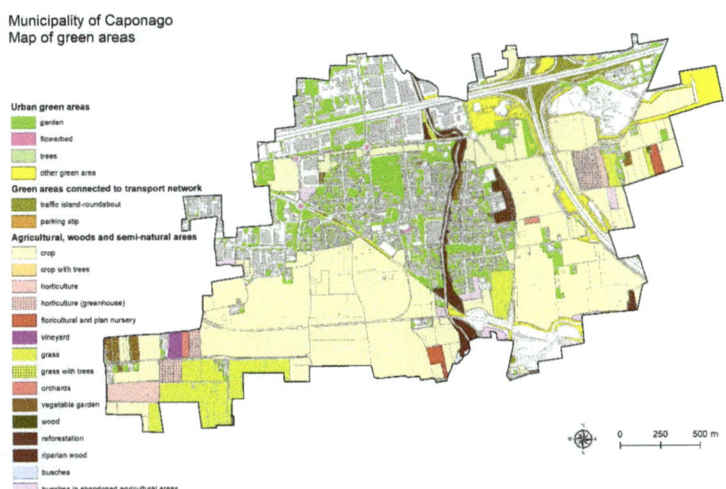

Figure 3. Map of green areas for the municipality of Caponago.

2.2.2. Map of Natural Surface Infiltration Potential

The second map of the procedure (phase II, map 2 of Figure 1) is related to the natural surface infiltration potential. It expresses the capacity of water to infiltrate through the most superficial layers of the soil. It is useful for the study of hydraulic risk and the evaluation of infiltration strategies. The map was built through a zoning of the territory into geological units that are "average homogeneous", from the point of view of infiltration, for each of which a saturated hydraulic conductivity value (m/s) was estimated. The zones were derived from the geological cartography available—in particular, from the Regional Geological Map at 1:10000 scale ("CARG" project)—integrated with other information from the geological cartography of the municipal urban plan.

The infiltration values were derived from an empirical estimation of the permeability of the different lithofacies, based on the available surveys and corrected by infiltration tests.

With regard to the study area, the geological units were derived from the geological map of the municipal urban plan. The Regional Geological Map (CARG) is not available at present. The delimited units were "average homogeneous", from the point of view of the surface infiltration potential and may present heterogeneity in different specific sites (Table 3).

Table 3. Characteristics of the geological units in Caponago.

Geological Unit	CARG Code	Description
Unità di Besnate	Bes	Fluvioglacial and glacial deposits, slightly weathered, up to 4 m. Sporadic Loess deposits.
Sintema del Po	Pg	Gravels, sands and silts from recent fluvial deposits, lacustrine deposits, slope and colluvial deposits, landslide deposits. Fresh upper surface, characterized by entisols and inceptisols.

Once zoning was carried out, a saturated hydraulic conductivity value was assigned to the units and appropriately reclassified into hydrofacies, thanks to the availability of infiltration data obtained by surveys at variable depth (Figure 4).

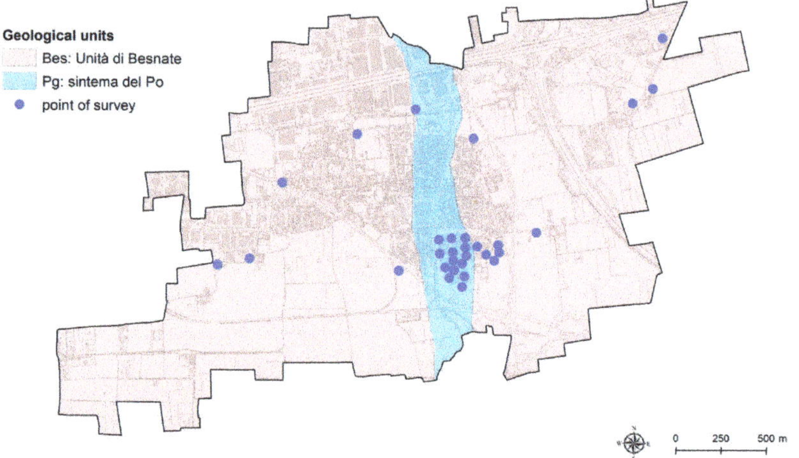

Figure 4. Geological units and localization of the points of survey for the municipality of Caponago.

The association of infiltration values derived from infiltration tests to the geological units allowed us to estimate a reference value for each unit and, thus, to proceed to the mapping of the infiltration potential. Being a parameter that varies over several orders of magnitude, it was considered appropriate to consider the logarithm of the infiltration potential and to average the available values.

Table 4 shows the reference values of saturated hydraulic conductivity (m/s) related to the different classes of infiltration potential, whereas Table 5 shows the attribution of the final class of surface infiltration potential to each geological unit in the study area.

The map of natural surface infiltration potential (phase II, map 2 of Figure 1) is represented in Figure 5.

Table 4. Classes of surface potential infiltration.

Classes of Surface Infiltration Potential	Reference Values of Saturated Hydraulic Conductivity (m/s)
Very high	$>10^{-2}$
High	10^{-2}–10^{-3}
Medium	10^{-3}–10^{-4}
Low	10^{-4}–10^{-5}
Very low	$<10^{-5}$

Table 5. Surface potential infiltration classes assigned to the geological units in Caponago.

Geological Unit	CARG Code	Surface Potential Infiltration Class
Unità di Besnate	Bes	Low
Sintema del Po	Pg	Medium

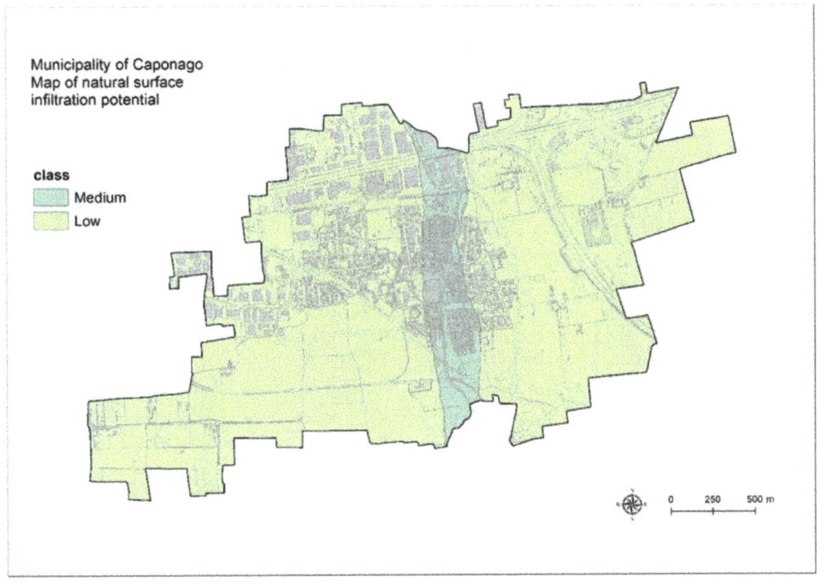

Figure 5. Map of natural surface infiltration potential.

2.2.3. Map of Exclusion Areas

The third map of the procedure (phase II, map 3 of Figure 1) is related to the exclusion areas. These are portions of the territory that have hydrogeological characteristics such that the infiltration of water could represent a risk to the safety of the population. These areas were identified on the basis of laws and regulations and were derived from various territorial plans, such as:

- The Flood Risk Management Plan (PGRA) of the Po river basin, as established from the EU Floods Directive (60/2007);
- The Hydrogeological Plan of the Po river (PAI), provided by the Po river basin authority;
- A geological feasibility map of the municipal urban plan;
- A geological general map of the municipal urban plan.

The layers, in shapefile format, were derived from these data and the map of exclusion areas was produced, as shown in Figure 6.

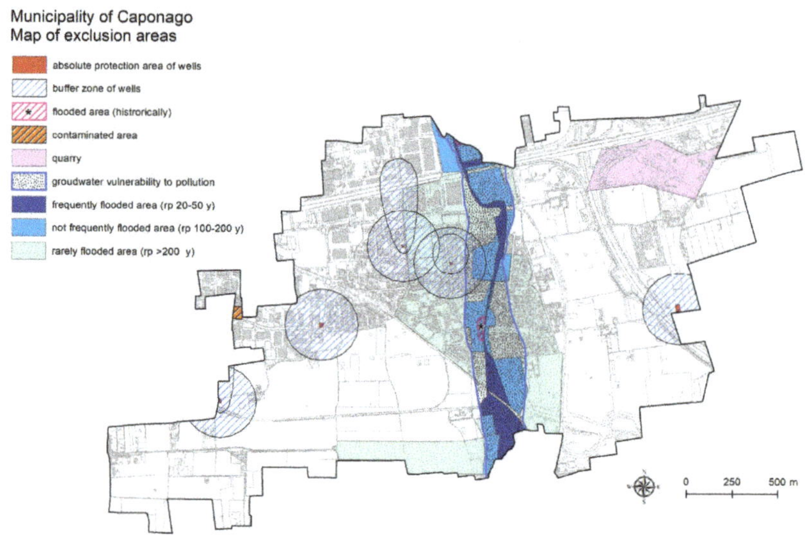

Figure 6. Map of exclusion areas for the municipality of Caponago. They represent the portions of the territory that have hydrogeological characteristics such that the infiltration of water could represent a risk to the safety of the population.

As already mentioned, land morphology must be also considered in order to exclude areas with inadequate slopes. The construction of a Digital Terrain Model (DTM) allowed us to identify areas with a slope greater than 15%. In the territory of Caponago, the slope never exceeded this value.

3. Results

In phase III of the procedure, first the "Green area compatibility map to realize the Green Infrastructure" (phase III, map 4 of Figure 1) was created. Then, using a GIS overlay procedure, this map was combined, after the assignment of appropriate weights, with the Map of natural surface infiltration potential (phase II, map 2 of Figure 1) and the map of exclusion areas (phase II, map 3 of Figure 1), in order to create the final Green Infrastructure Suitability Map (phase III, map 5 of Figure 1).

3.1. Green Area Compatibility Map to Realize the Green Infrastructure

We derived the "Green area compatibility map to realize the Green Infrastructure" by giving a compatibility score to each green area typology in the map of green areas. The compatibility score expresses how compatible each green area typology is with the realization of the Green Infrastructure for rainwater management.

In order to provide an appropriate compatibility score, we identified the "equipped green areas" (i.e., those with benches, playgrounds and so on) using municipal maps, satellite images and Google Street View (Figure 7).

The compatibility score was derived through the aggregation of different characteristics: Naturalness [N], Anthropic presence [A], Productive value [P] and Urban context [U]. Each characteristic was assessed by assigning a value ranging from 1 (low presence of the characteristic) to 5 (maximum presence of the characteristic).

Figure 7. Identification of "equipped green areas" using municipal maps, satellite images and Google Street View. In the example, only the area "A" is classified as an "equipped green area".

The compatibility score was directly proportional to naturalness and urban context while being inversely proportional to the anthropic presence and productive value. Every single score was then calculated as follows:

- Naturalness score [Ns] = [N];
- Urban context score [Us] = [U];
- Anthropic presence score [As] = 5 − [A] + 1;
- Productive value score [Ps] = 5 − [P] + 1.

The Total Compatibility score [TCs] for each green area (see Table 6) was finally calculated as follows:

$$[TCs] = [Ns] + [Us] + [As] + [Ps].$$

Green areas were finally reclassified, according to the TCs value, into three compatibility classes: High compatibility (TCs = 16–20), medium compatibility (TCs = 10–15) and low compatibility (TCs = 4–9). The Green area compatibility map to realize the GI was then produced, as shown in Figure 8.

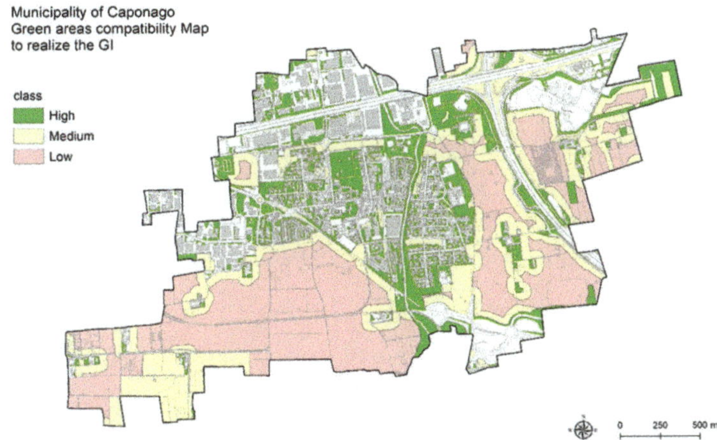

Figure 8. Green area compatibility map for realizing the Green Infrastructure for the municipality of Caponago.

Table 6. Compatibility scores for green area classes. H: high; M: medium; L: low.

Type of Green Area	Value of the Characteristic				Compatibility Score				Total Score	Compatibility Class
	Naturalness [N]	Ant. Pres. [A]	Prod. Value [P]	Urban Context [U]	Naturalness = [N]	Ant. Pres. [=(5 − A) + 1]	Prod. Value [=(5 − P) + 1]	Urban Context [=U]		
Equipped green area	1	5	1	5	1	1	5	5	12	M
Garden	4	1	1	1	4	5	5	1	15	M
Garden in urban context (uc)	4	1	1	5	4	5	5	5	19	H
Trees	4	1	1	1	4	5	5	1	15	M
Trees uc	4	1	1	5	4	5	5	5	19	H
Other green area	4	1	1	1	4	5	5	1	15	M
Other green area uc	4	1	1	5	4	5	5	5	19	H
Flowerbed	1	2	1	1	1	4	5	1	11	M
Flowerbed uc	1	2	1	5	1	4	5	5	15	M
Traffic island/roundabout	1	2	1	1	1	4	5	1	11	M
Traffic island/roundabout uc	1	2	1	5	1	4	5	5	15	M
Parking strip	1	5	1	1	1	1	5	1	8	L
Parking strip uc	1	5	1	5	1	1	5	5	12	M
Floricultural and plant nursery	1	5	5	1	1	1	1	1	4	L
Floricultural and plant nursery uc	1	5	5	5	1	1	1	5	8	L
Horticultural	1	5	5	1	1	1	1	1	4	L
Horticultural uc	1	5	5	5	1	1	1	5	8	L
Horticultural (greenhouse)	1	5	5	1	1	1	1	1	4	L
Horticultural (greenhouse) uc	1	5	5	5	1	1	1	5	8	L
Vegetable garden	2	3	3	1	2	3	3	1	9	L
Vegetable garden uc	2	3	3	5	2	3	3	5	13	M
Crop	2	3	4	1	2	3	2	1	8	L
Crop uc	2	3	4	5	2	3	2	5	12	M
Crop with trees	3	2	3	1	3	4	3	1	11	M
Crop with trees uc	3	2	3	5	3	4	3	5	15	M
Orchard	2	3	5	1	2	3	1	1	7	L
Orchard uc	2	3	5	5	2	3	1	5	11	M
Vineyard	2	3	5	1	2	3	1	1	7	L
Vineyard uc	2	3	5	5	2	3	1	5	11	M
Grassland	3	2	3	1	3	4	3	1	11	M
Grassland uc	3	2	3	5	3	4	3	5	15	M
Grassland with trees	4	2	2	1	4	4	4	1	13	M
Grassland with trees uc	4	2	2	5	4	4	4	5	17	H
Deciduous wood	4	2	2	1	4	4	4	1	13	M
Deciduous wood uc	4	2	2	5	4	4	4	5	17	H
Reforestation	4	2	2	1	4	4	4	1	13	M
Reforestation uc	4	2	2	5	4	4	4	5	17	H
Riparian wood	5	1	1	1	5	5	5	1	16	H
Riparian wood uc	5	1	1	5	5	5	5	5	20	H
Bushes	5	1	1	1	5	5	5	1	16	H
Bushes uc	5	1	1	5	5	5	5	5	20	H
Bushes in abandoned agricultural lands	5	1	1	1	5	5	5	1	16	H
Bushes in abandoned agricultural land uc	5	1	1	5	5	5	5	5	20	H

3.2. Weighting Procedure for Potential Infiltration and Exclusion Areas

The Green Infrastructure Suitability Map was produced by overlaying (with GIS) the Green area compatibility map to realize the GI, the map of natural surface infiltration potential and the map of exclusion areas.

The Green area compatibility map to realize the GI was, in fact, reduced by the potential infiltration of soil and the presence of areas where it is not possible to infiltrate.

For this reason, a weight (ranging from 0 to 1) was assigned to each infiltration potential class and to each exclusion area (Table 7) in order to take into account the reduction of the compatibility, which could remain unchanged (weight = 1) or decrease to a minimum (weight = 0).

Table 7. Weights assigned to potential infiltration and exclusion areas.

Potential Infiltration [PI]	Saturated Hydraulic Conductivity (m/s)	Weight [WPI]
High	$10^{-2}\_10^{-3}$	1.0
Medium	$10^{-3}\_10^{-4}$	0.7
Low	10^{-4}–10^{-5}	0.5
Very low	$<10^{-5}$	0.1
Exclusion Areas [EA]		**Weight [WEA]**
Absolute protection area of wells, buffer zone of wells, flooded area, contaminated area, quarry, frequently flooded areas (return period of 20–50 years)		0.0
Not frequently flooded areas (return period of 100–200 years), rarely flooded areas (return period of $p > 200$ years), groundwater vulnerability to pollution		1

3.3. Green Infrastructure Suitability Map

Once the three maps had been overlaid, the Green Infrastructure Suitability Map (Figure 9) was produced, classifying the total Green Infrastructure Suitability score into six classes [GI-Suit] (Table 8). [GI-Suit] was calculated as the product of the Total Compatibility score [TCs] and the weights [WPI] and [WEA], according to the following formula:

$$[GI - Suit] = [TCs] \cdot [WPI] \cdot [WEA].$$

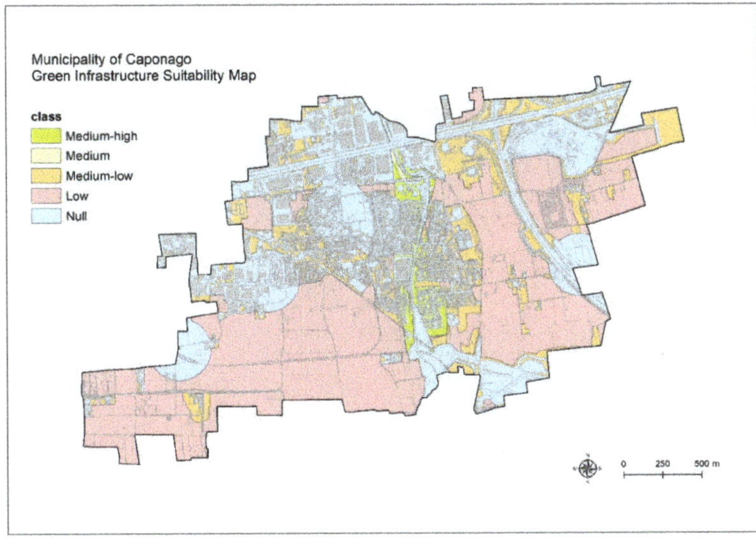

Figure 9. Green Infrastructure Suitability Map for the municipality of Caponago.

Table 8. GI-Suit Scores and Green infrastructure suitability classes.

Score of [GI-Suit]	Classes of Green Infrastructure Suitability
0	Null
1–7	Low
7–10	Medium–Low
10–13	Medium
13–15	Medium–High
16–20	High

4. Discussion and Conclusions

4.1. Discussion

The methodology applied to the study area allowed us to quantify the availability of green areas that are useful for the creation of GI at the municipal level.

The non-sealed areas in the municipality of Caponago cover 337.56 ha, equal to 67.28% of the municipal area (501.73 ha). The remaining areas are water (0.58%) and impervious surfaces (32.15%).

Most of these areas are agricultural, wooded and semi-natural areas (50.74% of the municipal area and 74.42% of the green areas), while urban green areas represent 15.13% of the municipal area and 22.48% of green areas. Significantly lower percentages concerned green areas connected to the transport network (Table 9).

Table 9. Municipal green areas: Values in ha, % of municipal area [ma] and % of total green areas [tga].

Green Area Class	Area (ha)	% of ma	% of tga
Agricultural, wooded and semi-natural areas	254.60	50.74	75.42
Green areas connected to the transport network	7.07	1.41	2.09
Urban green areas	75.89	15.13	22.48
Total	337.56	67.28	100.00

Despite the small size of the municipality of Caponago, the characteristics of the study area are those typical of an area with intermediate land-use intensity. Valtanen et al. [9] described three study areas in the city of Lahti (Finland) according to their land-use intensity and type: from high land-use intensity (80% of impervious area) to low land-use intensity (14% of impervious area). Yao et al. [5] and Du et al. [8] reported situations relating to large Chinese cities with impervious surfaces ranging from 20% to 50%. Surma [43] reported three case studies in Poland with impervious surfaces ranging from 19.8% to 47.5%.

With regards to the areas compatible with the construction of GI, the green areas with high compatibility are close to urban areas and road networks (Figure 8) and represent 17.64% of the municipal territory and 26.22% of the total green areas. This means that about three-quarters of the permeable areas of the municipality have from medium to null compatibility with the construction of GI (Table 10).

Among the green areas with high compatibility, the prevailing class was made up of Urban green areas (14.33% of the municipal area, 21.30% of total green areas). Among the areas with medium compatibility, the main class was represented by agricultural, wooded and semi-natural areas, with 18.19% of the municipal territory and 27.03% of the total green areas.

About green areas suitable for the construction of the GI, there are no areas with high suitability. This is due to the fact that the characteristics of the soils are such that the natural surface infiltration potential is medium or low (Figure 5) and the relative maximum weight is, therefore, 0.7 (Table 7). Charlesworth et al. [28] and Muthanna et al. [36] also reported very small percentages (2.5% and 3.2%) of areas suitable for infiltration SuDS in the city of Coventry (UK) and Trondheim (Norway) respectively.

The areas with medium-high suitability (10.14 ha) represent 2.02% of the municipal area and 3% of the total green areas (Table 11). 6.05 ha are urban green areas and 4.09 are agricultural, wooded and semi-natural areas. No green areas connected to the transport network are included in the medium-high suitability class.

Table 10. Municipal areas compatible with creating GI: Values in ha, % of municipal area [ma] and % of total green areas [tga].

Class of Green Area	High Compatibility			Medium Compatibility			Low Compatibility		
	ha	% of ma	% of tga	ha	% of ma	% of tga	ha	% of ma	% of tga
Agric., wooded and semi-natural	16.62	3.31	4.92	91.24	18.19	27.03	146.74	29.25	43.47
Green areas connec. transport	0.00	0.00	0.00	7.07	1.41	2.09	0.00	0.00	0.00
Urban green areas	71.9	14.33	21.30	3.99	0.80	1.18	0.00	0.00	0.00
Total	88.52	17.64	26.22	102.3	20.39	30.31	146.74	29.25	43.47

Table 11. Suitability of municipal areas for creating GI: Values in ha, % of municipal area [ma], and % of total green areas [tga].

Class of Green Area	Medium-High Suitability			Medium Suitability			Medium-Low Suitability			Low Suitability			Null Suitability		
	ha	% of ma	% of tga	ha	% of ma	% of tga	ha	% of ma	% of tga	ha	% of ma	% of tga	ha	% of ma	% of tga
Agric., wooded, and semi-natural	4.09	0.82	1.21	0.7	0.14	0.21	12.39	2.47	3.67	214.9	42.83	63.65	22.55	4.49	6.68
Green areas connec. transp.	0.00	0.00	0.00	0.00	0.00	0.00	6.02	1.20	1.78	0.00	0.00	0.00	1.04	0.21	0.31
Urban green areas	6.05	1.21	1.79	0.17	0.03	0.05	51.86	10.34	15.36	0.52	0.10	0.15	17.58	3.50	5.21
Total	10.14	2.02	3	0.87	0.17	0.26	70	14.01	20.82	215.4	42.93	63.81	44.06	8.78	12.20

The agricultural, wooded and semi-natural areas are very close to the urban area (Figure 10) so they are indeed interesting for the sustainable management of rainwater. Christman et al. [27] also reported the presence of a percentage (between 5% and 22%) of non-urban areas among the high-priority GSI (Green Stormwater Infrastructure) implementation sites in the city of Philadelphia (USA).

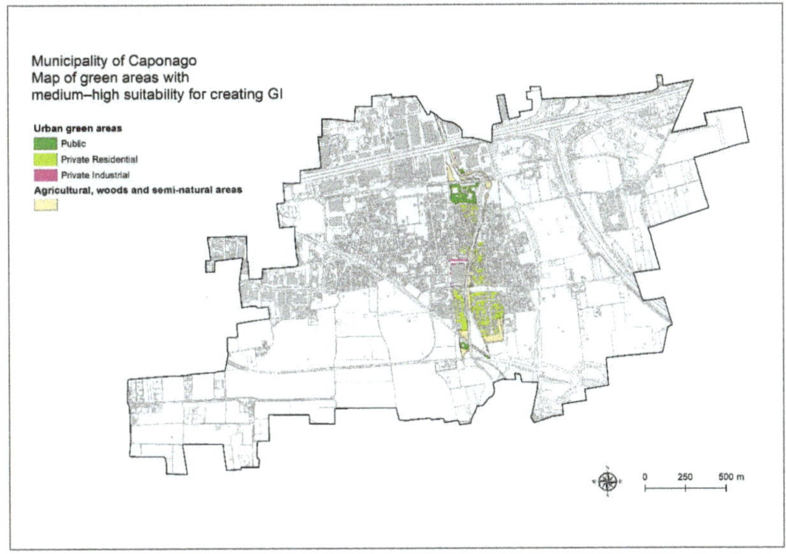

Figure 10. Typologies of green areas with medium-high suitability for creating GI in the municipality of Caponago.

Only 27% of urban green areas with medium-high suitability for the construction of the GI are public areas. The remaining 73% is private, 68% residential and 5% industrial (Figure 10). Dhakal and Chevalier [24] reported a similar average percentage (65–75%) of private land in five American cities (Portland, Seattle, Philadelphia, Chicago, and Syracuse), noting that incentives and other programs offered to private landowners have produced encouraging results.

The areas with medium-low suitability constitute 14.01% of the municipality and 20.82% of the green areas. The areas with low suitability represent 42.93% of the municipality and 63.81% of the green areas.

4.2. Conclusions

The proposed Green Infrastructure Suitability Map is a tool that municipal authorities can use as:

- An informative basis in the land-use planning process in order to set up or update the municipal plan (PGT) with reference to rainwater management, in accordance with the regulations of Lombardy Region;
- A necessary knowledge basis for the definition of municipal stormwater management plans, particularly related to the choice of the most appropriate NBS for each location.

The localization of the most appropriate intervention must be made on the basis of the assessed territorial characteristics, type of prevalent function required (e.g., detention, retention, flow control, infiltration, filtration, or evapotranspiration), context (urban or rural–natural), expected use (accessible to people or not), and maintenance needs.

Our work is still in progress. The Green Infrastructure Suitability Map is the first step towards the development of a more complete process of identifying the type of NBS which is best suited to address various specific local problems.

Author Contributions: Conceptualization, G.S. (Giulio Senes), P.S.F. and N.F.; methodology, G.S. (Giulio Senes), P.S.F., G.C. and N.F.; formal analysis, G.S. (Giulio Senes), P.S.F., G.C.; investigation, G.S., P.S.F. and G.C.; resources, P.S.F., G.C., P.F. and G.S. (Giulio Senes); data curation, P.S.F. and G.C.; writing—original draft preparation, G.S. (Giulio Senes) and P.S.F.; writing—review and editing, G.S. (Giulio Senes), P.S.F., N.F., P.F., G.S. (Giovanna Sacchi) and G.V.; visualization, P.S.F. and G.C.; supervision, G.S. (Giulio Senes) and N.F.; funding acquisition, G.S. (Giulio Senes) and G.S. (Giovanna Sacchi). All authors have read and agreed to the published version of the manuscript.

Funding: This research was funded by BriazAcque s.r.l., Monza (MB), Italy.

Institutional Review Board Statement: Not applicable.

Informed Consent Statement: Not applicable.

Data Availability Statement: Publicly dataset were analyzed in this study. This data can be found here: [www.geoportale.regione.lombardia.it]. The new data produced in this study are not publicly available, they are available on request from the corresponding author.

Acknowledgments: Authors are grateful to Marco Togni, Giacomo Redondi, Iuri Dino Tagliaferri.

Conflicts of Interest: The authors declare no conflict of interest.

Declaration: This work has been presented on Greening cities shaping cities symposium in October 2020.

References

1. Istituto Superiore per la Protezione e la Ricerca Ambientale. *Qualità Dell'ambiente Urbano. IX Rapporto. Focus su Acque e Ambiente Urbano*; ISPRA: Roma, Italy, 2013; ISBN 978-88-448-0622-4.
2. Davis, M.; Naumann, S. Making the Case for Sustainable Urban Drainage Systems as a Nature-Based Solution to Urban Flooding. In *Nature-Based Solutions to Climate Change Adaptation in Urban Areas. Theory and Practice of Urban Sustainability Transitions*; Kabisch, N., Korn, H., Stadler, J., Bonn, A., Eds.; Springer: Cham, Switzerland, 2017; pp. 123–137. [CrossRef]
3. Berndtsson, J.C. Green roof performance towards management of runoff water quantity and quality: A review. *Ecol. Eng.* **2010**, *36*, 351–360. [CrossRef]

4. Guan, M.F.; Sillanpaa, N.; Koivusalo, H. Modelling and assessment of hydrological changes in a developing urban catchment. *Hydrol. Process.* **2015**, *29*, 2880–2894. [CrossRef]

5. Yao, L.; Chen, L.D.; Wei, W. Assessing the effectiveness of imperviousness on stormwater runoff in micro urban catchments by model simulation. *Hydrol. Process.* **2016**, *30*, 1836–1848. [CrossRef]

6. Fairbrass, A.; Jones, K.; McIntosh, A.; Yao, Z.; Malki-Epshtein, L.; Bell, S. *Green Infrastructure for London: A Review of the Evidence. A Report by the Engineering Exchange for Just Space and the London Sustainability Exchange*; Natural Environmental Research Council: London, UK, 2018.

7. Chen, J.Q.; Theller, L.; Gitau, M.W.; Engel, B.A.; Harbor, J.M. Urbanization impacts on surface runoff of States the contiguous United States. *J. Environ. Manag.* **2017**, *187*, 470–481. [CrossRef]

8. Du, J.; Qian, L.; Rui, H.; Zuo, T.; Zheng, D.; Xu, Y.; Xu, C.Y. Assessing the effects of urbanization on annual runoff and flood events using an integrated hydrological modeling system for Qinhuai River basin, China. *J. Hydrol.* **2012**, *464*, 127–139. [CrossRef]

9. Valtanen, M.; Sillanpaa, N.; Setala, H. Effects of land use intensity on stormwater runoff and its temporal occurrence in cold climates. *Hydrol. Process.* **2014**, *28*, 2639–2650. [CrossRef]

10. Yang, G.X.; Bowling, L.C.; Cherkauer, K.A.; Pijanowski, B.C. The impact of urban development on hydrologic regime from catchment to basin scales. *Landsc. Urban Plan.* **2011**, *103*, 237–247. [CrossRef]

11. Tao, W.D.; Bays, J.S.; Meyer, D.; Smardon, R.C.; Levy, Z.F. Constructed wetlands for treatment of combined sewer overflow in the US: A review of design challenges and application status. *Water* **2014**, *6*, 3362–3385. [CrossRef]

12. Chocat, B.; Ashley, R.; Marsalek, J.; Matos, M.R.; Rauch, W.; Schilling, W.; Urbonas, B. Toward the sustainable management of urban storm-water. *Indoor Built Environ.* **2007**, *16*, 273–285. [CrossRef]

13. European Union. *Nature-Based Solutions for Climate Mitigation. Analysis of EU-Funded Projects*; European commission: Bruxelles, Belgium, 2020; ISBN 978-92-76-18200-9. [CrossRef]

14. Fletcher, T.D.; Shuster, W.; Hunt, W.F.; Ashley, R.; Butler, D.; Arthur, S.; Trowsdale, S.; Barraud, S.; Semadeni-Davies, A.; Bertrand-Krajewski, J.; et al. SUDS, LID, BMPs, WSUD and more—The evolution and application of terminology surrounding urban drainage. *Urban Water J.* **2015**, *12*, 525–542. [CrossRef]

15. Zhou, Q. A Review of Sustainable Urban Drainage Systems Considering the Climate Change and Urbanization Impacts. *Water* **2014**, *6*, 976–992. [CrossRef]

16. Keeler, B.L.; Hamel, P.; McPhearson, T.; Hamann, M.H.; Donahue, M.L.; Meza Prado, K.A.; Arkema, K.K.; Bratman, G.N.; Brauman, K.A.; Finlay, J.C.; et al. Social-ecological and technological factors moderate the value of urban nature. *Nat. Sustain.* **2019**, *2*, 29–38. [CrossRef]

17. European Union. *Nature-Based Solutions: State of the Art in EU-Funded Projects*; European commission: Bruxelles, Belgium, 2020; ISBN 978-92-76-17334-2. [CrossRef]

18. Senes, G.; Fumagalli, N.; Ferrario, P.S.; Rovelli, R.; Sigon, R. Definition of a land quality index to preserve the best territories from future land take. An application to a study area in Lombardy (Italy). *J. Agric. Eng.* **2020**, *51*, 43–55. [CrossRef]

19. US Environmental Protection Agency. *Low Impact Development (LID). A Literature Review*; EPA Office of Water (4203): Washington, DC, USA, 2000.

20. Ashley, R.; Illman, S.; Kellagher, R.; Scott, T.; Udale-Clarke, H.; Wilson, S.; Woods Ballard, B. *The SuDS Manual*; CIRIA: London, UK, 2015.

21. Walmsley, A. Greenways and the making of urban form. *Landsc. Urban Plan.* **1995**, *33*, 81–127. [CrossRef]

22. European Union. *Building a Green Infrastructure for Europe*; European Commission: Bruxelles, Belgium, 2013; ISBN 978-92-79-33428-3. [CrossRef]

23. Lähde, E.; Khadka, A.; Tahvonen, O.; Kokkonen, T. Can We Really Have It All?—Designing Multifunctionality with Sustainable Urban Drainage System Elements. *Sustainability* **2019**, *11*, 1854. [CrossRef]

24. Dhakal, K.P.; Chevalier, L.R. Urban stormwater governance: The need for a paradigm shift. *Environ. Manag.* **2016**, *57*, 1112–1124. [CrossRef]

25. Vogel, J.R.; Moore, T.L.; Coffman, R.R.; Rodie, S.N.; Hutchinson, S.L.; McDonough, K.R.; McLemore, A.J.; McMaine, J.T. Critical Review of Technical Questions Facing Low Impact Development and Green Infrastructure: A Perspective from the Great Plains. *Water Environ. Res.* **2015**, *87*, 849–862. [CrossRef]

26. Pappalardo, V.; Campisano, A.; Martinico, F.; Modica, C.; Barbarossa, L. A hydraulic invariance-based methodology for the implementation of storm-water release restrictions in urban land use master plans. *Hydrol. Process.* **2017**, *31*, 4046–4055. [CrossRef]

27. Christman, Z.; Meenar, M.; Mandarano, L.; Hearing, K. Prioritizing Suitable Locations for Green Stormwater Infrastructure Based on Social Factors in Philadelphia. *Land* **2018**, *7*, 145. [CrossRef]

28. Charlesworth, S.; Warwick, F.; Lashford, C. Decision-Making and Sustainable Drainage: Design and Scale. *Sustainability* **2016**, *8*, 782. [CrossRef]

29. Charlesworth, S.M.; Warwick, F. Sustainable drainage, green and blue infrastructure in urban areas. In *Sustainable Water Engineering*; Charlesworth, S.M., Booth, C., Adeyeye, K., Eds.; Elsevier: Amsterdam, The Netherlands, 2020; pp. 185–206. ISBN 9780128161203. [CrossRef]

30. Wang, X.; Shuster, W.; Pal, C.; Buchberger, S.; Bonta, J.; Avadhanula, K. Low Impact Development Design—Integrating Suitability Analysis and Site Planning for Reduction of Post-Development Stormwater Quantity. *Sustainability* **2010**, *2*, 2467–2482. [CrossRef]

31. Li, L.; Uyttenhove, P.; Vaneetvelde, V. Planning green infrastructure to mitigate urban surface water flooding risk—A methodology to identify priority areas applied in the city of Ghent. *Landsc. Urban Plan.* **2020**, *194*, 103703. [CrossRef]
32. Yau, W.K.; Radhakrishnan, M.; Liong, S.Y.; Zevenbergen, C.; Pathirana, A. Effectiveness of ABC waters design features for runoff quantity control in Urban Singapore. *Water* **2017**, *9*, 577. [CrossRef]
33. Dagenais, D.; Thomas, I.; Paquette, S. Siting green stormwater infrastructure in a neighbourhood to maximise secondary benefits: Lessons learned from a pilot project. *Landsc. Res.* **2017**, *42*, 195–210. [CrossRef]
34. Pappalardo, V.; La Rosa, D.; Campisano, A.; La Greca, P. The potential of green infrastructure application in urban runoff control for land use planning: A preliminary evaluation from a southern Italy case study. *Ecosyst. Serv.* **2017**, *26*, 345–354. [CrossRef]
35. Sun, Y.; Tong, S.; Yang, Y.J. Modeling the cost-effectiveness of stormwater best management practices in an urban watershed in Las Vegas Valley. *Appl. Geogr.* **2016**, *76*, 49–61. [CrossRef]
36. Muthanna, T.M.; Sivertsen, E.; Kliewer, D.; Jotta, L. Coupling field observations and Geographical Information System (GIS)-based analysis for improved Sustainable Urban Drainage Systems (SUDS) performance. *Sustainability* **2018**, *10*, 4683. [CrossRef]
37. Gallagher, K.V.; Alsharif, K.; Tsegaye, S.; Van Beynen, P. A new approach for using GIS to link infiltration BMPs to Groundwater Pollution Risk. *Urban Water J.* **2018**, *15*, 847–857. [CrossRef]
38. Kuller, M.; Bach, P.M.; Ramirez-Lovering, D.; Deletic, A. What drives the location choice for water sensitive infrastructure in Melbourne, Australia? *Landsc. Urban Plan.* **2018**, *175*, 92–101. [CrossRef]
39. Martin-Mikle, C.J.; de Beurs, K.M.; Julian, J.P.; Mayer, P.M. Identifying priority sites for low impact development (LID) in a mixed-use watershed. *Landsc. Urban Plan.* **2015**, *140*, 29–41. [CrossRef]
40. Gregory, J.H.; Dukes, M.D.; Jones, P.H.; Miller, G.L. Effect of urban soil compaction on infiltration rate. *J. Soil Water Conserv.* **2006**, *61*, 117–124.
41. Tredway, J.C.; Havlick, D.G. Assessing the Potential of Low-Impact Development Techniques on Runoff and Streamflow in the Templeton Gap Watershed, Colorado. *Profess. Geogr.* **2017**, *69*, 372–382. [CrossRef]
42. Tsegaye, S.; Singleton, T.L.; Koeser, A.K.; Lamb, D.S.; Landry, S.M.; Lu, S.; Barber, J.B.; Hilbert, D.R.; Hamilton, K.O.; Northrop, R.J.; et al. Transitioning from gray to green (G2G)—A green infrastructure planning tool for the urban forest. *Urban For. Urban Green.* **2018**, *40*, 204–214. [CrossRef]
43. Surma, M. Sustainable urban development through an application of green infrastructure in district scale—A case study of Wrocław (Poland). *J. Water Land Dev.* **2015**, *25*, 3–12. [CrossRef]
44. Foomani, M.S.; Malekmohammadi, B. Site selection of sustainable urban drainage systems using fuzzy logic and multi-criteria decision-making. *Water Environ. J.* **2019**, *34*, 584–599. [CrossRef]

sustainability

Article

Is Agent-Based Simulation a Valid Tool for Studying the Impact of Nature-Based Solutions on Local Economy? A Case Study of Four European Cities

Rembrandt Koppelaar [1], Antonino Marvuglia [2,*], Lisanne Havinga [3], Jelena Brajković [4] and Benedetto Rugani [2]

[1] EcoWise Ekodenge Ltd., London SE1 8ND, UK; rembrandt.koppelaar@eco-wise.co.uk
[2] Luxembourg Institute of Science and Technology (LIST), 41, rue du Brill, L-4422 Belvaux, Luxembourg; benedetto.rugani@list.lu
[3] Architectural History and Theory Unit, Built Environment Group, Eindhoven University of Technology, Postbus 513, 5600 MB Eindhoven, The Netherlands; L.C.Havinga@tue.nl
[4] Faculty of Architecture, University of Belgrade, Kralja Aleksandra Blvd. 73/II, 11120 Belgrade, Serbia; jelena.brajkovic@arh.bg.ac.rs
* Correspondence: antonino.marvuglia@list.lu; Tel.: +352-275-888-5036

Citation: Koppelaar, R.; Marvuglia, A.; Havinga, L.; Brajković, J.; Rugani, B. Is Agent-Based Simulation a Valid Tool for Studying the Impact of Nature-Based Solutions on Local Economy? A Case Study of Four European Cities. *Sustainability* **2021**, *13*, 7466. https://doi.org/10.3390/su13137466

Academic Editors: Israa H. Mahmoud, Eugenio Morello, Giuseppe Salvia and Emma Puerari

Received: 22 May 2021
Accepted: 29 June 2021
Published: 4 July 2021

Publisher's Note: MDPI stays neutral with regard to jurisdictional claims in published maps and institutional affiliations.

Abstract: Implementing nature-based solutions (NBSs) in cities, such as urban forests, can have multiple effects on the quality of life of inhabitants, acting on the mitigation of climate change, and in some cases also enhancing citizens' social life and the transformation of customer patterns in commercial activities. Assessing this latter effect is the aim of this paper. An agent-based model (ABM) was used to assess change in commercial activities by small and midsize companies in retail due to the development of parks. The paper focuses on the potential capacity of NBS green spaces to boost retail companies' business volumes, thus increasing their revenues, and at the same time create a pleasant feeling of space usability for the population. The type of NBS is not specified but generalized into large green spaces. The simulation contains two types of agents: (1) residents and (2) shop owners. Factors that attract new retail shops to be established in an area are simplified, based on attractor points, which identify areas such as large green spaces within and around which shops can form. The simulations provided insights on the number of retail shops that can be sustained based on the purchasing behavior of citizens that walk in parks. Four European cities were explored: Szeged (Hungary), Alcalá de Henares (Spain), Çankaya Municipality (Turkey) and Milan (Italy). The model allowed analyzing the indirect economic benefit of NBSs (i.e., large green spaces in this case) on a neighborhood's economic structure. More precisely, the presence of green parks in the model boosted the visits of customers to local small shops located within and around them, such as cafés and kiosks, allowing for the emergence of 5–6 retail shops (on average, for about 800 walking citizens) in the case of Szeged and an average 12–14 retail shops for a simulated population of 2900 persons that walk in parks in the case of Milan. Overall, results from this modeling exercise can be considered representative for large urban green areas usually visited by a substantial number of citizens. However, their pertinence to support for local policies for NBS implementation and other decision-making related activities of socioeconomic nature is hampered by the low representativeness of source data used for the simulations.

Keywords: urban green areas; nature-based solutions (NBSs); agent-based model (ABM); firmographics; market segmentation

1. Introduction

A recent analysis by the United Nations [1] highlighted that the current share of the world population living in cities and urban areas (around 55%) is expected to increase steadily to 68% by 2050. At the same time, in Europe, the number of people living in urban areas will considerably grow from approximately 73% today to over 80% by 2050.

New techniques and approaches are thus required to design sustainable cities for future dense populations in narrower areas [2]. Population growth has pushed cities to adopt new revitalization schemes, forcing market dynamics to reshape the landscape of retail businesses, with retailers attempting to adopt innovative strategies in order to keep up with the new patterns arising in a changing society [3]. In this framework, the concept of biophilia has emerged. This concept is based on the acknowledgement that connection with nature (not only on a physical level, but also mentally and at a social level) is an innate biological need of humans, and that this connection affects our personal wellbeing, productivity and societal relationships. In this sense, several studies show the stress relief properties derived from the ability to directly access nature and the many positive influences of the natural environment on various human psychological states [4,5]. In addition, the connection of natural stimuli coming from nature with significant and positive effects on emotional responses in retail-store settings has also been demonstrated [6].

Urban green and blue spaces, such as urban parks, forests, gardens, water streams, green roofs and green facades, not only result in multiple co-benefits for health, the economy, society and the environment but also provide habitats for several species [7]. These are all nature-based solutions (NBSs), which are defined by the European Commission (EC) as "solutions inspired and supported by nature, which are cost-effective, simultaneously provide environmental, social and economic benefits and help build resilience. Such solutions bring more and more diverse natural features and processes into cities, landscapes and seascapes, through locally adapted, resource-efficient and systemic interventions" (https://ec.europa.eu/info/research-and-innovation/research-area/environment/nature-based-solutions_en (accessed on 15 June 2021)) The EC has developed a framework to assist member states in mapping and assessing the "urban ecosystem", which consists of both built and green infrastructure and delivers a wide range of social, economic and ecological services [8].

The concept of NBSs has been extensively studied, with a broad range of literature reviews synthesizing existing knowledge on NBSs in relation to urban contexts being published in recent years [9,10]. According to a recent literature review on stakeholders' engagement in NBSs [11], most studies have focused on the social and environmental benefits of NBSs. The social benefits most often investigated are related to access or proximity to nature, physical and mental wellbeing and exercise. The most studied environmental aspects are related to air quality, climate regulation, biodiversity and wildlife. Studies focusing on economic benefits remain lacking, though some exist that include wood provision [12,13], increase in property value [14–16] and food provision [17–19]. This identified bias toward social and environmental benefits is also highlighted by Parker and co-authors [20]. In addition, a recent literature review on key enablers and barriers to the uptake and implementation of NBSs in urban settings [21] identified five publications [22–26] that included economic opportunities and jobs as part of the objectives evaluated in mostly integrated evaluations of case study applications.

More specifically, in Chen et al. [22], the socioeconomic benefits of farm ponds in Southern China are roughly estimated. These benefits stem from the hydrological regulation and flood alleviation services that ponds provide, contributing to social stability, and the improved nutrient cycling beneficial for agricultural production and the farm economy. Especially in the case of the city of Wuhan, the authors reported a generated profit from fish production between USD 1.96 and 2.55 per square meter and an estimated benefit for the entire city, coming from the industry of pond aquaculture, of approximately USD 145 million. In Liquete et al. [23], a multi-criteria analysis (MCA) is used to assess the environmental, social and economic benefits provided by the Gorla Maggiore water park in Northern Italy. Although an estimation of the costs associated with the alternatives assessed in the MCA is provided, conclusions from this study do not prove that economic benefits actually arise from the transition from an alternative grey infrastructure and from the previous situation (a poplar plantation) to the proposed green infrastructure. Relying on figures reported in Terrapin [27], Santiago Fink [24] provides instead some estimates of the potential economic benefits arising from access (even just visual) to green spaces and simply daylight. The benefits, which are only based on expert-based judgements, range from reduced stress and increased productivity of employees in businesses and offices to faster recovery after surgery in hospitals, improved test scores of students in schools and consequent reduction in school dropout rates, reduction in healthcare insurance premium expenses due to better physiological and mental state induction and finally reduction in criminal acts (violence and aggression). While the flood and coastal erosion risk management (FCRM) Partnership Calculator (https://www.gov.uk/guidance/partnership-funding (accessed on 15 June 2021)) developed by Short et al. [25] is used to estimate the economic benefits of protecting some river catchment properties in the United Kingdom from single rainfall events using natural flood management interventions, the approach used takes an economic-based perspective typical of the insurance sector. Finally, van der Jagt and co-authors [26] make the point that gardeners of large plots need to organize themselves in a company, meet food handling regulation and pay tax. This helps to create proper green jobs and revenue. However, no quantification of this job creation potential is given, even for the few European cities that are reported as examples of communal urban gardening (CUG) in the paper.

Despite this diversity of successful achievements and findings, none of these examples presents a computational tool to simulate and quantify the economic advantages associated with the future implementation of NBS scenarios. In other words, to the best of the authors' knowledge, the recent and growing NBS literature still does not offer examples of computational tools that allow simulating NBS implementation models with the possibility of tweaking parameters of interest. A gap regarding such aspects therefore exists in NBS research, which this paper has the ambition to, at least partially, address. Accordingly, the paper aims at creating quantitative simulations of the effects of developing large NBSs in cities, such as parks, on the change in commercial activities based on small and midsize companies in retail. The research hypothesis is that an NBS (or collection of NBSs), such as an urban park, can attract people in areas previously not frequented by outsiders. Such an increase (and the possible diversification) of daytime and sometimes nighttime human activities can potentially favor the growth of pre-existent commercial activities and attract new hospitality and retail activities in a certain urban area. An agent-based modeling approach, as illustrated in the section, is used to accomplish the objective of this study. The developed agent-based model (ABM) furthers NBS assessments from two perspectives: firstly, to explore the dynamics between NBSs and socioeconomic development as a deductive process. Beyond developing an understanding of what factors are relevant or not in this context, it also informs regarding what data to collect to establish inductive evidence on the relationship between the existence of NBSs in an area and socioeconomic indicators, using statistical analyses to validate relationships. Secondly, the model is used as a quantitative tool for urban planners to assess socioeconomic benefits once the relationships in the model have been validated by inference of relationships using real-world observations. This paper concerns the first exploratory aspect.

2. State of the Art

The economic attributes of NBSs have mostly been studied in relation to the economic effects of climate change mitigation and adaptation, e.g., in the recent report by the United Nations Environment Programme [28]. However, the potential economic co-benefits of NBSs can also go beyond these core objectives of NBSs, thereby providing even more economic reasons to implement them. Providing insight to practitioners and decision makers about these additional economic benefits is critical. As stressed in Babí Almenar et al. [10], in order to facilitate the operationalization of the NBS concept, the added value of NBSs and the potential co-benefits that they provide need to be easily understood by practitioners and decision makers. Those authors argue that practitioners and decision makers need more studies exploring the relationship between specific urban NBS and specific benefits.

Several studies [29–36] have underlined financial constraints and a lack of funding as part of the challenges identified by citizens and stakeholders in the implementation of NBSs. Additional studies [37–39] indicate that a lack of financial incentives and strong business models is an important barrier to the implementation of NBSs. As such, current literature reviews highlight that there is a lack of studies focusing on the economic benefits of NBSs [40,41]. Therefore, this paper contributes to the discourse on quantitatively evaluating and modeling the potential of NBSs from an economic perspective that goes beyond the economic benefits directly related to climate change mitigation or adaptation.

One of the benefits of NBSs, in particular urban green parks, is that new retail and hospitality companies/firms tend to move into such areas. With the foreseen increase in the number of people in urban areas, one of the possible boundary conditions for the willingness of companies to establish a presence in these areas is change in popularity. The rates of vacant spaces [42] and firm population can also be considered boundary conditions. There are several such factors mentioned in the literature. For instance, companies can move into an area in different ways: new companies may enter the market, or an existing chain may wish to open a new facility or relocate. Studies show that old firms prefer to stay at their home region or fixed locations [43], while new entrepreneurs often treat their hometown as a natural start-up location [44]. Van Dijk and Pellenbarg [45] used data from the Netherlands and showed that firm internal factors such as the economic sector, firm size and previous migration behavior are good predictors that can explain a firm's decision to relocate. Similar results can be found in other studies [43,46]. The relationship between a firm and other organizations in its environment has been shown to significantly enhance the explanatory power of firms' relocation models, and this effect varies depending on the strength and geographical distance of their relationships [47]. Bodenmann and Axhausen [48] showed that local taxes have a very positive effect on firm relocation; they also found that distance is an important indicator, and significant differences exist between sectors. Location selection models for firms in urban areas and details of the simulations can be found in De Bok and Bliemer [49] and Waddell et al. [42].

The design of ecological spaces and environments has a potential effect on visiting and shopping intentions [50]. This is in line with the increase in the likelihood of making a retail drink or food purchase for a person during a visit or recreational activity in large green spaces (either enroute in travel or at the NBS site). The effects of the ecological design of hotels on behavioral intentions and the resulting competitive advantage in terms of intention to revisit and willingness to pay more are discussed in Lee et al. [51]. Service and commercial environments offering natural settings have been studied and analyzed in terms of comfort, customer behavior and psychological responses in Purani and Kumar [52]. Some studies investigated nature-based applications as a variable of attraction [50], while other studies suggest that natural settings in retail areas support attention and bring cognitive benefits [53,54].

Biophilic design offers many possibilities through its attributes and elements [55], beyond the benefits of access to nature, for enhancing social life or commercial activities in urban environments. Within its element of place-based relationships [55], biophilic design recognizes attributes such as *geographical, historic, ecological* and *cultural* connection to a place, as well as *integration* of culture and ecology, *avoiding placelessness*, as well as creating a *spirit of place*. By recognizing these attributes, biophilic design emphasizes the importance of the issue of *identity of place* and unique site-specific values within urban locations. Integrating the abovementioned attributes in the planning of urban sites and NBS spaces can enrich the overall quality of space and create specific site values that offer unique, memorable and positive experiences to people. In terms of branding and marketing, a properly curated identity and spirit of a place (in line with biophilic guidelines) presents a unique site value which can be marketed and exploited, resulting in enhanced social and commercial activities and therefore boosting the place's attractiveness.

To examine the relationship between NBSs and socioeconomic development, this paper deploys a simulation approach. In particular, an ABM is used to assess change in commercial activities by small and midsize companies in retail due to the development of large NBSs, such as parks. This concept is based on the notion that an NBS (or collection of NBSs), such as an urban park, can attract people to areas previously not frequented by outsiders. The increase (and possible diversification) in daytime and sometimes nighttime activities potentially favors the growth of pre-existing commercial activities and attracts new hospitality and retail activities in an area.

The choice of simulation to study this relationship in a dynamic manner, as opposed, for example, to a static statistical evaluation of time–series datasets, was made.

An ABM (in its purest form) is made up of "objects", which from a computer science standpoint are "computational entities that encapsulate some state, are able to perform actions, or methods, on this state, and communicate by message passing" [56]. An ABM allows the representation of individual objects (agents) with their specific characteristics. Agent-based simulations are bottom-up: at the lowest level, agents interact and, as a result, the macro behavior of the system (not from superposition, but from the interaction of micro level behaviors) might emerge at a higher level [57]. This is why, in de Marchi and Page [58], ABMs are defined as consisting of "autonomous, interacting computational objects, called agents, often situated in space and time", and in Macy and Flache [59], it is stated that an ABM "replaces a single integrated model of the population with a population of models, each corresponding to an autonomous decision maker".

For this reason, ABMs make it possible to simulate agents belonging to different societal groups, which differ in terms of factors such as age, lifestyle, economic status, preferences and motivation. Moreover, in ABMs, spatial information and its interaction with agents can be readily incorporated based on a geographical information system (GIS) support map. In the context of the application described in this paper, this is meaningful for evaluating heterogeneity between NBS spaces, interactions between NBSs and local retailers and other structures associated with urban NBS networks. As described in Marvuglia et al. [60], an ABM comprises a set of agents which can belong to certain classes and are characterized by the asset of *attributes* defining their characteristics. These attributes can be static or vary in the course of the simulation as a result of the actions undertaken by the agents. The latter are regulated by a set of *decision rules* that steer agents' behavior. Agents operate within a certain *environment*, which not only is determined by the spatial context but, in a larger sense, also comprises *time* and *exogenous events* [60].

The property of emergence discussed above differentiates ABMs from other single-level simulation system models. In Laurenti et al. [61], it is clarified that system models can be either quantitative or qualitative: quantitative models are system dynamics (SD) models, while qualitative models are causal loop diagrams (CLDs). SD models are focused on dynamic behavior over time, have stocks and flows, represent feedback loop structures and require specific parameters, equations and computer simulations to run. A typical CLD, in contrast, consists of a set of symbols describing a dynamic system's causal structure: variables, causal links (between the variables) with a polarity and symbols that identify feedback loops with their polarity. Causal links have a direction and a polarity; they also sometimes have a delay mark. More details on CLDs and the concepts behind them can be found in Schaffernicht [62].

The SD society defines SD as "a computer-aided approach to policy analysis and design. It applies to dynamic problems arising in complex social, managerial, economic or ecological systems–literally any dynamic systems characterized by interdependence, mutual interaction, information feedback, and circular causality" (http://www.systemdynamics. org/what-is-s/ (accessed on 15 June 2021)).

Gilbert and Troitzsch [63] argued that SD simulations only have a single level: they model the individual, the firm, the organization or society; they cannot model interactions between scales or levels and so do not exhibit emergent behavior. Compared to SD, ABMs allow the modeling of more complex dynamics because system structure can also change during simulation. In SD, a fixed interaction structure is defined and maintained, which means that connections between different actors/elements have to be defined before starting the simulation. Using an ABM, one only defines an interaction space in the form of the types of interactions that agents are allowed to have [64]. Both SD and ABM approaches are capable of representing temporal aspects of dynamic systems, but ABM approaches are more appropriate for modeling spatially explicit complex systems because they can handle spatial heterogeneity in individual attributes and simulate their mobility [65]. ABM is also a better approach for modeling heterogeneity in individual attributes and in the network of interactions among population elements; however, this means that AB modeling requires the collection of more data at the level of individuals, which in turn leads to a slower modeling process, higher computational costs and more difficult calibration in ABM compared to the SD approach.

This paper investigates, via the agent concept, the number of retail shops that can be sustained through NBS green spaces and the potential revenues and profits that can be boosted by these spaces. The simulation utilizes a causal chain layering modeling approach [66] and draws from neoclassical economics theory about profit maximization. The type of NBS is not specified but generalized into large green spaces. In future versions of the model, this could be further subdivided into specific NBSs such as large urban parks, heritage gardens, green cemeteries, woods and so forth.

3. Materials and Methods

An ABM, implemented in the popular Netlogo (https://ccl.northwestern.edu/netlogo/ (accessed on 15 June 2021)) platform (version 6.0.4) [67,68], was applied in this paper. Figure 1 shows a screenshot of the user interface of the model. For an overview of ABM modeling platforms in the context of simulation modeling, please refer to Abar et al. [69]. The Netlogo platform was selected for its versatility and robustness in deploying a wide range of modeling rules for this use case. To this end, it is highly suitable for exploratory academic research such as the one carried out in this paper. Its limitations are its computational scalability and inability to be integrated in modern cloud computation, because of which it is not suitable for large-scale simulations or exploitation as a commercial service.

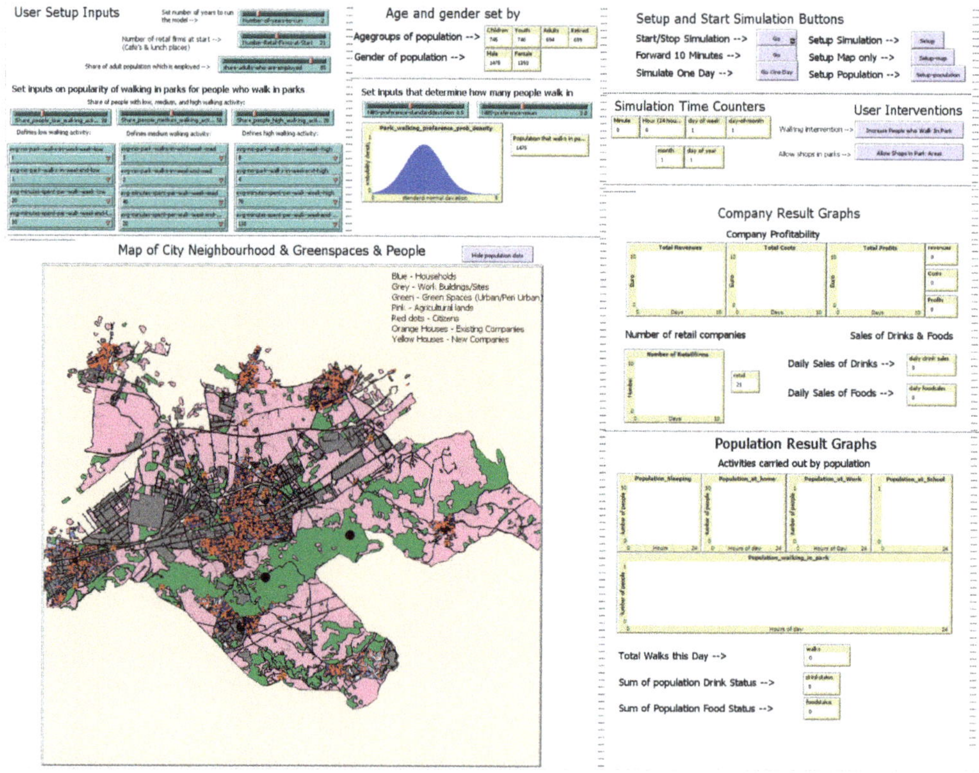

Figure 1. Screenshot of Netlogo model's interface for socioeconomic and commercial development resulting from NBS changes.

The simulation contains two types of agents: (1) residents, some of whom carry out activities in green spaces and as such sometimes purchase drinks and food items; and (2) shop owners, who establish shops across the city. The simulation is initialized with a number of initial shops as set by the user, and new shops are created while running the model. Factors that attract new retail shops to establish a presence in an area are simplified based on attractor points, which identify areas such as large green spaces within and around which shops can emerge.

Population agents are further segmented based on two characteristics: (1) the share of agents who walk in green spaces and (2) purchasing behavior, i.e., buying a drink and/or food item. Walking characteristics are initialized by a probability setting as an initial input. A Gaussian probability density function is used for this purpose. The user can set the mean and standard deviation to modify the population share that walks in the park.

The user can further divide those who walk in green spaces into:

- The proportion that has a low walking activity, with 1 walk during the workweek of 20 min on average and 1 during the weekend of 30 min on average;
- The proportion that has a medium walking activity, with 3 walks during the workweek of 30 min on average and 2 during the weekend of 60 min on average;
- The proportion that has a high walking activity, with 8 walks during the workweek of 40 min on average and 4 walks during the weekend of 80 min on average.

The share of the population that is employed can also be set to influence the population that stays at home during the day and thereby has more propensity to walk in green spaces (if that agent does walk in green spaces as described above). In addition, the number of initial shops and the number of years that the model should be run for (ranging from 1 to 10 years) can be set.

3.1. Agent Activities

Population agents in the simulation carry out three main categories of activities:

1. Daily routine activities (sleep, school or work or stay at home, leisure);
2. Walking in green spaces;
3. Purchasing a drink or food.

The population is initialized with a set of activities, such that not all agents have the same activity "set". The switch between activities as dependent variables takes place for each agent across the day based on an activity transition probability. At a set interval during which a transition can occur, such as between 6 a.m. and 9 a.m., a probability roll is made at a 10-minute timestep interval, resulting in 18 probability rolls during the time interval. A Gaussian probability distribution is assumed. Since it is not modeled what influences this probability, as this lies outside of the scope of the model, no additional independent variables are included.

Varying standard deviations can be provided for each transition. In total, the model contains eight transitions:

1. Sleeping *to* awake at home;
2. Awake at home *to* school (for population members who attend school);
3. School *to* home (for population members who attend school);
4. Awake at home *to* work (for population members who work);
5. Work *to* home (for population members who work);
6. Home *to* walk in park;
7. Walk in park *to* home;
8. At home *to* sleep.

Based on the set of transitions, a daily pattern emerges, which determines at what time residents are in green spaces or elsewhere. In addition, the difference between weekends and weekdays is considered, such that there are no to/from work and to/from school transitions during weekends.

3.2. Purchasing Behavior, Firm Financial Flows and Firm Disappearance

Resident agents are characterized by a daily food status and a daily drink status, which indicates the extent to which, for a given day, they need a food or drink item. If they have already consumed a food or drink item, the status of consumption is set to "saturated", and they will no longer seek to purchase any food or drinks. If their status is not yet "saturated", there is a probability that a food or drink item may be purchased. The occurrence of a purchase is based on a uniform probability roll from 0 to 1, with a 0.90 or higher probability threshold for drink item purchases and a 0.97 or higher probability threshold for food item purchases. These rolls are made in every 10-minute timestep, such that the longer the time a person spends in an NBS space, the higher the likelihood of a purchase. For example, a 50 min time spent in an NBS green space results in a 50% probability of a purchase. In the current version of the model, due to the lack of data, there is no distinction between those who never purchase any drinks and/or food items, those who rarely purchase drinks and/or food items and those who frequently purchase drinks and/or food items.

Once a drink purchase is made, it yields additional revenue for the shop where it is made. Prices are assumed based on a generic "drink" and generic "food item" and set to a randomized value per shop. The price value for drinks is set to vary between 2 and 3 euros, and food prices between 4 and 5 euros. Variable costs, including food costs and labor cost and other consumables, are assumed to be 50% of the price level, and fixed costs are introduced at 20,000 euros per year assuming space rental, financing, taxes and electricity and water charges at 1667 euros per month. The difference between revenues and costs results in the net profit or loss of the shop, which indicates its financial sustainability (or lack thereof).

It is assumed that a purchase is always made at the retail shop that is closest to the resident agent in terms of spatial distance. To this end, spatial distance is estimated for each purchase that is made to assign the purchase to the closest retail shop, which is critical in order to evaluate which retail shops survive because they make enough sales and which close down because of a lack of revenue.

If shops make a substantial loss that is equivalent to half of their fixed costs per annum, it is assumed that they disappear and are removed from the simulation. Thus, the only shops that survive are those that make a profit on a sustained basis, based on the number of customers they can obtain by being in a location close to a large number of customers who are local residents.

3.3. New Firm Appearance and Location Choice

The simulation also allows new shops to appear with similar rules for revenues, costs and profits. The locations in which new shops can appear are fixed as an initial input based on "attractor points". New shops can appear only near these attractor points, which denote either city center areas or the center of large green spaces in the simulation. The idea is that companies do not randomly set up shop somewhere but are attracted to particular locations based on their characteristics, such as population density and visitor popularity. The idea behind attractor points was developed by Arentze and Timmermans [70], who studied firm location choice in urban settings based on a specific value per spatial cell for a company.

The probability that a new shop will emerge is based on a uniform probability roll between 0 and 10, with a 5% probability that a new shop will establish a presence per probability roll. The periodicity at which a probability roll is made for a new shop to emerge can be on a daily, weekly or monthly basis.

4. Results

The model was run for four different European cities: Szeged (Hungary), Alcalá de Henares (Spain), the Metropolitan City of Milan (Italy) and Çankaya Municipality (Turkey). A total of six model runs were established for each city. Each of these runs covered a five-year period for which daily population activities at 10 min intervals were simulated, including park walking for weekdays and weekends. The input variations were the number of initial shops at the start of each model run, set to 1, 5 or 10 retail shops randomly allocated across the city. Each run was carried out twice to give a course understanding of variability across model runs.

4.1. Results for Szeged (Hungary)

The simulation area was selected to cover the center of Szeged with surrounding areas and satellite peri-urban areas. The model runs for Szeged showed that for a simulated population of approximately 800 citizens who walk in parks (out of 1583 citizens), a total of 5 to 6 retail shops can be supported based on drink and food purchases associated with time spent in green spaces (Table 1). The final number of shops at the end of the 5-year simulation varied substantially between 3 and 8. No substantial difference was found between different initial numbers of retail shops. The location of the shops was found to be closely related to green spaces. At least 50% of the shops were located within or at the edge of green space areas, and in most cases, 75% of shops were located in green spaces. In one simulation run, all retail shops were in green spaces (see Figure 2).

Table 1. Szeged: simulation results for socioeconomic and commercial development resulting from NBS changes.

Run Set #	Population (# People)	Population Walking in Parks (# People)	Number of Retail Shops		
			At Start of Model Run	At End of Model Run	Average across Model Run
1	1583	827	1	8	6
2	1583	783	1	3	5
3	1583	818	5	7	6
6	1583	873	5	6	5
5	1583	810	10	3	5
6	1583	845	10	5	5

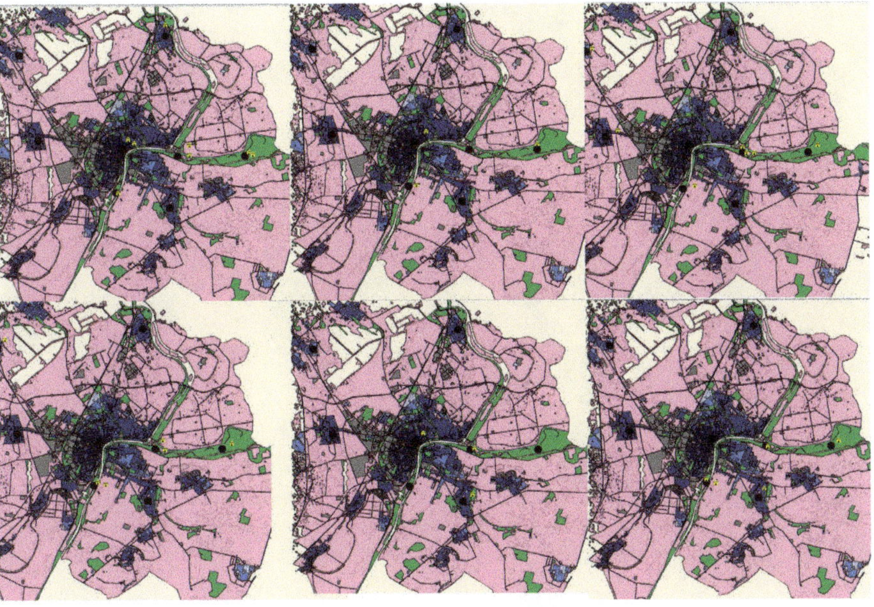

Figure 2. Final map location results for Szeged of retail shops across six model runs. Initial retail shops (if still existing) in orange; new retail shops in yellow. Land use is depicted based on green spaces (green), households (blue), agriculture (pink) and commercial spaces (grey).

4.2. Results for Alcalá de Henares (Spain)

The simulation area was selected to include the center of Alcalá de Henares with surrounding areas and satellite peri-urban areas. The model runs for Alcalá de Henares showed that for a simulated population of approximately 1500 citizens who walk in parks (out of 2900 citizens), a total of 7 to 9 retail shops can be supported based on drink and food purchases associated with time spent in green spaces (Table 2). The final number of shops at the end of the 5-year simulation was quite stable at 7 or 8. No substantial difference was found between different initial numbers of retail shops, which varied between 1 and 10 across the 6 model runs. The location of the shops was mostly in residential centers far from larger green spaces. Approximately 25% to 50% of shops were located within or at the edge of green space areas. At maximum, half of all retail shops were in green spaces (see Figure 3).

Table 2. Alcalá de Henares: simulation results for socioeconomic and commercial development resulting from NBS changes.

Run Set #	Population (# People)	Population Walking in Parks (# People)	Number of Retail Shops		
			At Start of Model Run	At End of Model Run	Average across Model Run
1	2868	1559	1	8	9
2	2868	1498	1	8	9
3	2868	1511	5	7	6
6	2868	1440	5	7	8
5	2868	1448	10	8	7
6	2868	1461	10	8	7

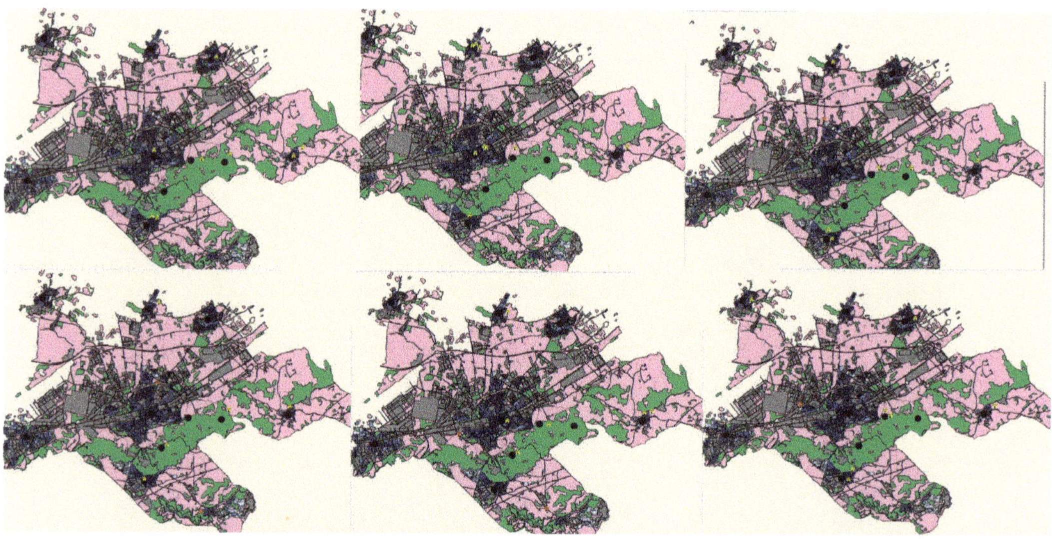

Figure 3. Final map location results for Alcalá de Henares of retail shops across six model runs. Initial retail shops (if still existing) in orange; new retail shops in yellow. Land use is depicted based on green spaces (green), households (blue), agriculture (pink) and commercial spaces (grey).

4.3. Results for the Metropolitan City of Milan (Italy)

The simulation area was based on a portion of the Northern Milan Metropolitan Area centered on the quarry restoration site in Parco Lago Nord selected with surrounding neighborhoods.

The model runs for the Metropolitan City of Milan showed approximately 3000 walks in parks for a simulated population of approximately 5600 citizens, and as a result, a total of 12 to 14 retail shops can be supported based on drink and food purchases associated with time spent in green spaces (Table 3). The final number of shops at the end of the 5-year simulation was quite stable at 13 or 14. No substantial difference was found between different initial numbers of retail shops, which varied between 1 and 10 across the 6 model runs. The location of the shops was mostly in residential centers at a reasonable distance from larger green spaces. Approximately 33% to 40% of shops were located within or at the edge of green space areas. At maximum, half of all retail shops were in green spaces (see Figure 4).

Table 3. Metropolitan City of Milan: simulation results for socioeconomic and commercial development resulting from NBS changes.

Run Set #	Population (# People)	Population Walking in Parks (# People)	Number of Retail Shops		
			At Start of Model Run	At End of Model Run	Average across Model Run
1	5592	2929	1	14	14
2	5592	2950	1	13	12
3	5592	2919	5	14	13
6	5592	2943	5	13	13
5	5592	2921	10	14	13
6	5592	2815	10	14	12

Figure 4. Final map location results of retail shops across six model runs for the Metropolitan City of Milan. Initial retail shops (if still existing) in orange; new retail shops in yellow. Land use is depicted based on green spaces (green), households (blue), agriculture (pink) and commercial spaces (grey).

4.4. Results for Çankaya Municipality (Turkey)

The simulation area was based on the southeast area of Ankara, where Çankaya Municipality is located.

The model runs for Çankaya Municipality showed close to 1600 walks in parks for a simulated population of approximately 3000 citizens, and as a result, a total of 7 to 9 retail shops can be supported based on drink and food purchases associated with time spent in green spaces (Table 4). The final number of shops at the end of the 5-year simulation varied significantly between 5 and 11. No substantial difference was found for the average number of retail shops, based on the initial number of retail shops, which varied between 1 and 10 across the 6 model runs. The location of the shops was primarily in residential centers quite far away from larger green spaces. In five out of six model runs, only approximately 12.5% of the shops were located within or at the edge of green space areas. At maximum, 25% of retail shops were in green spaces (see Figure 5).

Table 4. Çankaya Municipality: simulation results for socioeconomic and commercial development resulting from NBS changes.

Run Set #	Population (# People)	Population Walking in Parks (# People)	Number of Retail Shops		
			At Start of Model Run	At End of Model Run	Average across Model Run
1	3000	1569	1	8	9
2	3000	1596	1	11	8
3	3000	1545	5	7	7
6	3000	1553	5	6	8
5	3000	1608	10	5	8
6	3000	1610	10	11	8

Figure 5. Final map location results of retail shops across six model runs for Çankaya Municipality. Initial retail shops (if still existing) in orange; new retail shops in yellow. Land use is depicted based on green spaces (green), households (blue), agriculture (pink) and commercial spaces (grey).

5. Discussion

The results of the model can be interpreted to be valid for large NBS park spaces which are visited by a substantial number of citizens. Insights are thereby derived for such spaces based on how much revenue can be generated locally by allowing for retail places within or close to NBS park spaces.

The main limitation of the model relates to the absence of data on the preferences of park visitors in terms of frequency of purchases and the amount spent on purchases. At present, the model assumes a standard probability that does not change over time, while purchases will in reality differ depending on the type of visitor, weekday or weekend day, weather and other factors. The effect of this lack of information can cause a mismatch between the purchases predicted by the model and real ones. A survey among park visitors and/or observational results on purchases, if existing retail establishments are in place, would be required to fill this data gap.

The second limitation of the model lies in the fact that more factors, in addition to distance, can play a role in purchasing behavior, such as price and type of drink or food available at a retail shop, versus the generic drink and food item currently provided in the model. This is especially relevant for food items in cases where there are different dietary segments of the population, as non-alcoholic drinks are relatively universally available across retail shops. Moreover, a fixed ratio between profits, operational costs and a fixed investment cost per shop was assumed, to allow focusing entirely on how many shops emerged per size of the population that is walking and their location.

A third limitation consists in the selection of the number of locations where shops can be established, which was constrained to a few in order to understand differences between establishing shops in NBS park spaces or in city centers near or further away from NBS park spaces. The number of areas where retail shops can be established is much larger, and a much more complex locational option model would be needed if the aim is to predict specific locations. For the purposes of understanding the number of shops and the extent to which additional revenues can be generated by NBS green parks, this level of locational detail is instead unnecessary.

A fourth limitation is that only baseline data were available from the cities involved in the Nature4Cities project, due to which it was not possible to calibrate and parameterize the model based on real-world observations. In a future research design bespoke to examining NBSs and socioeconomic factors, significant funding and effort would be needed to further advance the field to also include on-site surveys. Such surveys would have to cover at least 1000 members of the local population in a city to establish activity–time–use patterns, walking propensity, purchasing behaviors and relevant socioeconomic drivers.

The model described in this paper is based on a much simpler structure, and the results are less detailed than those from other literature models. For example, Tsekeris and Vogiatzoglou [71] developed a model which takes into account various exogenous elements (migration and relocation, commuting, costs of households and firms, transport costs and agglomeration economics), whereas the model of De Bok and Bliemer [49] considers firm-specific behavior and quantifies the effects of spatial and transport planning on firm population and mobility. The model applied in this paper is further based on much lower information and data requirements and does not involve the use of complex economic or transport models. Moreover, it deals only with local retailer shops and not with large industrial firms, which thus makes it unnecessary to build an accurate model of the job market or of transport policies. Nonetheless, the model can serve as a preliminary screening tool to assess the possible impacts of urban green parks on the economic sustainability of neighborhood-level small shops.

6. Conclusions

The agent-based simulation model developed in this paper simulated the change in commercial activities by small and midsize companies in retail due to the development of large green parks, which are a specific example of an NBS. Simulations were performed for four European cities taken as case studies. The number of shops, scaled based on the size of the population, was simulated using a 5-year simulation with a 10-minute interval. In the case of Szeged (Hungary), approximately 5–6 retail shops emerged on average for approximately 800 walking citizens, with no variation if there were differences in new retail shops. In the cases of Alcalá de Henares (Spain) and Çankaya Municipality (Turkey), approximately 7–9 retail shops emerged for around 1500 walking citizens. The simulation with Alcalá de Henares showed a lower number of shops in the case of 10 initial shops (approximately 7 on average across the simulation) and higher in the case of 1 initial shop (9 across the simulation), which may be a random result given the limited number of simulations of 5 years and given that this result did not emerge in the simulations for other municipalities/cities. The simulation for the Metropolitan City of Milan (Italy) provided an average number of retail shops of 12–14 for a simulated population of 2900 that walks in parks.

The results are generated using a combination of the number of people walking in parks, the probability of a purchase, the number of purchases that need to be made for shops to run breakeven given investment cost and the ratio between revenues and operational costs. Since these were chosen based on educated estimates and not on local evaluations, the results demonstrate the potential of such a model but cannot be used for a real implementation of investment choices. If the model were built to concretely support decision making in that sense, it would need to be informed based on local surveys to better reflect the actual frequency of people who walk in parks and the segments they belong to (in the simulation set as low, medium and high park walkers), as well as their purchasing behavior in relation to socioeconomic factors, such as income. The simulation would then allow for testing different variations in segments and their changes over time, in terms of how many retail shops can be sustained.

Another established result was the location of retail shops in the simulations because of the proximity between the place where a person is walking and the place where the purchase happens.

It is further worth mentioning that the layout of the municipality and location of green spaces have a substantial influence on where retail shops emerge and are maintained because they can make sufficient profit to survive. The simulations showed that in cases where there are limited large continuous green spaces, such as in the case of Szeged, it is more likely that retail shops can be sustained in a concentrated green area, compared to situations where there is a large fairly continuous area of green NBS space across the city center, such as in the case of Alcalá de Henares and Çankaya Municipality. In the latter case, walking behavior is more scattered, and it is more likely that retail shops will emerge at the edge or closer to built-up residential areas. However, the simulations did not consider asymmetrical attractiveness of green areas. In other words, the model did not account for the possibility that certain green areas could be frequented more often than others for walking purposes. This could nonetheless have substantial influence on the outcome.

The added value of the model presented here essentially lies in the capacity to analyze the indirect economic benefit of NBSs on a neighborhood's economic structure—more specifically on the ability for large green spaces to provide revenue for their maintenance and upkeep, through cafés within and around the NBS. Through the model, this benefit in terms of commercial revenues that spring from NBSs can be quantified and thus provide a basis for establishing NBS infrastructure on a healthier financial basis. In future versions of the model, this could be further subdivided into specific NBSs such as large urban parks, heritage gardens, green cemeteries, woods and so forth.

Author Contributions: Conceptualization, A.M. and R.K.; methodology, A.M. and R.K.; software, R.K. and A.M.; validation, R.K.; formal analysis, R.K. and A.M.; investigation, R.K. and A.M.; resources, A.M. and R.K.; data curation, R.K.; writing—original draft preparation, A.M. and R.K.; writing—review and editing, A.M., R.K., L.H., J.B. and B.R.; visualization, R.K.; supervision, B.R.; project administration, B.R.; funding acquisition, B.R. All authors have read and agreed to the published version of the manuscript.

Funding: This work has received funding from the European Union's Horizon 2020 Research and Innovation Programme under grant agreement No 730468 (www.nature4cities.eu) for the project "Nature4Cities", which is gratefully acknowledged by R.K., A.M. and B.R.

Institutional Review Board Statement: Not applicable.

Informed Consent Statement: Not applicable.

Data Availability Statement: Not applicable.

Acknowledgments: The authors gratefully acknowledge the COST Action CA16114 "RESTORE: Rethinking Sustainability towards a Regenerative Economy", thanks to which the networking opportunities were created to develop this joint work.

Conflicts of Interest: The authors declare no conflict of interest.

References

1. United Nations. *World Urbanization Prospects: The 2018 Revision (ST/ESA/SER.A/420)*; United Nations, Department of Economic and Social Affairs, Population Division: New York, NY, USA, 2019; p. 126.
2. Pataki, D.E.; Carreiro, M.M.; Cherrier, J.; Grulke, N.E.; Jennings, V.; Pincetl, S.; Pouyat, R.V.; Whitlow, T.H.; Zipperer, W.C. Coupling biogeochemical cycles in urban environments: Ecosystem services, green solutions, and misconceptions. *Front. Ecol. Environ.* **2011**, *9*, 27–36. [CrossRef]
3. Ozuduru, B.H.; Guldmann, J.-M. Retail Location and Urban Resilience: Towards a New Framework for Retail Policy. *Surv. Perspect. Integr. Environ. Soc.* **2013**, *6*, 1–13.
4. Grahn, P.; Stigsdotter, U.K. The relation between perceived sensory dimensions of urban green space and stress restoration. *Landsc. Urban Plan.* **2010**, *94*, 264–275. [CrossRef]
5. Beute, F.; Andreucci, M.B.; Lammel, A.; Davies, Z.; Glanville, J.; Keune, H.; Marselle, M.; O'Brien, L.; Olszewska-Guizzo, A.; Remmen, R.; et al. Types and Characteristics of Urban and Peri-Urban Green Spaces Having an Impact on Human Mental Health and Wellbeing. EKLIPSE Expert Working Group. 2020. Available online: https://eklipse.eu/wp-content/uploads/website_db/Request/Mental_Health/EKLIPSE_HealthReport-Green_Final-v2-Digital.pdf (accessed on 15 June 2021).
6. Brengman, M.; Willems, K.; Joye, Y. The Impact of In-Store Greenery on Customers. *Psychol. Mark.* **2012**, *29*, 807–821. [CrossRef]
7. Niemelä, J. Ecology and urban planning. *Biodivers. Conserv.* **1999**, *8*, 119–131. [CrossRef]
8. European Commission. *Mapping and Assessment of Ecosystems and Their Services*; European Commission: Brussels, Belgium, 2016.
9. Li, L.; Cheshmehzangi, A.; Chan, F.K.S.; Ives, C.D. Mapping the Research Landscape of Nature-Based Solutions in Urbanism. *Sustainability* **2021**, *13*, 3876. [CrossRef]
10. Babí Almenar, J.; Elliot, T.; Rugani, B.; Philippe, B.; Navarrete Gutierrez, T.; Sonnemann, G.; Geneletti, D. Nexus between nature-based solutions, ecosystem services and urban challenges. *Land Use Policy* **2021**, *100*, 104898. [CrossRef]
11. Ferreira, V.; Barreira, A.P.; Loures, L.; Antunes, D.; Panagopoulos, T. Stakeholders' Engagement on Nature-Based Solutions: A Systematic Literature Review. *Sustainability* **2020**, *12*, 640. [CrossRef]
12. Meyer, M.A.; Schulz, C. Do ecosystem services provide an added value compared to existing forest planning approaches in Central Europe? *Ecol. Soc.* **2017**, *22*. [CrossRef]
13. Popoola, L.; Ajewole, O. Public perceptions of urban forests in ibadan, nigeria: Implications for environmental conservation. *Arboric. J.* **2001**, *25*, 1–22. [CrossRef]
14. Jim, C.; Chen, W. Perception and Attitude of Residents Toward Urban Green Spaces in Guangzhou (China). *Environ. Manag.* **2006**, *38*, 338–349. [CrossRef] [PubMed]
15. Panagopoulos, T.; Tampakis, S.; Karanikola, P.; Karipidou-Kanari, A.; Kantartzis, A. The Usage and Perception of Pedestrian and Cycling Streets on Residents' Well-being in Kalamaria, Greece. *Land* **2018**, *7*, 100. [CrossRef]
16. Yen, Y.; Wang, Z.; Shi, Y.; Soeung, B. An Assessment of the Knowledge and Demand of Young Residents regarding the Ecological Services of Urban Green Spaces in Phnom Penh, Cambodia. *Sustainability* **2016**, *8*, 523. [CrossRef]
17. Barau, A.S. Perceptions and contributions of households towards sustainable urban green infrastructure in Malaysia. *Habitat Int.* **2015**, *47*, 285–297. [CrossRef]
18. Guenat, S.; Dougill, A.J.; Kunin, W.E.; Dallimer, M. Untangling the motivations of different stakeholders for urban greenspace conservation in sub-Saharan Africa. *Ecosyst. Serv.* **2019**, *36*, 100904. [CrossRef]
19. Gwedla, N.; Shackleton, C.M. Perceptions and preferences for urban trees across multiple socio-economic contexts in the Eastern Cape, South Africa. *Landsc. Urban Plan.* **2019**, *189*, 225–234. [CrossRef]
20. Parker, J.; De Baro, M.E.Z. Green Infrastructure in the Urban Environment: A Systematic Quantitative Review. *Sustainability* **2019**, *11*, 3182. [CrossRef]
21. Ershad Sarabi, S.; Han, Q.; Romme, A.G.L.; De Vries, B.; Wendling, L. Key Enablers of and Barriers to the Uptake and Implementation of Nature-Based Solutions in Urban Settings: A Review. *Resources* **2019**, *8*, 121. [CrossRef]
22. Chen, W.; He, B.; Nover, D.; Lu, H.; Liu, J.; Sun, W.; Chen, W. Farm ponds in southern China: Challenges and solutions for conserving a neglected wetland ecosystem. *Sci. Total Environ.* **2019**, *659*, 1322–1334. [CrossRef] [PubMed]

23. Liquete, C.; Udias, A.; Conte, G.; Grizzetti, B.; Masi, F. Integrated valuation of a nature-based solution for water pollution control. Highlighting hidden benefits. *Ecosyst. Serv.* **2016**, *22*, 392–401. [CrossRef]
24. Santiago Fink, H. Human-Nature for Climate Action: Nature-Based Solutions for Urban Sustainability. *Sustainability* **2016**, *8*, 254. [CrossRef]
25. Short, C.; Clarke, L.; Carnelli, F.; Uttley, C.; Smith, B. Capturing the multiple benefits associated with nature-based solutions: Lessons from a natural flood management project in the Cotswolds, UK. *Land Degrad. Dev.* **2019**, *30*, 241–252. [CrossRef]
26. Van der Jagt, A.P.N.; Szaraz, L.R.; Delshammar, T.; Cvejić, R.; Santos, A.; Goodness, J.; Buijs, A. Cultivating nature-based solutions: The governance of communal urban gardens in the European Union. *Environ. Res.* **2017**, *159*, 264–275. [CrossRef]
27. Terrapin Bright Green. *Terrapin The Economics of Biophilia*; Terrapin Bright Green LLC: New York, NY, USA, 2015.
28. Simpson, D. *The Economics of Nature-Based Solutions: Current Status and Future Priorities*; United Nations Environment Programme (UNEP): Nairobi, Kenia, 2020.
29. Di Marino, M.; Tiitu, M.; Lapintie, K.; Viinikka, A.; Kopperoinen, L. Integrating green infrastructure and ecosystem services in land use planning. Results from two Finnish case studies. *Land Use Policy* **2019**, *82*, 643–656. [CrossRef]
30. Furlong, C.; Phelan, K.; Dodson, J. The role of water utilities in urban greening: A case study of Melbourne, Australia. *Util. Policy* **2018**, *53*, 25–31. [CrossRef]
31. Girma, Y.; Terefe, H.; Pauleit, S. Urban green spaces use and management in rapidly urbanizing countries:-The case of emerging towns of Oromia special zone surrounding Finfinne, Ethiopia. *Urban For. Urban Green.* **2019**, *43*, 126357. [CrossRef]
32. Keeley, M.; Koburger, A.; Dolowitz, D.P.; Medearis, D.; Nickel, D.; Shuster, W. Perspectives on the Use of Green Infrastructure for Stormwater Management in Cleveland and Milwaukee. *Environ. Manag.* **2013**, *51*, 1093–1108. [CrossRef] [PubMed]
33. Khoshkar, S.; Balfors, B.; Wärnbäck, A. Planning for green qualities in the densification of suburban Stockholm – opportunities and challenges. *J. Environ. Plan. Manag.* **2018**, *61*, 2613–2635. [CrossRef]
34. Lamichhane, D.; Thapa, H.B. Participatory urban forestry in Nepal: Gaps and ways forward. *Urban For. Urban Green.* **2012**, *11*, 105–111. [CrossRef]
35. Rall, E.L.; Kabisch, N.; Hansen, R. A comparative exploration of uptake and potential application of ecosystem services in urban planning. *Ecosyst. Serv.* **2015**, *16*, 230–242. [CrossRef]
36. Živojinović, I.; Wolfslehner, B. Perceptions of urban forestry stakeholders about climate change adaptation – A Q-method application in Serbia. *Urban For. Urban Green.* **2015**, *14*, 1079–1087. [CrossRef]
37. Davis, M.; Naumann, S. Making the Case for Sustainable Urban Drainage Systems as a Nature-Based Solution to Urban Flooding. In *Nature-Based Solutions to Climate Change Adaptation in Urban Areas: Linkages between Science, Policy and Practice*; Kabisch, N., Korn, H., Stadler, J., Bonn, A., Eds.; Springer International Publishing: Cham, Switzerland, 2017; pp. 123–137. ISBN 978-3-319-56091-5.
38. Li, C.; Peng, C.; Chiang, P.-C.; Cai, Y.; Wang, X.; Yang, Z. Mechanisms and applications of green infrastructure practices for stormwater control: A review. *J. Hydrol.* **2019**, *568*, 626–637. [CrossRef]
39. Zuniga-Teran, A.A.; Staddon, C.; De Vito, L.; Gerlak, A.K.; Ward, S.; Schoeman, Y.; Hart, A.; Booth, G. Challenges of mainstreaming green infrastructure in built environment professions. *J. Environ. Plan. Manag.* **2020**, *63*, 710–732. [CrossRef]
40. Raymond, C.M.; Frantzeskaki, N.; Kabisch, N.; Berry, P.; Breil, M.; Nita, M.R.; Geneletti, D.; Calfapietra, C. A framework for assessing and implementing the co-benefits of nature-based solutions in urban areas. *Environ. Sci. Policy* **2017**, *77*, 15–24. [CrossRef]
41. Lafortezza, R.; Sanesi, G. Nature-based solutions: Settling the issue of sustainable urbanization. *Environ. Res.* **2019**, *172*, 394–398. [CrossRef]
42. Waddell, P.; Borning, A.; Noth, M.; Freier, N.; Becke, M.; Ulfarsson, G. Microsimulation of Urban Development and Location Choices: Design and Implementation of UrbanSim. *Netw. Spat. Econ.* **2003**, *3*, 43–67. [CrossRef]
43. Brouwer, A.E.; Mariotti, I.; Van Ommeren, J.N. The firm relocation decision: An empirical investigation. *Ann. Reg. Sci.* **2004**, *38*, 335–347. [CrossRef]
44. Van Dijk, J.; Pellenbarg, P.H. Firm Migration. In *International Encyclopedia of Geography*; American Cancer Society: Atlanta, GA, USA, 2017; pp. 1–12. ISBN 978-1-118-78635-2.
45. Van Dijk, J.; Pellenbarg, P.H. Firm relocation decisions in The Netherlands: An ordered logit approach. *Pap. Reg. Sci.* **2000**, *79*, 191–219. [CrossRef]
46. De Bok, M.; van Oort, F. Agglomeration Economies, Accessibility, and the Spatial Choice Behavior of Relocating Firms. *J. Transp. Land Use* **2011**, *4*, 5–24. [CrossRef]
47. Knoben, J.; Oerlemans, L.A.G. Ties that Spatially Bind? A Relational Account of the Causes of Spatial Firm Mobility. *Reg. Stud.* **2008**, *42*, 385–400. [CrossRef]
48. Bodenmann, B.R.; Axhausen, K.W. Destination choice for relocating firms: A discrete choice model for the St. Gallen region, Switzerland. *Pap. Reg. Sci.* **2012**, *91*, 319–341. [CrossRef]
49. De Bok, M.; Bliemer, M.C.J. Infrastructure and Firm Dynamics: Calibration of Microsimulation Model for Firms in the Netherlands. *Transp. Res. Rec.* **2006**, *1977*, 132–144. [CrossRef]
50. Ortegón-Cortázar, L.; Royo-Vela, M. Attraction factors of shopping centers. *Eur. J. Manag. Bus. Econ.* **2017**, *26*, 199–219. [CrossRef]
51. Lee, J.-S.; Hsu, L.-T.; Han, H.; Kim, Y. Understanding how consumers view green hotels: How a hotel's green image can influence behavioural intentions. *J. Sustain. Tour.* **2010**, *18*, 901–914. [CrossRef]
52. Purani, K.; Kumar, D.S. Exploring restorative potential of biophilic servicescapes. *J. Serv. Mark.* **2018**, *32*, 414–429. [CrossRef]

53. Amérigo, M.; García, J.A.; Sánchez, T. Attitudes and Behavior towards Natural Environment. Environmental Health and Psychological Well-Being. *Univ. Psychol.* **2013**, *12*, 845–856.

54. Berman, M.G.; Jonides, J.; Kaplan, S. The Cognitive Benefits of Interacting With Nature. *Psychol. Sci.* **2008**, *19*, 1207–1212. [CrossRef]

55. Kellert, S. Dimensions, Elements, and Attributes of Biophilic Design. In *Biophilic Design: The Theory, Science, and Practice of Bringing Buildings to Life*; Kellert, S.R., Heerwagen, J., Mador, M., Eds.; John Wiley & Sons: Hoboken, NJ, USA, 2008.

56. Wooldridge, M. *An Introduction to MultiAgent Systems*, 2nd ed.; John Wiley & Sons: Chichester, UK, 2009.

57. Marks, R. Validating Simulation Models: A General Framework and Four Applied Examples. *Comput. Econ.* **2007**, *30*, 265–290. [CrossRef]

58. De Marchi, S.; Page, S.E. Agent-Based Models. *Annu. Rev. Political Sci.* **2014**, *17*, 1–20. [CrossRef]

59. Macy, M.; Flache, A. Social Dynamics from the Bottom-Up: Agent-based Models of Social Interaction. In *The Oxford Handbook of Analytical Sociology*; Oxford University Press: Oxford, UK, 2009.

60. Marvuglia, A.; Navarrete Gutiérrez, T.; Baustert, P.; Benetto, E. Implementation of Agent-Based Models to support Life Cycle Assessment: A review focusing on agriculture and land use. *AIMS Agric. Food* **2018**, *3*, 535–560. [CrossRef]

61. Laurenti, R.; Lazarevic, D.; Poulikidou, S.; Montrucchio, V.; Bistagnino, L.; Frostell, B. Group Model-Building to identify potential sources of environmental impacts outside the scope of LCA studies. *J. Clean. Prod.* **2014**, *72*, 96–109. [CrossRef]

62. Schaffernicht, M. Causal loop diagrams between structure and behaviour: A critical analysis of the relationship between polarity, behaviour and events. *Syst. Res. Behav. Sci.* **2010**, *27*, 653–666. [CrossRef]

63. Gilbert, N.; Troitzsch, K.G. *Simulation for the Social Scientist*, 2nd ed.; Open University Press: Buckingham, UK, 2005.

64. Davis, C.; Nikolić, I.; Dijkema, G.P.J. Integration of Life Cycle Assessment Into Agent-Based Modeling. *J. Ind. Ecol.* **2009**, *13*, 306–325. [CrossRef]

65. Ahmadi Achachlouei, M. *Exploring the Effects of ICT on Environmental Sustainability: From Life Cycle Assessment to Complex Systems Modeling*; KTH Royal Institute of Technology: Stockholm, Sweden, 2015.

66. Morgan, S.L.; Winship, C. *Counterfactuals and Causal Inference. Methods and Principles for Social Research*; Cambridge University Press: New York, NY, USA, 2015.

67. Tisue, S.; Wilensky, U. NetLogo: A Simple Environment for Modeling Complexity. In Proceedings of the International Conference on Complex Systems, Boston, MA, USA, 16–21 May 2004.

68. Tisue, S.; Wilensky, U. Design and Implementation of a Multi-Agent Modeling Environment. In Proceedings of the Agent 2004 Conference on Social Dynamics: Interaction, Reflexivity and Emergence, Chicago, IL, USA, 7–9 October 2004.

69. Abar, S.; Theodoropoulos, G.K.; Lemarinier, P.; O'Hare, G.M.P. Agent Based Modelling and Simulation tools: A review of the state-of-art software. *Comput. Sci. Rev.* **2017**, *24*, 13–33. [CrossRef]

70. Arentze, D.T.; Timmermans, D.H. A Multi-Agent Activity-Based Model of Facility Location Choice and Use. *disP Plan. Rev.* **2007**, *43*, 33–44. [CrossRef]

71. Tsekeris, T.; Vogiatzoglou, K. Spatial agent-based modeling of household and firm location with endogenous transport costs. *Netnomics Econ. Res. Electron. Netw.* **2011**, *12*, 77–98. [CrossRef]

sustainability

MDPI

Article

Multi-Level Perspective on Sustainability Transition towards Nature-Based Solutions and Co-Creation in Urban Planning of Belgrade, Serbia

Ana Mitić-Radulović [1,*] and Ksenija Lalović [2]

1   Centre for Experiments in Urban Studies—CEUS, Dalmatinska 70, 11000 Belgrade, Serbia
2   Faculty of Architecture, University of Belgrade, Bulevar kralja Aleksandra 73, 11000 Belgrade, Serbia; ksenija.lalovic@arh.bg.ac.rs
*   Correspondence: ana.mitic-radulovic@ceus.rs; Tel.: +381-65-8535-299

Abstract: In recent years, nature-based solutions have been increasingly promoted as a climate change adaptation instrument, strongly advocated to be co-created. Achieving clear, coherent, and ambitious urban greening strategies, embedded in urban planning and developed in a co-creative, participatory and inclusive manner, is highly challenging within the EU enlargement context. In this article, such challenges are studied through two recent urban development initiatives in Belgrade, the Capital of Serbia: the first initiative focuses on planning the new Linear Park, within the framework of the CLEVER Cities Horizon 2020 project; the second initiative envisages the transformation of the privatised Avala Film Complex in the Košutnjak Urban Forest, primarily led by private interests but supported by the local authorities. The multiple-case study research method is applied, with an exploratory purpose and as a basis for potential future research on evaluation of co-creation processes for NBS implementation. The theoretical basis of this article is founded in the research on sustainability transitions, focusing on multi-level perspective (MLP) framework. The urban planning system in Belgrade and Serbia is observed as a socio-technical regime of the MLP. In such framework, we recognize co-creative planning of the Linear Park as a niche innovation. We interpret opposition towards planning of the Avala Film Complex as escalation, or an extreme element of the socio-technical landscape, comprised of civic unrests and political tensions on one side, combined with the climate crisis and excessive pollution on the other side. Moreover, the article examines informal urban planning instruments that can be implemented by the practitioners of niche innovations, that could support urban planners and NBS advocates in the Serbian and EU enlargement contexts to face the challenges of motivating all stakeholders to proactively, constructively and appropriately engage in co-creation.

Keywords: co-creation; nature-based solutions; urban planning; multi-level perspective; sustainability transition

Citation: Mitić-Radulović, A.; Lalović, K. Multi-Level Perspective on Sustainability Transition towards Nature-Based Solutions and Co-Creation in Urban Planning of Belgrade, Serbia. *Sustainability* **2021**, *13*, 7576. https://doi.org/10.3390/su13147576

Academic Editors: Eugenio Morello, Israa H. Mahmoud, Giuseppe Salvia and Emma Puerari

Received: 18 March 2021
Accepted: 24 June 2021
Published: 7 July 2021

1. Introduction

Several environmental sustainability concepts have evolved through academic research and policy practice over the past two decades [1], ranging from ecosystem services, to over green infrastructure [2,3], to nature-based solutions (NBS) [4]. They nurture an interdisciplinary approach to urban ecosystems [1], emphasizing multi-level governance and strategic urban actions [5]. In recent years, the NBS have been increasingly promoted as a climate change adaptation instrument by International Union for Conservation of Nature (IUCN) [6–8] and the European Commission (EC) [9,10].

The EC defines NBS as dynamic and comprehensive "solutions that are inspired and supported by nature, which are cost-effective, simultaneously provide environmental, social and economic benefits and help build resilience; such solutions bring more, and more diverse, nature and natural features and processes into cities, landscapes and seascapes, through locally adapted, resource-efficient and systemic interventions". It further emphasizes that "nature-based solutions must benefit biodiversity and support the delivery of a range of ecosystem services" [10] (p. 3).

Following the EC's framework of Research and Innovation policy on "Re-Naturing Cities and Green Infrastructure" [9], since 2015 NBS have been strongly embedded in the Horizon 2020 Funding Programme [11] and advocated to be co-created in practice [10,11], implying citizen participation in all NBS development stages: in their planning, design, implementation, and monitoring and evaluation.

Although the notion of co-creation "arose from the business world" [12] (p. 205), in the academic literature it gained ground as a common framework where multiple stakeholders co-design research agendas, co-produce knowledge and co-disseminate it, in various practical dimensions: scientific, international and sectoral [13]. Reflection among all stakeholders is needed [13], but collaboration between researchers and policy officers is critical, both for "policy-relevance of research and its policy uptake", as well as for "new insights for research blind spots" [14] (p. 90).

According to IUCN's NBS Global Standards, one of the eight fundamental criteria of NBS are to be "based on inclusive, transparent and empowering governance processes", being further elaborated as that basic compliance of NBS with prevailing legal and regulatory provisions "need to be complemented with ancillary mechanisms that actively engage and empower local communities and other affected stakeholders" [8] (p. 14). This criterion or principle of NBS is commonly referred to as co-creation of NBS, and has been increasingly perceived as "a fundamental approach to address the impacts of global environmental changes and create new opportunities for all people" [12] (p. 205).

Since 2020, NBS have been promoted as the main instruments of the envisaged Urban Greening Plans—the new policy document recommended by the EC to all the cities bigger than 20,000 inhabitants, starting from the year 2021 [15]. With such policy and financial frameworks, the "promotion of healthy ecosystems, green infrastructure and nature-based solutions should be systematically integrated into urban planning, including in public spaces, infrastructure and the design of buildings and their surroundings" [15] (p. 13) within the cities of the European Union (EU) member states.

However, achieving clear, coherent and ambitious urban greening strategies, embedded in urban planning and developed in a co-creative, participatory and inclusive manner, is highly challenging within the EU enlargement context, particularly in the Candidate Countries and Potential Candidates of the Western Balkans.

Most of the countries/entities of the EU enlargement area belong to the post-socialist context of the ex-Yugoslavia: Serbia and Kosovo (for the European Union, this designation used is without prejudice to positions on status, and is in line with UN Security Council resolution 1244/99 and the International Court of Justice Opinion on the Kosovo declaration of independence), North Macedonia, Montenegro, and Bosnia and Herzegovina—namely, all except Albania and Turkey [16].

Urban planning legal frameworks and practices of ex-Yugoslav countries are strongly influenced by the Yugoslav socialist heritage and self-managed socialism as a unique ideological standpoint. From this perspective, participation was supposed to become part of day-to-day activities of citizens, which would lead to "authentic, free and creative self-fulfillment of the citizens and their community" [17] (p. 23).

However, recent historical research on socialist practice shows that participation in Yugoslav urban planning was rather declarative and poorly conducted, through limited techniques. Expert institutions substituted the power of the central state instead of delegating it to the citizens for self-management [18].

Post-socialist socio-economic transition imposed new challenges [19], with inadequate solutions for new plurality of interests within the market economy. The urban planning approach and methodological processes were practiced as technocratic and exclusively expert-based in most cases [20].

To strengthen and ensure citizen participation in urban planning, the legislative changes in Serbia in 2014 [21], introduced Early Public Consultation (EPC) as the first of the two milestones in the formal urban planning procedure where the government communicates the urban plan with the broader public. As a relatively new planning instrument in society with a long tradition of centralized planning, EPC did not have a significant role—until two years ago.

After realizing that a multitude of interest in post-socialist urban development significantly threatens natural resources, in particular the green infrastructure, leading to missed opportunities for NBS, the wider public favored raising pressure towards the local authorities and urban planning institutions [22].

In this new, often conflicting urban planning setting, the local authorities declaratively promote urban greening [23,24], but there is a significant discrepancy between policy and practice. Moreover, co-creation is rarely perceived as an added value, due to a still predominant rational top-down planning approach, the lack of facilitation expertise and the rise of social and political tensions. However, local administrations have started realizing the indispensability of communication with the wider public in recent months.

The challenge of inconsistency and duality of approaches towards nature-based solutions and co-creation in the EU enlargement context will be illustrated in this article with two recent urban development initiatives in Belgrade, the Capital of Serbia, more precisely with specific official planning phases of both of the initiatives: the Early Public Consultation procedures.

The first initiative focuses on the planning the new Linear Park, led by the City of Belgrade as a Follower City of the CLEVER Cities project (a European-Commission-funded project from the Horizon 2020 Innovation Action Programme under Grant Agreement No. 776604. See https://clevercities.eu/ (accesed on 20 May 2021)), and its supporting local partner Center for Experiments in Urban Studies (CEUS), represented by the authors of this paper. The second initiative envisages transformation of the privatized Avala Film Complex in the Košutnjak Urban Forest, primarily led by the private interest.

The theoretical basis of this article is founded in the research on sustainability transitions, as an overarching field of approaches and perspectives regarding large-scale, long-term and complex societal transformations toward sustainability [25–29]. Sustainability transitions represent "a threat to existing dynamically stable configurations facing persistent sustainability challenges, and they present opportunities for more radical, systemic, and accelerated change" [29] (p. 600).

There are various frameworks for analyzing and interpreting sustainability transitions [30,31], but this article will focus on multi-level perspective framework according to Geels [25,26], which recognizes three main analytical levels: (1) socio-technical regime, as a stable system of established practices and associated rules; (2) niches, as "protected spaces" of experimentation and emerging innovations which "provide the seeds for systemic change" (p. 27); and (3) the socio-technical landscape, as the wider, slowly-changing external context that creates pressure on the regime.

The main research question of this article explores in which manner various (and sometimes contradictory) urban planning cases can contribute to effective sustainability transition regarding the urban planning system towards NBS co-creation, and what can be learned from those cases if interpreted using the multi-level perspective framework.

Furthermore, we will examine which informal urban planning instruments and principles can be implemented by the practitioners of the niche innovations, in order to strengthen their impact on the current socio-technical regime and its' subsequent destabilization towards a sustainability transition.

2. Materials and Methods

This article represents an inquiry of mainstreaming NBS and co-creation in urban planning practice. The research focuses on analyzing how differences in planning initiatives' participation solicit various reactions and may be the cornerstones of long-term socio-technical transition. The multiple-case study research method is applied with an exploratory purpose to enable in-depth experiences, as contemporary phenomena within their "real-life contexts" [32] but with contradictory narratives and implications, where contextual conditions are highly pertinent to phenomena of the study. This may offer insights that might not be achieved with other approaches [33] as a basis for developing the "more structured" methodology that will be necessary for additional research strategies expected in the future. Since case-study allows "to *explain* the causal links in real-life interventions that are too complex for the survey or experimental strategies" [32] (p. 15), it will also serve as a basis for potential research on evaluation of co-creation for NBS.

The case study method was performed from a third-person perspective or observer position, even though the authors were directly involved in one of the examined planning initiatives, namely, they planned, designed and implemented all the co-creation actions related to the future Linear Park. Since these authors had very little control over the project initiative initiation and implementation process, it is considered that this will not influence the case study objectivity. On the contrary, their first- and second-person perspective insights will support the in-depth analysis.

This research strongly correlates with Action Research, since it associates and "combines theory and practice (and researchers and practitioners) through change and reflection in an immediate problematic situation" [34] (p. 94). Iterative processes include problem diagnosis, action intervention and reflective learning, performed jointly by researchers and practitioners. They even include close-up and detailed observations via direct involvement of researchers in the execution of the co-creation practice in real situations, namely, the participatory planning processes. Furthermore, this research has strong features of community-based participatory action research (PAR) as a distinct qualitative methodology, since it relies on democratic and liberating processes in which participants construct meaning [35].

The materials used in this research are official public documents and publications, critical research publications, public social media announcements and published surveys and interviews.

In the following subchapters, two major urban re-development initiatives in Belgrade will be presented in an integrated way, providing vital information on the formal and informal participation and planning processes. The Linear Park case-study with adopted CLEVER Cities co-creation methodology [36] will be described and explained in detail.

2.1. Overview of the Dual Urban Planning Practices Regarding Co-Creation for NBS in Belgrade in the Year 2020

Two major urban transformation initiatives raised the interest of both experts and the broader public in the City of Belgrade in the year 2020: the new Linear Park in the area of the former railway corridor from the Beton Hala until the Pančevo Bridge (4.5 km long) was perceived as a prominent practice example, while transformation of the privatized Avala Film Complex in the Košutnjak Urban Forest was perceived as problematic and ineffective practice.

There are significant similarities between those two planning initiatives: both are related to the green infrastructure of the City of Belgrade, as well as to investments and real estate development plans of the SEBRE company (Beograd, Serbia) from the Czech Republic. In both cases, the planning task is assigned to the Urban Planning Institute of Belgrade, the most experienced and the most resourceful public planning agency in Serbia. However, the planning initiatives' communication; their sensitivity for public participation, engagement and consultation; and the content and the ambition of the new proposals significantly differ. This research focuses on analyzinganalyzing how these differences solicit various reactions and may be the cornerstones of long-term socio-technical transition towards co-creation and mainstreaming of NBS in urban planning in the local context.

Both cases will be analyzed primarily via EPC procedure as the first official, formal planning milestone when the government communicates the urban plan's initial ideas under development with the broader public. However, by law, EPC does not impose any interactive communication: materials are exposed for public insight but without any presentations, discussions or workshops, although more significant and timely interaction with the public was recognized as a necessity among professionals [37] and was the initial goal of proposing and advocating for an additional instrument/step in the planning procedure. Moreover, inputs from the public, obtained in the EPC, are non-binding and do not require official feedback [21].

From 2018, the general public became more aware of urban planning enactment procedures. It started to react to announcements, media texts, official elaborations and related documents, and a significant number of community groups emerged in reaction to the government's urban plans intentions.

In the year 2020, just after the first peek of the COVID-19 pandemic, two interesting EPC occurred that are the subject of this research.

2.2. Linear Park

The railway corridor between the Beton Hala and the Pančevo Bridge was perceived as a zone for re-development from transit into a green space within the Plan of General Regulation of Belgrade [38]. In this context, it was proposed by Belgrade's City Authorities' Secretariat for Environmental Protection, as a testbed for introducing the NBS in urban planning practice within the CLEVER Cities project, initiated in June 2018 and funded from the European Union's Horizon 2020 innovation action program. In September 2018, the Assembly of the City of Belgrade adopted a Decision on Development of the Plan of Detailed Regulation (PDR) for Linear Park, Belgrade [39]. The main goal of this urban plan is transformation of app. 46.7 ha of the former railway corridor land into a healthy, inspiring and attractive space: a demonstration polygon for art and technology, using NBS and co-creation in a broad participatory process (Figure 1).

Figure 1. Elaborate for the Plan of Detailed Regulation for the Linear Park: planned land-use and contents [40].

Financing of the planning process is allocated from the local budget and the CLEVER Cities project, which is the first such case in the City of Belgrade. Additionally, financing PDR's implementation is expected from local and national budgets and private funds (of many real estate developments in the surrounding area but with unclear mechanisms), other R&D projects (e.g., EuPOLIS), resources from the Instrument of Pre-Accession (IPA funds) and loans of international financial institutions (e.g., European Bank for Reconstruction and Deveopment) [41]. The total investment value is roughly estimated at 50 million EUR [42].

In September 2019, the vacant city-owned land of the Marina Dorćol (4 ha of prime construction land, with the capacity of 76,000 m^2 over-ground new construction within the realm of the Linear Park plan) was sold to the company MD Investment, in property of the SEBRE company, as the only bidder of the public tendering process [43].

2.2.1. CLEVER Cities Co-Creation Methodology, Applied in Planning of the Linear Park

The conceptual setting of stakeholder engagement and co-creation critical issues is derived from the CLEVER Cities Co-Creation Methodology, developed by Politecnico di Milano and elaborated in detail within the CLEVER Cities guidance on co-creating nature-based solutions Part I [44] and Part II [45].

This particular methodology provides "a complete co-creation pathway that encourages decision-makers to embed citizen engagement methodologies as an approach to co-design and co-implement NBS" [46]. Two main mechanisms for implementing NBS in urban fabrics proposed by this methodology are: (1) urban innovation partnership (UIP), as a "city-wide or district-focused informal alliance of local and city authorities, community (groups), businesses, academics to promote the NBS for regeneration or urban transformation, facilitate and drive the cocreation process" [44] (p. 8), and (2) CLEVER Action Labs (CALs). According to this methodology, co-creation process is divided in five phases: urban innovation partnership (UIP) establishment, co-design, co-implementation, co-monitoring and co-development [12,44–46].

As this methodology was developed for the Front-Runner Cities (Front-Runner Cities of the CLEVER Cities project are Hamburg, Milan and London) of the CLEVER Cities project, which have been testing and demonstrating NBS in all the aforementioned phases over the five years, CLEVER Cities local partners from Belgrade had to adopt this methodology to the Follower City context, namely, to learn from project demonstrations, implement adequate solutions and integrate them in a specific urban plan (Follower Cities of the CLEVER Cities project are Belgrade, Larissa, Madrid, Malme, Sfantu George and Quito).

Therefore, co-creation process elaborated in this article focus on the first phase of the CLEVER Cities co-creation pathway: urban innovation partnership (UIP) establishment, comprised of fours specific steps/tools: (1) identification of CLEVER Cities project within the city local context; (2) mapping and engaging of stakeholders; (3) launch of the urban innovation partnership; and (4) design of the platform according to the local context [44]. The last step was optional, but in Belgrade it proved to be the crucial tool for establishing transparent and regular communication among the UIP members and interested citizens, as well as for building trust.

Since Belgrade does not demonstrate NBS implementation, the second, co-design phase was adopted to "co-planning", with appropriate adjustments of several specific steps: launcing of the CAL at local level was performed as a more expert process, resulting in the establishment and registration of BELLAB (BELgrade urban livig LAB, for further information visit: bellab.rs (accessed on 10 May 2021)) as the first urban living lab in the Western Balkans within the European Network of Living Labs (ENoLL). Co-design of the NBS has been performed two-fold: (1) as education of wide stakeholders about NBS via adjusted NBS Catalogue, selection of optimal NBS via polls and discussions, NBS community mapping technique—both physically during on-site workshops (Figure 2) and virtually on the miro board, as well as through NBS contest for conceptual design of the Linear park; and (2) as co-creation of urban parameters for the Plan of Detailed Regulation of the Linear Park, via online expert discussions and online public debate, that will be proposed in the following official Public Consultation process and, if accepted, will allow NBS design in the urban plan implementation phase.

Local Belgrade CLEVER Cities team of policy-makers, researchers and practitioners has been using the CLEVER Cities specific tools, templates and reports [44], which has been perceived as significant for the legitimacy of the process and wider acceptance and comprehension of the NBS planning. Within CLEVER Cities co-creation pathway, two main approaches to stakeholder engagement are recognized: non-participatory—one-sided methods, where knowledge is either imparted or extracted; and participatory—two-sided methods, which imply collaboration with others to generate change and new knowledge, where stakeholders lead the work and potentially take it forward (Table 1) [44].

Table 1. The citizen engagement approaches applied in the case of Linear park planning, according to the CLEVER Cities methodology [46].

Levels of Engagement	Nature of Approach	Description
Inform	non-participatory	The flow of information from the Local Belgrade CLEVER Cities team via bellab internet platform
Consult	non-participatory	Gaining the stakeholder's information and opinion through questionaries, interviews and academic research
Involve	participatory	Involvement of stakeholders in discussions about planning issues via focus groups, academic education, online expert discussions and online public debate
Collaborate	participatory	Full involvement of stakeholders in decision making via NBS conceptual design contest and co-creation of urban parameters
Empower	participatory	Stakeholders full involvement via supported official EPC propositions

In Belgrade context, involvement of multitude of stakeholders has been achieved as the pioneering success of the participatory planning approach.

During the summer and early autumn of 2019, several rounds of consultations between the Belgrade City Authority Office of the Chief Urban Planner and the Secretariat for Environmental Protection, and CEUS, as the local support partner to the City of Belgrade within the CLEVER Cities project, were organized. It was agreed that, in this case of the urban planning process that aims towards at least two innovative practices—mainstreaming of NBS and introducing the concept of co-creation—the public communication process should be tailored in accordance. This task beyond usual formal participation routines was entrusted to CEUS.

During Autumn 2019, CEUS with local partners conducted identification of CLEVER Cities project within the city local context, as well as Mapping and Engaging of Stakeholders. Within the field of community-based PAR [36], CEUS selected the urban living lab approach for further processes, in line with its distinct characteristics, as well as the CLEVER Cities Co-creation Methodology framework.

Urban living lab (ULL) is recognized as an emerging form of collective urban governance [47] but also as an approach or set of methods for reinforcing change in a co-creative way [48]. It is used for addressing complex urban development challenges, by collaborative innovation though involvement of diverse stakeholders [49]: citizens and community groups (enabled users), civil society, public administration, research and businesses. All the stakeholders actively contribute to co-creation in a real-life setting with territorial focus, in several phases: (1) research and joint exploration of challenges and needs from different perspectives, (2) development and experimentation in the real-life setting by prototyping, (3) testing and rebuilding the prototype, (4) evaluation and implementation and (5) commercialization [50–52].

In Europe, ULLs are increasingly seen as an explicit form of intervention delivering sustainability goals for cities [53] and "a tool or instrument to change mindsets, processes and material solutions" [48] (p. 18). They are used to bridge the gap between research and practice while achieving greater citizen participation and social cohesion.

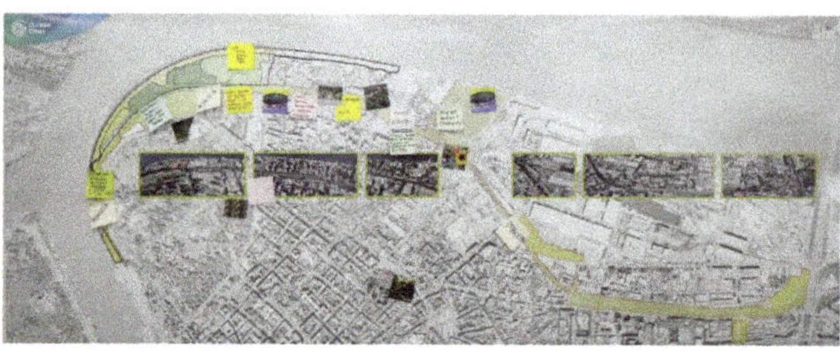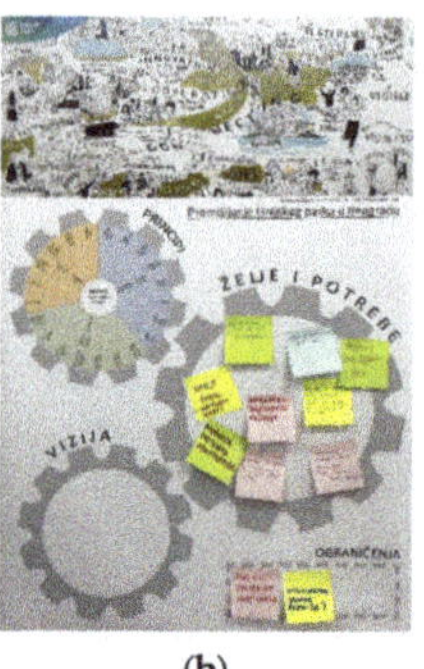

(a)   (b)

Figure 2. Results of the participatory mapping during the UIP launch: (**a**) community-mapping of possible NBS; (**b**) defining visions, SWOT and co-benefits of NBS [54] (Figure is produced by the Author).

2.2.2. Co-Creation in Practice of Planning the Linear Park

In November 2019, official launch of Belgrade's urban innovation partnership (UIP) was organized [54], and it was comprised of several introductory presentations on NBS and co-creation, and a discussion and initial mapping of challenges, opportunities and visions (Figure 2).

UIP was formalized with establishment of a task force by the Mayor's Decision [55]. It gathered representatives from 41 institutions (more information can be seen on bellab.rs, accessed on 16 March 2021), divided into seven distinct sections, which include the Core Project Team, educational and cultural institutions, sport and recreational institutions, public organizations and utility companies, private sector developers, academia, expert associations and SME's and (finally) national-level institutions (the last two among the institutions previously listed) [55].

Two focus groups followed UIP's establishment in December 2019 and January 2020, organized and facilitated by CEUS, when five new institutions joined the partnership and significantly contributed to its work. Particularly strong interest and proactive inputs and ideas were received from secondary schools and public cultural and educational institutions, namely, future beneficiaries of the Linear Park.

CEUS also developed a unique online platform to support its local ULL, named Belgrade urban living LAB (BELLAB) [56]. It is a comprehensive repository of all the actions taken, with very detailed minutes, questionnaire and polls results, individual inputs, illustrations, etc. It also serves as a medium for announcing future events, exchanging relevant news and establishing contacts with new UIP members and interested citizens.

Through the BELLAB platform and the UIP network, and in coordination with the City of Belgrade, in December 2019 CEUS launched an online questionnaire for citizens, on desired programs, content and activities, to be planned in the Linear Park [57]. The questionnaire was promoted in organized public events, via social media and TV reportage, and it was filled in by 570 citizens. Its results (Figure 3) revealed citizens' interest in using open public spaces: skate-parks, amphitheatres, multifunctional plateaus, community gardens, artistic pavilions, cultural-historical paths, green creative corridors, eco-educo centers, etc. The results also confirmed high interest in urban agriculture: 57.6% of examinees confirmed that they are interested in practicing urban agriculture, but almost 40% of them (22.4% of total responders) were concerned that urban agriculture would be too complicated. Answers to the "open question" about the park content revieled that responders highly appreciate urban biodiversity and simple green spaces and that they prefer landscape design over urban design. They even proposed "nature as the main creator" and renaturing of this urban corridor by ecological succession [57].

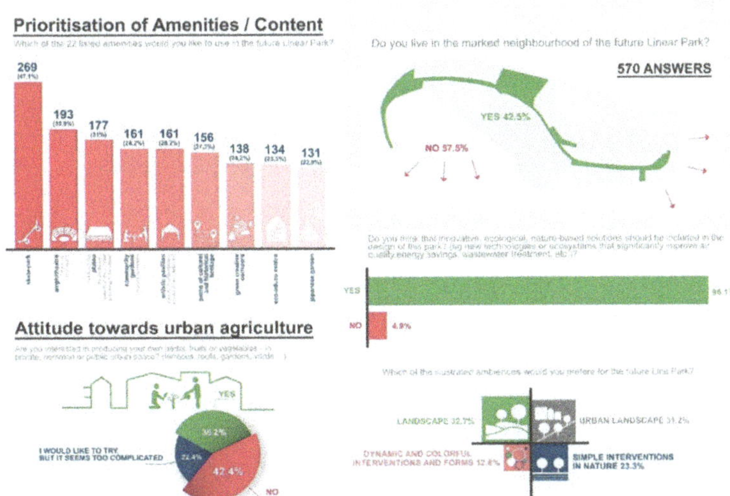

Figure 3. Results of the Linear Park Questionnaire [57] (Figure is produced by the Author).

In parallel, a call for young transdisciplinary teams was drafted by CEUS and finalized in collaboration with the Office of the Chief Urban Planner, Association of Belgrade Architects and EuPOLIS consortium. The call was announced in December 2019 [58]. A total of 145 young people in 28 teams answered this call, and criteria for selection of ten winning teams were: (1) proposed conceptual approach; (2) application of NBS; and (3) previous candidates' experiences, according to CVs and portfolios. At the end of February 2020, the ten teams were chosen by the Professional Committee comprised of 10 members (including the authors of this article) to develop conceptual designs for 10 sections of the Linear Park. They gathered 49 young authors (architects, landscape architects, civil engineers, electrotechnical engineers and chemical engineers) with 15 collaborators (additional transport engineers, mechanical engineers, engineers of urban planning and regional development, biologists, etc.). The programming basis of their designs was the results of the aforementioned questionnaire [57].

In February 2020, the first broader public event was organized, comprised of a panel of presentations of the CLEVER Cities partners, a discussion with citizens, a workshop for community mapping and an exhibition of NBS examples prepared by Master students of the University of Belgrade—Faculty of Architecture (UBFA) [59]. Over 130 people attended this event, and questions and comments from participants were pro-active, focused and constructive, which evidenced that the local community is dynamic and highly interested and motivated to (self) organize and invest their time, expertise and other resources for the future of this important public space and a new green oasis of Belgrade. Catalogue of NBS for Urban Regeneration [60], and the Co-creation Guidance [44,45] were translated into Serbian language and adapted to the local context, in collaboration with the students of the UBFA. Based on those materials, citizens could select the most desired NBS for the Linear Park and place them on the map of the area, as well as any other input from their own, local perspective (Figure 4). For public spaces interventions, citizens expressed interest, in particular, for NBS such as Infiltration Areas and Porous Paving, Community Gardens, Urban Bee-keeping, Facilities for Birds and Fauna, Butterfly Park, Urban Fruit Trees, Sensory Gardens, Urban Flower Fields, Usage of Treated Surface Water, The Living Garden Concept, Islands of Coolness, Green Noise Barriers, Eco-Urban Furniture, Shade provided by vegetation, etc. For new structures and complexes planning and design, such as Marina Dorćol, citizens believe that plans and technical documentation should integrate the following Building-Scale Interventions: Green Walls, Green Roofs, Urban Rooftop Farming, Rainwater Collection, etc. [59].

Figure 4. Results of the community mapping during the public workshop [59] (Figure is produced by the Author).

In April 2020, it was announced in the media that SEBRE Marina Dorćol company would donate resources for young architectural teams to develop conceptual designs for the Liner Park, from the beginning of June, by the end of August 2020 [61]. All those processes occurred before any formal citizens engagement procedure, namely before the EPC.

The EPC for the Linear Park was announced on the first day of the Planning Committee's work following the COVID 19 lock-down and conducted in the period 13–27, May, 2020. Due to prohibition of public gatherings in May, CEUS organized an online consultation process for the members of the UIP and the broader citizenry using the ZOOM application, with the possibility of sending comments and questions in advance. Since the first COVID 19 pandemic wave had a significant impact on people and shifted their focus and interest, and due to still insufficient general skills for public discussions in an on-line realm at that moment, a total of 25 participants attended this meeting. However, those who did attend the meeting very clearly expressed their doubts, wishes and suggestions. A poll for prioritising NBS was organized as well, and Community Gardens were voted as the most desired NBS. On 27 May, CEUS submitted suggestions regarding the Draft Plan development in 27 points to the Secretariat for Urban Planning and Construction, based on the results of all the consultative processes conducted before and during the EPC process [62]. No significant objections were received by the CEUS team nor communicated in the media, including social media.

However, there are still significant challenges in planning and implementation of the Linear Park that can be expected: construction of the cultural, educational, commercial and sports facilities (37,250 m^2) within the green areas [40], relocation of the sub-standard settlements along the railroad [63], etc. However, the CLEVER Cities team believes in the initial sense of ownership created and good synergies among institutions, and will advocate for a transparent, collaborative approach in the subsequent steps of this urban plan development to implement NBS and green infrastructure.

2.3. Avala Film Complex (Košutnjak Urban Forest)

The Avala Film Complex privatization process occurred in April 2015 and caused dissatisfaction and distrust among citizens [64]. The Avala Studio's d.o.o. (70% owned by the SEBRE company), Avala Film Complex's new owner [65], obtained the rights over valuable cultural heritage (producer's right over 600 movies) and the *right of land use* over 37 ha of urban land. According to Serbian Law on Planning and Construction [21] and Law on Conversion of Right of Use into the Right of Ownership Over the Construction Land with a Fee [66], in cases of privatization, the *right of land use* can be converted into the *right of ownership* without any cost.

In December 2017, following the Avala Studio's d.o.o. initiative, the Assembly of the City of Belgrade adopted a Decision on developing the Plan of Detailed Regulation (PDR) for Avala Film Complex [67] for an app. 86.8 ha of land. From the moment of enactment of the Decision on the plan development, until the formal EPC announcement, no communication was initiated with the expert or broader public regarding this initiative (Table 2).

Table 2. The EPC citizen engagement in the case of the Avala Film Complex (Košutnjak Urban Forest).

Levels of Engagement	Nature of Approach	Description
Inform	non-participatory	Media announcements on urban planning initiation and EPC
Consult	non-participatory	No activities
Involve	participatory	No activities
Collaborate	participatory	No activities
Empower	participatory	No activities Spontaneous civil engagement supported by NGOs and academia

EPC for the Avala Film Complex was announced on 29 June 2020 and lasted until 13 July. The EPC Elaborate [68] revealed that surprising land-use changes had been proposed. Along with the construction of the app. 80,000 m^2 of public facilities and complexes, the construction of 422,000 m^2 of residential space, 147,000 m^2 of commercial space and 42,000 m^2 of sports facilities were planned (Figure 5). According to the EPC Elaborate, such real estate development would destroy over 16 ha of the Košutnjak urban forest [68].

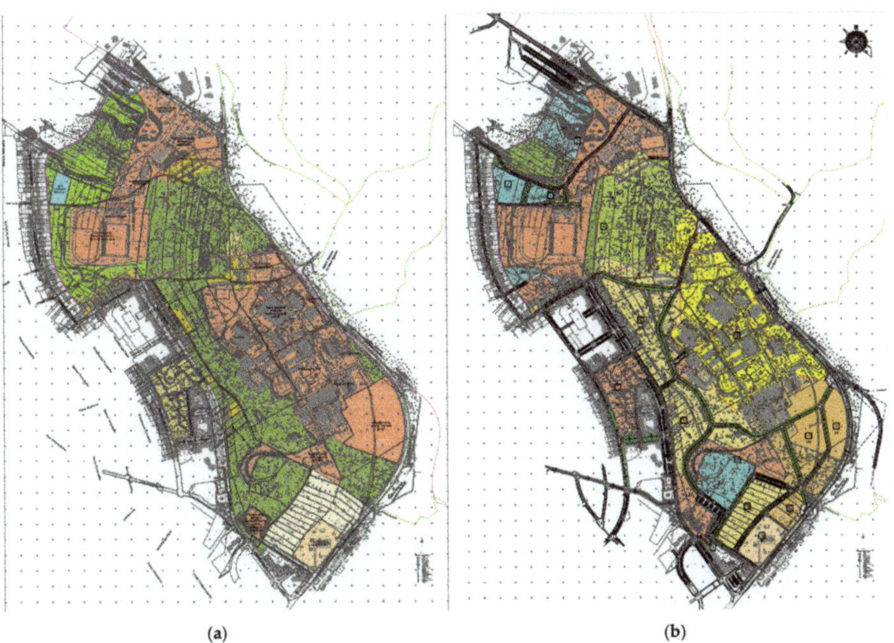

(a)   (b)

Figure 5. PDR for Avala Film Complex: (**a**) left—existing land-use, and (**b**) right—planned land-use [68].

As soon as this information was made public, a strong civil engagement swiftly got organized against the planning initiative. An online petition against forest destruction got almost 15,000 signatures in just one day and over 30,000 signatures until the deadline for submission of complaints on EPC Elaborate [69]. Signatures are still being collected, and currently there are over 76,000 electronic signatures [70]. Furthermore, over 7500 elaborated hard copy, signed official complaints were submitted by citizens [71], collected within a week. Numerous professional associations, ecology movements, students' unions, and four representatives of the academia (three Deans of the University of Belgrade: Faculties of Forestry, Biology and Sport, and Physical Education, and Director of the Institute for Biological Research) publicly criticized the project [72].

Following the EPC process, the Chief Urban Planner, as the president of the Planning Committee, announced in early September that the city had decided to completely stop development of the PDR for the Košutnjak area, in line with the Planning Committee's Conclusion [73]. Although many media and civic groups celebrated this information as the citizens' victory, it soon became apparent that the planning process is just delayed but not terminated. The urban parameters and capacities will be reconsidered within the same development concept [74]. This was confirmed with the Secretariat for Urban Planning and Construction's official answer regarding free access to public interest information, clarifying that due to numerous complaints, it was not possible to presume when the Planning Committee will adopt the Report on EPS and conclude that the EPC process is finalised [75]. Urban Planning Institute's Working Plan for 2021 includes the continuation of the PDR for Avala Film Complex.

The citizen group "Pozdrav sa Košutnjaka" published their manifesto, requiring a change of the plan's title, protecting the forest, developing the new elaborate, and initiating procedures for the protection of the area as natural and cultural heritage [75]. This group also conducted analyses of the Avala Film Complex's property rights, concluding that only over 11 cadastre plots, namely, 2.7 ha (out of 87 ha within the scope of the plan), Avala Studios d.o.o. have ownership rights. However, the company does have the privilege of use over a much larger area (app. 40 ha), and they can efficiently conduct conversion of rights in the National Cadastre. Moreover, the Group analyzed land use and land cultures and noted that some of the plots (covering over 30 ha) are marked as prime forest land regarding the "culture". However, they are also marked as "urban construction land" regarding land use [76]. Thus, there is a lot of research and interpretation, as well as a lot of misinterpretation and misleading information. That is why expert facilitation and professionally-led co-creation is crucial.

The other citizen group, "Bitka za Košutnjak", submitted the initiative on 31 July 2020, to the Assembly of the City of Belgrade to terminate this plan in the legal procedure. Following several media texts in which the Serbian President criticized the current Elaborate for Košutnjak Urban Forest, the group "Bitka za Košutnjak" sent him an official invitation to sign the Petition against this plan: first as the citizen, and a month later as the President of the Republic of Serbia and all its citizens [77]. As expected, no answer was obtained. Finally, this group collected additional 2420 signatures, for initiating the procedure of obligatory public voting of all councillors at the next session of the Assembly, with media coverage, so that everybody could see who in the City Assembly had defended the public interest [71].

3. Results

This article explores how various urban planning cases can be understood and analyzed as different forces with potential to induce the long-term systemic change, and thus analyzed and interpreted as different analytical levels of the multi-level perspective framework (MLP). The urban planning formal process, although common for both cases, is observed as a socio-technical regime of the MLP. In such framework, we recognize co-creative planning of the Linear Park as a niche innovation, namely, the "protected space" provided by the CLEVER Cities project. We interpret opposition towards planning of the Avala Film Complex as escalation, or extreme element of the socio-technical landscape, comprised of civic unrests and political tensions on one side, combined with climate crisis and excessive pollution on the other side. Those interpretations will be further elaborated in the following sub-chapters.

3.1. Linear Park as the Niche Innovation: Follow-Up and Expected Results

CLEVER Cities project provided a niche and allowed introduction of novelties in the formal planning procedure for the Linear Park. Citizen engagement from the very initial moment of the plan development, active public participation, careful *expectations* management, articulation of *visions* [26] (p. 28) and gradual building of trust make this planning practice the first example of co-creation in Belgrade and have the potential to become a role model for future co-creative NBS and greening strategies.

Following the formal EPC for the PDR of the Linear Park, CEUS team kept regular communication with the expert and the wider pubic and kept building social *networks* [26] (p. 28) by organizing several online events: Online Discussion with presentation of the Zone 4, during the EU Green Week in October 2020 (85 participants); Educational Session of Presentations and Discussions between Designing Teams and Students in November 2020 (60 participants); Expert Discussion on Urban Parameters for NBS in November 2020 (35 participants); and Discussion on Instruments for Co-creation of NBS in Serbia and Ecological Index, during The Nature Of Cities global online festival in February 2021 (65 participants). All those interactions allowed for *"learning and articulation processes* on various dimensions" [26] (p. 28). CEUS also established a quarterly Newsletter, which has 600 subscribers, and prepared a miro board for online community mapping [78].

Public Consultation for this urban plan is expected in May and June 2021, and in order to achieve a higher sense of belonging, ownership of spaces, and citizen-centered solutions, as well as the legitimacy of the procedures from the European perspective, the following subsequent activities are planned: Open Public Discussion about the Draft Urban Plan, with review of Conceptual Designs and final NBS selection, intensive communication among stakeholders, and Peer Review of the Public Consultation/Draft Urban Plan among the CLEVER Follower Cities.

In order to keep the citizens engaged and keep the momentum going, it is planned to strengthen collaboration with local schools and cultural and educational institutions around the Linear Park, organize local volunteering actions (e.g., demonstration urban farming) and promote NBS via youth contests and challenge prizes.

The only (but significant) risk factor for this specific urban plan at the moment is the strong distrust of citizens towards the investor of the Marina Dorćol and Linear Park conceptual designs—the SEBRE company, created by the unclear intentions regarding the Avala Film Complex. On the other hand, this crisis is also an opportunity for both the government and the investor to realize how different approaches towards the citizens create radically different results and may be highly beneficial and informative for future planning practice.

Authors of this article believe that Linear Park planning example has a strong potential to "provide the seeds for systemic change" [26] (p. 27); thus, systematization of the utilised tools and principles will be conducted, for possible future use in similar situations in the Western Balkans and the EU enlargement context.

Informal Instruments for Co-Creation of NBS in Niche Innovations

It is critical to reflect on communication strategy and informal urban planning instruments applied in this case, that can be implemented by the practitioners of the niche innovations, while avoiding conflicts, and even benefiting from the existing civic pressure and political tensions, as the socio-technical landscape which contributes to sustainability transition of the urban planning system.

In co-creative planning of the Linear Park, several tools and instruments have been introduced (Figure 6):

- A wide urban innovation partnership was established;
- An online platform for information exchange was created;
- A questionnaire regarding the content was conducted;
- Thematic focus groups were organized;
- Public presentations and a workshop were held;
- Community mapping was introduced—both on site and online;
- Open discussion during EPC was held;
- Joint cumulative remarks in the official planning procedure were prepared;
- Several complementary online discussions were organized;
- Catalogue of NBS examples was prepared;
- Design competition for multidisciplinary teams was organized;
- Conceptual design was co-created in collaboration with 10 teams.

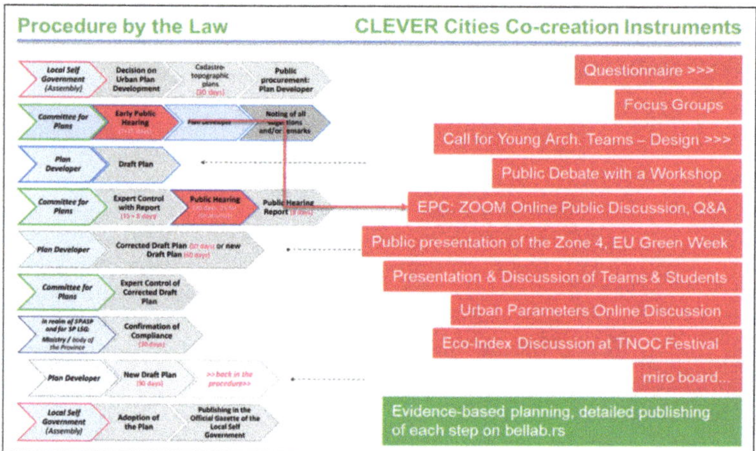

Figure 6. Illustration of the stakeholder participation and citizen engagement procedures defined by the Law (interpreted by the Author), and complementary, informal co-creation instruments tested within the Niche of the CLEVER Cities project and planning of the Linear Park (Figure is produced by the Author).

These instruments allowed empowered stakeholders to take an active role in urban planning, e.g., by (1) preparation of cumulative, joint remarks in the official planning procedure; (2) nature-based solutions introduction to the wider citizenry by the Catalogue of good NBS practices, prepared within the academic collaboration; and (3) urban design capacity building of the young practitioners in the winning teams, etc. More critical assessment of the effects of these instruments will be conducted in the subsequent research, following this urban plan adoption and clear evaluation of co-creation for integrating NBS in urban planning. Although these instruments cannot be claimed as relevant for all the innovation niches, as conceived in general MLP middle-range theory, it is believed that these or similar tools can be highly beneficial in urban planning practice, if further exploited in the Serbian and the EU enlargement context.

3.2. Avala Film Complex as an Extreme Element of the Socio-Technical Landscape: Follow-Up and Expected Results

The goal of the Avala Film Complex re-development, at least declaratively, was and is the enhancement of the Serbian film industry [68]. However, according to the PDR Elaborate and planning proposition, it is evident that the new owner is a real estate developer, with the intention to build a new city quarter, mainly residential and commercial, instead of protecting the current "forest within public facilities" [68]. Concerning these ambitions, significant public opinion emerged, assuming that Belgrade's green infrastructure network is seriously threatened and NBS mainstreaming is highly undermined.

It is highly unexpected that the expert and wider public will accept the existing urban development concept for the Avala Film Complex. Civic groups announced mass protests as soon as the COVID-19 pandemics allows it. Citizens claim their readiness to defend the forest with their own bodies if such radical measures become necessary.

While many authors in early September described protection of the Košutnjak Urban Forest as the "outline of new local politics, which achieved an important victory" [65] and "success of public participation" [79], it seems to be a complete failure regarding urban planning processes, co-creation and NBS mainstreaming.

Nevertheless, authors of this article believe that civic pressure created around Avala Film Complex represents the escalation of dissatisfaction with the current urban planning system and argue that it can be interpreted as an extreme element of the socio-technical landscape, which will be highly significant for the establishment of substantially better planning praxis. "Stopping of the process for [. . .] plan is maybe the critical moment in which participatory urban planning may become a regular model, which will be used from the early stage until its full implementation [. . .] with reducing the chances of "duplicating the work" for the Urban Planning Institute of Belgrade if Draft Plans can be developed without negative feedback from citizens" [79].

Although Avala Film Complex obviously does not represent the participatory urban planning, nor paves the road for such pro-active practice, its re-active strength, expressed by defending the public interest in the street and in a rebellious manner, reflects societal values and significantly influences regime dynamics. It is highly valuable "in interaction with processes at different levels" [26] (p. 29), namely, with dynamics of the Linear Park niche innovation.

4. Discussion

Performed multiple-case study results indicate the possibility of sustainability transition in Belgrade context by destabilization of the socio-technical regime regarding the urban planning system, by "active participation, struggle and negotiation" [80].

Since the socio-technical regime represents the "deep structure" [81], comprised of "cognitive routines and shared beliefs, capabilities and competences, lifestyles and user practices, favourable institutional arrangements and regulations, and legally binding contracts [. . .] characterized by lock-in" [26], it is necessary that niche innovation, such as Linear Park example, "build up internal momentum" that "changes at the landscape level create pressure on the regime", such as Avala Film Complex opposition, so that regime's destabilization "creates window of opportunity for niche-innovations" [26] (p. 29).

This multiple-case study illustrates how niche innovation of co-creation in planning of the Linear Park, and socio-technical landscape of the civil pressure escalated around planning of the Avala Film Complex, are "processes in multiple dimensions and at different levels which link up with, and reinforce, each other ("circular causality")" [26] (p. 29) (Figure 7).

In Belgrade, there are indications that the two case-studies described in this article already have significant although indirect impact on systemic change of the planning practice, interpreted as the socio-technical regime in this research.

In particular, development pathway of the General Urban Plan (GUP) for Belgrade 2041 can be encouraging, for several reasons: (1) at the online discussion during The Nature Of Cities festival, organized by CEUS, representatives of the Urban Planning Institute for the first time communicated their strategic framework regarding the GUP with the public, and for the first time they interacted with criticism of well-established urban planning activists (Collective "Ministry of Space"); (2) subsequently, the second author of this article joined the GUP expert team and supported preparation of the citizen engagement strategy; (3) initial questionnaire for preparation of the GUP EPC Elaborate was published in March 2021, promising a more inclusive and participatory procedure than the usual practice.

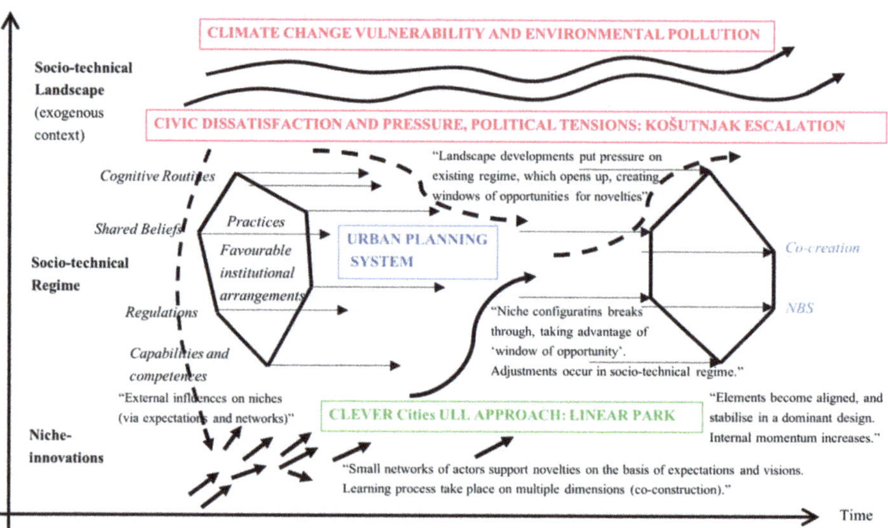

Figure 7. Interpretation according to Geels [26]: the multi-level perspective on sustainability transition regarding urban planning in Belgrade (Figure is produced by the Author).

The MLP framework on sustainability transition has been commonly used for interpreting long-term changes in the energy sector [80], GMO and food production [27], organic food and eco-housing [82], etc. This research, however, focuses on urban planning system as a complex formal procedure and uses its' own elements, namely, particular planning cases, as illustrations for different analytical levels of the framework. Although bearing the risk of ambiguity, this approach is considered innovative and appropriate to the socio-political and geographical context of the study. Moreover, the focus of co-creation, as a social innovation rather than a technological one, brings added value to this research.

5. Conclusions

The MLP framework for sustainability transition regarding urban planning system can be further analyzed, with unfolded interactions sub-divided into several phases. However, the initial conclusion is that urban sustainability transition in this context is possible and should be expected.

By comparison of the two case studies of citizens' engagement regarding urban planning of green infrastructure in Belgrade (Table 3), it is evident that the reactions of citizens against the unwanted proposal at the moment are much easier, prompter and more numerous than proactive citizens' engagement. This is understandable, considering the long tradition of centralized planning in a post-socialist context and lack of planning dialogue experience in the past. This is also the major challenge for urban planners and NBS advocates in the Serbian and the EU enlargement context. This article highlighted the importance of tailored and contextually sensitive early communication of the planning initiatives as the cornerstone of successful and sustainable urban development. When planning significant urban transformations related to urban greening resources or potentials, it is critical to perform high sensitivity for the site, and socio-economic and environmental specificities and values. Initiation of the co-creation process before the formal urban planning procedure (EPC) is essential, as well as the utilization of complementary, informal co-creation tools to enable trustworthy and constructive communication.

Table 3. Summary of the MLP framework interpretation, describing the case-studies at various analytical levels and in relation to NBS and co-creation. (Table is produced by the Author).

MLP Framework	Interpretation of the Three Analytical Levels and Their Reinforcements	NBS Considerations?	Co-Creation Processes?
Opposition to planning of the **Avala Film Complex** redevelopment as an extreme element of the **Socio-Technical Landscape**	- Accumulated dissatisfaction with non-transparent procedures and introverted urban planning; - Civic pressure, and political tensions - Concerns regarding climate vulnerability and air pollution - Extreme manifestation with mass mobilization against the concrete redevelopment initiative	- Reduction of 16 ha of urban forest in the land use overview - None NBS considerations - Evidence on the lack of systemic approach to NBS	- No communication prior to EPC - No co-creation - Misleading communication after the EPC
Urban planning system in Belgrade, Serbia, as a **Socio-Technical Regime**	- Post-socialist transition introduced multiple interest in urban development, but there are no adequate tools for addressing this variety of interests - Strong political influence; - Modest financial instruments	- Declarative care for green infrastructure, rather than practical - Ecological index or green urban parameters not yet in usage	- Post-socialist heritage influence perception of participation - Lack of professional capacities for facilitation discourage co-creation
Linear Park ULL as a **Niche Innovation**	- Developed in the protected space of the CLEVER Cities Horizon 2020 project - Careful *expectations* management was conducted, as well as articulation of *visions* - Via UIP, social *networks* were built up and *learning and articulation processes* were facilitated on various dimensions - This ULL has the potential to "provide the seeds for systemic change" [27]	- NBS have been mainstreamed in urban planning, via advocating green urban parameters - NBS Catalogue has been prepared and communicated to expert and wider public - NBS have been elaborated and promoted via community mapping, polls and collaborative prioritization - Co-design and prototyping of NBS via design competition	- Cross-sectoral collaboration - Citizen engagement and community empowerment - Introduction of new, informal planning instruments - Building of trust via regular and transparent communication

From the experience of the Linear Park planning, several important principles of **co-creation** in the EU enlargement context have been recognized:

I. **Formalization** of stakeholders' engagement is important, in the form of a task force, such as urban innovation partnership;

II. **Building of trust** is critical, and can be obtained by external facilitation (by engaging representatives of professional association or academia, rather than the local self-government); regular communication and transparent information about processes are highly appreciated (e.g., via an online platform—web interface, Newsletter, etc.);

III. **Direct communication** is indispensable, in various forms: interviews, focus groups, workshops, presentations, discussions, etc.

IV. **Co-design and prototyping** give stakeholders and citizens a strong sense of ownership; examples include a design competition for young transdisciplinary teams and conceptual designs of multiple teams;

V. **Empowerment of stakeholders** during the process is crucial for long-term constructive changes.

This research also paves the road for future evaluation of the impacts of co-creation in the subsequent phases of the two urban transformations, as well as further exploration of the ULL approach and its impact on urban sustainability transition in the EU enlargement context.

Author Contributions: Conceptualization, A.M.-R. and K.L.; methodology, A.M.-R. and K.L.; validation, A.M.-R. and K.L.; formal analysis, A.M.-R.; investigation, A.M.-R.; writing—original draft preparation, A.M.-R. Writing—review and editing, K.L.; visualisation, A.M.-R.; supervision, K.L.; project administration, A.M.-R.; All authors have read and agreed to the published version of the manuscript.

Funding: This research was funded by the European Union's Horizon 2020 innovation action program under grant agreement number 776604.

Institutional Review Board Statement: Not applicable.

Informed Consent Statement: Not applicable.

Data Availability Statement: Data presented in this study are available on request from the corresponding author.

Acknowledgments: This work has been presented at the Greening Cities Shaping Cities Symposium in October 2020 (https://www.greeningcities-shapingcities.polimi.it/ accessed on 15 March 2021). The work was partly prepared in the framework of the European Horizon 2020 project Clever Cities. Herewith, the authors would like to thank Israa Mahmoud and Eugenio Morello (Politecnico di Milano) as editors of this Special Issue. The authors would like to thank Marija Martinović for her feedback on the last version of this paper. They are also grateful to the three anonymous reviewers of this journal for their helpful comments and suggestions.

Conflicts of Interest: The authors declare no conflict of interest. The funders had no role in the design of the study; in the collection, analyses, or interpretation of data; in the writing of the manuscript; or in the decision to publish the results.

References

1. Escobedo, F.J.; Giannico, V.; Jim, C.Y.; Sanesi, G.; Lafortezza, R. Urban forests, ecosystem services, green infrastructure and nature-based solutions: Nexus or evolving metaphors? *Urban For. Urban Green.* **2019**, *37*, 3–12. [CrossRef]
2. Benedict, M.A.; McMahon, E.T. Green Infrastructure: Smart Conservation for the 21st Century. *Renew. Resour. J.* **2002**, *20*, 12–17.
3. Chatzimentor, A.; Apostolopoulou, E.; Mazaris, A.D. A review of green infrastructure research in Europe: Challenges and opportunities. *Landsc. Urban Plan.* **2020**, *198*. [CrossRef]
4. Kabisch, N.; Korn, H.; Stadler, J.; Bonn, A. *Nature-based Solutions to Climate Change Adaptation in Urban Areas—Linkages between Science, Policy and Practice*; Springer: Cham, Switzerland, 2017; pp. 1–337.
5. Ahern, J. Green Infrastructure for cities. The spatial dimension. In *Cities of the Future. Towards Integrated Sustainable Water and Landscape Management*; Novotny, V., Ed.; IWA Publications: London, UK, 2007; pp. 267–283.

6. IUCN. The IUCN Programme 2013–2016. Available online: https://portals.iucn.org/library/node/10320 (accessed on 16 March 2021).
7. IUCN. The IUCN Programme 2017–2020. Available online: https://portals.iucn.org/library/node/46366 (accessed on 16 March 2021).
8. IUCN. *IUCN Global Standard for Nature-Based Solutions*, 1st ed.; IUCN: Gland, Switzerland, 2020; ISBN 978-2-8317-2058-6. Available online: https://portals.iucn.org/library/node/49070 (accessed on 16 March 2021)
9. European Commission. Towards an EU Research and Innovation Policy Agenda for Nature-Based Solutions & Re-Naturing Cities. *Final Report of the Horizon 2020 Expert Group (Full Version)*. 2015. Available online: https://op.europa.eu/en/publication-detail/-/publication/fb117980-d5aa-46df-8edc-af367cddc202 (accessed on 16 March 2021).
10. European Commission. Nature-Based Solutions: State of the Art in EU-funded Projects. 2020. Available online: https://ec.europa.eu/info/publications/nature-based-solutions-state-art-eu-funded-projects_en (accessed on 30 March 2021).
11. European Commission. Nature-Based Solutions Research Policy: EU Research Policy, What Nature-Based Solutions Are, Background, News and Documents. Available online: https://ec.europa.eu/info/research-and-innovation/research-area/environment/nature-based-solutions/research-policy_en (accessed on 10 April 2021).
12. Mahmoud, I.; Morello, E. Co-Creation Pathway as a catalyst for implementing Nature-based Solution in Urban Regeneration Strategies Learning from CLEVER Cities framework and Milano as test-bed. *Urban. Inf. Spec. Issue* **2018**, 204–2010.
13. Mauser, W.; Klepper, G.; Rice, M.; Schmalzbauer, B.S.; Hackmann, H.; Leemans, R.; Moore, H. Transdisciplinary global change research: The co-creation of knowledge for sustainability. *Curr. Opin. Environ. Sustain.* **2013**, *5*, 420–431. [CrossRef]
14. Frantzeskaki, N.; Kabisch, N. Designing a knowledge co-production operating space for urban environmental governance—Lessons from Rotterdam, Netherlands and Berlin, Germany. *Environ. Sci. Policy* **2016**, *62*, 90–98. [CrossRef]
15. European Commission. EU Biodiversity Strategy, Bringing Nature back into Our Lives. EC COM(2020) 380 Final. Available online: https://ec.europa.eu/environment/strategy/biodiversity-strategy-2030_en (accessed on 21 April 2021).
16. European Commission. Candidate Countries and Potential Candidates. Available online: https://ec.europa.eu/environment/enlarg/candidates.htm (accessed on 10 April 2021).
17. Aleksić, B. Mesna zajednica—Osnove za razmatranje prostorno-funkcionalnih svojstava, u: Mate Bylon, (ur.). In *Prilog Poslediplomskog Kursa Stanovanje Naucnom Skupu 'Mesna Zajednica' u Beogradu, Juna 1978*; Arhitektonski Fakultet: Belgrade, Serbia, 1978; p. 23.
18. Martinović, M. Jugoslovensko Samoupravljanje U Arhitekturi Centara Mesnih Zajednica U Beogradu Od 1950. DO 1978. Ph.D. Thesis, Faculty of Architecture, University of Belgrade, Belgrade, Serbia, 26 September 2020.
19. Vujošević, M.; Zeković, S.; Maričić, T. Post-Socialist Transition in Serbia and Its Unsustainable Path. *Eur. Plan. Stud.* **2012**, *20*, 1707–1727. [CrossRef]
20. Zekovic, S.; Vujošević, M.; Maričić, T. Spatial regularisation, planning instruments and urban land market in a post-socialist society: The case of Belgrade. *Habitat Int.* **2015**, *48*, 65–78. [CrossRef]
21. Law on Planning and Construction (Zakon o planiranju i izgradnji), "Official Gazette of the Republic of Serbia" No. 72/2009, 81/2009-rev., 64/2010—Decision of the CC, 24/2011, 121/2012, 42/2013—Decision of the CC, 50/2013—Decision of the CC, 98/2013—Decision of the CC, 132/2014, 145/2014, 83/2018, 31/2019, 37/2019—Other Laws and 9/2020. Available online: https://www.paragraf.rs/propisi/zakon_o_planiranju_i_izgradnji.html (accessed on 13 May 2020).
22. Združene Inicijative i Pokreti. Available online: http://www.zdruzeno.org/o-nama/ (accessed on 10 April 2020).
23. Plan of General Regulation of the System of Green Areas of Belgrade, Official Gazette of the City of Belgrade No. 110/2019. Available online: http://www.sllistbeograd.rs/pdf/2019/110-2019.pdf#view=Fit&page=1 (accessed on 13 May 2020).
24. City of Belgrade—Secretariat of Environmental Protection, Climate Change Adaptation Action Plan and Vulnerability Assessment. 2015. Available online: http://www.beograd.rs/images/data/c83d368b72364ac6c9f9740f9cda05ed_6180150278.pdf (accessed on 10 April 2021).
25. Geels, F.W. Ontologies, socio-technical transitions (to sustainability), and the multi-level perspective. *Res. Policy* **2010**, *39*, 495–510. [CrossRef]
26. Geels, F.W. The multi-level perspective on sustainability transitions: Responses to seven criticisms. *Environ. Innov. Soc. Transit.* **2011**, *1*, 24–40. [CrossRef]
27. Lawhon, M.; Murphy, J.T. Socio-technical regimes and sustainability transitions: Insights from political ecology. *Prog. Hum. Geogr.* **2011**, *36*, 354–378. [CrossRef]
28. Ernst, L.; de Graaf-Van Dinther, R.E.; Peek, G.J.; Loorbach, D.A. Sustainable urban transformation and sustainability transitions; Conceptual framework and case study. *J. Clean. Prod.* **2015**, *112*, 2988–2999. [CrossRef]
29. Loorbach, D.A.; Frantzeskaki, N.; Avelino, F. Sustainability Transitions Research: Transforming Science and Practice for Societal Change. *Annu. Rev. Environ. Resour.* **2016**, *42*, 599–626. [CrossRef]
30. Markard, J.; Truffer, B. Technological innovation systems and the multi-level perspective: Towards an integrated framework. *Res. Policy* **2008**, *37*, 596–615. [CrossRef]
31. Jørgensen, U. Mapping and navigating transitions—The multi-level perspective compared with arenas of development. *Res. Policy* **2012**, *41*, 996–1010. [CrossRef]
32. Yin, R.K. *Case Study Research: Design and Methods*, 4th ed.; SAGE Publications, Inc.: Thousand Oaks, CA, USA, 2009.
33. Rowley, J. Using case studies in research. *Manag. Res. News* **2002**, *25*, 16–27. [CrossRef]
34. Avison, D.E.; Lau, F.; Myers, M.D.; Nielsen, P.A. Action Research. *Commun. ACM* **1999**, *42*. [CrossRef]

35. MacDonald, C. Understanding participatory action research: A qualitative research methodology option. *Can. J. Action Res.* **2012**, *13*, 34–50.
36. Guidance on Co-Creating NBS. Available online: https://clevercitiesguidance.wordpress.com/ (accessed on 15 April 2021).
37. Danilović Hristić, N.; Stefanović, N. The role of public insight in urban planning process: Increasing efficiency and effectiveness. *Spatium Int. Rev.* **2013**, *30*, 1–7. [CrossRef]
38. Plan of General Regulation of the City of Belgrade (Plan Generalne Regulacije Građevinskog Područja Sedišta Jedinice Lokalne Samouprave—Grad Beograd) "Official Gazette of the City of Belgrade" No. 20/16, 97/16, 69/17 and 97/17. Available online: https://www.beoland.com/planovi/pgr-beograda/ (accessed on 1 October 2020).
39. The City Assembly of Belgrade, Decision on Development of the Plan of Detailed Regulation for Linear Park—Belgrade, Urban Municipalities of Old City and Palilula, Official Gazette of the City of Belgrade No. 88/2018. Available online: http://www.sllistbeograd.rs/pdf/2018/88-2018.pdf#view=Fit&page=14 (accessed on 13 May 2020).
40. Urban Planning Institute of Belgrade—URBEL. Plan of Detailed Regulation for the Linear Park—Belgrade, Urban Municipalities of Old Town and Palilula—Elaborate for Early Public Participation. Available online: http://mapa.urbel.com/publish/PDR_za_linijski_park%E2%80%93Beograd/01_tekst/PDF/01_tekst%20RJU_Linijski%20park.pdf (accessed on 13 May 2020).
41. City of Belgrade—Secretariat of Environmental Protection, Draft Green City Action Plan for the City of Belgrade. 2021. Available online: https://www.beograd.rs/images/file/741ccefb8afef0372f993fee441c847f_9117346413.pdf (accessed on 15 March 2021).
42. Mučibabić, D. Rešenje za Linijski Park u Septembru. 14 May 2020. Available online: http://www.politika.rs/scc/clanak/454105/Resenje-za-linijski-park-u-septembru (accessed on 1 July 2020).
43. Radonjić, M. Marina na Dorćolu Prodata za 32 MILIONA EVRA, Stigla Samo Jedna Ponuda, evo ko je KUPAC, 2019. Blic, 13 September 2019. Available online: https://www.blic.rs/vesti/beograd/marina-na-dorcolu-prodata-za-32-miliona-evra-stigla-samo-jedna-ponuda-evo-ko-je-kupac/8j6yjqr (accessed on 1 October 2020).
44. Morello, E.; Mahmoud, I.; Gulyurtlu, S. CLEVER Cities Guidance on Co-Creating Nature-Based Solutions: PART I—Defining the Co-Creation Framework and Stakeholder Engagement, Deliverable 1.1.5, CLEVER Cities, H2020 Grant No 776604. 2018. Available online: https://clevercitiesguidance.files.wordpress.com/2018/12/cocreation-guidance_part-i.pdf (accessed on 16 March 2021).
45. Morello, E.; Mahmoud, I.; Gulyurtlu, S. CLEVER Cities Guidance on Co-Creating Nature-Based Solutions: PART II—Running CLEVER Action Labs in 16 Steps. Deliverable 1.1.6, CLEVER Cities, H2020 Grant No 776604. 2018. Available online: https://clevercitiesguidance.files.wordpress.com/2019/09/20190913_cocreation-guidance_part-ii.pdf (accessed on 15 March 2021).
46. Mahmoud, I.; Morello, E. Co-creation Pathway for Urban Nature-Based Solutions: Testing a Shared-Governance Approach in Three Cities and Nine Action Labs. In *Smart and Sustainable Planning for Cities and Regions; Green Energy and, Technology*; Bisello, A., Vettorato, D., Ludlow, D., Baranzelli, C., Eds.; Springer: Cham, Switzerland, 2019. [CrossRef]
47. Voytenko, Y.; McCormick, K.; Evans, J.; Schliwa, G. Urban living labs for sustainability and low carbon cities in Europe: Towards a research agenda. *J. Clean. Prod.* **2015**, *123*, 45–54. [CrossRef]
48. Bylund, J.; Riegler, J.; Wrangsten, C. Are urban living labs the new normal in co-creating places? In *Co-Creation of Public Open Places. Practice-Reflection-Learning*; Smaniotto Costa, C., Mačiulienė, M., Menezes, M., Goličnik Marušić, B., Eds.; C3Places Project; Lusófona University Press: Lisbon, Portugal, 2020; ISBN 978-989-757-125-1. [CrossRef]
49. Bergvall-Kåreborn, B.; Holst, M.; Ståhlbröst, A. Concept Design with a Living Lab Approach. In Proceedings of the Hawaii International Conference on Systems Science (HICSS'42), Big Island, HI, USA, 5–8 January 2009.
50. Steen, K.; van Bueren, E. *Urban Living Labs—A Living Lab Way of Working*; AMS Research Report: Amsterdam, The Netherlands, 2017.
51. Steen, K.; van Bueren, E. The Defining Characteristics of Urban Living Labs. *Technol. Innov. Manag. Rev.* **2017**, *7*, 10–22. [CrossRef]
52. UNaLAB. Living Lab Handbook for Urban Living Labs Developing Nature-Based Solutions. 2020. Available online: https://unalab.eu/system/files/2020-07/living-lab-handbook2020-07-09.pdf (accessed on 20 January 2021).
53. Bulkeley, H.; Coenen, L.; Frantzeskaki, N.; Hartmann, C.; Kronsell, A.; Mai, L.; Marvin, S.; McCormick, K.; van Steenbergen, F.; Palgan, Y.V. Urban living labs: Governing urban sustainability transitions. *Curr. Opin. Environ. Sustain.* **2016**, *22*, 13–17. [CrossRef]
54. CEUS. Partnerstvo za Urbane Inovacije. Available online: http://bellab.rs/partnerstvo/ (accessed on 20 July 2020).
55. Mayor of Belgrade. Decision No. 020-7998/19-G from 27 November 2019. Available online: http://bellab.rs/wp-content/uploads/2019/12/RESENJE_UIP-Radna-grupa_CLEVER-Cities-projekat.pdf (accessed on 20 July 2020).
56. BELgrade Urban Living LAB. Available online: http://bellab.rs/ (accessed on 20 July 2020).
57. CEUS. Upitnik za Građane. Available online: http://bellab.rs/upitnik/ (accessed on 20 July 2020).
58. Služba Glavnog Gradskog Urbaniste. Poziv za Mlade Arhitektonske i Transdisciplinarne Timove za Kvalifikaciju za Izradu Idejnog Rešenja Dela Područja Linijskog Parka u Beogradu, DAB 18.12.2020. Available online: http://www.dab.rs/downloads/%D0%9F%D0%BE%D0%B7%D0%B8%D0%B2%20%D0%B7%D0%B0%20%D0%BA%D0%B2%D0%B0%D0%BB%D0%B8%D1%84%D0%B8%D0%BA%D0%B0%D1%86%D0%B8%D1%98%D1%83%20%D0%B7%D0%B0%20%D0%BF%D1%80%D0%BE%D1%98%D0%B5%D0%BA%D0%B0%D1%82%20%D0%BF%D0%BE%D0%B4%D1%80%D1%83%D1%87%D1%98%D0%B0%20%D0%9B%D0%B8%D0%BD%D0%B8%D1%98%D1%81%D0%BA%D0%B8%20%D0%BF%D0%B0%D1%80%D0%BA%2018.11.2019..pdf (accessed on 20 July 2020).
59. CEUS. Radionica za Građane 12 February 2020. Available online: http://bellab.rs/radionica/ (accessed on 20 July 2020).

60. Morello, E.; Mahmoud, I.; Colaninno, N. (Eds.) Catalogue of Nature-Based Solutions for Urban Regeneration, Energy & Urban Planning Workshop, School of Architecture Urban Planning Construction Engineering, Politecnico di Milano. 2020. Available online: http://www.labsimurb.polimi.it/nbs-catalogue/ (accessed on 3 October 2020).
61. Mučibabić, D. Politika. Za "Marinu Dorćol—150 Miliona Evra", 8 April 2020. Available online: http://www.politika.rs/scc/clanak/451815/Za-Marinu-Dorcol-150-miliona-evra (accessed on 1 July 2020).
62. CEUS. Rani Javni Uvid. Available online: http://bellab.rs/elaborat/ (accessed on 20 July 2020).
63. Mladenović, M. Gradnja. Linijski Park: Šta da Očekujemo i od čega da Strepimo, 29 May 2020. Available online: https://www.gradnja.rs/linijski-park-sta-da-ocekujemo-i-od-cega-da-strepimo/ (accessed on 10 September 2020).
64. Momčilović, P. Slučaj Košutnjak: Tragedija Uništavanja Javnog Dobra u tri čina, 7 July 2020. Available online: https://www.masina.rs/?p=14121&fbclid=IwAR0G2K2kFFj9pJU09VTIfIUryMxYk-ivkPtyk_Mil8IO_TyZI7n67LSA64I (accessed on 8 July 2020).
65. Ilić, S. Peščanik. Bitka za Košutnjak, 10 August 2020. Available online: https://pescanik.net/bitka-za-kosutnjak/ (accessed on 2 September 2020).
66. Law on Conversion of Right of Use into the Right of Ownership over the Construction Land with a Fee. Official Gazette of the Republic of Serbia; No. 64/2015 and 9/2020. Available online: https://www.paragraf.rs/propisi/zakon_o_pretvaranju_prava_koriscenja_u_pravo_svojine_na_gradjevinskom_zemljistu_uz_naknadu.html (accessed on 13 July 2020).
67. The City Assembly of Belgrade. Decision on Development of the Plan of Detailed Regulation for Avala Film Complex, Urban Municipality of Čukarica, Official Gazette of the City of Belgrade No. 97/2017. Available online: http://www.sllistbeograd.rs/pdf/2017/97-2017.pdf#view=Fit&page=13 (accessed on 13 July 2020).
68. Urban Planning Institute of Belgrade—URBEL. Plan of Detailed Regulation for the "Avala Film" Complex, Urban Municipality of Čukarica—Elaborate for Early Public Participation. Available online: http://mapa.urbel.com/publish/PDR%20za%20kompleks%20Avala%20filma/01_TEKST/01_tekst.pdf?fbclid=IwAR0Yh_RmYGWlgt0C5MyfmeQHYCHIoQQnZZ9KP0aXMf6dX1tMj0blA1TQmfQ (accessed on 8 July 2020).
69. N1info, N1 Beograd, NDM BGD: Sakupićemo 100.000 Potpisa za Odbranu Košutnjaka, ne Damo Pluća Grada, 13 July 2020. Available online: http://rs.n1info.com/Vesti/a619286/NDM-BGD-Sakupicemo-100.000-potpisa-za-odbranu-Kosutnjaka-ne-damo-pluca-grada.html (accessed on 1 October 2020).
70. Inicijativa za odbranu Košutnjaka/Initiative for Defending Košutnjak. Available online: https://www.peticije.online/inicijativa_za_odbranu_kosutnjaka (accessed on 1 October 2020).
71. Bitka za Košutnjak/Battle for Košutnjak, Facebook Private Group. Available online: https://www.facebook.com/BitkazaKosutnjak/ (accessed on 10 October 2020).
72. Danas. Stručnjaci protiv seče Košutnjaka. Available online: https://www.danas.rs/drustvo/strucnjaci-protiv-sece-kosutnjaka/?fbclid=IwAR1z0CxRw89ZbdA9fu7i6sOmMHyHaFqolrzTMYk0W9wJtTdpTfe_bO71Y3o (accessed on 12 September 2020).
73. RTS, Gradski Urbanista: Neće Biti Seče Šume na Košutnjaku, 4 September 2020. Available online: https://www.rts.rs/page/stories/sr/story/125/drustvo/4068850/urbanista-suma-kosutnjak-seca.html (accessed on 1 October 2020).
74. NIN, Košutnjak—Nije Još Sve Završeno, 4 October 2020. Available online: http://www.nin.co.rs/pages/article.php?id=102354757&fbclid=IwAR0J5Ei_hp2efntpVoH4QxCzp5C3qRt8ZmZi8MJvj0ADWrD5y0OYSvTWtrY (accessed on 5 October 2020).
75. Pozdrav sa Košutnjaka/Regards from Košutnjak, Facebook Private Group. Available online: https://www.facebook.com/pozdravsakosutnjaka (accessed on 10 October 2020).
76. E-Cadastre. Available online: https://katastar.rgz.gov.rs/eKatastarPublic/FindParcela.aspx?KoID=704083 (accessed on 10 September 2020).
77. Jovanović, D. "Bitka za Košutnjak" Predala Zahtev Vučiću: Sprečite Seču Šume, 22 September 2020. Available online: http://rs.n1info.com/Vesti/a652451/Bitka-za-Kosutnjak-predala-zahtev-Vucicu.html (accessed on 22 September 2020).
78. CEUS. Učestvuj. Available online: http://bellab.rs/ucestvuj/ (accessed on 14 March 2020).
79. Mladenović, M. Gradnja. Uspeh Participacije Građana: Avala Film i Kosančićev Venac Ostaju Kakvi Jesu (Za Sada), 8 September 2020. Available online: https://www.gradnja.rs/uspeh-participacije-gradjana-avala-film-i-kosancicev-venac-ostaju-kakvi-jesu-za-sada/ (accessed on 9 September 2020).
80. Wieczorek, A.J.; Berkhout, F. Transitions to sustainability as societal innovations. In *Principles of Environmental Sciences*; Springer: Berlin/Heidelberg, Germany, 2009; pp. 503–512. Available online: https://link.springer.com/chapter/10.1007/978-1-4020-9158-2_27 (accessed on 1 October 2020).
81. Geels, F.W. Understanding system innovations: A critical literature review and a conceptual synthesis. In *System Innovation and the Transition to Sustainability: Theory, Evidence and Policy*; Elzen, B., Geels, F.W., Green, K., Eds.; Edward Elgar: Cheltenham, UK, 2004; pp. 19–47.
82. Smith, A. Translating Sustainabilities between Green Niches and Socio-Technical Regimes. *Technol. Anal. Strat. Manag.* **2007**, *19*, 427–450. [CrossRef]

sustainability

MDPI

Article

Setting the Social Monitoring Framework for Nature-Based Solutions Impact: Methodological Approach and Pre-Greening Measurements in the Case Study from CLEVER Cities Milan

Israa H. Mahmoud [1,*], Eugenio Morello [1,*], Chiara Vona [2], Maria Benciolini [2], Iliriana Sejdullahu [3], Marina Trentin [3] and Karmele Herranz Pascual [4]

[1] Laboratorio di Simulazione Urbana Fausto Curti, Department of Architecture and Urban Studies (DAStU), Politecnico di Milano, 20133 Milan, Italy

[2] Cooperativa Sociale Eliante Onlus, via San Vittore 49, 20123 Milan, Italy; vona@eliante.it (C.V.); benciolini@eliante.it (M.B.)

[3] Ambiente Italia Società a Responsabilità Limitata, Department of Adaptation and Resilience, 20129 Milan, Italy; iliriana.sejdullahu@ambienteitalia.it (I.S.); marina.trentin@ambienteitalia.it (M.T.)

[4] TECNALIA, Energy and Environment Division, Parque Tecnológico de Bizkaia, Astondo bidea, Edificio 700, E-48160 Derio-Bizkaia, Spain; karmele.herranz@tecnalia.com

[*] Correspondence: israa.mahmoud@polimi.it (I.H.M.); eugenio.morello@polimi.it (E.M.)

Citation: Mahmoud, I.H.; Morello, E.; Vona, C.; Benciolini, M.; Sejdullahu, I.; Trentin, M.; Pascual, K.H. Setting the Social Monitoring Framework for Nature-Based Solutions Impact: Methodological Approach and Pre-Greening Measurements in the Case Study from CLEVER Cities Milan. *Sustainability* **2021**, *13*, 9672. https://doi.org/10.3390/su13179672

Academic Editors: Miguel Amado and Juan Miguel Kanai

Received: 2 July 2021
Accepted: 20 August 2021
Published: 27 August 2021

Publisher's Note: MDPI stays neutral with regard to jurisdictional claims in published maps and institutional affiliations.

Abstract: Nature-based solutions (NBS) are currently being deployed in many European Commission Horizon 2020 projects in reaction to the increasing number of environmental threats, such as climate change, unsustainable urbanization, degradation and loss of natural capital and ecosystem services. In this research, we consider the application of NBS as a catalyst for social inclusivity in urban regeneration strategies, enabled through civic participation in the co-creation of green interventions with respect to social cohesion and wellbeing. This article is focused on a social monitoring framework elaborated within the H2020 CLEVER Cities project, with the city of Milan as a case study. Firstly, we overviewed the major regeneration challenges and expected co-benefits of the project, which are mainly human health and wellbeing, social cohesion and environmental justice, as well as citizen perception about safety and security related to the NBS implementation process. Secondly, we examined the relevance of using NBS in addressing social co-benefits by analyzing data from questionnaires against a set of five major indicators, submitted to citizens and participants of activities during pre-greening interventions: (1) Place, use of space and relationship with nature, (2) Perceived ownership and sense of belonging, (3) Psychosocial issues, social interactions and social cohesion, (4) Citizen perception about safety and security, and lastly, we analyzed (5) knowledge about CLEVER interventions and NBS benefits in relation to socio-demographics of the questionnaires' respondents. Thirdly, we cross-referenced a wind-rose multi-model of co-benefits analysis for NBS across the regeneration challenges of the project. Because of the COVID-19 emergency, in this research we mainly focused on site observations and online questionnaires, as well as on monitoring pre-greening scenarios in three Urban Living Labs (ULLs) in Milan, namely CLEVER Action Labs. Lastly, this study emphasizes the expected social added values of NBS impact over long-term urban regeneration projects. Insights from the pre-greening surveys results accentuate the importance of the NBS interventions in citizens' perceptions about their wellbeing, general health and strong sense of neighborhood belonging. A wider interest towards civic participation in co-management and getting informed about NBS interventions in the Milanese context is also noted.

Keywords: nature-based solutions; social monitoring; social cohesion; co-creation; urban living lab; CLEVER Cities

1. Introduction

While many scientific contributions discuss the definitions and the theoretical frameworks of monitoring environmental impacts related to nature-based solutions (NBS) [1–3]

hands-on experiences and evidence-based effects from cities are still required to improve our understanding of the range of social, wellbeing and general health benefits provided by NBS. This is a key first step for promoting their introduction in urban planning policies and decision-making processes in cities [4,5]. Not only the development of conceptual models of social impacts, but evidence-based monitoring frameworks related to NBS in urban environments are also a relatively new topic in academic research and fairly peripheral [6,7]. In theory, the original definition of NBS derives from the International Union for Conservation of Nature (IUCN) 2013–2016 Programme as: *"actions to protect, sustainably manage and restore natural or modified ecosystems, which address societal challenges (e.g., climate change, food and water security or natural disasters) effectively and adaptively, while simultaneously providing human well-being and biodiversity benefits"* [8–10]. The European Commission [11] gives a broader definition of NBS, as *"actions inspired by, supported by or copied from nature that aim to help societies address a variety of environmental, social and economic challenges in sustainable ways"*. According to the European Commission scopes, NBS can transform environmental and societal challenges into innovation opportunities, by turning natural capital into a source for green growth and sustainable development for application in urban areas [12–15].

In practice, implementing NBS concepts exceeds the boundaries of traditional urban regeneration approaches that aim to "protect and preserve nature" by also considering the enhancement and restoration of urban ecosystem services [16,17] in addition to the enhancement of social impacts generated from NBS [14]. Specifically, relative to the topics of social justice and social cohesion, NBS have been linked to the notion of environmental justice across studies that explore the role that providing equal access to neighborhood green spaces has in the fostering of social cohesion. Such spaces bridge and bond social capital and support the cultural integration of typically marginalized and fragile social groups (vulnerable groups) such as the elderly, immigrants, persons with disabilities, chronic diseases, etc. (i.e., recognition-based justice) [7,18–21].

It is critical to note that NBS are believed to enhance levels of social inclusivity in urban planning "only if" they are supported by citizen engagement and public participation practices throughout the implementation [22–24]. Haase et al. [25] stress the potential for NBS to generate positive impacts on social inclusion whenever implemented. This aspect will depend on: (1) respect for local urban and institutional contexts, (2) the type of NBS to be implemented; as well as (3) the different actors and stakeholders who are to be involved in the project execution. In a similar manner, Dumitru et al. [1] emphasizes the optimal performance of NBS depending on their social uptake and continued use overtime.

Moreover, scientific research pinpoints the potential of NBS to deliver social-ecological justice in urban planning [26]. In the latest publication by Beute et al. [27] they emphasize the positive impact NBS have on human health and wellbeing, which also further strengthens social equity through the accessibility to green and blue infrastructures. The COVID-19 pandemic painfully pointed out the lack of regular use of green spaces while emphasizing the increased interest in connectedness with nature and the critical role proximity to green spaces plays in improving mental and physical health and wellbeing [28,29].

In addition, NBS are not simply 'just' green; rather, they are considered to be essential urban design measures for green and blue infrastructure, capable of providing multiple environmental purposes. For instance, scientific evidence highlights the role of policies at local and metropolitan scales to promote the use of NBS as multiple-benefit solutions for climate-change-related effects on health, wellbeing and citizens' sense of ownership. Broader general evidence discusses that just the environmental-related impact of NBS could be related to a deeper, more widespread knowledge of their co-benefits, connectedness to nature in relation with sociodemographic aspects as well as increasing community engagement and place-based ownership [28,30–32].

Connection and relationship with nature is associated with an improvement in people's general health and wellbeing. This is supported by scientific evidence and well-established theories, such as Attention Restoration Theory [33,34] and the Stress Recovery

Theory [33]. However, in recent years, studies are emerging that support the need to bring the psychological restorative capacity of nature to urban environments [35–37]. This is where urban interventions integrating NBS, such as those being carried out in the CLEVER Cities project in Milan as explained later in this article, play an important role in providing evidence of the benefits of natural elements over health and well-being. In addition, NBS in the urban environment are associated with another co-benefit, which is the increase and improvement of social relations because of the positive impact they have on social cohesion and the feeling of belonging to a place. Hence, the integration of NBS in urban public spaces, with their associated co-benefits, allows for the recovery of the cultural functions of these spaces and their consideration as socio-ecosystems. The CLEVER Cities project focuses on implementing NBS using a pathway of co-creation that is community-driven through the monitoring of the physical and social effects of NBS experimentations. Special attention is given to Milan city context and the selection of relevant regeneration challenges to specifically address, according to the areas of intervention.

In this article, we aim to shed some light on the gap in knowledge between the theoretical models of NBS, social monitoring and experiences from real world case studies such as the CLEVER Cities application in Milan The theoretical models of NBS promote them as problem solvers to climate and social challenges; however, the real experience of using NBS through Horizon 2020 projects and beyond is still lagging behind on evidence to showcase whether they really solve all the problems they are touted to solve, especially regarding intersections with gender equity, accessibility to green areas with respect to social cohesion aspects, etc. (Nonetheless, a quick Scopus and Science Direct databases' review of the literature reveals a major lack in monitoring methodologies specifically related to NBS pre- and post-greening implementation and their impacts on wellbeing in general terms, as well as psychosocial aspects connected to social cohesion specifically. The query included "Social Monitoring" OR "Social perception" AND "nature-based solutions" in two datasets by keywords AND title, always revealed less than 100 publications after a schematic check of relevance on the impact from a human-centred approach. See https://www.sciencedirect.com/search?qs=social%20monitoring%3B%20nature-based%20solutions&years=2022%2C2021&lastSelectedFacet=publicationTitles&publicationTitles=271784 (accessed on 20 April 2021)).

In the CLEVER Cities project (For more information on CLEVER Cities project, see https://clevercities.eu/the-project/), which started in 2018, the physical medium for the implementation of NBS is the ULL (Urban Living Labs, hereafter CLEVER Action Labs, CALs), and all the pilot projects' results in this article are referring to the social co-monitoring activities happening during the pre-greening phase of the project. Moreover, data are analyzed according to a co-designed methodological pathway initially developed by the responsible partner POLIMI, supported by ELIANTE and then shared with all the local partners in Milan (The stakeholders involved in this collaborative process were mainly a university partner (DAStU—Politecnico di Milano, hereafter POLIMI), a facilitating partner (Eliante), the Municipality of Milan (CDM), Ambiente Italia Srl. (AMBIT), the Mobility and Environmental Agency of Milan (AMAT), Rete Ferroviaria Italiana (RFI) and Italferr (Società Italferr Spa—Gruppo Ferrovie dello Stato italiane).) For more on the co-creation of CLEVER Cities see [38,39].

2. Materials and Research Context

The reflections in this research article connect the social influences generated from the co-design activities for integrating NBS in urban regeneration processes carried out with a wide array of public stakeholders in the city of Milan. In CLEVER Cities, the co-design activities are considered the first phase of a complete co-creation pathway that encompasses other phases of co-implementation, co-monitoring and co-development of NBS [39,40]. In particular, the co-creation phases and tools were conceived with some flexibility, in order to take into account the different opportunities that diverse NBS types and actors involved (e.g., in terms of scale, ownership, localization) offer regarding shared decision-making.

This was done by monitoring and analyzing a set of established indicators related to the social impacts of NBS during the pre-greening phase of the project. Specifically, the methodology presented in this article is related to three main urban regeneration challenges identified by the CLEVER Cities project consortium and locally by the Milanese team: (1) human health and wellbeing, (2) social cohesion and environmental justice and (3) citizen safety and security perception.

The social monitoring impact framework falls within the project activities and Work Package 4, "Assessing NBS impact through the CLEVER Monitor", related to the monitoring and impact measurement of NBS implementation generally [41]. Focus on these specific problems has been highlighted by the municipality for Milan, in order to ensure resilience related to heat waves and water management issues generated within dense urbanized areas. This challenge can cause health and safety risks to vulnerable targets such as the chronically diseased, young children, and elders.

Throughout the two and half years of the project, a set of Key Performance Indicators (KPIs) were identified and divided into two main sets by category of measurement (environmental and social KPIs). Within the project's wider monitoring plans, the methodological framework presented in this article is only related to the social KPIs utilized and is based on the need to evaluate and monitor the advancements of the social impacts related to NBS co-implementation in the city of Milan. The Local Monitoring Team (LMT) started by identifying the main environmental and social aspects to be evaluated. Next, the team analyzed them with respect to the specific CALs in Milan and, finally, verified them in different team meetings starting in February 2019 and onwards.

In March 2019, three collaborative workshops were conducted, one per CAL. A Theory of Change (ToC) collaborative activity was carried out in order to forecast the possible expected outcomes in each CAL context. A first version of the Local Monitoring Plan (LMP) was developed afterwards in June 2019. The social monitoring methodology was developed collaboratively with all the interested stakeholder groups that were part of the Milan LMT. The initial idea was to develop a mixed methodological framework using a variety of quantitative and qualitative measurement tools such as: surveys, on site observations, interviews with stakeholders, focus groups and online questionnaires. The scientific validation (in this sense: scientific validation refers to verifying actual needs from site visits and focus groups to concretize the methodological framework) of the LMP and social monitoring methodological framework initially started in September 2019 during the Milan Green Week festival by conducting site visits to the three CALs, including a guided tour to Milan's existing green roofs and walls for CAL 1, a tour of Giambellino Park 129 for CAL 2, and the Tibaldi train stop for CAL 3. The Project coordinator and other Front Runner Cities' leaders were also invited to site visits and observations within events occurring at the festival, (For more information on CLEVER Cities Milan, see https://milanoclever.net/).

From October 2019 until February 2020, a first tailored methodology was drafted and shared with CAL leaders to check on the scope and the set of indicators, including the feasibility of measuring a pre-greening baseline built on place-based criterion. Later on, the arrival of the COVID-19 pandemic constrained the number of tools available to the team, leading to the choice of submitting online questionnaires starting in February and March 2020 when emergency levels of sickness hit Milan and blocked all activities in a hard lockdown [42–44]. The complete LMP for the pre-greening phase of each CAL, including the social KPIs, was then co-designed and approved by all the involved partners based on their specific interests.

For each CAL, a lead partner is currently guiding the co-implementation of the NBS and is therefore responsible for following up the data-collection process and refining the overlap between the execution and the monitoring process. For CAL 1, Ambiente Italia (AMBIT) is responsible for conducting the co-design processes in four pilot green roof projects as well as online workshops, which helped to collect initial pre-greening data from November 2020 onward. For CAL 2, Eliante (ELI) together with MiloLab (a community association) were responsible for survey dissemination as well as conducting interviews and collecting data from site visits and co-design participants, which started in September 2019. For CAL 3, the Municipality of Milan (CDM) in conjunction with jurisdiction 5 and 6 Municipalities, conducted a public consultation with RFI and Italferr about the Tibaldi train stop in December 2019. This work later moved to online platforms and focused mainly on an online co-design survey that commenced in June 2020.

3. Methodology

The co-production process of this mixed-method social monitoring framework was based on several steps, see Figure 1: (1) scoping and gathering information: what aspects are to be measured related to the social impact of NBS in a Milanese context; (2) developing the theoretical model and scientific triangulation: why specific aspects are measured; (3) verification workshops with partners: how to measure specific impacts for each case; (4) scientific iteration and testing of the methodology through questionnaire development: developing a baseline and database for an online depository; (5) launching the questionnaires to a wider public and collecting pre-greening data from questionnaires; (6) data elaboration for specific CALs' place-based situations.

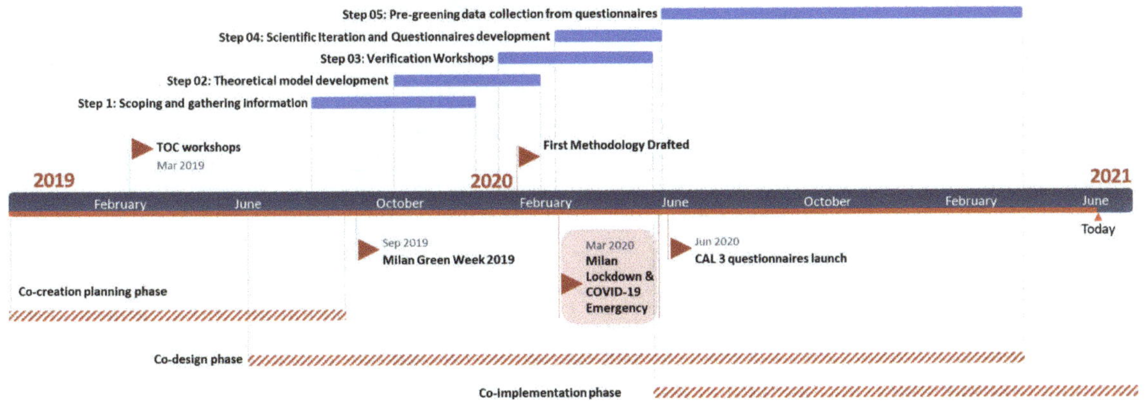

Figure 1. Timeline of the development for the Social Monitoring Methodology and the different steps presented in the work. Source: the first two authors.

Since the topic of this article is quite new amongst to the deduced similar methodologies [45–47] produced in this area of academic research, the efforts in selecting indicators were mainly related to the general aim of the project in using NBS to increase inclusiveness and strengthen collaboration between cities and citizens as seen in CLEVER Cities guidelines. The first step of the scoping activity of this methodological framework included gathering information on the three regeneration challenges related to social co-benefits of NBS.

The second step was complemented with a grey literature search for analogue indicators that have possible links with place-based connectedness to nature, NBS co-benefit measurements, mainly addressing wellbeing, psychosocial issues and social cohesion, but also considering safety and security, see Table 1. A series of internal team validation workshops were necessary to focus on regeneration challenges 1, 3 and 4 collectively in all three CALs between March and September 2019. In this step, the transformation of the ToC results into possible KPIs relative to each CAL place-based context was carried out by restructuring a logic and coherent chain from assumptions of the current situation into outputs and expected impacts that could be measured. Particular attention was paid to the Milanese context from earlier ToC workshops held for each CAL in March 2019, as well as site observations and visits during co-design workshops in relation to the main impacts identified, (Figure 2a–d).

During the third step, progress towards the development of the theoretical model and the scientific validation of possible similar research methodologies was achieved. The resulting indicators were in the majority divided into one of these five macro categories.

Table 1. A summary of identified challenges and considerations from literature review carried out by POLIMI, DAStU.

Regeneration Challenges Identified by the Project	Topics (Macro Category Resulting from ToC in Milan Social Monitoring Framework)	Integrating Approaches Linking Social Impacts and NBS Co-Benefits (Leading to Micro Indicators in the Milan Framework)	Relevant Literature *
Regeneration challenge 1: Human Health and wellbeing	Relationship with nature and well-being related to NBS	Human wellbeing and general health Positive impact of greenery on environmental values and general aesthetics	[48–53]
	Use of space (leisure, sport, relax, outdoor activity, etc.)	Connectedness to nature and wellbeing Frequency and use of spaces Effect of COVID-19 change to use of space *	[54–63] *
Regeneration challenge 3: Social cohesion and environmental justice	Perceived ownership of space and place satisfaction	Satisfaction with the building characteristics and proximity to green areas relationships Perceived ownership of green areas	[13,64–67]
	Psychosocial issues and social cohesion	Social interactions, support and cohesion Place identity and sense of belonging Civic participation and willingness to participate in co-design activities	[25,68–72]
Regeneration challenge 4: Citizen safety **	Citizen perception about safety and security	Increase in safety and security perception related to lighting, accessibility, maintenance, aesthetics, and interactions in places with the presence of other people	[73–75]

* These references were mainly identified during the course of the social monitoring methodology development timeframe from March 2019 till June 2020. Afterwards, some relevant literature also evidenced the social impacts generated from NBS, following the COVID-19 pandemic period. Henceforth, an additional set of micro indicators and survey questions were added to measure the use of green spaces during the lockdown period and its impact on perception related to relationship with nature and wellbeing, as well as to measure interest in participation in the co-maintenance aspects on the CALs of Milan. That was after the launch of the CAL 3 questionnaire in June 2020 which was the first pilot project. Hence in CAL 3, the indicators related to COVID-19 use of space were not measured for the pre-greening phase but will be monitored in post greening. ** POLIMI was the responsible partner for developing the framework of possible KPIs related to this regeneration challenge for the three CLEVER Cities project Front Runner Cities, see Appendix 02 in Supplementary materials. A Scientific master thesis was developed under the responsibility of Morello and Mahmoud in 2020, see [75].

During the fourth step, the scientific iteration and testing of the methodology through progress questionnaires took place. Two main partners carried out this process, POLIMI and ELI, working collectively on the three different CALs. The development of the questionnaires took place across different formats (online and offline) and in two languages (Italian and English) initially. All versions and elaborations on the questionnaires were collaboratively shared with other partners in the LMT such as AMBIT for CAL 1, MiloLab for CAL 2, RFI and Italferr for CAL 3. A testbed carried out by the local CAL 2 team and MiloLab together with a small group of local stakeholders helped develop a baseline (19 answers), which was stored in a POLIMI database (online repository). In order to correlate the spatial impact of NBS on the beneficiaries of each CAL, two other sections (macro categories) were added to the social monitoring methodology and the questionnaires after this iteration, looking specifically at the relative knowledge about CLEVER interventions and expectations related to NBS co-benefits and socio-demographic data.

The fifth step started with launching the CAL 3 questionnaires online to the public in June 2020, together with an online campaign that was created with the help of the CLEVER Milan social media team and the website was prepared by the local team. This wider public launch and data collection step helped the scientific triangulation of some indicators and questions that, afterwards, were considered of critical importance in the other CAL 1 and

CAL 2 questionnaires. This helped ensure some cross-comparability in the local Milan context.

The last step is demonstrated in Sections 5 and 6 in data collection, analysis, results and discussions.

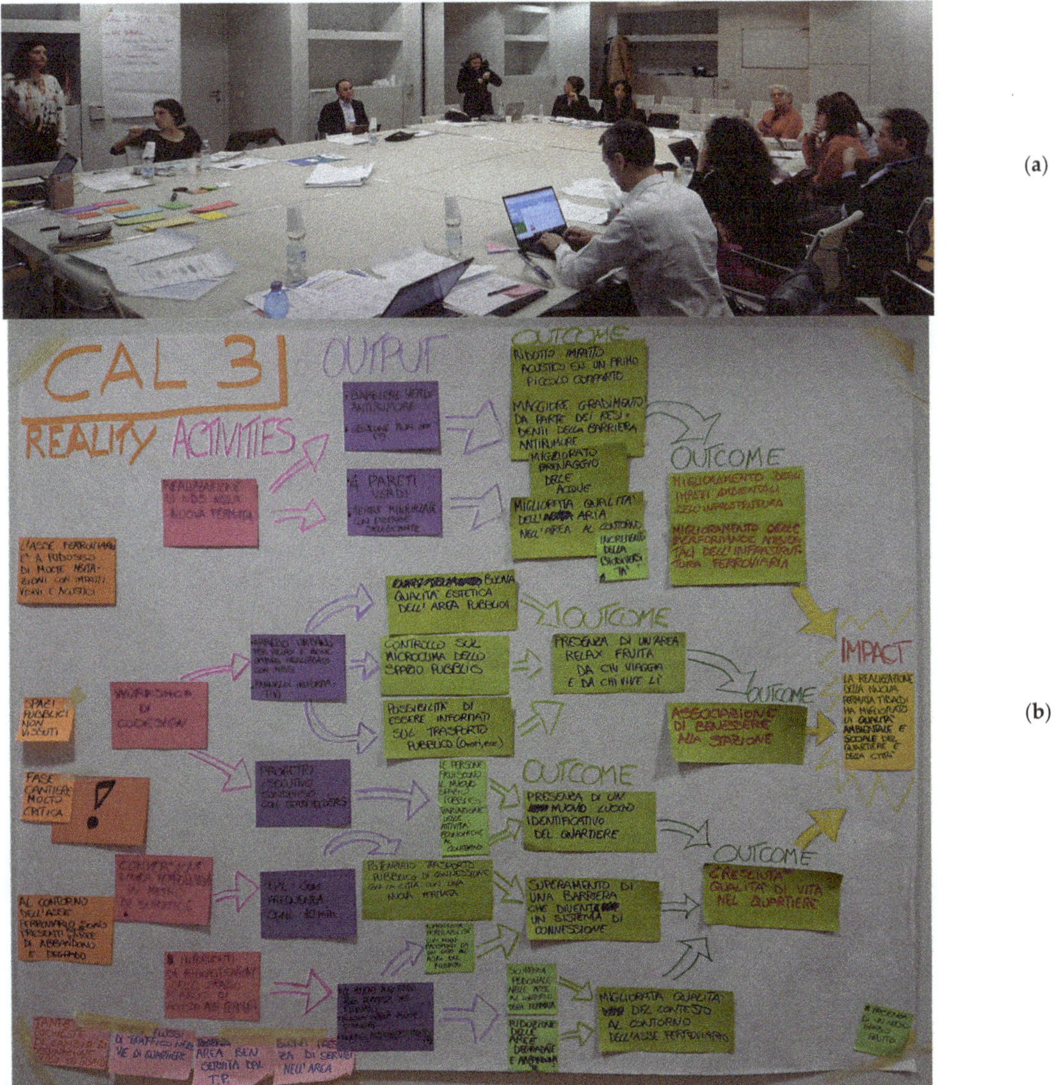

Figure 2. *Cont.*

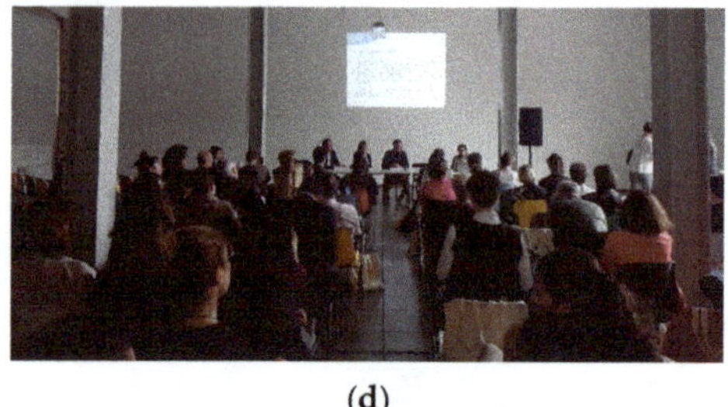

(c)   (d)

Figure 2. Different focus groups and workshops to validate the needs to monitor specific aspects in the three different CALs. From top to bottom: (**a**) ToC workshop with local stakeholders, March 2019; (**b**) A typical panel of ToC, specifically here CAL 3; (**c**) Co-design by immersion activity in CAL2, September 2019; (**d**) Milan Green Week press conference by CDM, FPM, POLIMI, ELI, AMBIT and WWF, September 2019.

3.1. Implementing the Methodology in Practice

In order to create a mixed-method approach for this evaluation process, the assessment framework is structured as a matrix. *Horizontally*, it is based on the macro-and micro indicators that are relevant to the three general regeneration challenges previously mentioned. Then, the macro categories that relate to the main outcomes to be measured in each specific CAL were added. *Vertically*, the framework is divided in different sections, as follows:

1. Who: the target groups of the analysis that will benefit from the NBS intervention,
2. How: the measurement tools (quantitative surveys, and qualitative interviews),
3. What: the needs of each CAL (if the indicator itself will be evaluated in specific CAL),
4. When: the stage this measurement should be addressed (pre-greening or post-greening), and
5. The type of questions: descriptions of the type of questions to be utilized (binary, ranking using Likert scale, multiple choice questions, or open-ended).

Following horizontally, each macro-category has micro-indicators underneath that correspond to a specific section transferred from the survey template developed from April 2020 onwards (Appendix 01 in the supplementary materials). In the "What" columns, the options given to measure each micro indicator in each CAL were given by adding a drop-down button. This will ensure that the same question is being elaborated and the question number is added next to it for easier reading of the matrix.

In the following Figure 3, a simulation using this methodological tool was run for the CALs of Milan, see original tabular tool in supplementary material. Taking into consideration the different timelines of the application of the questionnaires in the three CALs and the timeline of step 05 as explained in Figure 1, a set of indicators was identified on the horizontal axes in order to facilitate the cross-comparability between the results obtained and the data analyzed. The results from this simulation have shown the most important micro indicators to focus on, as below:

- Relationships with nature, wellbeing related to NBS and the use of space.
- The perceived ownership of space by different groups together with place satisfaction.
- Psychosocial issues, such as social cohesion, place identity and the focus on a sense of belonging towards the NBS in area of intervention.
- Knowledge about CLEVER Interventions and participation in community activities related to NBS.
- Citizens perceptions about the interventions in terms of safety and security related aspects.
- Socio-demographic data related to the area of intervention.

MACRO/MICRO indicators	Target groups			Measurement tool				where						when		Question
analysis of demographic data will consider: age, sex, residence vicinity, interest in the area of intervention in all monitoring periods.	these are indications for the sample of the respondents to be gathered during all monitoring periods.			quantitative		qualitative		CAL 1		CAL 2		CAL 3		#	#	type
These target groups should be diversified and balanced in age, gender, race (if needed) and include sample of social-level outcomes and all different vulnerable groups	stakeholder	residents	other specific groups	surveys on/off	on site observations	interviews stakeholders	focus groups	GRW	qn#	GIAM129	qn#2	TIBALDI	qn#	pre	post	type
1. Relationship with nature and well-being related to NBS in the area of intervention																
1.1 Relationship with nature and well-being related to NBS																
importance of the green as a priority in the neighborhood/ area of intervention	yes	yes	NA	yes	no	yes	yes	yes	19	yes	13	yes	8	yes	yes	scale
positive impact of the green areas in your neighborhood/area of intervention (health and well-being, environmental values, air quality, biodiversity, heat in summer, aesthetics of surroundings, social cohesion and relationships)	yes	yes	NA	yes	no	yes	yes	yes	20	yes	14	yes	12	yes	yes	ranking
did the recent emergency crisis change the use and perception of green spaces	no	yes	NA	yes	no	yes	yes	yes	22	no		no		no	yes	binary
2. Place, use of space and connectedness to Nature																
use the green areas around you or in your neighborhood (frequency of visits)	no	yes	NA	yes	yes	yes	yes	yes	21	yes		no		yes	yes	binary
type of use for the green space (leisure, sport, social, relax, outdoor activity, etc.)	no	yes	NA	yes	yes	yes	yes	yes	7	yes	6	yes	1	yes	yes	multiple choice
Time of use / work/ living in building/ neighborhood / area of interest (COVID)		yes	NA	yes	no	no	no	yes	10	yes	22	no		no		binary
Frequency relationship time with building/ neighborhood/ area of intervention	yes	yes	NA	yes	no	no	no	yes	9	yes	8	yes	34	yes		binary
Activities usually carried out in the place (multiple answer)	no	yes	NA	yes	no	no	yes	no		yes		yes	3	yes		multiple choice
3. Perceived ownership of space and Place satisfaction																
3.1 Place Satisfaction (general residential, open space or building)																
satisfaction with the neighborhood where you live (in case of large scale intervention)	yes	yes	NA	yes	no	yes	yes	no		yes		no		yes	yes	binary
overall satisfaction with the building where you live (increase of green roofs and walls)	yes	yes	residents of same building in CAL 1 in pre-post	yes	no	yes	yes	yes	14	no		no		yes	yes	binary
General satisfaction with the NBS/green area of intervention around where you live (in case of urban gardening, urban parks and green noise barriers)	yes	yes	commuters in CAL 3	yes	no	yes	yes	yes	23	yes	24	no		yes	yes	binary
Place Satisfaction with the building characteristics (thermal comfort, landscape, aesthetics [of buildings] sound environment, lighting , availability of common spaces, local services and amenities, quality of public areas, accessibility to green spaces)	yes	yes	residents of same building in CAL 1 in pre-post	yes	no	no	no	yes	15	yes	12	yes	7	yes	yes	scale
4.Psychosocial issues and Social cohesion																
4.1.Social Interaction and cohesion																
Social interaction, support, and cohesion (asking a favor, trust people in neighborhood, asking for help, getting along, people bond from different backgrounds, happy with relationships)	yes	yes	same participants in pre-post	yes	yes	yes	yes	yes	13	yes	11	yes	6	yes	no	scale
talk with neighbours apart greetings	no	yes	same participants in pre-post	yes	no	yes	yes	yes	11	yes	10	yes	4	yes	no	binary
4.2.Place identity and sense of belonging																
evaluate sense of belonging to the building/ neighborhood/ area of intervention	yes	yes	NA	YES	no	no	yes	yes	8	yes	7	yes	2	yes	yes	binary
5. Citizen perception about safety and security																
5.1 Lighting and clear visibility																
the area is lightened, visually clear paths, no sense of fear is perceived	yes	yes	NA	yes	yes	yes	yes	yes	29	yes	27	yes	26	yes	yes	ranking
5.2 Accessibility to green area																
increase of accessibility means in the area (walkability, bikeability, physical activities, etc.)	yes	yes	NA	yes	yes	yes	yes	no		yes		yes	26	yes	yes	ranking
5.3 Maintenance of green area																
status of the green area (litter, green condition, furniture, etc.)	yes	yes	NA	yes	yes	yes	yes	yes	34	yes	27	yes	26	yes	yes	ranking
Difficulty of maintenance (high cost and technical errors) / vandalism, degradation	yes	yes	no	yes	no			yes	34	yes	28	yes	27	yes	no	scale
5.4 Aesthetics																
green increase aesthetic quality of the area (green roof, walls, parks, etc.)	yes	yes	NA	yes	no	yes	yes	yes	34	yes	27	yes	26	yes	yes	ranking
5.5 Activities and presence of other people																
interaction in spaces, variety of activities, stickiness to places help you stay	yes	yes	NA	yes	yes	yes	yes	yes	29	yes	27	yes	27	yes	yes	ranking
6.Knowledge about CLEVER Cities project and interventions																
Knowledge about clever project	yes	yes	NA	yes	no	no	no	yes	24	yes	30	yes	9	yes	no	binary
what do you know about clever interventions	yes	yes	NA	yes	no	no	no	yes	25	yes	31	yes	10	yes	no	open ended
knowledge about NBS in general	yes	yes	NA	yes	no	no	no	yes	26	yes	32	yes	11	yes	no	open ended
Knowledge about Milan green roofs / shared gardens / green train stations	yes	yes	NA	yes	no	no	no	yes	27-28	yes	26	yes	23	yes	no	multiple choice
participation to co-design and co-management of intervention	yes	yes	NA	yes	no	no	no	yes	35	yes	29	yes	35	yes	no	ranking
7.Socio-demographic data Characteristics																
sex/gender	yes	yes	NA	yes	no	no	no	yes	2	yes	1	yes	28	yes	yes	binary
age	yes	yes	NA	yes	no	no	no	yes	3	yes	2	yes	29	yes	yes	binary
Laboral situation	yes	yes	NA	yes	no	no	no	yes	5	yes	4	yes	31	yes	yes	binary
Education	yes	yes	NA	yes	no	no	no	yes	6	yes	5	yes	32	yes	yes	binary
Legend		Macro indicator		Micro indicator		NA		yes	d as is			no	osed			
	QWERTY	sample of question explanation														

Figure 3. Simulation of the Methodological tool for social monitoring framework, source: the first author. A copy from the tabular tool is provided in supplementary material (methodology Excel sheet).

On the vertical axes, the most common target groups are the residents for all indicators. The most common measurement tool turned out to be the surveys, both online and offline, followed by interviews with local stakeholders and focus groups. However, the latter two instruments are less relevant since they might be not resident or completely familiar with the context with respect to the Macro categories related to place satisfaction, CLEVER interventions, and socio-demographic data analysis. In general, the macro indicators and most measurement tools are mainly to be used collectively in CAL 1, CAL 2 and to some degree in CAL 3. Hence, a combination of quantitative surveys and qualitative interviews were considered for complementary assessment, also, as in the ToC approach, to set a socio-economic framework that can support final decision-making for NBS.

The highlighted areas (red rectangles) in the Figure 3 below show how the overall selection of the target groups for the questionnaire's distribution and the measurement tool for the analysis was evaluated. In addition, the highlighted areas show the exact CALs where the simulation of the overall methodology was built-upon and on which indicators this was focused.

3.2. Some Notes on the Methodology and the Questionnaires' Form

The methodology is meant to be transversal to all three CALs in Milan in order to coordinate and better understand if some of the survey structures could be identical and allow some comparability in results between different CALs. However, we understand that the rest of the Front Runner cities do not necessarily have the same macro thematic categories for social interaction and cohesion related to NBS interventions impact, and that the results are not comparable to the other frontrunner cities. However, it is also a flexible tool that has the ability to change the macro categories in order to replace them with whatever other themes or macro indicators are needed for the specific context.

An on-site, visual observation tool is also considered highly important in providing more insight on the actual status. However, it may not be used for some indicators in order to avoid bias of the observers, as much of the observation work is referred by CLEVER Cities team and not easily transferred to outsiders. Nonetheless, it is highly relevant to the type of green space use and the activities people carry out in the space itself. It is then recommended in the post-greening phase evaluation as a key measurement.

For the pre-greening phase, we started drafting an online survey that has the same macro-category and then translated each micro-indicator into a type of question as indicated in the last column, respectively. Some questions have then incorporated a more complete list of elements to be evaluated based on the status of the CALs. The survey was initially pre-tested with people from the local community and residents not involved in the methodology design to assure the questions are convenient to respond to, clear, and easy to understand.

The CALs in Milan then required a more in-depth interview form using the same methodology as the macro/micro-indicators structure; however, the queries have more open-ended questions with relation to pre-greening and post-greening phases. The analysis of these interviews is still to be completed and will be included in the future research undertaken after the post-greening phases.

4. The Case Study of Milan CLEVER Action Labs

Before the analysis of the collected data from the surveys, in the following section we give an overview of the three CLEVER Action Labs (CALs) [76]. The CLEVER Cities Milan project area is situated in the south of Milan. It has three CALs, two spot interventions (CAL 2 and CAL 3) and one extended area (CAL1) mainly in the south part of the city, (Figure 4). In the local context, CdM, AMAT, AMBIT, ELI and POLIMI are responsible for collaboratively promoting urban greening measures such as NBS in terms of policy, planning, design and implementation.

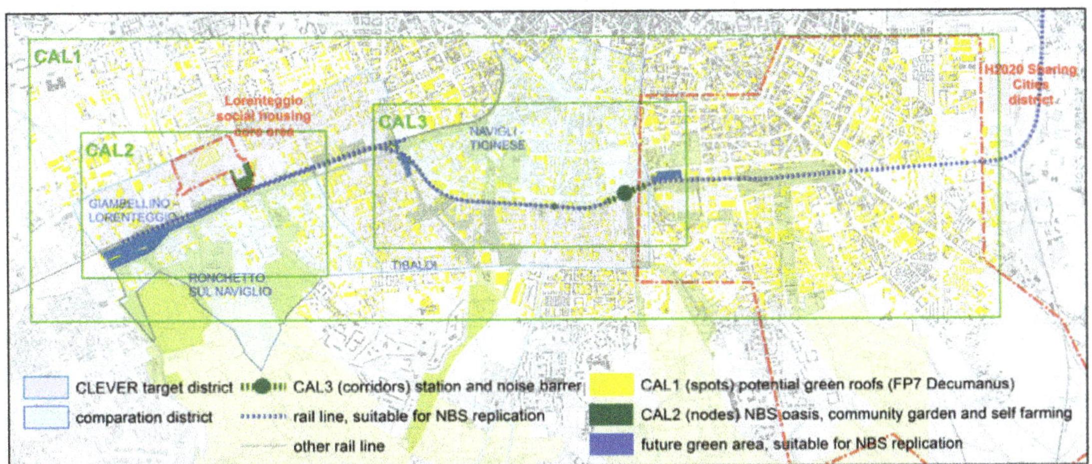

Figure 4. Territorial operating area of the CALs of Milan, southern transect. Source: the CLEVER Milan team, GA—June 2018.

The work being conducted in CAL1 has influenced policymakers to incorporate these urban greening measures (NBS) into the New Building Code of Milan (Regolamento Edilizio del Comune di Milano 2021, see https://www.comune.milano.it/aree-tematiche/urbanistica-ed-edilizia/sportello-unico-edilizia/regolamento-edilizio-del-comune-di-milano accessed on 19 July 2021). Moreover, the CAL 1 is focused on mainstreaming green roofs and walls to raise wider awareness of their benefits, to increase the overall amount being installed, and to encourage professionals and companies to embrace their use as part of their own approach [77].

The CAL 2 and CAL 3 are located in deprived areas, heavily affected by the railway infrastructure that crosses them. The CAL 2 is situated in a densely built-up area that is mostly residential, and it is focused on the neighborhood Lorenteggio Giambellino. Whereas CAL 3 comprises the area where the new railway stop Tibaldi is being constructed.

In CAL 2, ELI with CdM and AMAT are transforming the fragmented neglected areas near railway tracks into spaces for community farming that will serve as natural oases to increase community cohesion and improve storm water management. In CAL 3 with RFI, ITALFERR, CdM, AMAT and ELI, the local partners are developing new types of noise barriers using NBS that include interventions to strengthen biodiversity aspects and mitigate environmental impacts from the Tibaldi railway station.

Each CAL applies different modalities to mobilize public and private resources. CAL 1 is testing how the co-creation process can help raise private funds to complement and facilitate municipal funding. Based on the current development of activities, CAL 1 has progressed slightly at an advanced rate in terms of planning and co-implementing activities. This is because their successful implementation was subject to a complex structure of arrangements that involved different stakeholders, respectively their time availability, organizational capacity and technical assistance provided for CAL 1 activities.

4.1. CAL 1: Regreening Milan Green Roofs and Walls

The focus of CAL 1 (Figure 5) lies on the design and promotion of innovative NBS, such as the experimental and multifunctional green roofs and walls. To promote NBS, CAL1 has been developing an awareness-raising campaign. Its goal is not solely to increase knowledge about the importance of greenery in our buildings, but also to adopt a more strategic approach to public interest communication. It aims for the translation of this awareness into action, such as triggering a legislative change or supplementation, and helping drive professionals to employ green roofs and green walls in their building practices.

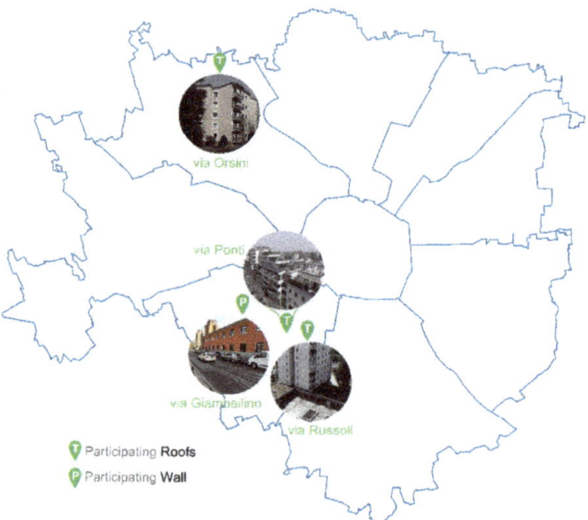

Figure 5. CAL 1, operating four green roofs and walls in Milan, source: AMBIT, May 2021.

Given the aforementioned objectives of this CAL, the activities were defined in two main tasks on which they will work:

- Increasing the knowledge through engagement and dissemination activities; i.e., the awareness-raising campaign.
- Turning knowledge into action in the form of the CLEVER pilot projects (green roofs and walls).

As part of the awareness-raising campaign, the activities carried out were related to knowledge exchange. These included two guided tours to discover green roofs and walls in Milan (one right after the CAL 1 launch on the 14th of June 2019 and the other one during the Milano Green Week, September 2019), three training courses on "Green Roofs and Walls" (in October 2019), and to disseminate knowledge during Milano Green Week, including a mobile exhibition that travels to different events and explains the important benefits of the CLEVER Cities project.

The second task of CAL1 initiated with the procurement process. The role of AMBIT, together with the CDM, was to advance the co-financing schemes for the implementation of the CLEVER pilot projects. In the subsidy scheme set for this purpose, two public calls have been launched. One for the selection of 10 potential green roofs and walls that will apply for 35% subsidy and technical support for NBS co-implementation, and another for the identification of experts skilled in designing green roofs and walls, who will provide the technical support through a co-creation process in the CLEVER Cities framework [78].

Due to the many consequences caused by the COVID-19 emergency, mostly in relation to the financial availability of resources, six of these projects that initially confirmed their interest in co-creating green roofs and walls have withdrawn their applications. With some delays, the co-design for the remaining four projects has started, which have also been subject to social monitoring in the pre-greening phase as in this article scope. The questionnaires for social monitoring of the four participating projects (via Russoli, via Orsini, via Giambellino, and via Ponti) have been submitted simultaneously during the co-design workshops. As explained below (Section 5.1—data collection), the compilation was carried out online and with technical assistance from co-design teams.

4.2. CAL 2: A Community Park in Giambellino, 129. Milan

The focus of CAL 2 lies on creating a new community garden, in the public area located in Giambellino 129, previously abandoned and with contaminated soil (Figure 6). The green area in G129 could be considered a steppingstone in the green ecological network on the Milan Circle line railways side [79]. Surrounding the social housing neighborhood of Lorenteggio-Giambellino is a dense residential area with a strong need for (and lack of) green and shared spaces. The old social housing block needs to be rehabilitated, which will be realized in the coming years by the Lorenteggio Suburban Rehabilitation Programme's Masterplan and the Peripheries Rehabilitation Plan of the Municipality. The social context is complex: Lorenteggio's population is mainly composed of elderly residents and migrants from different countries (the latter category 40% of residents in 2015). The degradation of some social houses causes crimes and conflicts; hence, a lot of people perceive a sense of insecurity and lack of safety.

Figure 6. CAL 2, Giambellino 129 Community public park, co-designed with residents from the neighborhood, authorized use from the CDM—published on CLEVER Milan website, April 2020.

However, Lorenteggio has a strong local community that is active in many initiatives to promote social cohesion and citizen engagement, working together towards a better use of urban spaces. The co-creation process implemented by MiloLab, and the CLEVER Cities project aims to encourage citizens and local organizations to co-design, co-manage and co-monitor the new community garden. Different types of NBS have been designed to create a high-quality multifunctional green space that focuses on enhancing connectedness and relationship to nature as well as social cohesion. Examples of these projects include a bird garden, an orchard, a community garden and a butterfly garden. The overall aim of this CAL is to provide a high-quality multifunctional green infrastructure in Giambellino 129 that can enhance presidium, social cohesion and ecological values. In particular, the social monitoring activities of CAL 2 include the evaluation of the impact associated with the new area on wellbeing and quality of life, social cohesion and sense of belonging.

4.3. CAL 3: A New Train Stop in Tibaldi

CAL 3 focuses on the opportunities arising from the construction of the new Tibaldi railway stop (See https://www.tibaldiscarl.it/presentazione/2.html [in Italian, accessed on 20 April 2021]) by working on a threefold program: improving the stations' environmental performance (rainwater management, microclimate and thermal comfort for the travelers), allowing the continuity of the ecological corridor for biodiversity and introducing groundbreaking standards that would incorporate NBS for noise mitigation. The experimental integration of NBS in railway infrastructure is in line with the principles of the European Union strategy on Green Infrastructures to help enhance health, wellbeing, provide jobs and deliver many benefits from nature to citizens [80].

In line with the program and within the CLEVER co-creation framework, the co-design of the public square, in front of Tibaldi's railway stop, has been supported by several activities. An internal focus group has initiated the design of the project (September 2019) which afterwards was presented and opened for public discussion (July 2020). Due to the pandemic context, a questionnaire on co-design was conducted (10 November 2020–31 December 2020) in which a significant number of local citizens participated (no. 325) (https://milanoclever.net/2021/04/28/risultati-sondaggio-cal3-tibaldi/ [in Italian, accessed om 20 April 2021]). This questionnaire aimed to engage the local citizens in the co-design process by giving them the opportunity to choose the functions, urban furniture, tree and plant species and paving materials. The co-design phase also foresees the engagement of technical NBS experts, with whom a workshop was organized in January 2021. Besides the public space that serves as an "open-air waiting area" of the railway station, the project also encompasses a number of NBS such as green walls, green railway embankments, and green noise barriers (Table 2).

Table 2. ToC Summary table for CALs interventions in Milan related to social monitoring framework, source: the first two authors, elaborated from ELI and AMB.

	CAL 1	**CAL 2**	**CAL 3**
Brief Description	Green Roofs and Walls	A Community Public Park	An Open-Air Waiting Area
CLEVER identified Regeneration Challenge	Regeneration challenge 1: Human Health and wellbeing	Regeneration challenge 3: Social cohesion and environmental justice	Regeneration challenge 4: Citizen safety and security
Aims and expected outputs related to ToC	Better training of citizens in workshops New financial partnerships	Soil restoration Citizen Engagement in co-design activities	Changes to planning policies related to NBS
Expected Outcomes	Higher availability of green roof spaces Increased sense of belonging and social wellbeing Increased quality of built environment	Increased Biodiversity * Increase of citizens awareness through co-monitoring of Nature-based solutions	Reduction in Crime Reduction of acoustic noise from the stationIncrease sense of belonging towards the neighborhood of interventions
Specific Micro Indicators	Increase connectedness to Nature and aesthetics	Increased social cohesion and support	Increase in sense of safety and security
Expected Measured impact from social monitoring framework	Greener urban spaces generate higher wellbeing for residents and better environmental quality	A higher quality multifunctional green infrastructure with community involvement and social presidium	A new railway stop, with higher social and environmental quality for the surrounding neighborhood and city

* Beginning in summer of 2021, biodiversity measurements in CAL 2 will be collected with similar methods of observation, community walks and focus groups, but will use separate sets of indicators in LMP, apart from these social monitoring framework purposes.

5. Data Collection, Analysis and Results

To simplify the process of adapting this social monitoring framework for the comparative analysis of this research article, the following micro indicators (Figure 7) were selected transversally from the three CALs to be analyzed commonly to build on the different aspects of the Milano context case study. They are as follows:

- Relationship with nature and wellbeing related to the NBS intervention (Regeneration. Challenge. 1)
- Positive impact of greenery on environmental values related to the neighborhood (Regeneration. Challenge. 1)
- Connectedness to Nature and use of space (leisure, sport, relaxation, outdoor activity, etc.) (Regeneration. Challenge. 1)
- Place satisfaction (general residential, open space or building), (Regeneration. Challenge. 3)
- Social interaction and cohesion within the place (Regeneration. Challenge. 3)
- Place-identity and sense of belonging (Regeneration. Challenge. 3)
- Citizen perceptions and concerns on safety and security of NBS interventions (Regeneration. Challenge. 4).

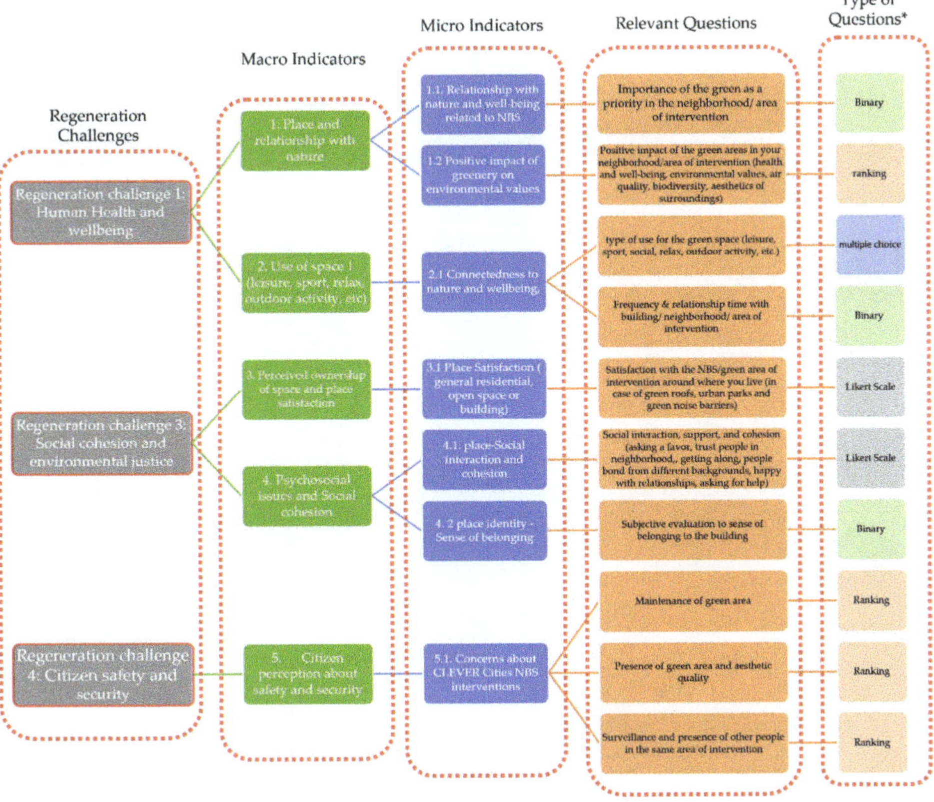

Figure 7. Tree map of the selection of relevant macro categories and micro indicators for the analysis of questionnaires, as well as the type of questions *. Source: the first author. * Referring to the type of questions: binary is mainly yes/no questions; Likert scale is mainly based on exhaustive mutual scale of preferences; ranking questions are prioritized ranking to questions with relevance or preferences; multiple choice questions refer to different possible choices within the available answers to select.

Another two sets of micro indicators were added to these previous ones not uniquely related to the urban regeneration challenges of the CLEVER Cities project, but rather to the city-specific CALs context, after the iteration described in Step 04 (Figure 1). They measure the following points:

- Knowledge about the CLEVER interventions and NBS in general in the city of Milan, in addition to the willingness to participate in co-design and co-management of CLEVER interventions.
- Socio-demographic data (gender, age, labor situation, and educational level).

All relevant questions recorded in Figure 7, have been tested and checked with local stakeholder groups from CAL 2 (19 answers) before the official launch of the questionnaires to confirm common question types (binary, Likert scale, ranking, multiple choice or open-ended questions). In the online Italian questionnaires, the order of the sections and certain relevant questions were alternated in order to avoid bot fraud and to lessen online monotony for the respondents due to its total length (average 35 questions).

5.1. Data Collection

The data collection was divided into a few phases, and it lasted approximately one year, from May 2020 until May 2021. Table 3 summarizes the initial start and end dates of the data collection as well as the status of the pre-greening questionnaires. According to the aforementioned methodology, all pre-greening questionnaires were designed to contain approximately 35 questions in total, with an expected maximum of filling-in time of 20 min. These constraints will also be considered for data collection during the post-greening phase since they have been developed in concordance with other Front Runner cities of the CLEVER Cities project.

Another relevant note on data collection between March 2020 and March 2021 is that the general use of online questionnaires by different municipality departments was gaining popularity from a wider public consensus; hence its major use during the pandemic emergency. Nonetheless, to avoid digital divide and marginalization of vulnerable populations, a dedicated team from ELI and AMBIT was following onsite data collection through paper questionnaires and assisted in compilation of the forms.

Table 3. Data collection target groups, timelines and methods of dissemination.

#	Target Groups	Timeline	Methods of Dissemination	Average Time Elapsed	Number of Respondents
CAL 1	People who live or work in the buildings where the green roof/wall will be built	November 2020–May 2021	Online + on site assisted compilation (in Via Russoli and Via Ponti)	36:45 * Min	79
CAL 2	Stakeholders who took part in the participatory process of co-design of G129	May 2020–October 2020 (Limited distribution within the MiloLab and co-design participants)	Online + on site assisted compilation	23:36 Min	19 ***
	Inhabitants or frequenters of Giambellino neighborhood	March 2021–April 2021 ** (Wider distribution with municipality newsletter)	Online + on site assisted compilation	19:07 Min	167
CAL 3	Inhabitants or frequenters of Tibaldi neighborhood	June 2020–September 2020	Online	19:36 Min	92
			Total		338 ***

* In CAL 1, one answer was recorded during an extensive elapsed time (24 h) due to a human error and it artificially raised the average elapsed time, substantially. ** In CAL 2, a wider online and offline campaign was carried out between March and April 2021 in order to include a younger age range in the analysis. This was in response to the predominance of older age categories noticed during the initial phases of data collection. *** The initial testbed questionnaires are not analyzed in this research article since the need for this analysis is obsolete; it was needed to test the questionnaires flow, logical chain and progress time but does not add major statistical information to the results since it was conducted with local team members and a small group of stakeholders. Hence the total is 357 − 19 = 338 questionnaires analyzed. No sensitive data was collected during the questionnaire's submission. The LMT decision was to cover the ethical issues regarding the participation of people and their data, taking the consideration not to collect any personal information unless participants gave consent.

The population sample of the questionnaire's respondents were equally distributed among the residents and frequenters of the neighborhoods and eventually across possible age ranges and gender; however, in CAL 2, major interest from female residents was noted.

5.2. Data Analysis and Results

Cross-comparative analysis to the exclusively selected micro-indicators in this research, as explained before, was used to identify correlations between NBS interventions and perceptions related to NBS social impacts. The authors have related only the positive responses recorded from each question (Table 4). The reason for this decision is that the final aim of this research article is to provide insight into simple quantitative analysis and methods to support NBS pre-greening procedures and the co-implementation phase [25,28,63,81–89]. Hence, the percentages or numbers reported below refer to the highest positive value recorded in each category: very important or very satisfied on Likert scale questions; yes (or only one category) in Binary questions; for multiple choice or ranking scales, the first four priorities are considered in the matrix.

Table 4. Cross-comparative analysis results from social monitoring impacts and questionnaires on perceptions. In bold, the highest % in each CAL vertically, the last column shows an averaged % evaluation for each indicator; in red, the most relevant. = <45%   > 100% Positive high relation between Micro indicator and questionnaires results.

	MACRO Categories	MICRO Indicators	CAL 1	CAL 2	CAL 3	Indicator Evaluation
Regeneration Challenge 1: Human Health and wellbeing	**1. Relationship with nature and well-being related to NBS in the area of intervention**	**1.1. Importance of the green as a priority in the building or neighborhood of intervention in personal opinion**				
		Very important	80%	86%	87%	High
		1.2. Positive impact of the green areas in your neighborhood/area of intervention in personal opinion				
		Aesthetics of the neighborhood or buildings	55.7%	65.3%	48.9%	High
		Citizen's health	53.2%	60.5%	52.2%	Medium
		Citizen's well-being	49.4%	65.9%	57.6%	High
		Perceived temperature and thermal comfort	45.6%			Low
		Air quality	45.6%	70.7%	53.3%	High
	2. Place, use of space and connectedness to Nature	**2.1 Type of use for the building or neighborhood relationship**				
		Living in the same building or Neighborhood	65.8%	80.2%	72.8%	High
		Working in the same building or Neighborhood	26.5%	4.1%		Low
		Frequenting cultural activities in the neighborhood		8.3%	8.6%	Low
		Visiting for green areas or physical activity in the neighborhood		23.3%	15.20%	Low
		Other or personal reasons (family or friends)	7.5%	11.9%	35.8%	Low
		2.2. Frequency relationship time with building/ neighborhood/area of intervention				
		More than 5 years	84%	82%	84%	High
Regeneration challenge 3: Social cohesion and environmental justice	**3. Perceived ownership of space and place satisfaction**	**3.1. Place Satisfaction with the building or neighborhood characteristics**				
		Accessibility to parks and green areas	82%			High
		Maintenance and Cleaning of the area	67.2%			Medium
		Availability of common spaces	63.3%			Medium
		Economic accessibility and services prices		43.7%		Low
		Public services availability		62.90%	64.1%	Medium
		Environment and Landscape attributes		19.8%	48.9%	Low
		Transportation and logistics			64.2%	High
		Aesthetics of the neighborhood or buildings	62.5%			Low
		The neighborhood in general		25.7%	49%	Low
	4. Psychosocial issues and social interactions	**4.1. Place Social interaction, support and Cohesion**				
		Staying Long in this Building /Neighborhood	74.3%	71.8%	67.1%	High
		Happy with relationships and vicinity in this building/neighborhood	76.%	64.70%	69.70%	High
		Exchange favors and things with the residents	59.50%	49.70%	52.80%	Medium
		I know people that I can ask for help and support			64.80%	Medium
		I trust people in my neighborhood	53.90%	38.30%	51.10%	Medium
		4.2 Place identity and sense of belonging				
		Very strong sense of belonging	81%	71%	76%	High
Regeneration Challenge 4: Citizen security	**5. Citizen perception about safety and security**	**5.1. Concerns about CLEVER Cities NBS interventions related to the building or the neighborhood**				
		Lighting and clear visibility		56.30%	71.70%	High
		Accessibility pedestrian and Cycling		34.10%	67.40%	Medium
		Maintenance	42.40%	64.70%	81.50%	High
		Presence of green areas	84.80%		48.90%	High
		Aesthetics	84.80%	29.30%	41.30%	High
		Presence of other people in space	62%	29.90%	45.70%	Medium
		Presence of security personnel and surveillance	36.80%	37.10%	69.60%	Medium

Knowledge of the CLEVER Cities Project and socio-demographic analysis are presented afterwards (Table 5), hence providing evidence of a clear relationship between the three main regeneration challenges of the project with evidence-based data on general knowledge of NBS and social structures in the three CALs specific contexts.

Table 5. Cross-comparative analysis results from the socio-demographic data in three CALs and knowledge about CLEVER Cities interventions. In red, the most prominent categories.

	MACRO Categories	MICRO Indicators	CAL 1	CAL 2	CAL 3	Indicator Evaluation
	Knowledge about CLEVER Cities interventions	**Information about CLEVER Cities project and NBS**				
		Knowledge about CLEVER Cities project generally before the questionnaire	29.0%	20.0%	23.0%	Low
		Knowledge about Milan green roofs/shared gardens/green stations respectively	48.6%	47.0%	68.5%	Medium
		Willingness to participate in co-design and co-management of intervention				
		I want to be more informed about how the roof/wall will be built in the building or Neighborhood where I live/work	64.6%	80.0%	84.4%	High *
		I want to collaborate in the co-management and co-maintenance of the green roof/wall in the building or Neighborhood where I live/work	39.5%	20.0%	28.6%	low
City specific CALs context	**Socio-demographic data Characteristics**	**Gender**				
		Male	57.0%	26.0%	46.0%	Medium
		Female	42.0%	74.0%	53.0%	High
		I prefer not to say	01.0%	0%	01.0%	low
		Age Range (% calculated over all respondents in each CAL)				
		16–24	1.3%	2.4%	4.3%	Low
		25–34	0.0%	9.0%	8.7%	Low
		35–49	21.5%	16.8%	21.7%	Low
		50–64	39.2%	32.9%	0.0%	Medium
		65–79	32.9%	36.5%	42.4%	High
		I prefer not to say	5.1%	2.4%	21.7%	Low
		Labor Situation				
		Unemployed	5.1%	2.4%	1.1%	Low
		Employee or self-employed/freelancer without employees	48.1%	44.3%	62.0%	High
		Self-employed with employees	1.3%	1.8%	3.3%	Low
		Retired	38.0%	37.7%	22.8%	Medium
		Household	1.3%	5.4%	2.2%	Low
		Not working—disability or long-term sick leave	1.3%	0.6%	0.0%	Low
		Student	1.3%	2.4%	3.3%	Low
		I prefer not to answer	1.3%	1.8%	3.3%	Low
		Education				
		PhD./Master	2.53%	6.59%	2.17%	Low
		University degree/Bachelor	10.13%	35.33%	51.09%	Medium
		High School Diploma	49.37%	46.11%	43.48%	High
		Middle School	25.32%	8.98%	2.17%	Low
		Elementary School	10.13%	1.80%	0.00%	Low
		No educational qualification	0.00%	0.00%	0.00%	NA
		I prefer not to answer	2.53%	1.20%	1.09%	Low

* A noticeable high willingness to participate in co-design and co-management of the activities and interest in information about the NBS interventions. Even though the initial knowledge about the CLEVER Cities project results are low, there is remarkable interest in information about Milan NBS. That interest is also reflected in a high number of subscriptions to social media channels and the local CLEVER Milan website, as respondents were invited to subscribe after submitting their questionnaires, in order to receive updates from the project.

In general, the cross analysis between the three different CALs gives insight into the Milanese territorial cohesion and stability in the relationship with the neighborhood where they live. Socio-demographic data reveal a major interest in public participation in co-creation activities as well as higher response values from females, generally in the age range of 50–79. Specifically, a noticeable percentage of the respondents were part of the mature population of 35–49 years (21.5%), 50–64 (32.9%) and 65–79 (36.5%). In addition, high rates of employees and self-employed (or freelance without employers) and retirement categories were noticed, as well as a high rate of high school diploma respondents (46.32%), followed by university degree holders (32.18%).

6. Discussions and Conclusions

The results from the questionnaires give indications of the different social impacts of NBS interventions in urban environments and the correlation of the human relationship to nature. These impacts are related to the main co-benefits of improving general health and wellbeing, social interactions and cohesion, and an increase in the use of space, place satisfaction, connectedness to nature and safety perception. With particular focus on each regeneration challenge raised in the project, we can summarize the following on each indicator (see supplementary material):

- **Relationship to nature and improved wellbeing related to NBS intervention (Reg. Ch. 1)**

This indicator shows a collective consensus about green areas as a priority for all respondents (all CALs $\geq$ 80%). CAL 2 showed an internal correlation with the neighborhood or building since these 80% are all residents or daily frequenters of the same building for more than 5 years. CAL 3 showed an external correlation as a majority of respondents did not participate in the public introductory event by the municipality regarding the Tibaldi station in December 2019.

- **Positive impact of greenery on environmental values related to the neighborhood (Reg. Ch. 1)**

Noticeably, this indicator highlights the synergies between individualistic preferences such as health and wellbeing of citizens in comparison to general preferences related to neighborhood aesthetics or air quality and pollution in all the three CALs. The percentage shows the cumulative prioritization of the higher four selections in each CAL from the "strongly agree" response, with percentages $\geq$ 45% (In social studies, the general consensus is that correlation percentage is considered positive if above 47%.)

- **Connectedness to nature and use of space (leisure, sport, relaxation, outdoor activity, etc.) (Reg. Ch. 1)**

The answers reported in this indicator are mostly from respondents that have either a residential or labor relationship with the building (or both) and neighborhood where the NBS are built or realized. A high correlation between neighborhood residency and place satisfaction related to usage of green areas for leisure or physical activity is also noted in CAL 2 and CAL 3, respectively. In other words, the majority of the questionnaire's respondents are also from the same neighborhood, which is also due to the exclusivity in the questionnaire's distribution either online or offline, since the target population was the users of the buildings or neighborhoods where the CLEVER intervention will be carried out. While in all CALs the majority of participants have a residential relationship to the place, the second most frequent relationship is specific to each CAL: work in CAL 1 buildings, visit green area or do physical activity in CAL 2 community garden, and family and friends in CAL3 station.

- **Place satisfaction (accessibility to parks and green areas, maintenance and cleaning status), (Reg. Ch. 3)**

All high percentages in this indicator are referring to people with more than 5 years stable relationship with the same building or neighborhood. In CAL 1, 92% of these stable relationships have been either residents or high frequenters that visit the building at least once daily. In CAL 2, 86% of these stable relationships have selected the green areas in the neighborhood as very important for them from the first indicator on relationship with nature. In CAL 3, 94% of these stable relationships think the green areas of the neighborhood are very important.

- **Place-social interaction and cohesion (Reg. Ch. 3)**

A high satisfaction with their social interaction was noticed in all CALs. In CAL1 and CAL 3, valorization of social bonds, trust and support is also remarkable. Contrarily, in CAL 2, people in the neighborhood show a lower general satisfaction; nonetheless, residents are content with their relationships and plan to stay in the same neighborhood.

- **Place-identity and sense of belonging (Reg. Ch. 3)**

In CAL 1, the value on sense of belonging was slightly higher, which is possibly due to the perception of a higher personal attachment to a building rather than the larger neighborhood, which is the case in CAL 2 or CAL 3.

- **Citizen perceptions and concerns on safety and security of NBS interventions (Reg. Ch. 4).**

General concerns of citizen perceptions on safety and security were highly recorded in CAL 3, mainly related to lighting and clear visibility (71.7%), accessibility (67.4%), maintenance (81.5%) and the presence of security personnel and surveillance (69.6%), presence of green areas (48.9%) and presence of other people in the space (45.7%)

The following graphical representation of the wind-rose (Figure 8) aims to give evidence from the previous analysis on the most relevant categories of interest, hence correlating between social impacts from NBS and outcomes from the methodological analysis of the questionnaires' data. The legend indicates if the resulting percentage is representing results from all the three CALs or just one or two of them. For each sub-indicator, data was averaged and elaborated according to a new percentage scale (green <60%, yellow >60% and <70%, Orange >70% and <80%, Red >80%) to visually showcase the most important macro categories and micro indicators by consequences.

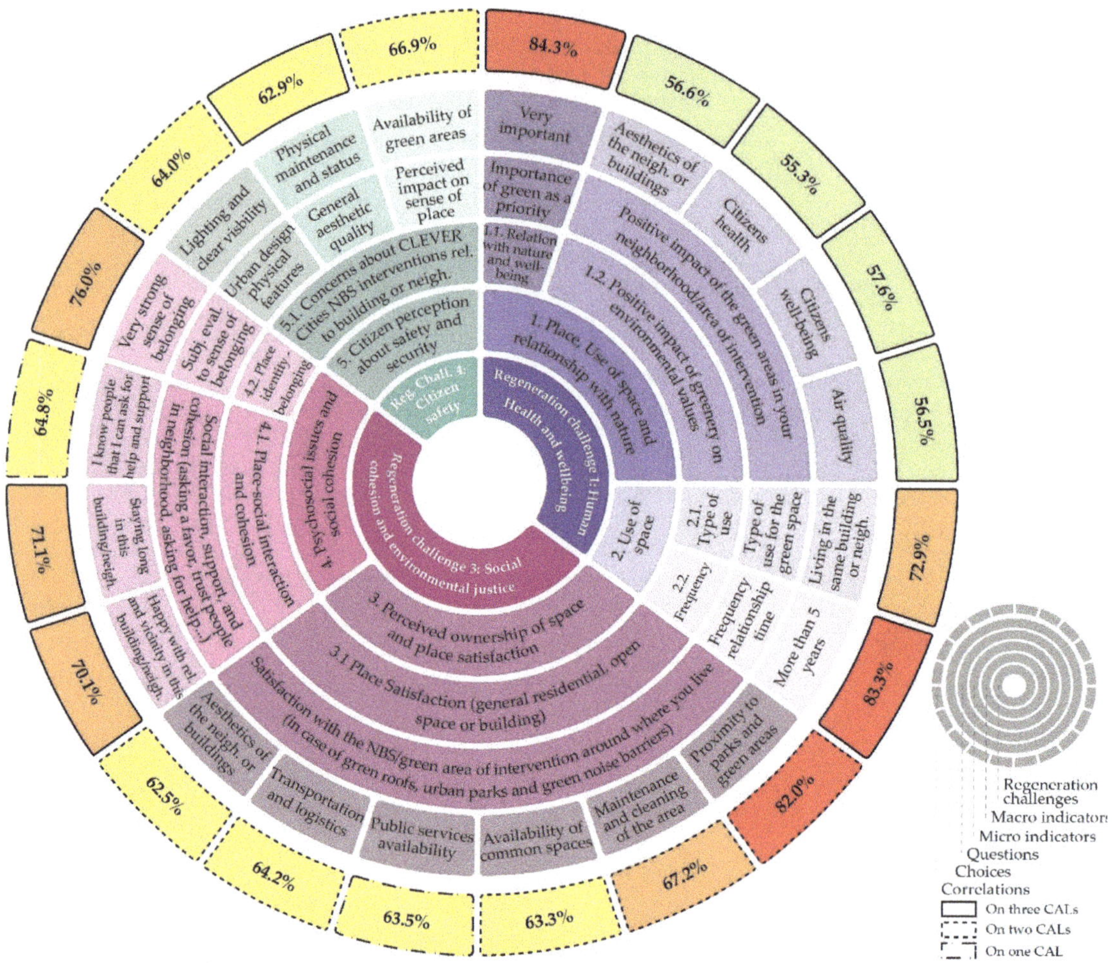

Figure 8. Wind-rose multi-model of co-benefits of NBS according to CLEVER Cities methodology. Considered data is the average percentage score for each indicator. Source: the first authors.

In relation to human health and wellbeing, it emphasizes the high importance of green infrastructure as a priority, medium positive impact from green areas on aesthetics, air quality and general wellbeing in residents' opinions. The model also reflects on the high connection of the relationship between residents and their permanence stability with the building and/or neighborhood where the CLEVER Cities interventions are taking place.

Reflecting on social cohesion and environmental justice, the model specifically investigates the clear high value of measuring aspects related to proximity to parks and green areas, maintenance and cleaning of the area with perception on general satisfaction and place ownership of one's building or neighborhood of residence. Commonly, the survey results guide a high social interaction in terms of happiness with relationship to vicinity and significant trust and support among the neighbors. Increased sense of belonging also results as an important aspect to focus on throughout the interventions in the CALs context.

Reflecting, then, on regeneration challenge 4 regarding safety and security, citizens' perceptions reveal high interest on maintenance, aesthetics and presence of other people in the green areas towards lowering their concerns on the areas of interventions related to safety and security. However, general reflection on safety and security did not result as a priority in all three CALs equally but were overlooked on average in two CALs only.

The most striking result of this study is the high and widespread priority given by participants to proximity to green and natural elements within their urban environment, especially related to CLEVER interventions. This is irrespective of whether the interventions are carried out in buildings, train station environments or in urban public spaces itself. This result contrasts with the trend observed in recent decades in our cities of soil sealing and land consumption in our environments, eliminating green or blue elements, both in public spaces (elimination of trees, gardens, fountains...) and in our residential buildings, where flowerpots and small vegetation on balconies have been noticeably disappearing. What the public seems to be calling for is a return to greening and bluing our spaces of coexistence with nature. During the COVID-19 pandemic and, especially, during the period of confinement, the windows and balconies of our residential buildings have recovered their function as public spaces for enjoyment and social interaction.

To conclude, this research article aims to give evidence on the gap between methodological approaches towards measuring NBS social impacts. From the data analysis, it is clear that relevant KPIs from the practice carried throughout questionnaires emphasize the need to have a coherent simulation model from pre-greening and post-greening phases in order to cross-compare the increased or decreased social impacts of NBS. Moreover, the cross-comparability between the three different CLEVER Action Labs in Milan reflects on social inclusivity as the main aim of the CLEVER Cities project. Nonetheless, positive impact from proximity to green areas and connectedness to nature relate to an increased general wellbeing and satisfaction with one's building or neighborhood. It is valid to consider the application of NBS as a driver and catalyst in terms of social cohesion and wellbeing, but equally important is the engagement of citizens and voiceless groups in the implementation of NBS through a co-creation dimension.

Our aim from this methodological approach carried out throughout a year and a half of research on the theme of co-creation and co-implementation of such complex work is to reflect on the place-based needs emerging from social impacts related to NBS co-benefits. The evidence from literature is quite prominent, yet the evidence from practice-based on implemented projects is more valuable and quite remarkable. Future research will include implementing the same cross-comparative analysis on the post-greening phase after the implementation of the NBS interventions by the end of the year 2023.

Limitations

The research results also highlight the drawbacks of the long-term process of monitoring aspects related to social cohesion that make the results outdated by the end of the project lifetime. Another relevant drawback is the lack of unified measurement methodological framework when compared to other similar H2020 sister projects. The finding is emphasized from the work of Task Force II established on evaluating the NBS impact in place [21,90].

Another limitation on the general methodological approaches to social impacts related to project implementing NBS are the place-based constraints and relation to specific contextual attributes. In the case of CLEVER Cities, the project focuses on social inclusivity, which was emphasized by positive relationships in the different neighborhoods and pilot project areas.

Last, other noticeable limitations are the impossibility to measure accurate social benefits in quantitative terms except after the finalized project implementation and the conclusion of process evaluation. Meanwhile, the readings of the questionnaires and other instruments remain perceptual and are considered guidelines for the real implementation pathways.

Supplementary Materials: The methodological instrument and the data analysis details are available online at https://www.mdpi.com/article/10.3390/su13179672/s1. In addition, the following

Appendices are included as supplementary materials: **Appendix 01:** The Questionnaire templates in English. **Appendix 02:** Table of safety and security methodological analysis in CLEVER Cities Milan.

Author Contributions: Conceptualization, I.H.M. and E.M.; methodology, I.H.M.; questionnaire implementation: I.H.M., E.M., C.V. and K.H.P.; data analysis, I.H.M.; validation, I.H.M., E.M. and C.V.; formal analysis, I.H.M.; data investigation, I.H.M. and C.V.; resources on CAL1, I.S. and M.T.; resources on CAL 2, I.H.M., C.V. and M.B.; resources on CAL3, I.H.M. and I.S.; data curation, I.H.M.; writing—original draft preparation, I.H.M. and E.M.; writing—review and editing, I.H.M. and E.M.; visualization, I.H.M.; supervision, E.M.; project administration, E.M.; funding acquisition, E.M. All authors have read and agreed to the published version of the manuscript.

Funding: This document has been prepared in the framework of the European project CLEVER Cities. This project has received funding from the European Union's Horizon 2020 innovation action program under grant agreement no. 776604. The sole responsibility for the content of this publication lies with the authors. It does not necessarily represent the opinion of the European Union. Neither the EASME nor the European Commission are responsible for any use that may be made of the information contained herein.

Institutional Review Board Statement: Ethical review and approval were waived for this study, due to the authors' decision on not collecting personal information from questionnaires respondents. Hence, all data were collected and stored anonymously with no ethical requirements to fulfill.

Informed Consent Statement: Informed consent was obtained from all subjects involved in the research study.

Data Availability Statement: Original data can be obtained by contacting the first author. Authorization for publication consent was obtained from all authors and involved partners.

Acknowledgments: The authors also want to thank Claudia Carlini (Responsabile Gestione Operativa—Direzione Investimenti Nord-Est) from RFI—Rete Ferroviaria Italiana, Gloria Dajelli (Coordinatore Settore Studi e Progettazione Ambientale—Direzione Tecnica—U.O. Architettura Ambiente e Territorio—S.O Ambiente) from Italferr Spa—Gruppo Ferrovie dello Stato italiane, and Emilia Barone (Direzione Urbanistica) from Comune di Milano for the CLEVER CITIES local Milan team support and follow-up. Martin Krekeler, CLEVER Cities project coordinator from the Senate of the Free and Hanseatic City of Hamburg—Senate Chancellery, Sean Bradley from Groundwork London—CLEVER Cities London team, and Alessandro Arlati from the Department of Urban Planning and Regional Development, HafenCity University Hamburg—CLEVER Cities Hamburg team for the graphical representation support.

Conflicts of Interest: The authors declare no conflict of interest.

References

1. Dumitru, A.; Frantzeskaki, N.; Collier, M. Identifying principles for the design of robust impact evaluation frameworks for nature-based solutions in cities. *Environ. Sci. Policy* **2020**, *112*, 107–116. [CrossRef]
2. Ershad Sarabi, S.; Han, Q.; L Romme, A.G.; de Vries, B.; Wendling, L. Key Enablers of and Barriers to the Uptake and Implementation of Nature-Based Solutions in Urban Settings: A Review. *Resources* **2019**, *8*, 121. [CrossRef]
3. Kabisch, N.; Qureshi, S.; Haase, D. Human-environment interactions in urban green spaces—A systematic review of contemporary issues and prospects for future research. *Environ. Impact Assess. Rev.* **2015**, *50*, 25–34. [CrossRef]
4. European Commission. *Towards an EU Research and Innovation Policy Agenda for Nature-Based Solutions & Re-Naturing Cities*; European Commission: Brussels, Belgium, 2015; ISBN 9789279460500.
5. Frantzeskaki, N.; Vandergert, P.; Connop, S.; Schipper, K.; Zwierzchowska, I.; Collier, M.; Lodder, M. Examining the policy needs for implementing nature-based solutions in cities: Findings from city-wide transdisciplinary experiences in Glasgow (UK), Genk (Belgium) and Poznań (Poland). *Land Use Policy* **2020**, *96*, 104688. [CrossRef]
6. Tzoulas, K.; Galan, J.; Venn, S.; Dennis, M.; Pedroli, B.; Mishra, H.; Haase, D.; Pauleit, S.; Niemelä, J.; James, P. A conceptual model of the social–ecological system of nature-based solutions in urban environments. *Ambio* **2021**, *50*, 335–345. [CrossRef]
7. Cousins, J.J. Justice in nature-based solutions: Research and pathways. *Ecol. Econ.* **2021**, *180*, 106874. [CrossRef]
8. IUCN French Committee. Nature-based Solutions for Climate Change Adaptation & Disaster Risk Reduction. 2019. Available online: https://uicn.fr/wp-content/uploads/2019/07/uicn-g20-light.pdf (accessed on 29 March 2021).
9. IUCN. *IUCN Global Standard for Nature-Based Solutions: A User-Friendly Framework for the Verification, Design and Scaling Up of NbS: First Edition*; IUCN: Gland, Switzerland, 2020.
10. IUCN. The IUCN Programme 2013–2016. In Proceedings of the IUCN World Conservation Congress, Jeju, Korea, 6–15 September 2012; pp. 1–30.

11. European Commission. Nature-Based Solutions. Available online: https://ec.europa.eu/research/environment/index.cfm?pg=nbs (accessed on 29 August 2019).
12. ICLEI. Nature-Based Solutions for Sustainable Urban Development. 2017. Available online: https://unfccc.int/files/parties_observers/submissions_from_observers/application/pdf/777.pdf (accessed on 19 June 2018).
13. Nesshöver, C.; Assmuth, T.; Irvine, K.N.; Rusch, G.M.; Waylen, K.A.; Delbaere, B.; Haase, D.; Jones-walters, L.; Keune, H.; Kovacs, E.; et al. The science, policy and practice of nature-based solutions: An interdisciplinary perspective. *Sci. Total Environ.* **2017**, *579*, 1215–1227. [CrossRef]
14. Fink, H.S. Human-nature for climate action: Nature-based solutions for urban sustainability. *Sustainability* **2016**, *8*, 254. [CrossRef]
15. Kabisch, N.; Korn, H.; Stadler, J.; Bonn, A. *Nature Based Solutions to Climate Change Adaptation in Urban Areas: Linkages between Science, Policy and Practice*; Springer OPEN: Berlin, Germany, 2017; ISBN 9783319537504.
16. Dushkova, D.; Haase, D. Not Simply Green: Nature-Based Solutions as a Concept and Practical Approach for Sustainability Studies and Planning Agendas in Cities. *Land* **2020**, *9*, 19. [CrossRef]
17. Bulkeley, H.; Kok, M.; Xie, L. *Realising the Urban Opportunity: Cities and the Post-2020 Biodiversity Governance*; PBL Netherlands Environmental Assessment Agency: The Hague, The Netherlands, 2021.
18. Raymond, C.M.; Frantzeskaki, N.; Kabisch, N.; Berry, P.; Breil, M.; Nita, M.R.; Geneletti, D.; Calfapietra, C. A framework for assessing and implementing the co-benefits of nature-based solutions in urban areas. *Environ. Sci. Policy* **2017**, *77*, 15–24. [CrossRef]
19. Dick, J.; Jones, J.C.; Carver, S.; Dobel, A.J.; Miller, J.D. How are nature-based solutions contributing to priority societal challenges surrounding human well-being in the United Kingdom: A systematic map. *Environ. Evid.* **2020**, *9*, 1–21. [CrossRef]
20. Ferreira, V.; Barreira, A.P.; Loures, L.; Antunes, D.; Panagopoulos, T. Stakeholders' engagement on nature-based solutions: A systematic literature review. *Sustainability* **2020**, *12*, 640. [CrossRef]
21. Skodra, J.; Connop, S.; Tacnet, J.-M.; Van Cauwenbergh, N.; Almassy, D.; Baldacchini, C.; Basco Carrera, L.; Caitana, B.; Cardinali, M.; Feliu, E.; et al. Principles guiding NBS performance and impact evaluation. In *Evaluating the Impact of Nature-Based Solutions*; Dumitru, A., Wendling, L., Eds.; European Commission: Brussels, Belgium, 2021; pp. 47–70, ISBN 9789276229612.
22. Řešení Inspirovaná Přírodou v Horizontu. Available online: https://www.h2020.cz/files/cejkova/NBS-Echo.pdf (accessed on 29 August 2019).
23. Bourguignon, D. Nature-based solutions concept, opportunities and challenges. *Environ. Res.* **2017**, *159*, 509–518.
24. European Commission. *Nature-Based Solutions: State of the Art in EU-Funded Projects*; Wild, T., Freitas, T., Vandewoestijne, S., Eds.; European Commission: Brussels, Belgium, 2020; ISBN 9789276181989.
25. Haase, D.; Kabisch, S.; Haase, A.; Andersson, E.; Banzhaf, E.; Baró, F.; Brenck, M.; Fischer, L.K.; Frantzeskaki, N.; Kabisch, N.; et al. Greening cities—To be socially inclusive? About the alleged paradox of society and ecology in cities. *Habitat Int.* **2017**, *64*, 41–48. [CrossRef]
26. Pineda-pinto, M.; Frantzeskaki, N.; Nygaard, C.A. The potential of nature-based solutions to deliver ecologically just cities: Lessons for research and urban planning from a systematic literature review. *Ambio* **2021**. [CrossRef] [PubMed]
27. Beute, F.; Andreucci, M.B.; Lammel, A.; Davies, Z.; Glanville, J.; Keune, H.; Marselle, M.; O'Brien, L.; Olszewska-Guizzo, A.; Remmen, R.; et al. *Types and Characteristics of Urban and Peri-Urban Green Spaces Having an Impact on Human Mental Health and Wellbeing. Report Prepared by an EKLIPSE Expert Working Group*; UK Centre for Ecology & Hydrology: Wallingford, UK, 2020; ISBN 9781906698751.
28. Bayulken, B.; Huisingh, D.; Fisher, P.M.J. How are nature based solutions helping in the greening of cities in the context of crises such as climate change and pandemics? A comprehensive review. *J. Clean. Prod.* **2021**, *288*, 125569. [CrossRef]
29. Winch, R.; Moss, C. *Principles for Delivering Urban Nature-Based Solutions*; Green Building Council: London, UK, 2021.
30. Mayer, F.S.; Frantz, C.M.P. The connectedness to nature scale: A measure of individuals' feeling in community with nature. *J. Environ. Psychol.* **2004**, *24*, 503–515. [CrossRef]
31. Boley, B.B.; Strzelecka, M.; Yeager, E.P.; Ribeiro, M.A.; Aleshinloye, K.D.; Woosnam, K.M.; Mimbs, B.P. Measuring place attachment with the Abbreviated Place Attachment Scale (APAS). *J. Environ. Psychol.* **2021**, *74*, 101577. [CrossRef]
32. van der Jagt, A.P.N.; Smith, M.; Ambrose-Oji, B.; Konijnendijk, C.C.; Giannico, V.; Haase, D.; Lafortezza, R.; Nastran, M.; Pintar, M.; Železnikar, Š.; et al. Co-creating urban green infrastructure connecting people and nature: A guiding framework and approach. *J. Environ. Manag.* **2019**, *233*, 757–767. [CrossRef] [PubMed]
33. Kaplan, R.; Kaplan, S. *The Experience of Nature: A Psychological Perspective*; Cambridge University Press: New York, NY, USA, 1989; ISBN 0-521-34139-6.
34. Kaplan, S. The restorative benefits of nature: Toward an integrative framework. *J. Environ. Psychol.* **1995**, *15*, 169–182. [CrossRef]
35. Subiza-Pérez, M.; Vozmediano, L.; San Juan, C. Restoration in urban settings: Pilot adaptation and psychometric properties of two psychological restoration and place bonding scales/Restauración en contextos urbanos: Adaptación piloto y propiedades psicométricas de dos escalas de restauración psicoló. *PsyEcology* **2017**, *8*, 234–255. [CrossRef]
36. Aletta, F.; Kang, J. Promoting healthy and supportive acoustic environments: Going beyond the quietness. *Int. J. Environ. Res. Public Health* **2019**, *16*, 4988. [CrossRef]
37. Staats, H.; Jahncke, H.; Herzog, T.R.; Hartig, T. Urban options for psychological restoration: Common strategies in everyday situations. *PLoS ONE* **2016**, *11*, 1–24. [CrossRef]

38. Mahmoud, I.; Morello, E. Co-creation Pathway for Urban Nature-Based Solutions: Testing a Shared-Governance Approach in Three Cities and Nine Action Labs. In *Smart and Sustainable Planning for Cities and Regions*; Bisello, A., Ed.; Springer International Publishing: Berlin, Germany, 2021; pp. 259–276, ISBN 9783030577643.

39. Mahmoud, I.; Morello, E. Are Nature-based solutions the answer to urban sustainability dilemma? The case of CLEVER Cities CALs within the Milanese urban context. In *L'Urbanistica Italiana di Fronte all'Agenda Portare Territori e Comunità Sulla Strada Della Sostenibilità e Della Resilienza, Proceedings of the Atti della XXII Conferenza Nazionale SIU*; SIU Società Italiana degli Urbanisti: Matera, Italy, 2020; pp. 1322–1327.

40. Morello, E.; Mahmoud, I.; Gulyurtlu, S. Guidance on Co-Creating Nature-Based Solutions PART II—Running CLEVER Action Labs in 16 Steps. Deliverable 1.1.6. 2018. Available online: http://guidance.clevercities.eu/ (accessed on 28 April 2019).

41. CLEVER Cities. D4.3 Monitoring Strategy in the FR Interventions. 2020. Available online: https://clevercities.eu/fileadmin/user_upload/Resources/CLEVER_D4.3_Monitoring_Strategy_in_the_FR_interventions_vF2.pdf (accessed on 10 February 2021).

42. Italy Announces Restrictions Over Entire Country in Attempt to Halt Coronavirus. Available online: https://www.nytimes.com/2020/03/09/world/europe/italy-lockdown-coronavirus.html (accessed on 21 April 2021).

43. Italy Goes Into Nationwide Lockdown as Virus Numbers Spiral. Available online: https://www.bloomberg.com/news/articles/2020-03-09/italy-to-extend-lockdown-nationwide-after-virus-spreads (accessed on 21 April 2021).

44. Coronavirus Italy: PM Extends Lockdown to Entire Country. Available online: https://www.theguardian.com/world/2020/mar/09/coronavirus-italy-prime-minister-country-lockdown (accessed on 21 April 2021).

45. Gaber, J.; Gaber, S.L. Utilizing Mixed-Method Research Designs in Planning: The Case of 14th Street, New York City. *J. Plan. Educ. Res.* **1997**, *17*, 95–103. [CrossRef]

46. Gaber, J.; Gaber, S.L. *Qualitative Analysis for Planning and Policy: Beyond the Numbers*; American Planning Association: Chicago, IL, USA, 2007.

47. Flyvbjerg, B. Five Misunderstandings About Case-Study Research. *Qual. Inq.* **2006**, *12*, 219–245. [CrossRef]

48. Eggermont, H.; Balian, E.; Azevedo, J.M.N.; Beumer, V.; Brodin, T.; Claudet, J.; Fady, B.; Grube, M.; Keune, H.; Lamarque, P.; et al. Nature-based Solutions: New Influence for Environmental Management and Research in Europe. *GAIA-Ecol. Perspect. Sci. Soc.* **2015**, *24*, 243–248. [CrossRef]

49. Cohen-Shacham, E.; Walters, G.; Janzen, C.; Maginnis, S. *Nature-Based Solutions to Address Global Societal Challenges*; IUCN: Gland, Switzerland, 2016; ISBN 9782831718125.

50. Andersson, E.; Langemeyer, J.; Borgström, S.; McPhearson, T.; Haase, D.; Kronenberg, J.; Barton, D.N.; Davis, M.; Naumann, S.; Röschel, L.; et al. Enabling Green and Blue Infrastructure to Improve Contributions to Human Well-Being and Equity in Urban Systems. *Bioscience* **2019**, *69*, 566–574. [CrossRef]

51. Carrus, G.; Scopelliti, M.; Lafortezza, R.; Colangelo, G.; Ferrini, F.; Salbitano, F.; Agrimi, M.; Portoghesi, L.; Semenzato, P.; Sanesi, G. Go greener, feel better? The positive effects of biodiversity on the well-being of individuals visiting urban and peri-urban green areas. *Landsc. Urban Plan.* **2015**, *134*, 221–228. [CrossRef]

52. Braubach, M.; Egorov, A.; Mudu, P.; Wolf, T.; Thompson, C.W.; Martuzzi, M. Effects of Urban Green Space on Environmental Health, Equity and Resilience. In *Nature-Based Solutions to Climate Change Adaptation in Urban Areas*; Kabisch, N., Korn, H., Stadler, J., Bonn, A., Eds.; Springer: Cham, Switzerland, 2017; pp. 51–64, ISBN 978-3-319-53750-4.

53. Marselle, M.R.; Hartig, T.; Cox, D.T.C.; De Bell, S.; Knapp, S.; Lindley, S.; Triguero-mas, M.; Böhning-Gaese, K.; Cook, P.A.; de Vries, S.; et al. Pathways linking biodiversity to human health: A conceptual framework. *Environ. Int.* **2021**, *150*, 106420. [CrossRef] [PubMed]

54. Escobedo, F.J.; Giannico, V.; Jim, C.Y.; Sanesi, G.; Lafortezza, R. Urban forests, ecosystem services, green infrastructure and nature-based solutions: Nexus or evolving metaphors? *Urban For. Urban Green.* **2018**, *37*, 3–12. [CrossRef]

55. Nisbet, E.K.; Shaw, D.W.; Lachance, D.G. Connectedness With Nearby Nature and Well-Being. *Front. Sustain. Cities* **2020**, *2*. [CrossRef]

56. Ghahramani, M.; Galle, N.J.; Ratti, C.; Pilla, F. Tales of a city: Sentiment analysis of urban green space in Dublin. *Cities.* **2021**, 103395. [CrossRef]

57. Hunter, A.J.; Luck, G.W. Defining and measuring the social-ecological quality of urban greenspace: A semi-systematic review. *Urban Ecosyst.* **2015**, *18*, 1139–1163. [CrossRef]

58. Rice, W.L.; Mateer, T.J.; Reigner, N.; Newman, P.; Lawhon, B.; Taff, B.D. Changes in recreational behaviors of outdoor enthusiasts during the COVID-19 pandemic: Analysis across urban and rural communities. *J. Urban Ecol.* **2020**, *6*, 1–7. [CrossRef]

59. Rousseau, S.; Deschacht, N. Public Awareness of Nature and the Environment During the COVID-19 Crisis. *Environ. Resour. Econ.* **2020**, *76*, 1149–1159. [CrossRef] [PubMed]

60. Larcher, F.; Pomatto, E.; Battisti, L.; Gullino, P.; Devecchi, M. Perceptions of Urban Green Areas during the Social Distancing Period for COVID-19 Containment in Italy. *Horticulturae* **2021**, *7*, 55. [CrossRef]

61. Ugolini, F.; Massetti, L.; Calaza-Martínez, P.; Cariñanos, P.; Dobbs, C.; Ostoic, S.K.; Marin, A.M.; Pearlmutter, D.; Saaroni, H.; Šaulienė, I.; et al. Effects of the COVID-19 pandemic on the use and perceptions of urban green space: An international exploratory study. *Urban For. Urban Green.* **2020**, *56*, 126888. [CrossRef]

62. European Commission. *Nature-Based Solutions A Thematic Collection of Innovative EU-Funded Research Results—Unlocking Nature's Potential*; European Commission: Brussels, Belgium, 2020.

63. Baldwin, C.; Vincent, P.; Anderson, J.; Rawstorne, P. Measuring Well-Being: Trial of the Neighbourhood Thriving Scale for Social Well-Being Among Pro-Social Individuals. *Int. J. Community Well-Being* **2020**, *3*, 361–390. [CrossRef]
64. Laage-Thomsen, J.; Blok, A. Varieties of green: On aesthetic contestations over urban sustainability pathways in a Copenhagen community garden. *Environ. Plan. E Nat. Sp.* **2020**. [CrossRef]
65. Hörschelmann, K.; Werner, A.; Bogacki, M.; Lazova, Y. Taking Action for Urban Nature: Citizen Engagement Handbook. 2019. Available online: https://naturvation.eu/result/taking-action-urban-nature-citizen-engagement (accessed on 4 May 2020).
66. Frantzeskaki, N. Seven lessons for planning nature-based solutions in cities. *Environ. Sci. Policy* **2019**, *93*, 101–111. [CrossRef]
67. Langemeyer, J.; Camps-Calvet, M.; Calvet-Mir, L.; Barthel, S.; Gómez-Baggethun, E. Stewardship of urban ecosystem services: Understanding the value(s) of urban gardens in Barcelona. *Landsc. Urban Plan.* **2018**, *170*, 79–89. [CrossRef]
68. Faivre, N.; Fritz, M.; Freitas, T.; de Boissezon, B.; Vandewoestijne, S. Nature-Based Solutions in the EU: Innovating with nature to address social, economic and environmental challenges. *Environ. Res.* **2017**, *159*, 509–518. [CrossRef] [PubMed]
69. Pauleit, S.; Zölch, T.; Hansen, R.; Randrup, T.B.; Konijnendijk van den Bosch, C. Nature-Based Solutions and Climate Change—Four Shades of Green. In *Nature Based Solutions to Climate Change Adaptation in Urban Areas: Linkages between Science, Policy and Practice*; Springer: Berlin, Germany, 2017; pp. 29–49.
70. van der Jagt, A.P.N.; Szaraz, L.R.; Delshammar, T.; Cvejić, R.; Santos, A.; Goodness, J.; Buijs, A. Cultivating nature-based solutions: The governance of communal urban gardens in the European Union. *Environ. Res.* **2017**, *159*, 264–275. [CrossRef] [PubMed]
71. Dadvand, P.; Bartoll, X.; Basagaña, X.; Dalmau-Bueno, A.; Martinez, D.; Ambros, A.; Cirach, M.; Triguero-Mas, M.; Gascon, M.; Borrell, C.; et al. Green spaces and General Health: Roles of mental health status, social support, and physical activity. *Environ. Int.* **2016**, *91*, 161–167. [CrossRef]
72. Nared, J.; Bole, D. *Participatory Research and Planning in Practice*; The Urban; Springer: Ljubljana, Slovenia, 2020; ISBN 9783030280130.
73. Machielse, W. Perceived Safety in Public Spaces: A Quantitative Investigation of the Spatial and Social Influences on Safety Perception among Young Adults in Stockholm. 2015. Available online: https://www.semanticscholar.org/paper/Perceived-safety-in-public-spaces-%3A-A-quantitative-Machielse/6a9bafd76af6a3da8b0d9678681dc098def88b12 (accessed on 25 April 2021).
74. Hashim, N.H.M.; Thani, S.K.S.O.; Jamaludin, M.A.; Yatim, N.M. A Perceptual Study on the Influence of Vegetation Design Towards Women's Safety in Public Park. *Procedia-Soc. Behav. Sci.* **2016**, *234*, 280–288. [CrossRef]
75. Hosseinalizadeh, S. Safer Green Cities A Study about Vegetation Impacts on Perception of Safety in Green Spaces Case Study: Biblioteca Degli Alberi di Milano (BAM). Master's Thesis, Politecnico di Milano, Milan, Italy, 2020.
76. Cantergiani, C.; Herranz, K.; Murphy-Evans, N.; Bradley, S.; Pastoors, J.; Menny, M.; Robert, J.; Casagrande, S.; Barone, E.; Berrini, M.; et al. Co-Creation Plan and Co-Design of Solutions in CALs. CLEVERCities Deliverable 2.2. 2019. Available online: https://clevercities.eu/fileadmin/user_upload/Resources/D2.2_Co-creation.pdf (accessed on 10 May 2021).
77. Konjaria-Christian, S.; Pastoors, J.; Arlatti, A.; Rödl, A.; Berghausen, M.; Quanz, J.; Robert, J.; Rinsch, F.; Lüders, B.; Schmalzbauer, A.; et al. CAL Specific co Implementation Plan. CLEVER Cities. Deliverable 2.3. 2019. Available online: https://ec.europa.eu/research/participants/documents/downloadPublic?documentIds=080166e5d12c3540&appId=PPGMS (accessed on 6 July 2020).
78. Mahmoud, I.; Morello, E. Co-Creation Pathway as a catalyst for implementing Nature-based Solution in Urban Regeneration Strategies Learning from CLEVER Cities framework and Milano as test-bed. *Urban. Inf.* **2018**, *278*, 204–210.
79. AdP SCALI FERROVIARI—Milano. Available online: https://www.comune.milano.it/-/adp-scali-ferroviari-milano (accessed on 9 May 2021).
80. The EU Strategy on Green Infrastructure. Available online: https://ec.europa.eu/environment/nature/ecosystems/strategy/index_en.htm (accessed on 7 May 2021).
81. Barton, M. Nature-Based Solutions in Urban Contexts: A Case Study of Malmö, Sweden. Available online: https://lup.lub.lu.se/luur/download?func=downloadFile&recordOId=8890909&fileOId=8890910 (accessed on 15 October 2019).
82. Haase, A. The Contribution of Nature-Based Solutions to Socially Inclusive Urban Development—Some Reflections from a Social-environmental Perspective. In *Nature-Based Solutions to Climate Change adaptation in Urban Areas*; Springer: Cham, Switzerland, 2017; pp. 221–236. [CrossRef]
83. English Partnerships. *Additionality Guide: A Standard Approach to Assessing the Additional Impact of Projects*, 2nd ed.; 2004. Available online: https://assets.publishing.service.gov.uk/government/uploads/system/uploads/attachment_data/file/191511/Additionality_Guide_0.pdf (accessed on 22 January 2021).
84. Raymond, C.M.; Berry, P.; Breil, M.; Nita, M.R.; Kabisch, N.; de Bel, M.; Enzi, V.; Frantzeskaki, N.; Geneletti, D.; Cardinaletti, M.; et al. *An Impact Evaluation Framework to Support Planning and Evaluation of Nature-Based Solutions Projects*; Centre for Ecology and Hydrology: Lancaster, UK, 2017; ISBN 9781906698621.
85. Schönfeld, K.C. Von Urban Planning and European Innovation Policy: Achieving Sustainability, Social Inclusion, and Economic Growth. *Sci. Public Policy* **2019**, *46*, 772–783.
86. Shams, I.; Barker, A. Urban Forestry & Urban Greening Barriers and opportunities of combining social and ecological functions of urban greenspaces—Users' and landscape professionals' perspectives. *Urban For. Urban Green.* **2019**, *39*, 67–78. [CrossRef]
87. UNaLab. *Performance and Impact Monitoring of Nature-Based Solutions*. 2019. Available online: https://unalab.eu/system/files/2020-02/d31-nbs-performance-and-impact-monitoring-report2020-02-17.pdf (accessed on 22 January 2021).
88. Perrin, M. Impact-Driven Financing and Investment Strategies For Urban Regeneration: Types of NBS Financing Sources. 2018. CLEVER Cities Project. Available online: https://clevercities.eu/fileadmin/user_upload/Resources/D1.1_Theme_3_financing_urban_regeneration_EBN_12.2018.pdf (accessed on 11 April 2019).

89. European Commission. *Nature-Based Solutions Learning Scenario*; European Commission: Brussels, Belgium, 2020.
90. European Commission. *Evaluating the Impact of Nature-Based Solutions A Handbook for Practitioners*; Dumitru, A., Wendling, L.A., Eds.; European Commission: Brussels, Belgium, 2021.

sustainability

Article

Municipal Practices for Integrated Planning of Nature-Based Solutions in Urban Development in the Stockholm Region

Peter Brokking [1,*], Ulla Mörtberg [2] and Berit Balfors [2]

[1] Department of Urban Planning and Environment, KTH Royal Institute of Technology, SE-100 44 Stockholm, Sweden
[2] Department of Sustainable Development, Environmental Science and Engineering, KTH Royal Institute of Technology, SE-100 44 Stockholm, Sweden; mortberg@kth.se (U.M.); balfors@kth.se (B.B.)
*   Correspondence: peter.brokking@abe.kth.se

Abstract: Urban planning is assumed to play an important role in developing nature-based solutions (NBS). To explore how NBS is addressed in urban development, municipal planning practices are analyzed based on three case studies in the Stockholm region of Sweden. Through focus group discussions, interviews and document studies, the planning and implementation of NBS and their intended contribution to regional green infrastructure (GI) and social and ecological qualities are investigated. The results show that the planning and design of urban green spaces engages the local community. Moreover, different conceptual frameworks are used to strengthen an ecological perspective and nurture expected outcomes, in particular ecosystem services and GI. Through competence development and collaborative approaches, the co-creation of innovative solutions for public and private green spaces is promoted. However, institutional conditions, e.g., legal frameworks and landownership shape the planning process and can challenge the ability to enhance social and ecological qualities. An assessment of the planning processes indicates a strong focus on ecosystem services and local GI, while the potential to contribute to regional GI differs widely between cases. The study concludes that a knowledge-driven and integrative planning process can foster the potential of NBS for green and sustainable cities.

Keywords: green infrastructure; municipal planning; ecosystem services; shared governance; co-creation; public-private collaboration; competence development; land development

Citation: Brokking, P.; Mörtberg, U.; Balfors, B. Municipal Practices for Integrated Planning of Nature-Based Solutions in Urban Development in the Stockholm Region. *Sustainability* **2021**, *13*, 10389. https://doi.org/10.3390/su131810389

Academic Editors: Israa H. Mahmoud, Eugenio Morello, Giuseppe Salvia and Emma Puerari

Received: 14 July 2021
Accepted: 14 September 2021
Published: 17 September 2021

Publisher's Note: MDPI stays neutral with regard to jurisdictional claims in published maps and institutional affiliations.

1. Introduction

Worldwide, urban green spaces are under pressure as a result of the expansion and densification of urbanized areas, the exploitation of land for the development of buildings and roads, and the altering of landscapes and ecosystems [1]. The transformation of urban areas poses a threat to the social and ecological qualities that urban green spaces provide in terms of ecosystems services, biodiversity and wellbeing [2–4]. Hence, preserving green spaces is a pressing global challenge [5] that calls for urban responses that can invert the trends and accelerate change towards both local and global sustainability [6]. From a policy perspective, the crucial role of urban green space in future urban development is recognized in the UN 2030 Agenda, which is particularly addressed in Sustainable Development Goal 11, *Make cities and human settlements inclusive, safe, resilient and sustainable* [7].

In response to the decline of urban green space, nature-based solutions (NBS) has emerged as a concept to operationalize an ecosystem services approach within spatial planning [8]. NBS is a relatively new concept, but gained momentum when it was launched as a major research area by the European Commission (EC) in 2015 [9] to improve the implementation capacity through research and innovation activities [10]. This engagement has yielded a diversity of results, but revealed certain key challenges e.g., the refinement of the NBS concept in relation to other established concepts, a deeper understanding of potential conflicts with investment interests, and the risk for gentrification [11].

The reading of the NBS concept vis-à-vis related concepts (Table 1) is a recurring theme in the scientific literature and an important aspect in the communication with stakeholders [12]. As a clarification of the definition, the EC added that NBS must benefit biodiversity and support the delivery of a range of ecosystem services [13], which implies a call for added social and ecological qualities. As an umbrella concept, NBS is intended to 'sweep up' all other concepts for sustainability interventions that employ nature [14]. Hence, NBS integrates existing approaches, e.g., 'ecosystem services' and 'green-blue infrastructure', with assessments of the social and economic benefits of resource-efficient and systemic solutions that combine technical, governance, regulatory and social innovation [15]. This means that NBS embraces all types of measures that aim to foster social and ecological qualities of urban green spaces, strengthens green infrastructure (GI) and/or supports urban resilience.

Table 1. Definitions of core concepts.

Concept	Definition
Nature-based solution (NBS)	Solutions that are inspired and supported by nature which are cost-effective, simultaneously provide environmental, social and economic benefits, and help build resilience; such solutions must benefit biodiversity and support the delivery of a range of ecosystem services and bring more, and more diverse, nature and natural features and processes into cities, landscapes and seascapes, through locally adapted, resource-efficient and systemic interventions [13]
Green Infrastructure (GI)	A strategically planned network of natural and semi-natural areas, designed and managed to enhance biodiversity as well as deliver ecosystem services [16]
Ecosystem Services	Ecological characteristics, functions, or processes that directly or indirectly contribute to human well-being [17]
Green Space Ratio	The ratio between the "eco-efficient surface" and the entire surface of the plot or property [18]
Biodiversity	The variability among living organisms from all sources including, inter alia, terrestrial, marine and other aquatic ecosystems and the ecological complexes of which they are part; this includes diversity within species, between species and of ecosystems [19]

GI can be defined as a strategically planned network of natural and semi-natural areas, designed and managed to enhance biodiversity as well as to deliver ecosystem services [16,20]. In an urban context, NBS can be integrated in urban landscapes, e.g., by mitigating the loss of green spaces in order to further GI and thereby contribute to multiple dimensions of urban sustainability [21]. However, there is no consensus on how to design GI for promoting biodiversity and ecosystem services [22–24], but major efforts on developing GI are currently being undertaken and methodological frameworks for selecting appropriate green space designs are suggested [25]. Therefore, integrating NBS to nurture biodiversity and ecosystem services is associated with major uncertainties.

Although NBS has been endorsed to contribute to sustainable communities [26], translating the concept into legal and institutional systems to support implementation remains challenging [11]. Since the design of green spaces is contingent upon local circumstances, NBS practice is embedded in local settings and often connected to new urban development. Therefore, urban planning is reckoned to play an important role in achieving the integration of NBS in cities [27] and to merge social and ecological systems [2]. The integrative approach fosters the multifunctional nature of NBS [28] and requires collaboration across disciplines and governmental domains [14]. While the focus may initially be on developing a joint vision and a design that meets diverse and often contradictory objectives for the NBS, the collaboration also needs to align activities, financial commitments and responsibilities for the development and maintenance of urban green spaces. This implies that institutional conditions, e.g., legal frameworks, governmental responsibilities and land ownership, are important in developing NBS, which shapes the planning process and the ability to deliver social and ecological qualities. Hence, planning administrations need to adapt to prevailing conditions to ensure a collaborative and integrated planning trajectory that is broader than statutory planning procedures and advances to the anticipated goals for the NBS and the urban development project. Accordingly, urban planning should be understood as the governance of place, which necessitates a collaborative and deliberate approach that includes both the qualities of place and of process [29]. In this manner, municipal planning processes can provide a framework that enables shared governance for the development of NBS [30]. In the collaborative approach, active involvement of the local community and NGOs should be encouraged to empower citizens in the development of their local environment and equip them with knowledge about developing, operating and maintaining NBS [31]. Grounded on these preconditions, urban planning can enhance the merger of competencies and perspectives in the design and implementation of green spaces and leverage the potentials of NBS.

In response to the identified need to integrate NBS in institutional systems, this paper explores how NBS is addressed in municipal planning and urban development, and in what way the siting and design of green spaces benefit from policy frameworks for NBS, GI, ecosystem services and biodiversity. In addition, the paper focuses on the collaboration between municipal authorities and other stakeholders aiming to promote the integration of high-quality NBS in urban development. The paper is based on three empirical studies in the Stockholm region of Sweden. From an international perspective, the Swedish planning system is highly decentralized, with a planning monopoly for municipalities [32]. This means that Swedish municipalities have an important role in integrating environmental and sustainability issues in local planning [33] and enabling NBS.

This paper aims to analyze municipal practices for sustaining and developing GI, biodiversity and ecosystem services through NBS as part of urban development and to identify expedient pathways for the planning and implementation of NBS. More specifically, the aim is to understand the drivers for NBS, i.e., what qualities the planning strives for and how these benefit GI and support the delivery of ecosystem services. The following research questions are addressed in this paper:

1. What social and ecological qualities does the development of green space aim at and how are these qualities embedded in the planning process?
2. How does the planning process govern the design and implementation of green spaces in urban development projects to ensure the desired qualities of NBS?
3. How are the NBS-related conceptual frameworks of green infrastructure, biodiversity and ecosystem services used in urban planning and development and how do they enhance the social and ecological qualities of green spaces?

2. Materials and Methods

The research is based on a study of urban development projects in three municipalities in the Stockholm region, i.e., Stockholm, Täby and Upplands Väsby, see Figure 1. The City of Stockholm is the capital of Sweden, and with almost one million inhabitants the largest among 26 municipalities in the region [34]. During the last two decades, the city has focused on developing central areas through infills and extensive brownfield developments close to the inner city [35], and this has raised concerns about the loss of green spaces in central areas. A study of changes in nonurban land cover in the City of Stockholm shows that the quantity of green spaces decreased by 2% between 2003 and 2018 [36]. Täby and Upplands Väsby are both commuter municipalities located north of Stockholm. Within the municipal borders, there are large coherent rural areas and peri-urban green spaces with social and ecological qualities which contribute to the regional GI. Urban development takes place primarily in the central parts of both municipalities where green space is fragmented [37,38].

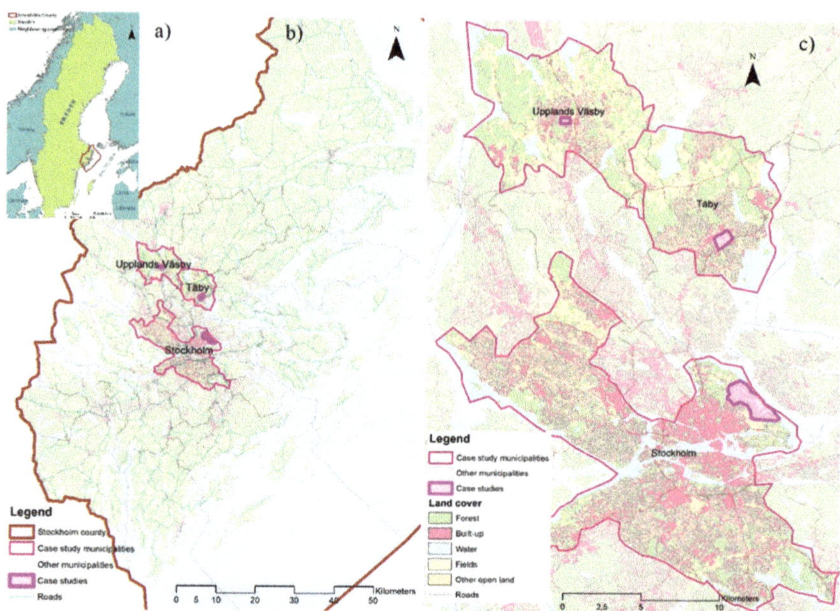

Figure 1. Study area with case studies: (**a**) Stockholm county in Sweden, (**b**) case studies and their municipalities in Stockholm county, and (**c**) zoom-in on case studies and their municipalities. Spatial data © Lantmäteriet and © EuroGeographics.

To achieve the aims of the paper, the planning processes of the three urban development projects are analyzed to gain an understanding of how municipal agencies run the process of developing NBS within urban development in collaboration with other public and private stakeholders. The results of the analyses are used to examine how the detailed design of the NBS evolves as part of the overall project planning to identify approaches for interdisciplinary collaboration and conceptualization of NBS that contribute to the drafting of the project and to single out mechanisms that are employed to ensure the implementation of the NBS. The study covers a planning period of 10 years and does not include an evaluation of the final NBS because the development of green spaces is still in progress.

The paper is based on the results from two research projects. The first, ISSUE (Integrated Sustainable Strategies for Urban Environments), is a transdisciplinary research project that studied local planning practices for sustainable development in urban and peri-urban areas. Through the collaboration between researchers and practitioners, knowledge was collected in a series of focus group discussions [39] regarding existing challenges and preconditions for novel planning practices that enhance sustainable urban development [40]. These focus group discussions took place in different thematic think-tanks, among which one focused on social sustainability and another on planning practices for sustainable development. The transdisciplinary think tanks included urban planners and environmental planners from different units of the municipal administration of Täby, Stockholm and Upplands Väsby, as well as sustainability experts from consultancy firms, sustainability specialists and business developers from private developers, and researchers in urban planning and sustainability analysis from KTH. The think tanks met three to four times per year over a period of three years to discuss predefined questions on the basis of ongoing sustainable urban development projects within the participating organizations. The co-production of knowledge was based on an incremental and iterative approach that included a sequence of research activities to deepen the understanding of key issues related to sustainable urban development [40]. In addition, a series of semi-structured interviews [41] were conducted with 17 representatives from the municipalities of Täby, Upplands Väsby and Nacka (i.e., environmental planners, urban planners, development engineers and building permit officers) and private developers (i.e., business developers and specialists), to collect individual opinions on tools and approaches to strengthen the ecosystem services in detailed development planning [42].The focus group discussions and interviews were documented by participating researchers. Within ISSUE, the case studies from the municipalities of Täby and Upplands Väsby that are presented in this paper were discussed in the thinktanks.

The other research project that feeds results into this paper is 'Sustainability, regulation and roles from detailed development plan to building permit'. This research studied opportunities and obstacles for promoting sustainable urban development in the planning and land development process [43]. Four urban development projects were studied in the research, including Stockholm Royal Seaport, which is included in this paper. For each project, one environmental planner and one urban planner that were significantly involved in the municipal planning process were interviewed. In addition, a group interview was conducted with four representatives from the real estate industry to collect their experiences and perspectives on contemporary practices for sustainable urban development, in particular in Stockholm [44].

Besides experiential knowledge that was obtained from the focus group discussions and interviews, studies of literature, and official documents (such as planning documents, investigations and online resources), were conducted to collect factual information on the three urban development projects.

3. Regional Green Infrastructure Initiatives in Stockholm County

Stockholm County is the fastest-growing region in Sweden with 2.3 million inhabitants and is expected to increase its population by 50% until 2050 (Stockholm County Council 2018). To meet the ongoing regional growth, a large number of urban developments are initiated in the municipalities. To guide local development, the Regional Development Plan (RUFS 2050) for Stockholm presents a vision that has been prepared and discussed with the municipal authorities and other actors. The latest regional development plan was adopted in 2018 by Stockholm County Council [45] that in 2019 changed its name to Region Stockholm.

Within Stockholm County, two different initiatives for regional GI coexist, each with a different focus and responsible authority. The first is the Stockholm green wedges that are promoted by Region Stockholm and have been part of regional planning and policy since the 1990s [46]. These ten green wedges create contiguous green spaces that extend from the countryside in Stockholm County (ca 30–50 km from the city center) to the center of the City of Stockholm. In addition, a number of large green-blue areas with high recreational, ecological and cultural qualities are outlined, as seen in Figure 2. As a planning concept, the green wedges have been important for the planning of the spatial development of the region since maintaining a coherent regional GI has been in focus. Some parts of the green wedges are vulnerable and defined as weak connections. Strengthening these weak connections is deemed to be vital for binding together the green wedges and their green core areas to secure recreational paths, creating access to larger strolling areas, and maintaining ecological connectivity [45,47,48].

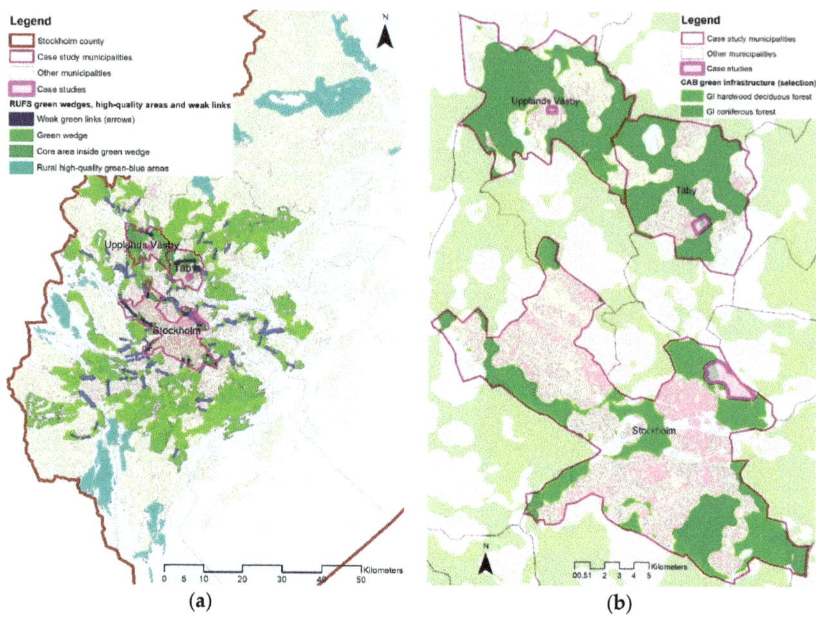

Figure 2. The location of the case studies in relation to the GI initiatives in Stockholm County. (**a**) RUFS GI comprising green wedges, rural high-quality areas and weak links. (**b**) an illustration of parts of the Stockholm CAB GI, showing main corridors for hardwood deciduous forest (light green) and coniferous forest (dark green) within and around the three case studies and their municipalities. Spatial data © Lantmäteriet, [45,49].

The second regional GI is based on the EU Biodiversity Strategy [50,51] and the EU strategy and guidelines on GI [16,20] for which guidelines were developed by the Swedish Environmental Protection Agency [52]. The Stockholm County Administrative Board (Stockholm CAB), which is the State representative in Stockholm County, is responsible for developing the regional GI that consists of ecologically functional networks of habitats, structures, natural areas and landscape elements that are designed, used and managed to maintain biodiversity and promote ecosystem services [49]. In 2019, the Stockholm CAB published a regional action plan for GI that sets priorities among different nature conservation activities, and concretizes goals and approaches to the different areas.

As shown in Figure 2a, the case studies in Upplands Väsby and Täby are situated in urban central positions within the municipalities, and not in direct contact with the RUFS GI or its weak links. However, the Stockholm case is situated within or very close to a weak link and can therefore directly be used to strengthen it. Regarding the Stockholm CAB GI (Figure 2b), the situation is different. Upplands Väsby is not connected to either of the selected GI components, i.e., coniferous forest and hardwood deciduous forest, while the Täby case study is situated within the coniferous forest link. The Stockholm case is very strategically located to strengthen the hardwood deciduous forest link.

Both regional GI initiatives value ecological qualities in terms of biodiversity and ecosystem services, but the Stockholm green wedges include a wider range of functions [53], e.g., recreation, health and attractiveness [45], which are linked to the conurbation of the Stockholm metropolitan area. In this regard, the Stockholm green wedges differ from the regional GI that is coordinated by the Stockholm CAB, which is based on a mapping of ecological qualities for the entire county while social qualities are subordinated. Accordingly, the action plan focuses on measures that foster biodiversity and ecosystem services, both in protected areas and in everyday landscapes [49]. Both regional GIs overlap geographically and the regional authorities collaborate in the planning of measures, gaining from the longstanding work with the green wedges and the actor networks that have been set up for many of the wedges.

This shows that both Stockholm CAB and Region Stockholm have important roles in conceptualizing regional GI and sharing knowledge on the social and ecological qualities of the different components that are part of it. Several protected areas are included in the regional GIs and cannot be exploited. However, for the remaining areas, regional authorities rely heavily on the municipalities for the maintenance, planning and implementation of NBS. Hence, the municipal responses affect to a large extent the long-term development of the regional GIs.

4. Planning for NBS through Urban Development

To control the right to develop or change land use, all EU countries use different planning instruments to balance development and the protection of land in the public interest [54]. In Sweden, the Planning and Building Act [55] regulates land use planning and provides legal tools, such as detailed development plans and building permits that allow municipalities to control land use and thereby protect green spaces from development or allocate sites for NBS. The municipal practices for sustaining and developing ecological and social qualities in urban development through NBS are contingent upon the location and conditions of the development site, the solutions that are considered, and the terms for implementing selected measures.

An important condition is land ownership. Private landowners have the right to develop their property in line with the provisions in the detailed development plan. In the plan, the municipality can articulate the need to foster ecological and social qualities on the development site, but it cannot stipulate binding detailed instructions related to ecosystem services, green roofs, or other types of NBS on private land [56].

The detailed development plan also applies to the development of municipal property. A large share of green spaces in urban areas, such as parks, common land and roadsides, is developed on publicly owned land, which allows municipalities to fully control land use and cater for ecosystem services in urban settings. However, municipalities also use land ownership to facilitate and control urban development, and thus public property is sold to private actors for housing purposes through land allocation [57]. When selling land to private developers, municipalities can attach conditions to the land transfer that have far-reaching requirements in a civil agreement. In this way, municipalities can put demands on developers to create green spaces with ecological and social qualities on the building plot [43]. Hence, land ownership enables municipalities to push for NBS in projects that are developed by the private actors and raise the sustainability targets beyond the ambitions of the detailed development plan.

Most of the urban development in the Stockholm region takes place in or adjacent to built-up areas through densification at locations with good accessibility to public transportation [45]. In central locations, this may affect weak links within the GI when parks and green pockets are transformed into housing areas [47,49] but the development of sites close to weak links also provides opportunities for strengthening GI using NBS. To nurture the planning of NBS in urban development projects there is also a need for expertise and political commitment at the municipal level as well as for local policies for green spaces that are embedded in other policies and within the municipal organization. Such policies may primarily address public space, but can also include domestic gardens that contribute to urban GI [58].

Numerous urban development projects that involve NBS have been initiated in the last two decades in the Stockholm region. To gain a deeper understanding of the planning trajectories for NBS as part of urban development, three projects have been selected that represent recent urban developments with a focus on sustainability in different municipalities in Stockholm County, see Figure 1 and Table 2. All three projects are centrally located in their respective municipality. The first project is the Stockholm Royal Seaport, which is a development area in the City of Stockholm, located adjacent to the Royal National City Park and 5 km from the inner city. It is the largest ongoing urban development in Sweden, which accommodates at least 12,000 new homes and 35,000 workplaces and serves as a model of good practice for sustainable urban development [59].

Table 2. Key characteristics of project areas included in the case studies.

Development Area	Stockholm Royal Seaport	Täby Park	Fyrklövern
Municipality	Stockholm	Täby	Upplands Väsby
Type of development	brownfield development	new development	urban renewal/densification
Key figures	12,000 dwellings, 236 ha	6000 dwellings, 70 ha	2000 dwellings, 27 ha
Milestones	2009 Detailed development plan first stages	2010 Municipality starts planning with landowners	2011—Väsby Labs: broad 2013 dialogue
	2010 First version of sustainability program	2013 Structure plan: shared vision landowners	2014 Land allocation using the point system
	2011 Launching of the competence development program	2015 Municipal plan program Täby Park	2015 Detailed development plan public space
	2012 First residents move into new dwellings	2015 Sustainability program: shared program landowners	2016 Development plan for ecosystem services
	2019 Implementation connectivity link from Hjorthagens park	2017 Detailed development plans first stages for housing development	2021 Start implementation of park
Land ownership	100% municipal	80% private, 20% municipal	100% municipal

The second project is Fyrklövern (Four Leaf Clover) in the municipality of Upplands Väsby, in which a mixed housing area with 2000 dwellings will be developed next to the city center. It is the largest development project in the municipality of Upplands Väsby in 30 years and aims to build "a modern small town" [60].

The third case study is Täby Park in the municipality of Täby. This site is a former horse race track that is being transformed into a new urban district with 6000 dwellings and green links. According to the vision for Täby Park "*Everything, from the design of the district, to the construction phase and the operation must be characterized by sustainability*" [61].

The development of Fyrklövern and Täby Park started in 2010 and should be framed in the wider context of the debate on the planning ideal that emphasizes density, multifunctionality and city life. In this debate, city centers that represent the ideal of the mixed city are depicted as attractive, while suburbs are portrayed as dull [62]. Against this background, Täby municipality decided in 2010 to shift from suburban to city-like development [63] and Upplands Väsby municipality initiated a broad public dialogue within Väsby Labs to discuss the transformation of Fyrklövern [64]. The planning of the Stockholm Royal Seaport started after the turn of the millennium and was built on the experiences from the development of Hammarby Sjöstad [65].

5. Three Case Studies of NBS in Urban Development

5.1. Stockholm—Stockholm Royal Seaport

Since the 1990s the city of Stockholm has been developing several new multi-family housing districts with a sustainability profile on former brownfield areas around the inner city, such as Hammarby Sjöstad and the Stockholm Royal Seaport. Through the years, the municipality has broadened the sustainability focus for these developments from waste and water management to climate neutrality, social sustainability and ecosystems services.

As the municipality owns the land in both project areas, it can control the development and ensure the fulfillment of set sustainability goals.

The municipal sustainability strategies and goals for the ongoing projects in the Stockholm Royal Seaport area are specified in the Program for sustainable urban development [66]. The goals are used to define the requirements and criteria in land allocation competitions that are announced for each of the construction phases. Developers are invited to submit bids that were assessed on basis of the criteria. Because of the attractive location of the development site, the municipality usually receives a number of competitive bids and can select the contribution that fulfills the requirements and criteria in the best way.

As part of the competence development program for the Stockholm Royal Seaport, the municipality organizes a variety of activities such as workshops, seminars and innovations projects [67]. These activities intend to increase and exchange knowledge on sustainability challenges and best practices among developers, suppliers, consultants and public administrations, but also to develop new solutions in collaboration with academia and industry. In such a way, these open dialogues foster the introduction of sustainable solutions in urban development and enhance the innovative capacity of the entire sector in the Stockholm region.

Since the Stockholm Royal Seaport is located next to a narrow passage that connects the northern and southern part of the Royal National City Park, the new urban district is expected to provide ecological qualities to strengthen ecological connectivity [66]. Hence, the urban design, the public parks, and private courtyards are assessed on their merits related to ecological connectivity, the provision of ecosystems services, and the green space ratio (see Table 1) of the development plots.

The comprehensive approach that is applied in the development of green spaces in the Stockholm Royal Seaport includes multiple urban design principles to enhance ecosystems services and NBS, e.g., multifunctional green areas, strengthening ecosystems and their connectivity, local stormwater management, urban gardening, and green buildings and roofs. The progress and goal fulfillment is monitored and the latest annual sustainability report highlights that all inhabitants have access to a park within 200 m and that all developments meet the required green space ratio [67]. Moreover, the report presents some of the green spaces that have been completed, which are designed as green corridors to nourish the local ecosystem by focusing on oak and amphibian habitats (see Figure 3). This NBS was already mentioned in the detailed development plan from 2009. Although the site was affected by soil pollution, its role as an ecological link between the Royal National City Park and Hjorthagens Park was highlighted [68].

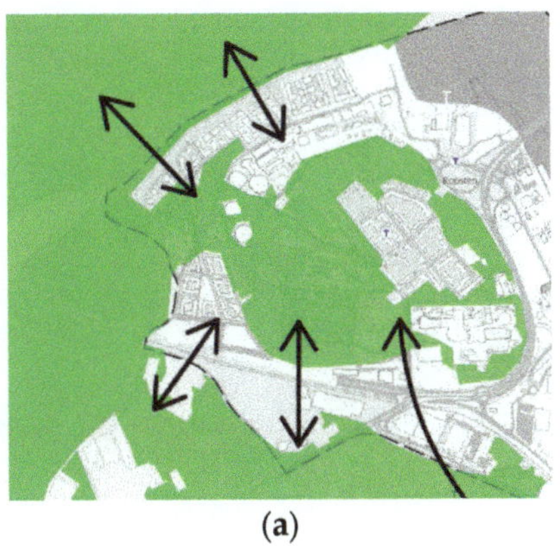

(a)

(b)

Figure 3. (**a**) Main connectivity links between Hjorthagens Park within the development area of Stockholm Royal Seaport and the Royal National City Park. (**b**) Example of a green space in the development area that is designed as a connectivity link. Source: (**a**) City of Stockholm and (**b**) City of Stockholm/Flickr 2019, Creative Commons license.

Citizens are involved in the planning and development of the Stockholm Royal Seaport, partly through social media and consultation meetings and partly through participation in joint activities such as urban gardening and pop-up parks [67]. These activities increase knowledge and promote engagement and cooperative responsibility among residents for maintaining the ecological qualities of these sites.

5.2. Upplands Väsby—Fyrklövern

In 2011, the municipality of Upplands Väsby initiated Väsby Labs to conduct a broad dialogue with citizens, developers and other stakeholders with the aim to provoke innovative ideas for the renewal of a centrally located area called Fyrklövern [64] In the area, a moisture-damaged school building was torn down, which opened up new development in the area. The dialogue generated many ideas that were used to define quality criteria for the development of Fyrklövern [69]. The municipality owns the land, but due to its peri-urban location in the Stockholm region, it tends to be more difficult to attract developers to invest in the municipality of Upplands Väsby compared to centrally located municipalities. However, a number of invited developers participated in the co-creation activities within Väsby Labs.

To kickstart the development process, the municipality set up a point system for the assessment of project proposals through which developers could gain a rebate on the land price for projects that meet the quality criteria that were identified in Väsby Labs. In the prospect of a discount on the land price, 14 developers that submitted a bid for housing development in Fyrklövern signed contracts [69]. The discount created room for developers to test innovative solutions related to urbanity, energy efficiency, co-creation and citizen involvement. A discussion on the detailed planning, which encompassed both the individual projects and the development of Fyrklövern, was initiated with the developers of the winning bids. This process of co-production and joint commitment among parties facilitated the development of the area in line with the quality criteria.

In the detailed development plan, the municipality designated land for the development of a new park at a central location in the area using grown-up trees from other parts of the development site [70]. Besides the new park, a walking passage will be created through the housing area (see Figure 4b). These green spaces are presumed to contribute to the overall ambitions for the area related to social cohesion and safety by creating a permeable structure with different public spaces to accommodate diverse social qualities [71]. Although not explicitly addressed in the point system, adding ecosystem services into the area is a focal issue for the development of both public and private green spaces in Fyrklövern. The public green space and the private courtyards make up a coherent NBS that supports ecosystem services with fruit trees, rain beds, opportunities for gardening and arrangements for social activities.

(a)

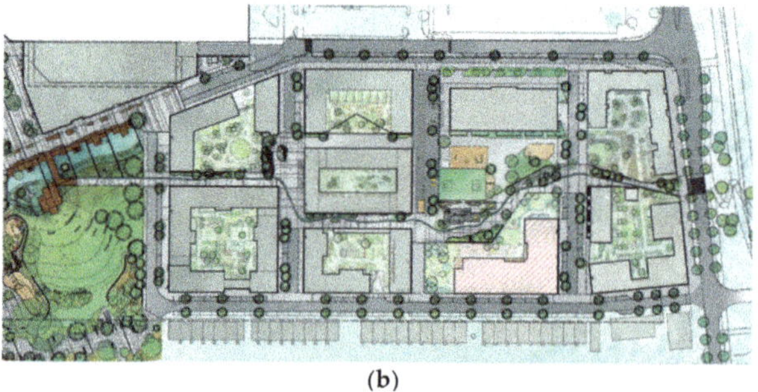

(b)

Figure 4. (**a**) Development area of Fyrklövern. Source: Upplands Väsby municipality. (**b**) Illustration of public and private green spaces, which are designed to deliver a variety of ecosystem services and connected by a walking passage through the housing area Illustration: Funkia.

Due to its location in the central urban area of Upplands Väsby, the NBS in Fyrklövern is poorly connected to the regional GI. However, the Municipal Development Plan for Ecosystem Services [72], which describes proposals for priorities of ecosystem services for the entire municipality, marks out the central urban area as being in a high need for green investments, specifically those related to risks for flooding. The survey of ecosystem services also identifies Fyrklövern as a potential area for developing ecosystem services, e.g., for measures to strengthen the local pollinators network [38]. Hence, the Development Plan for Ecosystem Services is a valuable resource in the planning of NBS in urban development.

5.3. Täby—Täby Park

The municipal comprehensive plan for Täby municipality from 2010 marks a shift from suburban towards a city-like development [63]. Following this plan, the municipality aims to achieve a cohesive urban development in the central parts of Täby, among others in Täby Park, which is a former racecourse that will be transformed into an urban district [73]. The land is mostly owned by four private developers with the exception of 20% of the area that is owned by the municipality. Accordingly, the area is developed on the landowners' terms within the constraints of the municipal detailed development plan. To coordinate the planning of the site, the municipality and the other landowners collaboratively developed a structure plan that describes a shared vision for the area and was approved in 2013 but was not binding on the parties. The plan expresses the intention to develop a city park to enhance ecological qualities and connect to surrounding areas, which was supported by the 3000 participants in dialogue that followed the presentation of the structure plan [73]. In addition, the partners developed a sustainability program, which describes a joint policy for social and environmental aspects of the development. The program is updated prior to the detailed development plan for each of the planning stages. Hence, the planning of Täby Park is to a large degree a collaborative achievement of the municipality and private partners.

In preparation for the detailed development plans, the municipality drafted a plan program, which presents an elaborated plan for Täby Park and links the proposed development to different local and regional policy documents. The program presents a plan for local GI that consists of an interlinked network of parks (see Figure 5b). The local GI is supposed to nurture multiple ecological qualities but is also considered to offer social qualities such as playgrounds for schools, outdoor experiences, and sports [73]. However, the environmental impact assessment of the program points at the isolated location of the development area and foresees that the natural environments that are saved will be impacted by high recreation pressure [73]. This means that existing ecological qualities will to a limited extent contribute to the local and regional GI.

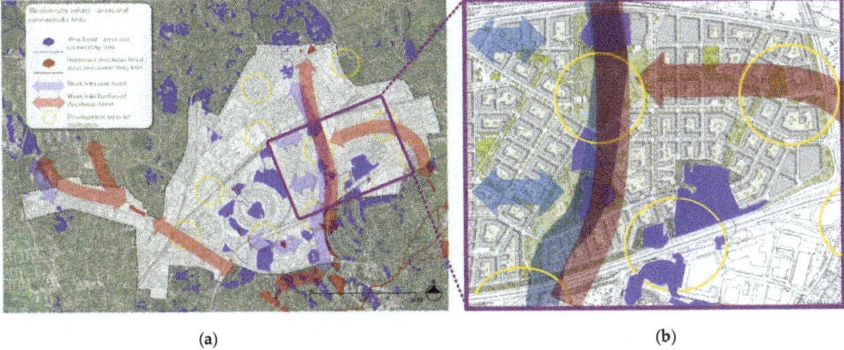

(a) (b)

Figure 5. (**a**) Habitats (filled shapes) dominated by pine forest (blue) and hardwood deciduous forest (red) with connectivity links (dashed lines in corresponding colors) in the central parts of Täby after implementation of existing plans. Arrows indicate a potential for strengthened connectivity for pine forest (blue), hardwood deciduous forest (red) and wild bees (yellow rings) [37]. (**b**) Illustration plan for Täby Park with biodiversity values added [73].

While some of the existing natural environments are incorporated in the planned city park and other parks, most of the green space will be newly-created. The sustainability program underlines the prominence of a coherent local GI as a major quality of the area, which is developed on municipal land and on private land that is transferred to the municipality [74]. The sustainability program specifies the green space ratio for both public space and private development sites. Based on the analysis of connectivity within the local GI, the municipality anticipates that the created NBS in Täby Park will facilitate the dispersal of species tied to pine forest and hardwood deciduous forest, as well as wild bees [37], see Figure 5a.

6. Governance for NBS in Urban Planning and Development

The three projects that were presented in the previous section describe municipal governance practices for the planning of green spaces as part of urban development in the Stockholm region. Based on the scope and the intended outcomes as to biodiversity and ecosystem services, these projects are examples of NBS, although the concept is not explicitly mentioned in the planning documents. All three projects share an ambition to create multifunctional green spaces that contribute to social and ecological qualities, both in a very local context and from a municipal or regional perspective. An analysis of the planning trajectories of the three projects in the Stockholm region provides some significant insights as to what factors enable the integrated planning approach that caters to NBS.

6.1. Agenda for NBS in Urban Planning and Development

Urban development is generally considered a threat to urban green spaces as it mainly focuses on residential development while the planning of green spaces ends up in a subsidiary role [75]. Even though the projects that are presented in this paper also focus on housing development, the results show that urban development also offers opportunities to preserve and create green spaces. In the case studies, the intention to create green spaces within the new developments was expressed early in the process, but initially as part of the urban design. The concept of the multi-functional mixed city that guided the developments of Täby Park and Fyrklövern presupposes access to parks and green pockets in the neighborhood as a necessary function for integration [76,77]. This is reflected in the point system that was used in Fyrklövern for the assessment of the developers' bids, where green space is evaluated as part of urbanity [78]. Besides urban design principles, public opinion played a role in putting green space on the agenda, either through consent, as in the case of Täby Park, where the proposed city park was endorsed in a public dialogue [73], or through criticism, as in the case of Stockholm Royal Seaport, where the detailed development plan from 2005 was substantially revised after public consultation to minimize ecological impacts [68]. In Fyrklövern the public discussion within Väsby Labs was initiated before a plan was drafted. This points to a strong public engagement in the development of NBS. In addition, it demonstrates the enabling role of the urban planning process to enhance public dialogue, either within the scope of formal procedures or as part of local initiatives, although the contribution of citizen participation to the democratization of urban governance is subject to debate [79,80].

The conceptualization and design of NBS in urban planning requires multi-disciplinary cooperation to bridge different fields of expertise [15], and to let ecology become a frame for decision making [81], relevant expertise needs to be represented in the planning process. When and how different competencies or municipal departments are involved in urban planning depends on the internal routines, which vary between municipalities [40]. In the municipalities of Täby and Upplands Väsby, these routines ensure that environmental planners partake in the detailed development planning, but their involvement in early and more informal phases of the planning process is in most projects not secured [42]. As a consequence, ecological and other types of expertise may be overlooked when the structure and the scope of the urban development project are framed. This can explain why GI, ecosystem services, and biodiversity are more thoroughly addressed in the later planning stages of Fyrklövern and Täby Park. For the Stockholm Royal Seaport the situation is different due to its location next to the Royal National City Park. The first version of one of early detailed developments was heavily criticized by experts and NGOs, which resulted in increased attention to actions to mitigate impacts on the local and regional GI.

6.2. Enabling the Implementation of NBS

Implementing NBS on public property is normally not a problem as long as formal requirements are met and funding is secured. All case studies have examples of such green space developments, in particular along streets and pathways, but also public parks and green pockets within the residential areas. Although NBS often focus on public space, research has shown that domestic gardens also can support ecosystems services and biodiversity, provided that individual owners adopt sustainable garden practices [82–84]. In addition, gardeners expand their understanding of and attachment to their local environment and enhance involvement in stewardship [85]. Therefore, combined planning of green space on public and private property in urban development can generate synergies and increase the ecological and social qualities of green space in the area.

Municipal authorities in Sweden do not have the regulatory means to demand a specific design of buildings or green spaces through building permits. Thus the implementation of NBS that is not part of public green space can only be guaranteed through voluntary commitments of developers that are formalized in agreements. In Täby Park this was achieved through negotiations between the private landowners and the municipality. In the cases of Fyrklövern and Stockholm Royal Seaport, developers were invited to submit competitive bids in land allocation competitions that met stated quality and sustainability criteria. The bids presented in most cases included innovative solutions for green building, stormwater management, urban gardening, etc., but they were merely business offers. Depending on the additional cost that the criteria entail, the project calculations require an attractive location and a low land price or a high density, as in the case studies, but on less profitable locations the number and quality of the bids decreases as well as the probability for the implementation of NBS on private property. This implies that a market-driven hybrid governance approach where for-profit actors participate in the realization of sustainable urban development cannot assure an equal distribution of urban NBS benefits across the country [86].

In line with common practices, the municipalities of Stockholm and Upplands Väsby drafted a detailed development plan and conducted public consultation prior to the land allocation. However, following the neo-performative model, the plan was not approved until after the reconciliation of the plan and the development projects [87]. Accordingly, the municipal authorities and the developers of the winning bids discuss the proposed projects to align them with the municipal goals and policies. These discussions are normally completed within two years and offer an arena for the co-creation of NBS and other actions that foster sustainable urban development. The discussion involves only participants from the municipal organization and the housing developers and can thus be described as a formal co-creation process with the primary purpose of value creation and sustainable practice [88]. Although the process encompasses a lot of negotiations, the meeting between the parties involved can open new perspectives or solutions, such as in Fyrklövern, where the co-creation process resulted in actualizing semi-public gardens that are designed to deliver ecosystem services and the walking passage connecting these gardens, which is regulated in the land allocation agreement between the municipality and the developers. In the planning of the Stockholm Royal Seaport, research shows that conflicts over sustainability requirements emerge during the negotiations, but through conflict resolution the parties co-create sustainable value [89], e.g., the implementation of a local GI on public and private property that is included in the land allocation agreement. In Täby Park the process also focuses on value creation, but due to private land ownership, the terms for the formal co-creation process differ. The drafting of the detailed development plans, which is based on the joint vision in the structure plan, runs parallel to the discussions on the joint sustainability plan. This means that the new neighborhoods are designed through collaborative planning within a context of deliberations on sustainable urban development.

Aside from value creation, the co-creation processes involve important elements of learning together [88]. A customized program for knowledge building fosters commitment and a shared understanding as part of a collaborative process [90]. The drafting of the sustainability plan for Täby Park was supported with seminars and activities to foster knowledge sharing and learning. Within the competence development program in the Stockholm Royal Seaport, a large number of seminars and meetings were organized and several research projects were initiated, e.g., a state-funded project with other stakeholders to develop methods for integrating ecosystem services in urban development, which generated knowledge for the planning and realization of a multifunctional local GI that is designed to enhance connectivity [91]. Competence development supports the co-production of situated knowledge among participants, which increases their ability to impact societal change processes [92].

6.3. Framing the NBS in Urban Planning and Development

In the case studies, NBS is embedded in a vision of developing multifunctional public and private green spaces in urban areas that deliver ecosystem services and strengthen local and regional GI. To design the green spaces and analyze their potential impacts, the municipalities primarily use three conceptual frameworks, i.e., GI, ecosystem services, and green space ratio. NBS is commonly associated with ecosystem service and GI [9]. To what extent these frameworks can create NBS that deliver the anticipated social and ecological qualities in the three case study areas is too early to assess, since only parts of the projects have been realized. Hence, a detailed assessment of the generated qualities after implementation that is supported with quantitative analyses would be needed to gain a thorough understanding of the actual contribution of the conceptual frameworks to foster social and ecological qualities. Nonetheless, the role of the three frameworks in the planning of NBS in urban development can be evaluated.

Ecosystem services is highlighted for its ability to bridge communication challenges between different stakeholders and to provide an integrated framework to adapt complexity to local planning practice [93]. In this manner, ecosystem services have served as a tool to identify tangible local measures in Fyrklövern and the Stockholm Royal Seaport that respond to challenges and needs in urban development, e.g., climate change adaption and stormwater management. By emphasizing the connection between measures and expected social and ecological qualities, ecosystem services prove to provide a valuable framework for ensuring the integration of ecological knowledge into local spatial settings. In Fyrklövern, ecosystem services are also used as an analytical framework for mapping ecological development needs [72] that is integrated into the municipal comprehensive plan and used to motivate the NBS in Fyrklövern [42]. However, as a regulatory tool in urban planning, municipal authorities in Stockholm and Täby use green space ratio to define green space requirements for building plots that are developed by private actors, which leaves the selection of tangible measures to developers. Although the green space ratio does not cope with spatial and ecological relations, it promotes social and ecological qualities, including ecosystem services, by giving higher weights to certain types of green space, e.g., the preservation of grown-up trees and sensitive biotopes [94]. The City of Stockholm monitors the use of green space ratio and ecosystem services in the development of the Stockholm Royal Seaport.

GI is another central framework that can offer guidance on practices to integrate NBS into urban planning [14]. In the case studies, GI was primarily employed to describe interconnected local green spaces, e.g., along the walking passage in Fyrklövern. Although the green spaces are not linked to the regional GI initiatives, Fyrklövern can contribute as a hub for pollination between local green spaces [38]. In Täby Park and the Stockholm Royal Seaport, GI focuses not only on interconnected local green spaces but also on connections to surrounding areas and regional GI initiatives, see Figure 2. The strategic location of the

Stockholm Royal Seaport entails a potential and a need to strengthen connectivity links with both the Stockholm CAB GI as well as the RUFS GI. However, to create durable links, more than a physical connection is required because the green spaces that compose the link need to offer specific habitat conditions as in the Stockholm Royal Seaport, where investments are made that aim to strengthen connectivity for amphibians and oak related species. Hence, it remains unclear whether the local GIs contribute to the regional GIs and biodiversity. For Täby Park, the environmental impact assessment of the plan program expressed concerns regarding the impact of recreation on ecological qualities [73], and the planned local GI suffers from different barriers that give rise to fragmentation. In the Stockholm Royal Seaport, the project organization monitors connectivity and the dispersal of oak-living insects but the ongoing land development will continue to put pressure on existing green spaces that may counteract the potential benefits of the NBS. Nevertheless, GI provides an important framework for the planning of NBS that allows for addressing connections between green spaces and regional GI, but ecological relations that enhance connectivity and biodiversity are associated with great uncertainty.

The potential benefits of green spaces for delivering social qualities and enhancing well-being are well documented, e.g., [3,95–97]. In the case studies, these benefits are often implied, as the anticipated social qualities are not specified in the planning documents, which focus more on functions, i.e., recreation, playgrounds and meeting points for social interaction. In addition, proximity is highlighted, e.g., in the Stockholm Royal Seaport where access to green space within 200 m is used as an indicator to follow up the sustainability goal of "ecosystem services for a resilient and healthy urban environment" [66]. In Fyrklövern the goals and conditions for the local GI connect to the urban quality objectives for urban development, which manifests in the significant role of the green spaces in creating social qualities in urban neighborhoods.

The analysis of the case studies reveals the complementarity of the three conceptual frameworks in the planning and development of NBS. Ecosystem services are used in the planning and design of green spaces while GI provides a tool to address local and regional connectivity in planning as a means to enhance biodiversity and spatial structures. The greenspace ratio plays, above all, a role as a regulatory tool to promote ecosystems services on private property. The need for all of these frameworks has been demonstrated in the case studies.

7. Conclusions

The planning for dense cities has focused attention on multifunctional land use from city to site-level planning across Europe [76], which calls for an integrated approach to urban development to capture the complex relations and interactions in urban socio-ecological systems. Although the role of urban planning to address multifunctionality is widely recognized, there is a need for planning approaches that accommodate urban complexity and are oriented towards providing solutions for urban sustainability [14]. The results from the case studies in the Stockholm region provide empirical insights into the planning and development of multifunctional and interconnected urban green spaces through the collaboration between municipal agencies and for-profit housing developers. The municipal practices that were studied include both public and private green spaces and focus on the search for NBS that integrate biodiversity, climate change adaptation and social qualities. Although the results of the research are based on the Swedish planning context, the findings can be transferred to countries with a similar planning system in North-Western Europe [87]. Moreover, there are components of the municipal planning practices that apply to situations in different planning settings.

The results from the empirical studies in Stockholm County clarify the role of urban planning in actualizing NBS to ensure the connection with and adaptation to local knowledge, conditions, and needs, which embeds the NBS in the local context. As the results from the research show, citizens express a large interest in the development of green spaces, and the planning process offers an institutional context to involve the local community in the design, development and stewardship of NBS. Through an area-wide perspective on urban development, a comprehensive approach to multifunctionality and urban sustainability can be applied. However, urban planning and development entail the balancing of interests, which is not always in favor of NBS.

In accordance with the results of the study, municipal authorities express in an early stage of the planning process the intention to create social and ecological qualities by developing NBS in the form of a local GI, thereby increasing biodiversity and fostering ecosystem services. The municipal practices to meet these ambitions are mainly based on three pillars. First, they involve a collaboration between public and private actors, e.g., landowners and developers, by establishing a dialogue on goals, challenges and alternative solutions to gain a joint understanding of how urban development can promote anticipated qualities of green spaces. However, the incentives to set up a dialogue and agree on a coherent vision differ between projects depending on land ownership and land price. The second pillar of the municipal strategy is to pursue activities for competence development that support the collaborative planning process with insights from other projects and research, and establish an arena for knowledge exchange and innovation. These activities may involve consultants or academics, but can also be part of a research project that is aligned with aspects of ongoing projects. Third, they enable the integration of ecological knowledge that is supported by conceptual frameworks that facilitate the planning of green space. Among these, GI and ecosystem services play an important role in urban planning by providing a common language for analyzing and communicating how proposals for NBS foster social and ecological qualities of green space. In the case studies, municipal agencies benefit from existing regional GI initiatives that provide knowledge on existing ecological qualities and a structure to which local green spaces can be linked, but the potential to contribute to regional GI differs widely between cases. From a longer-term perspective, experiential knowledge and a continued focus on developing GI can enhance biodiversity and create a resilient GI in the Stockholm region. Municipalities that include these three pillars in their urban planning processes gain better preconditions for developing NBS that contribute to a green and climate-resilient urban development.

Author Contributions: Conceptualization, all; methodology, P.B.; validation, all; formal analysis, P.B.; investigation, P.B.; writing—original draft preparation, P.B.; writing—review and editing, U.M and B.B.; visualization, U.M. All authors have read and agreed to the published version of the manuscript.

Funding: This research was funded by the Swedish Research Council for Sustainable Development (FORMAS), grant number 2015-00133.

Institutional Review Board Statement: Not applicable.

Informed Consent Statement: Not applicable.

Acknowledgments: The authors would like to thank the editors and the anonymous reviewers for their constructive comments on the earlier drafts of this paper.

Conflicts of Interest: The authors declare no conflict of interest.

Notes

An earlier version of this work was presented at the Greening Cities Shaping Cities Symposium in October 2020 (https://www.greeningcities-shapingcities.polimi.it/, accessed on 10 July 2021).

References

1. Haaland, C.; van den Bosch, C. Challenges and strategies for urban green-space planning in cities undergoing densification: A review. *Urban For. Urban Green.* **2015**, *14*, 760–771. [CrossRef]
2. Niemelä, J. Ecology of urban green spaces: The way forward in answering major research questions. *Landsc. Urban Plan.* **2014**, *125*, 298–303. [CrossRef]
3. van den Bosch, M.; Nieuwenhuijsen, M. No time to lose—Green the cities now. *Environ. Int.* **2017**, *99*, 343–350. [CrossRef] [PubMed]

4.   Dobson, J.; Dempsey, N. Known but not done: How logics of inaction limit the benefits of urban green spaces. *Landsc. Res.* **2021**, *46*, 390–402. [CrossRef]

5.   Colding, J.; Gren, Å.; Barthel, S. The Incremental Demise of Urban Green Spaces. *Land* **2020**, *9*, 162. [CrossRef]

6.   Wolfram, M.; Frantzeskaki, N. Cities and Systemic Change for Sustainability: Prevailing Epistemologies and an Emerging Research Agenda. *Sustainability* **2016**, *8*, 144. [CrossRef]

7.   United Nations. *Transforming Our World: The 2030 Agenda for Sustainable Development*; UN: New York, NY, USA, 2015.

8.   Scott, M.; Lennon, M.; Haase, D.; Kazmierczak, A.; Clabby, G.; Beatley, T. Nature-based solutions for the contemporary city/Re-naturing the city/Reflections on urban landscapes, ecosystems services and nature-based solutions in cities/Multifunctional green infrastructure and climate change adaptation: Brownfield greening as an adaptation strategy for vulnerable communities?/Delivering green infrastructure through planning: Insights from practice in Fingal, Ireland/Planning for biophilic cities: From theory to practice. *Plan. Theory Pract.* **2016**, *17*, 267–300.

9.   Hanson, H.; Wickenberg, B.; Alkan Olsson, J. Working on the boundaries—How do science use and interpret the nature-based solution concept? *Land Use Policy* **2020**, *90*, 104302. [CrossRef]

10.   Faivre, N.; Fritz, M.; Freitas, T.; de Boissezon, B.; Vandewoestijne, S. Nature-Based Solutions in the EU: Innovating with nature to address social, economic and environmental challenges. *Environ. Res.* **2017**, *159*, 509–518. [CrossRef]

11.   Davies, C.; Chen, W.; Sanesi, G.; Lafortezza, R. The European Union roadmap for implementing nature-based solutions: A review. *Environ. Sci. Policy* **2021**, *121*, 49–67. [CrossRef]

12.   Nesshöver, C.; Assmuth, T.; Irvine, K.; Rusch, G.; Waylen, K.; Delbaere, B.; Haase, D.; Jones-Walters, L.; Keune, H.; Kovacs, E.; et al. The science, policy and practice of nature-based solutions: An interdisciplinary perspective. *Sci. Total Environ.* **2017**, *579*, 1215–1227. [CrossRef]

13.   Directorate-General for Research and Innovation, European Commission. *Nature-Based Solutions State of the Art in EU-Funded Projects*; Publications Office of the European Union: Luxembourg, 2020.

14.   Dorst, H.; van der Jagt, A.; Raven, R.; Runhaar, H. Urban greening through nature-based solutions—Key characteristics of an emerging concept. *Sustain. Cities Soc.* **2019**, *49*, 101620. [CrossRef]

15.   Raymond, C.; Frantzeskaki, N.; Kabisch, N.; Berry, P.; Breil, M.; Nita, M.; Geneletti, D.; Calfapietra, C. A framework for assessing and implementing the co-benefits of nature-based solutions in urban areas. *Environ. Sci. Policy* **2017**, *77*, 15–24. [CrossRef]

16.   European Commission. *Guidance on a Strategic Framework for Further Supporting Deployment of EU-Level Green the and Blue Infrastructure*; Commission Staff Working Document 24.5.2019 SWD (2019) 193 Final; European Commission: Brussels, Belgium, 2019; p. 102.

17.   Millennium Ecosystem Assessment. *Ecosystems and Human Well-Being. Synthesis*; Island Press: Washington, DC, USA, 2005.

18.   National Board of Housing, Building and Planning. Grönytefaktor—Räkna Med Ekosystemtjänster. Available online: https://www.boverket.se/sv/PBL-kunskapsbanken/Allmant-om-PBL/teman/ekosystemtjanster/verktyg/gronytefaktor/ (accessed on 13 May 2021).

19.   Secretariat of the Convention on Biological Diversity. *Handbook of the Convention on Biological Diversity Including Its Cartagena Protocol on Biosafety*, 3rd ed.; Convention on Biological Diversity: Montreal, QC, Canada, 2005.

20.   European Commission. *Green Infrastructure (GI) Enhancing Europe's Natural Capital*; Communication from the Commission to the European Parliament, the Council, the European Economic and Social Committee and the Committee of the Regions; European Commission: Brussels, Belgium, 2013; p. 11.

21.   Kabisch, N.; Korn, H.; Stadler, J.; Bonn, A. *Nature-Based Solutions to Climate Change in Urban Areas—Linkages of Science, Policy and Practice*; Theory and Practice of Urban Sustainability Transitions; Springer: Cham, Switzerland, 2017; pp. 1–11.

22.   Chatzimentor, A.; Apostolopoulou, E.; Mazaris, A. A review of green infrastructure research in Europe: Challenges and opportunities. *Landsc. Urban Plan.* **2020**, *198*, 103775. [CrossRef]

23.   Badiu, D.; Nita, A.; Iojă, C.; Niță, M. Disentangling the connections: A network analysis of approaches to urban green infrastructure. *Urban For. Urban Green.* **2019**, *41*, 211–220. [CrossRef]

24.   Hostetler, M.; Allen, W.; Meurk, C. Conserving urban biodiversity? Creating green infrastructure is only the first step. *Landsc. Urban Plan.* **2011**, *100*, 369–371. [CrossRef]

25.   Gavrilidis, A.; Niță, M.; Onose, D.; Badiu, D.; Năstase, I. Methodological framework for urban sprawl control through sustainable planning of urban green infrastructure. *Ecol. Indic.* **2019**, *96*, 67–78. [CrossRef]

26.   Directorate-General for Research and Innovation, European Commission. *Nature-Based Solutions towards Sustainable Communities. Analysis of EU-Funded Projects*; Independent Expert Report; Publications Office of the European Union: Luxembourg, 2020.

27.   Pauleit, S.; Hansen, R.; Rall, E.; Zölch, T.; Andersson, E.; Luz, A.; Szaraz, L.; Tosics, I.; Vierikko, K. *Urban Landscapes and Green Infrastructure*; Oxford Research Encyclopedia of Environmental Science: Oxford, UK, 2017.

28.   Somarakis, G.; Stagakis, S.; Chrysoulakis, N. (Eds.) *ThinkNature Nature-Based Solutions Handbook*; European Union: Maastricht, The Netherlands, 2019.

29.   Healey, P. Collaborative planning in perspective. *Plan. Theory* **2003**, *2*, 101–123. [CrossRef]

30.   Mahmoud, I.; Morello, E. Co-creation Pathway for Urban Nature-Based Solutions: Testing a Shared-Governance Approach in Three Cities and Nine Action Labs. *Green Energy Technol.* **2021**, 259–276. [CrossRef]

31.   Frantzeskaki, N. Seven lessons for planning nature-based solutions in cities. *Environ. Sci. Policy* **2019**, *93*, 101–111. [CrossRef]

32. Högström, J.; Balfors, B.; Hammer, M. The role of small-scale planning projects in urban development: A case study in the metropolitan Stockholm region, Sweden. *Land Use Policy* **2019**, *84*, 294–304. [CrossRef]
33. Khoshkar, S.; Hammer, M.; Borgström, S.; Dinnétz, P.; Balfors, B. Moving from vision to action- integrating ecosystem services in the Swedish local planning context. *Land Use Policy* **2020**, *97*, 104791. [CrossRef]
34. Statistics Sweden. Folkmängd i Riket, Län Och Kommuner 31 Mars 2020 Och Befolkningsförändringar 1 Januari—31 Mars 2020. Totalt. Available online: https://www.scb.se/hitta-statistik/statistik-efter-amne/befolkning/befolkningens-sammansattning/befolkningsstatistik/pong/tabell-och-diagram/kvartals--och-halvarsstatistik--kommun-lan-och-riket/kvartal-1-2020/ (accessed on 9 May 2021).
35. Littke, H. Planning the Green Walkable City: Conceptualizing Values and Conflicts for Urban Green Space Strategies in Stockholm. *Sustainability* **2015**, *7*, 11306–11320. [CrossRef]
36. Furberg, D.; Ban, Y.; Mörtberg, U. Monitoring Urban Green Infrastructure Changes and Impact on Habitat Connectivity Using High-Resolution Satellite Data. *Remote Sens.* **2020**, *12*, 3072. [CrossRef]
37. Täby Municipality. *Täby Stadskärna 2050 Fördjupning av Översiktsplan (Antagen KS 2019-11-04)*; Täby Municipality: Stockholm, Sweden, 2019.
38. Ekologigruppen. *Kartläggning av Ekosystemtjänster i Upplands Väsby Kommun. Underlag Till Utvecklingsplan för Ekosystemtjänster*; Upplands Väsby Municipality: Stockholm, Sweden, 2015.
39. Nyumba, T.O.; Wilson, K.; Derrick, C.; Mukherjee, N. The use of focus group discussion methodology: Insights from two decades of application in conservation. *Methods Ecol. Evol.* **2018**, *9*, 20–32. [CrossRef]
40. Högström, J.; Brokking, P.; Balfors, B.; Hammer, M. Approaching Sustainability in Local Spatial Planning Processes: A Case Study in the Stockholm Region, Sweden. *Sustainability* **2021**, *13*, 2601. [CrossRef]
41. Bryman, A. *Social Research Methods*, 5th ed.; Oxford University Press: Oxford, UK, 2016.
42. Jacobsson, A.; Kofoed Schröder, J.; Balfors, B. *Verktyg Och Arbetssätt för Hantering av Ekosystemstjänster i Detaljplanering*; Institution för Utveckling, Miljövetenskap Och Teknik, KTH: Stockholm, Sweden, 2020.
43. Brokking, P.; Liedholm Johnson, E.; Paulsson, J. Implementation Strategies for Sustainable Urban Development: Examples from Swedish practice. In *Methods and Concepts of Land Management Diversity, Changes and New Approaches*; Hepperle, E., Paulsson, J., Maliene, V., Mansberger, R., Auzins, A., Valciukiene, J., Eds.; European Academy of Land Use and Development (EALD); vdf Hochschulverlag AG an der ETH Zürich: Zürich, Switzerland, 2020; pp. 223–234.
44. Brokking, P.; Liedholm Johnson, E.; Paulsson, J. *Hållbarhet, Regelverk Och Roller Från Detaljplan Till Bygglov*; Avdelningen för Fastighetsvetenskap, KTH: Stockholm, Sweden, 2020.
45. Stockholm County Council. *Regional Utvecklingsplan för Stockholm Regionen. RUFS 2050. Europas Mest Attraktiva Storstadsregion*; Trafik Och Regionplaneförvaltningen: Stockholm, Sweden, 2018.
46. Stockholm County Council. *Regionplan 1991 för Stockholms Län 1990—2020 (Utställningsförslag)*; Office of Regional Planning and Urban Transportation: Stockholm, Sweden, 1991.
47. Stockholm County Council. *När, vad Och Hur? Svaga Samband i Stockholmsregionens Gröna Kilar*; Tillväxt, Miljö Och Regionplanering: Stockholm, Sweden, 2012.
48. Stockholm County Council. *Regional Utvecklingsplan för Stockholmsregionen. Så Blir vi Europas Mest Attraktiva Storstadsregion (Antagen av Landstingsfullmäktige 11 Maj 2010)*; Regionplanekontoret: Stockholm, Sweden, 2010.
49. Stockholm County Administrative Board. *Grön Infrastruktur. Regional Handlingsplan för Stockholms Län (Fastställd November 2018)*; Stockholm County Administrative Board: Stockholm, Sweden, 2019.
50. European Commission. *Our Life Insurance, Our Natural Capital: An EU Biodiversity Strategy to 2020*; Communication from the Commission to the European Parliament, the Council, the European Economic and Social Committee and the Committee of the Regions; European Commission: Brussels, Belgium, 2011; p. 17.
51. European Commission. *EU Biodiversity Strategy for 2030. Bringing Nature Back into Our Lives*; Communication from the Commission to the European Parliament, the Council, the European Economic and Social Committee and the Committee of the Regions; European Commission: Brussels, Belgium, 2020.
52. Swedish Environmental Protection Agency. *Vägledning om Regionala Handlingsplaner för Grön Infrastruktur i Prövning Och Planering. Med Ett Känt Nätverk av Natur Kan vi Planera Effektivare*; Swedish Environmental Protection Agency: Stockholm, Sweden, 2017.
53. Lemes de Oliveira, F. *Green Wedge Urbanism: History, Theory and Contemporary Practice*; Bloomsbury Academic: London, UK, 2017.
54. Nadin, V.; Fernández Maldonado, A.M.; Zonneveld, W.; Stead, D.; Dabrowski, M.; Piskorek, K.; Sarkar, A.; Schmitt, P.; Smas, L.; Cotella, G.; et al. *COMPASS—Comparative Analysis of Territorial Governance and Spatial Planning Systems in Europe: Applied Research 2016–2018: Final Report*; ESPON: Luxembourg, 2018.
55. *Legislation Planning and Building Act (2010:900) Planning and Building Ordinance (2011:338)*, 3rd ed.; Swedish National Board for Housing, Building and Planning: Karlskrona, Sweden, 2018.
56. Swedish National Board for Housing, Building and Planning. Säkerställ Ekosystemtjänster i Detaljplan. 2019. Available online: https://www.boverket.se/sv/PBL-kunskapsbanken/Allmant-om-PBL/teman/ekosystemtjanster/metod_planering/dp/sakerstall-dp/ (accessed on 20 June 2021).
57. Olsson, L. The Neoliberalization of Municipal Land Policy in Sweden. *Int. J. Urban Reg. Res.* **2018**, *42*, 633–650. [CrossRef]
58. Cameron, R.; Blanuša, T.; Taylor, J.; Salisbury, A.; Halstead, A.; Henricot, B.; Thompson, K. The domestic garden—Its contribution to urban green infrastructure. *Urban For. Urban Green.* **2012**, *11*, 129–137. [CrossRef]

59. City of Stockholm. Stockholm Royal Seaport. Available online: https://vaxer.stockholm/omraden/norra-djurgardsstaden/in-english/ (accessed on 13 May 2021).
60. Upplands Väsby Municipality. Fyrklövern—Den Moderna Småstaden. Available online: http://www.upplandsvasby.se/fyrklovern#language (accessed on 28 May 2021).
61. Täby Park. Här Kan du Leva Gott Och Hållbart. Available online: https://taby-park.se/om-taby-park/ (accessed on 22 May 2021).
62. Tunström, M. *På Spaning Efter Den Goda Staden. Om Konstruktioner av Ideal Och Problem i Svensk Stadsbyggnadsdiskussion*; Örebro Studies in Human Geography, Örebro University: Örebro, Sweden, 2009.
63. Täby Municipality. *Det Nya Täby. Översiktsplan 2010–2030 (Antagen KF 14-12-2009)*; Täby Municipality: Stockholm, Sweden, 2010.
64. Upplands Väsby Municipality. Väsby Labs—Att Våga Och Göra. 2015. Available online: http://masarin.se/wp-content/uploads/2015/01/vasby_labs_att_vaga_och_gora_2014.pdf (accessed on 4 May 2021).
65. Svane, Ö.; Wangel, J.; Engberg, L.A.; Palm, J. Compromise and learning when negotiating sustainabilities: The brownfield development of Hammarby Sjöstad, Stockholm. *Int. J. Urban Sustain. Dev.* **2011**, *3*, 141–155. [CrossRef]
66. City of Stockholm. *Program för Hållbar Stadsutveckling. Norra Djurgårdsstaden Visar Vägen Mot en Hållbar Framtid (2017-03-22)*; City of Stockholm: Stockholm, Sweden, 2017.
67. City of Stockholm. Norra Djurgårdsstaden Hållbarhetsredovisning. 2019. Available online: https://vaxer.stockholm/globalassets/omraden/-stadsutvecklingsomraden/ostermalm-norra-djurgardsstaden/hallbar-stadsutveckling/resultat-2019/hallbarhetsredovisning_norradjurgardsstaden_2019.pdf (accessed on 10 October 2020).
68. City of Stockholm. *Detaljplan för del av Norra Djurgårdsstaden (Västra Delen) i Stadsdelarna Hjorthagen Och Norra Djurgården i Stockholm Dp 2008-12203-54 (Planbeskrivning- Godkänd 15-05-2009)*; Stadsbyggnadskontoret: Stockholm, Sweden, 2009.
69. Drotte, F.; von Hofsten, A. Mer stadskvalitéer till lägre pris. In *PLAN May 18*; Föreningen av Samhällsplanering: Stockholm, Sweden, 2016; Available online: http://www.planering.org/plan-blog/2016/5/17/mer-stadskvaliter-till-lgre-pris (accessed on 4 May 2021).
70. Upplands Väsby Municipality. *Detaljplan för Fyrklövern 1- Allmän Platsmark i Upplands Väsby Kommun. 1385 (Planbeskrivning)*; Kontoret för Samhällsbyggnad: Upplands Väsby, Sweden, 2015.
71. Upplands Väsby Municipality. *Planprogram för Fyrklövern (Godkänt KS Mars 2012)*; Upplands Väsby Municipality: Upplands Väsby, Sweden, 2012.
72. Ekologigruppen. *Utvecklingsplan för Ekosystemtjänster i Upplands Väsby Kommun. Översiktliga Prioriteringar Inför Fortsatt Planarbete (Slutversion 05-02-2016)*; Upplands Väsby Municipality: Stockholm, Sweden, 2016.
73. Täby Municipality. *Täby Park Planprogram (Godkänt KF 151102)*; Täby Municipality: Täby, Sweden, 2015.
74. Täby Municipality, Täby Park AB, Riksbyggen. *Hållbarhetsprogram för Täby Park för Detaljplan Del av Hästen 4 m fl (Område 5)*; Täby Municipality: Täby, Sweden, 2020.
75. Jim, C. Sustainable urban greening strategies for compact cities in developing and developed economies. *Urban Ecosyst.* **2012**, *16*, 741–761. [CrossRef]
76. Hansen, R.; Stahl Olafsson, A.; van der Jagt, A.P.N.; Rall, E.; Pauleit, S. Planning multifunctional green infrastructure for compact cities: What is the state of practice? *Ecol. Indic.* **2019**, *96*, 99–110. [CrossRef]
77. Bellander, G. *Blandstaden Ett Planeringskoncept för en Hållbar Bebyggelse Utveckling?* Boverket, Miljödepartementet, Formas: Karlskrona, Sweden, 2005.
78. Upplands Väsby Municipality. *Poängsystem för Fyrklövern. Stadsutveckling Kopplat Till Flexibla Priser (Version 2016-05-25)*; Upplands Väsby Municipality: Upplands Väsby, Sweden, 2016.
79. Nadin, V.; Stead, D.; Dabrowski, M.; Fernandez-Maldonado, A.M. Integrated, adaptive and participatory spatial planning: Trends across Europe. *Reg. Stud.* **2021**, *55*, 791–803. [CrossRef]
80. Zakhour, S. The democratic legitimacy of public participation in planning: Contrasting optimistic, critical, and agnostic understandings. *Plan. Theory* **2020**, *19*, 349–370. [CrossRef]
81. Randrup, T.; Buijs, A.; Konijnendijk, C.; Wild, T. Moving beyond the nature-based solutions discourse: Introducing nature-based thinking. *Urban Ecosyst.* **2020**, *23*, 919–926. [CrossRef]
82. Goddard, M.; Dougill, A.; Benton, T. Scaling up from gardens: Biodiversity conservation in urban environments. *Trends Ecol. Evol.* **2010**, *25*, 90–98. [CrossRef]
83. Dewaelheyns, V.; Jakobsson, A.; Saltzman, K. Strategic gardens and gardening: Inviting a widened perspective on the values of private green space. *Urban For. Urban Green.* **2018**, *30*, 207–209. [CrossRef]
84. Diduck, A.; Raymond, C.; Rodela, R.; Moquin, R.; Boerchers, M. Pathways of learning about biodiversity and sustainability in private urban gardens. *J. Environ. Plan. Manag.* **2019**, *63*, 1056–1076. [CrossRef]
85. Raymond, C.; Diduck, A.; Buijs, A.; Boerchers, M.; Moquin, R. Exploring the co-benefits (and costs) of home gardening for biodiversity conservation. *Local Environ.* **2018**, *24*, 258–273. [CrossRef]
86. Toxopeus, H.; Kotsila, P.; Conde, M.; Katona, A.; van der Jagt, A.P.N.; Polzin, F. How 'just' is hybrid governance of urban nature-based solutions? *Cities* **2020**, *105*, 102839. [CrossRef]
87. Berisha, E.; Cotella, G.; Janin Rivolin, U.; Solly, A. Spatial governance and planning systems in the public control of spatial development: A European typology. *Eur. Plan. Stud.* **2021**, *29*, 181–200. [CrossRef]
88. Puerari, E.; de Koning, J.I.J.C.; von Wirth, T.; Karré, P.M.; Mulder, I.J.; Loorbach, D.A. Co-creation dynamics in Urban Living Labs. *Sustainability* **2018**, *10*, 1893. [CrossRef]

89. Candel, M.; Karrbom Gustavsson, T.; Eriksson, P.E. Front-end value co-creation in housing development projects. *Constr. Manag. Econ.* **2021**, *39*, 245–260. [CrossRef]
90. Ansell, C.; Gash, A. Collaborative Governance in Theory and Practice. *J. Public Adm. Res. Theory* **2007**, *18*, 543–571. [CrossRef]
91. C/O City. *Gröna Lösningar Ger Levande Städer*; C/O City: Stockholm, Sweden, 2017.
92. Westberg, L.; Polk, M. The role of learning in transdisciplinary research: Moving from a normative concept to an analytical tool through a practice-based approach. *Sustain. Sci.* **2016**, *11*, 385–397. [CrossRef]
93. Frantzeskaki, N.; Kabisch, N. Designing a knowledge co-production operating space for urban environmental governance—Lessons from Rotterdam, Netherlands and Berlin, Germany. *Environ. Sci. Policy* **2016**, *62*, 90–98. [CrossRef]
94. City of Stockholm. *GYF—Grönytefaktor för Kvartersmark (2021-02-01)*; Exploateringskontoret: Stockholm, Sweden, 2021.
95. Dobson, J.; Birch, J.; Brindley, P.; Henneberry, J.; McEwan, K.; Mears, M.; Richardson, M.; Jorgensen, A. The magic of the mundane: The vulnerable web of connections between urban nature and wellbeing. *Cities* **2021**, *108*, 102989. [CrossRef]
96. Palliwoda, J.; Kowarik, I.; von der Lippe, M. Human-biodiversity interactions in urban parks: The species level matters. *Landsc. Urban Plan.* **2017**, *157*, 394–406. [CrossRef]
97. Jennings, V.; Bamkole, O. The relationship between social cohesion and urban green space: An avenue for health promotion. *Int. J. Environ. Res. Public Health* **2019**, *16*, 452. [CrossRef]

sustainability

Article

Green and Compact: A Spatial Planning Model for Knowledge-Based Urban Development in Peri-Urban Areas

Patricia Sanches [1], Fabiano Lemes de Oliveira [2,*] and Gabriela Celani [1]

[1] Department of Architecture and Construction, University of Campinas (UNICAMP), Campinas 13083-889, SP, Brazil; sanchesm@unicamp.br (P.S.); celani@unicamp.br (G.C.)
[2] Department of Architecture and Urban Studies (DASTU), Politecnico di Milano, 20133 Milan, Italy
* Correspondence: fabiano.lemes@polimi.it

Abstract: A seemingly unresolved debate in urban planning is the call for compactness and the provision of intra-urban green spaces. This article defines a multi-scalar spatial planning model for peri-urban areas and urban voids able to reconcile medium to high building densities with the provision of ecosystem services. The research is framed within design science research, and the theoretical definition of the model was followed by its application to the International Hub for Sustainable Development (HIDS) proposed by the University of Campinas, Brazil. The model's parameters and indicators derive from a literature review, case studies, and GIS spatial analyses. A series of expert workshops and a survey were carried out to test and validate the model. The results show that the model can support knowledge-based development in peri-urban areas with high levels of population density while ensuring good accessibility to green spaces and productive landscapes. The model can serve as a planning and design tool and support the development of public policies for other contexts committed to more resilient and sustainable development.

Keywords: planning models; spatial planning; green infrastructure; nature-based solutions (NBS); knowledge-based urban development

Citation: Sanches, P.; Lemes de Oliveira, F.; Celani, G. Green and Compact: A Spatial Planning Model for Knowledge-Based Urban Development in Peri-Urban Areas. *Sustainability* **2021**, *13*, 13365. https://doi.org/10.3390/su132313365

Academic Editor: Wen-Hsien Tsai

Received: 12 November 2021
Accepted: 29 November 2021
Published: 2 December 2021

Publisher's Note: MDPI stays neutral with regard to jurisdictional claims in published maps and institutional affiliations.

1. Introduction

Since the post-war period, urban sprawl has been on the increase worldwide. In fact, most urban areas have expanded their areas beyond their population growth rates [1,2]. Sprawl poses a range of challenges to urban management, with increased needs for expanding infrastructure, transport, and other services; for residents, in terms of lack of access to those; and to ecology due to land fragmentation, degradation and destruction of habitats. Additionally, increased rates of urbanisation are linked to higher urban temperatures and greenhouse gas emissions. While climate change is our most pressing issue, not a single G20 country is in line with the Paris Agreement [3]. Controlling land take remains one of the main difficulties for local governments in developing countries [4]. Competing agendas are at play since the real state sector is a significant employer, a motor of economic growth, and housing needs are still to be met. Yet, equating urban development with local needs in a sustainable and resilient manner is critical.

The call for dense, compact, mixed-use and traditional urban structures were at the core of the severe criticism placed against modernist planning at the birth of urban design as a discipline in the 1960s. The 1990s saw a renewed interest in compactness as a potential counter option to urban sprawl towards a more sustainable way of urban living [5]. Principles of the compact city model include compactness, density, diversity, mixed uses, sustainable modes of transportation and green spaces [6]. Although many authors claim this model to be capable of absorbing urban population while also protecting the environment [7–9], revisions of the model question its capacity to effectively bring nature close to residents [10,11]. Haaland and Konijnendik showed that the provision of urban green space in compact cities is a major challenge and that densification processes tend to pose threats to existing and planned urban green spaces [12]. Within the frame of the compact city, the loss of urban green spaces is seen across typologies [13], including allotment areas [14] and recreational sites [15].

In turn, since the emergence of the garden city idea, approaches that put good access to green spaces and their generous provision at the core are posited as contrasting to the compact city model. Popular preference to live in detached houses, in privacy and close to nature, has fuelled the splendid growth that suburban development has had in the last 70 years [16] and may have a resurgence given the recent mobility patterns suggesting a potential move away from urban areas seen during the COVID-19 pandemic. Planning debates have shown a persisting dichotomist approach in considering the benefits and shortfalls of such models [17,18].

While principles of urban sustainability such as compactness, medium-high density, walkability and liveability continue to be significant, the question of the presence and the roles of nature in urban planning has recently taken centre stage. The concept of green infrastructure (GI), defined as a strategically planned network of natural and seminatural areas aimed at delivering a range of ecosystem services [19], and more recently that of nature-based solutions (NBS), as actions inspired by, supported by or copied from nature that aim to help societies address a variety of environmental, social and economic challenges [20], have been used in the planning, delivery and stewardship of ecosystem services in and around urban areas. There is considerable evidence that green spaces bring a range of benefits to people [21,22], such as stress reduction [23], beauty, places for active and passive recreation and direct contact with nature [24]. During the recent pandemic, it was noted how their use increased, despite lockdowns and restrictions to mobility [25,26]. In cities, green spaces reduce temperatures, help prevent flooding and improve air quality, among other advantages [27,28]. More and more, nature is considered a key ally in combatting climate change [29,30].

Sustainable cities can arguably have many forms [31,32]. Planning debates on the roles of urban morphology on indicators related to sustainability often consider distinct planning models potentially appropriate to face current global challenges. Research on planning models has shown progress on integrating ecosystem services into planning [4,33] and on seeking to overcome the distance between compactness and the provision of ecosystem services in cities through new frameworks and methods [34,35]. Artmann et al., for instance, proposed a framework combining the concepts of smart growth with that of green infrastructure [36], and Ritcher and Behnisch put forward a methodology for multicriteria assessment of environmental concerns [37]. There has been too modelling research into aspects of density and accessibility to green space [38–40]. Despite advancements, research on linking compactness and the green city often treats density as a parameter instead of a central element of the equation. Furthermore, there is a lack of multi-scalar models dealing with the peri-urban areas and in the context of developing countries [41].

This article aims to present, test and validate a spatial conceptual model able to balance density and green spaces in the planning and design of peri-urban contexts, particularly for knowledge-based urban development areas. It addresses the following research questions: (1) how can medium/high densities be reconciled with the provision of green spaces in a peri-urban context for knowledge-based urban development? (2) in which way can the definition of a spatial planning model help in the proposition of solutions for a new development area? (3) How can the model contribute to a circular economy? (4) How can both urbanity and access to nature in a liveable and healthy environment be provided?

The model shows that achieving compactness and high densities do not exclude the adequate provision of green spaces. The multi-scalar approach employed allowed for the integration of ecosystem services into design thinking, and the integration of urban, peri-urban and hinterland areas.

The next sections describe the background of this research and the methods utilised. The presentation of the spatial planning model in its three scales: micro, meso and macro follows. Subsequently, the model is applied to the case of the International Hub for

Sustainable Development (HIDS) in Campinas, Brazil. The discussion brings together the key findings, puts them in relation to the state of the art, and suggests further research directions. We conclude with reflections on how this research addresses the question of compactness and greenery in cities today for knowledge-based urban development in peri-urban areas.

2. Background

The starting point of this research was the current planning of HIDS. Considering the need to prevent further land fragmentation due to sprawl, infilling with the generation of compact and multifunctional urban spaces has become crucial for creating liveable and sustainable environments [9].

In Brazil, as in other Latin American countries, new urban patterns mostly associated with the upper-middle classes, such as gated communities, have emerged in the past decades as a reaction to increasing urban violence. This resulted in low-density, scattered urbanization around metropolitan areas [42], leading to even higher social segregation and negative environmental impacts [43–45]. As a result, the peri-urban circles around large cities in these countries often show an asymmetrical occupation of poor and affluent populations. Different factors led to this uneven distribution, such as the presence of landfills, in the former case, or amenities such as natural parks or university campuses, in the latter.

The city of Campinas, in the state of São Paulo, is the fifth largest urbanised area in Brazil and displays a typical pattern of uneven peri-urban areas. Favelas were formed in the western skirts of the city, while upper-class residential neighbourhoods and gated communities were implemented in the north, close to former prominent coffee plantations and to the University of Campinas' (UNICAMP) campus, created in the 1960s. The university's presence stimulated the subdivision of existing farms into residential neighbourhoods and gated communities, increasing land value and resulting in real estate speculation and a discontinuous urban fabric interspersed with natural forest patches and protected areas.

The case of Campinas is a recurring pattern. During the 20th century, many university campuses were founded in the peri-urban areas of Brazilian cities, where land was affordable. However, these universities are now interested in implementing a more bustling and diverse urbanization model around them to attract high-technology firms and establish innovation hubs. Yigitcanlar has defined the new "knowledge-based urban developments" as "a place containing economic prosperity, environmental sustainability, just socio-spatial order and good governance" [46]. They support the production of knowledge through interactions between the interested parties that make this process possible. The current view of such areas has been represented by a quintuple helix model of innovation, in which society and the environment extend the triple helix model of academia (the university), industry and government. The built environment is thus an essential part of the system, shaping and accommodating the production of knowledge [47]. Knowledge districts bring together universities, research centres, and companies to promote the transfer of the most advanced scientific knowledge to the productive sector. They depend on dense and active urban spaces to promote social interaction and foster innovation. Their desirable qualities include not only scientific facilities and services but also social amenities, such as accessible public spaces with unique surroundings, such as waterfront locations, national parks or historical sites [48].

In face of the growing importance of the knowledge economy, and with the lack of adequate models for creating the desired ambience and at the same time dealing with the natural fragilities of peri-urban regions, city planners often find themselves stuck between gated communities' developers' expectations and the need to implement knowledge-based urban areas, which could be more sustainable both socially, environmentally and economically.

3. Materials and Methods

This research is framed within Design Science Research, which advocates the reduction of the gap between theory and practice, proposing solutions to real problems. By applying the solutions and evaluating their results, knowledge and new theoretical frameworks for science can be generated through an evidence-based approach, not only in an exploratory and descriptive way, but also prescriptive [49–51]. It too draws from research through design, as the design process is an integral part of the research [52,53]. A spatial planning model articulating density parameters with the provision of greenery was developed and subsequently applied to the planning and design of HIDS.

We built on Nassauer and Opdam's analytic framework [54] for knowledge innovation (Figure 1), which goes from process analysis to a phase of translating the acquired knowledge into planning and design rules and, finally, their application to a site-specific case. Hence, the first stage of the work involved the definition of parameters of compactness and landscape metrics and indicators (Figure 1's green rectangle). This was done through an expert workshop, literature review and case studies analyses [55]. In the second stage (blue rectangle), the spatial concept model was developed across three main scales: the urban block, the district, and the city, considering the area of HIDS. Further details of specific methods employed in the definition of the model are presented in the respective sections. In the third stage (Figure 1's pink rectangle), policy analyses and regulations relevant to the HIDS' site were undertaken, and spatial data collected and generated. Subsequently, the model was presented and discussed in focus groups with experts in environmental sciences and urban planning, and in a public event with experts in agroforestry. Finally, the model was applied to the HIDS area through a series of design events. As Madureira and Monteiro showed, the relationships between density and green spaces are strongly mediated by the quality of the latter [56]. Hence, this research brings together quantitative data, its analysis and evaluation, and qualitative design. These events involved academics, the local authority planning department and other stakeholders. A questionnaire on the usability of the model was completed by the design teams at the end of the process, which helped to evaluate and validate the model. They were asked whether and, if so, how the model was useful in the design process and for feedback on its potential improvements.

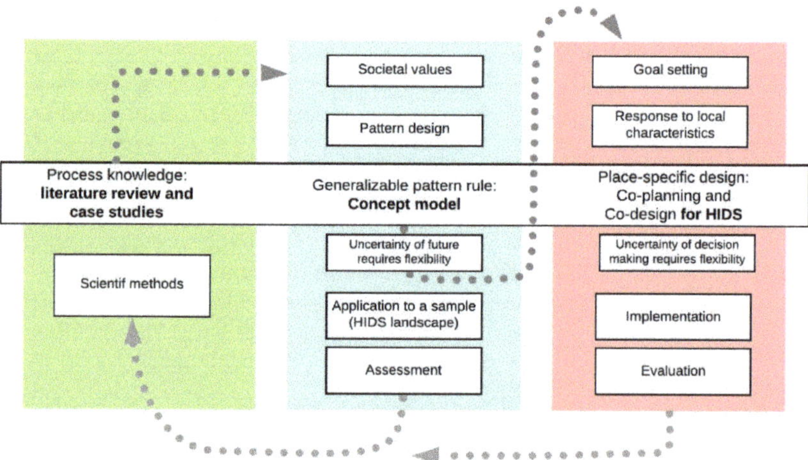

Figure 1. Diagram of the methodological framework.

4. The Spatial Planning Model

This section presents the spatial planning model proposed to reconcile density with the presence of nature, in a context of knowledge-based urban developments. It was developed to find a balance between the delivery of a range of ecosystem services, while concomitantly allowing for medium-high densities. Density is here considered as the net housing density and net building density. Net housing density refers to the number of dwelling units per hectare of net residential area (considering only predominantly residential plots). The net building density is represented by the Floor Area Ratio (FAR), which means the relationship between the total amount of the building's floor area (gross floor area) and the total area of the plot on which it stands.

The following subsection introduces the scale of the urban block, which will be followed by the meso and macro scales. The work in the various scales was cyclically revised after the results from the various stages were produced. The whole model was evaluated, finetuned and validated through the process of its application to the planning of HIDS.

4.1. The Urban Block

This initial stage of development sought to explore the concept of an "ideal urban block". It was developed from the guidelines and parameters established by Sanches [55] and complemented by the analyses of more than 400 case studies worldwide and a literature review. The following metrics were selected to maximise the provision of green spaces while providing the highest density. For green spaces, they are a minimum of 35% of green space area, a maximum of 26 green spots, the minimum average size of the green areas of 240 m^2, and the maximum average distance between green spots of 7 m. With regards to the built-form and density, they are as follows: intra-block parking area of up to 10% to minimise its presence, the maximum Building Coverage Ratio (BCR) of 57%, the minimum Floor Area Ration (FAR) of 1.5, the minimum of 220 density of dwelling units per hectare (du/ha), and the minimum size of apartments of 90 m^2 (Figure 2).

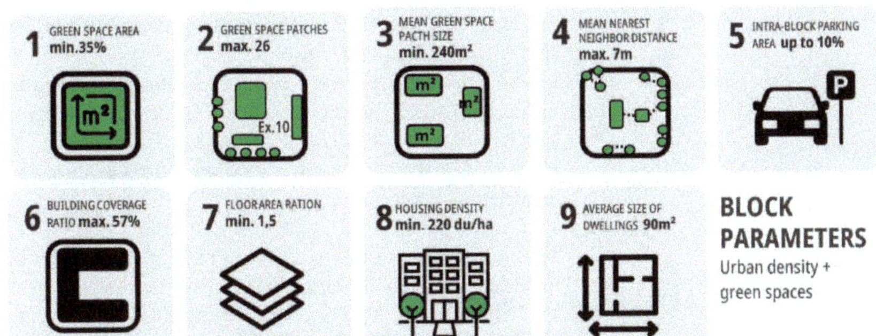

Figure 2. The urban block metrics.

A design workshop was subsequently organised with a multidisciplinary group of experts (see Section 4.3) who were asked to employ and test the selected metrics in proposals for a generic urban block.

Besides the metrics adopted in the workshop, types of intra-block green spaces were categorised based on accessibility and use. This stage was too based on literature review and analyses of case studies [55]. Four types were defined (Figure 3):

- Squares and other green public spaces;
- Privately-owned public open spaces (POPOS 1) [57], which are private green spaces where mixed uses prevail on the ground floor, with active facades. The inner area of the block is accessible, with entrances on at least two sides of the block;
- Privately-owned public open spaces—2 (POPOS 2). Predominant in residential areas, in open or semi-open blocks, these green spaces allow access to non-residents but are more secluded;
- Privately-owned green spaces of restricted access. These are residential courtyards and gardens with resident-only access.

Figure 3. Types of intra-urban green spaces defined for the spatial planning model ordered by level of accessibility and use, from public to private.

4.2. The Meso Scale: The Neighbourhood and the District

The second stage addressed the neighbourhood and district scales. By defining the typology of green spaces at this scale, we sought to equate urban and ecological connectivity issues and explore different degrees of urban density (housing and general building). The types of green spaces selected were linear, given their ability to extend throughout the territory and link to existing natural patches and agricultural areas (Figure 4). They are:

- Tree-lined streets: street, avenues, boulevards that are predominantly tree-lined and, when possible, with permeable pavements and flowerbeds on the sidewalks.
- Greenways: green corridors with pedestrians and bicycle paths, in which active mobility is the primary function.
- Linear parks: linear green spaces, providing multiple ecosystem services.
- Ecological corridor and parks (buffer strip): forested corridors connected to natural patches of relevant ecological value. They allow for the movement of fauna and provide a buffer strip with uses akin to linear parks to minimise edge effects (whether anthropic or natural) [58–61].
- Ecological corridor and productive green spaces (buffer strip): similar to the type above, but with buffer strips for productive uses, which can be areas of urban agriculture, agroforestry or forestry with native species.

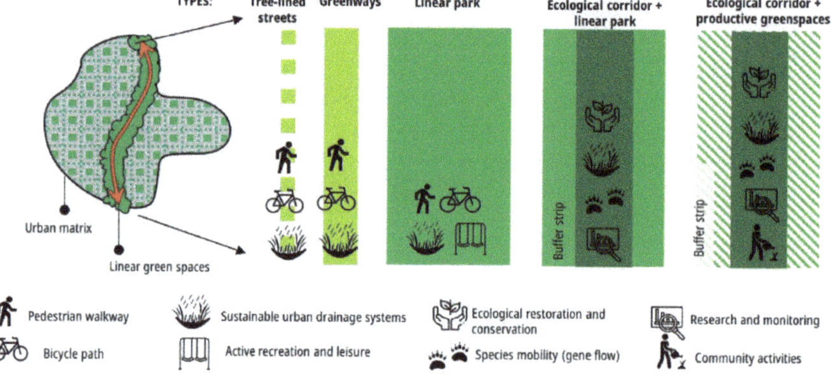

Figure 4. Types of linear green spaces in the spatial planning model at the neighbourhood and district scales.

The minimum widths of these green spaces must be adapted to their local context. Since Campinas, Brazil, is the location of our test case, we adopted the minimum requirements as stipulated by its local legislation.

Productive green spaces associated with ecological corridors take into account circular economy precepts [62]. Food and forestry products produced are intended to be primarily consumed and used on site. This approach supports the development of community-driven urban allotments, which provide direct access to zero-kilometre food and encourage healthier food consumption habits [63]. Locally sourced timber would be employed in the building construction. Other benefits include the reduction of CO_2 emissions related to the reduced need for the transportation of goods and the use of renewable materials. The model can thus be a thriving force to foster circular economy.

Following a density gradient logic, based on a series of studies on the impact of urbanisation on urban biodiversity, such as due to noise levels, traffic, number of people circulating, housing density and the height of buildings [61,64], we established three intervals (high, medium, low): the closer to a natural area with high ecological value (ecological corridors and natural patches) the lower the housing density and the floor to area ratio (FAR).

The linear green areas with a vocation for recreation and leisure (greenways and linear parks), and therefore destined for the intense use of the population, do not fit into this logic of gradients. As such, placing medium and high-density areas close to them would allow for the greatest number of people to benefit from their use.

The indicators of built-up and housing densities (high, medium and low ranges) were obtained with the aid of Principal Component Analysis (PCA), a multivariate analysis visualisation technique, from the raw data from Sanches [55]. The high-density intervals we use in the model are the ones previously defined for the urban block scale (Figure 2). In turn, the medium and low-density ranges were extracted from the data from sample blocks that distanced from the vectors of the PCA corresponding to the built-up and housing density variables (FAR and DOMIC) towards the vectors corresponding to the variables of the percentage of vegetation cover and the average size of green spots (PLAND and PATCH). Therefore, the lower the density, the greater the tree cover; that is, there is a negative relationship between these two parameters. The sample blocks selected as reference are indicated by the dotted circles in the PCA of Figure 5.

The model assumes a feasible scenario of implementation. Hence, we set housing density ranges (high, medium and low), and their respective FAR and maximum BCR, and the lowest minimum percentage of vegetation within the blocks, as seen in Figure 6.

4.3. The Regional Scale

The third stage addressed the regional scale, and how to articulate the meso scale to the peri-urban landscapes, often marked by the coexistence of sprawl, agricultural areas, natural patches and traffic infrastructure, as is the case of HIDS. The main premises were to minimise the negative effect of highways as barriers to ecological and urban connectivity and the edge effect in the transition of contrasting land uses. At the same time, the aim was to push for compactness in order to support sustainable urban development, minimising sprawl over protected natural and productive areas.

Greenways were identified as a type of green space that would link the development area to the surroundings supporting active mobility. Ecological corridors were too employed at this scale to promote ecological connectivity of natural patches, natural areas of permanent protection, and natural reserves within the surrounding agricultural areas.

4.4. The Spatial Planning Model across Scales

Figure 7 presents the general framework of the development of the spatial planning model. It articulates the concepts, guidelines and theoretical principles explained above regarding the combined planning of green areas in a scenario of increasing urban density.

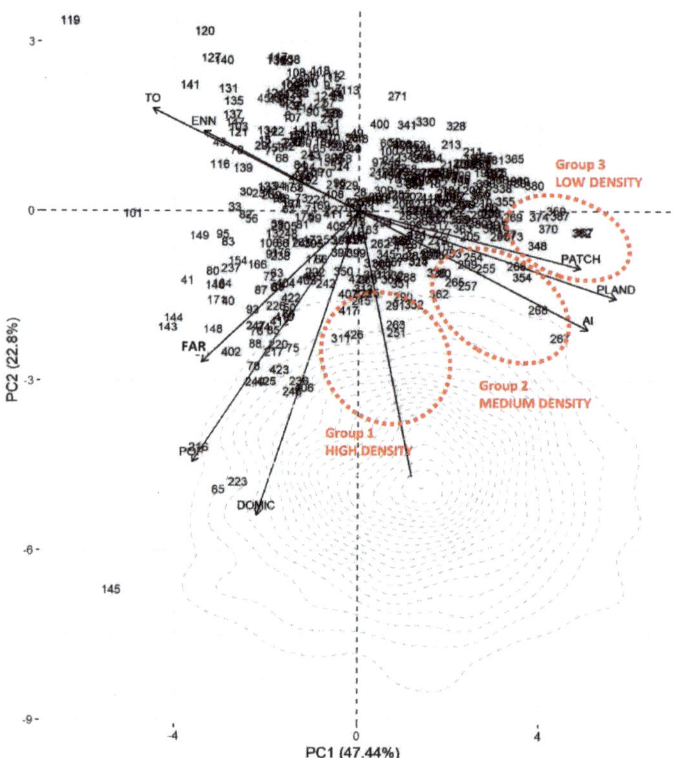

Figure 5. Principal component analysis (PCA) ordination (triplot) for urban block data with density (DOMIC and FAR) and landscape ecology (PATCH and PLAND) variables. The PCA was used to set parameters of density intervals and percentage of vegetation cover. DOMIC and FAR are housing density and built-up density, respectively. PATCH and PLAND are average size of green spots and percentage of vegetation cover, respectively.

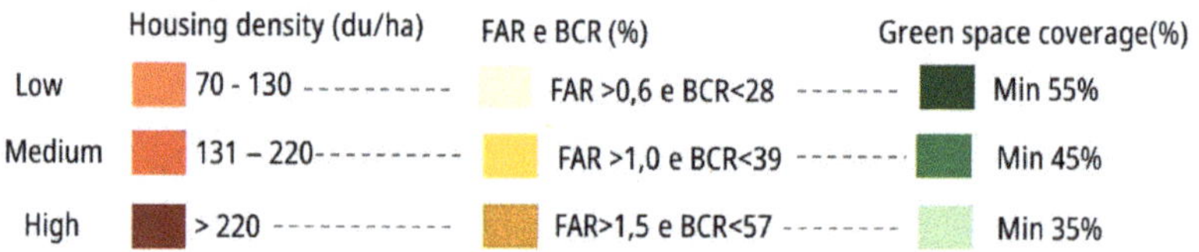

Figure 6. Parameters of net housing density, net built-up density (FAR), building coverage rate (BCR) and vegetation cover rates with intervals defined from the PCA.

The model is broken down into layers in Figure 8 to clarify its structure further and enhance its readability. Prior to implementing the density and green space layers, it is necessary to identify significant pre-existences such as urbanised areas and natural patches in the vicinity or on the edges of the site.

As a first layer, ecological corridors protected by buffer strips, which could be productive green spaces (agriculture, agroforestry, and forestry of native species) or parks are established as a priority. The second layer comprises the other types of linear green spaces: tree-lined roads, greenways and parks, which fulfil the role of ecological connectivity, active mobility and leisure. It is worth noting that the greenways are meant to go beyond the intervention area, linking across scales. The third layer comprises the green areas within the blocks, here represented by nodes. They are intended to be articulated to the tree-lined roads and the greenways. The fourth layer consists of the occupation of the territory according to a density gradient (distributed in three intervals of housing density and three intervals of building density) in which the areas closest to ecological corridors must present the lowest density rates, which gradually increase towards the centre of the urbanised area. The minimum percentages of intra-block vegetation cover also change depending on the density range in which the block is located, as explained in Figure 6. With all the layers overlapping, we have the complete model that can be replicated in similar contexts to HIDS.

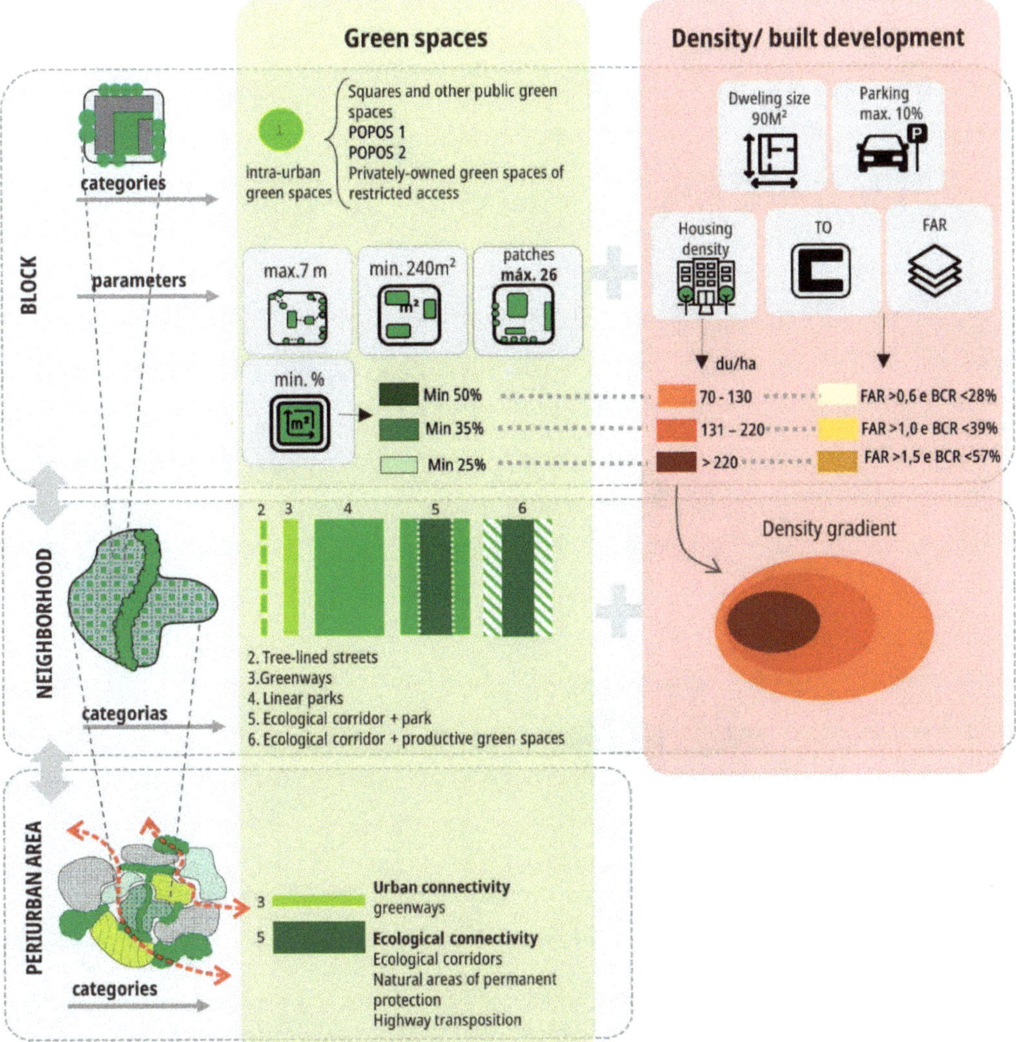

Figure 7. Framework of the definition of the spatial planning model. The two themes, green areas and density, run across the scales (urban block, neighbourhood/district and the city/region).

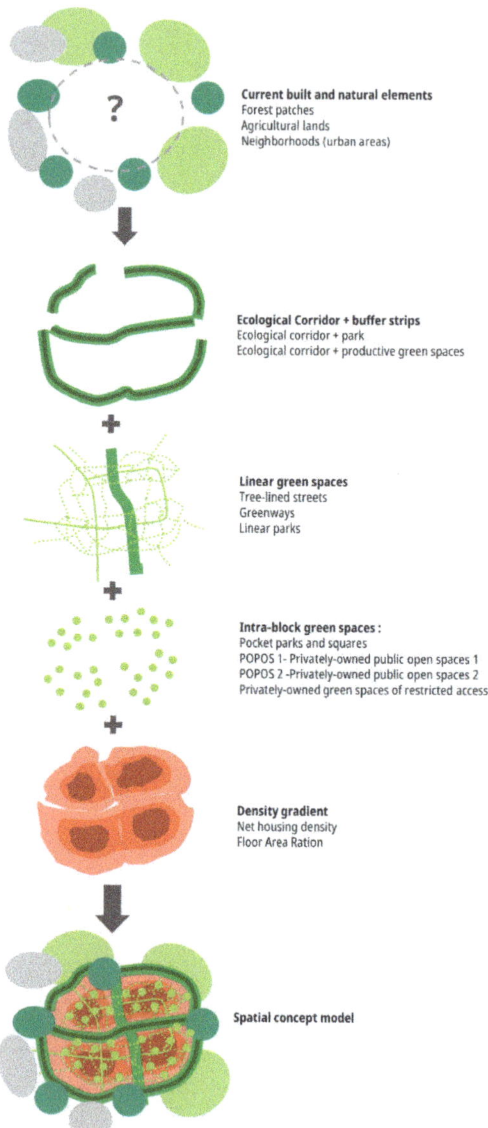

Figure 8. The model broken down into green space layers and density gradients.

5. The Case Study: The International Hub for Sustainable Development (HIDS)

5.1. Background

Campinas' High Technology Hub Development Company (*Companhia de Desenvolvimento do Polo de Alta Tecnologia de Campinas*—CIATEC) was conceived in Campinas in the late 1970s, as a second-generation park next to the University of Campinas. Located 15 km north of the city centre, it was then expected to become the Brazilian Silicon Valley. In the 1980s and 1990s, a few private companies and two large public research institutes settled in the area, which, however, lacked the necessary infrastructure investments, and most of the land remained undeveloped.

According to Gyurkovics e Lukovics [65], first and second generation technology parks were based, respectively, on the "science push"and "market pull" models of innovation, and they were typically located inside or next to university campuses. Third generation parks, on the other hand, tend to be located in bustling city centres, and they are based on a different innovation approach, known as interactive or feedback-based, typically developed through living labs. They look less like parks and more like hubs, districts, or simply knowledge-based urban areas, and their aim is to "improve the welfare of the local community" and "the development of their regions", connecting "the regional economy to the processes of knowledge-based economy".

After acquiring a large land parcel in CIATEC in 2014, the University of Campinas proposed converting this second-generation science park into a third-generation innovation area, named International Hub for Sustainable Development (HIDS in the Portuguese acronym) (Figure 9). This conversion not only involves introducing new and updated infrastructure but also a renovation in terms of culture and objectives. The most significant change is the focus on a new economy that explores opportunities related to sustainable development, such as clean energy and lower impact urbanization. In order to achieve this goal, housing, services and public spaces must be introduced to generate density and a diverse and active urban environment, attracting innovative firms and creative researchers. At the same time, the area encloses historical heritage and natural forest fragments that must be protected and connected through ecological corridors. A careful environmental plan must be carried out in order to set up an exemplary scheme that is in harmony with the hub's new guidelines. Moreover, the literature shows that many of the characteristics searched and valued by third-generation innovation hubs' users relate to the quality of green areas and open public spaces. For this reason, the site seemed a perfect opportunity for the application of our model.

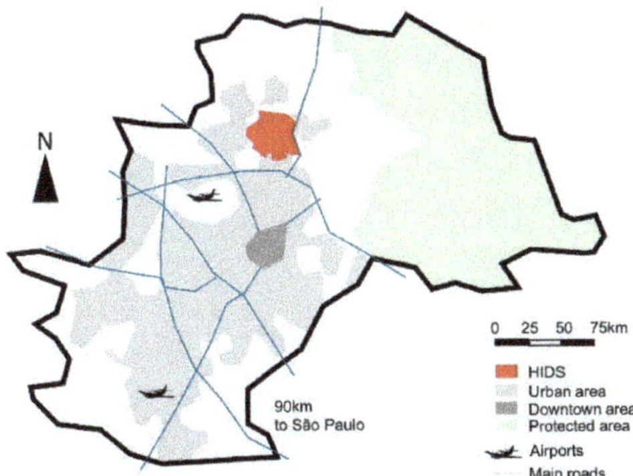

Figure 9. Location map of HIDS, in Campinas.

5.2. The Site-Specific Data Collection

Geo-referenced data was collected from the municipality of Campinas data portal and statistical data from the IBGE website (Brazilian Institute of Geography and Statistics). The collection of secondary data covered themes such as environmental, socio-economic, spatial characteristics and planning legislation. Primary data comprising spatial, tabular, and textual data relating to HIDS was produced. It included land cover, the quantification of each land cover class, and the calculation of metrics, such as FAR and BCR. To obtain the land cover data, supervised classification was used. This remote sensing technique classifies the land cover from satellite images based on spectral and texture characteristics. Satellite images were obtained from Planet's Skysat satellite sensor, with a spatial resolution of 3m and RGB and Near Infrared (NIR) spectral bands (bands 1,2,3 and 4—imaging date from 8 September 2020). The classifier selects the sampling units of pixels representing each land cover class, and the software generates an automatically classified image. The final product is a raster image, whose values for each pixel correspond to a certain cover class. We use the open source QGIS software and the Dzetsaka classifier (plugin) with the Random Forest classification algorithm. The land cover categories defined were: (1) tree and shrub, (2) herbaceous, grassland and agricultural land (3) impermeable areas, (4) river or lake and (5) bare soil.

The existing buildings in the HIDS urbanised cores were manually mapped, and the total built area was surveyed to determine the FAR and the BCR. This allowed for an understanding of the initial conditions related to density and availability of green spaces. Accordingly, "buildings" were added to the land cover map. For the calculation of the FAR and the BCR the total land area of each nucleus was considered, excluding the road system; that is, only the area of the blocks, which eventually houses, in addition to buildings, parking lots and common green areas, was used. After the mapping, the land cover classes were quantified in hectares and percentage in relation to the total area of HIDS and the area of each urbanised nucleus within it.

The result of the supervised classification obtained a Kappa index of 99.74% accuracy. The amount of each land cover class is shown in the pie chart below (Figure 10).

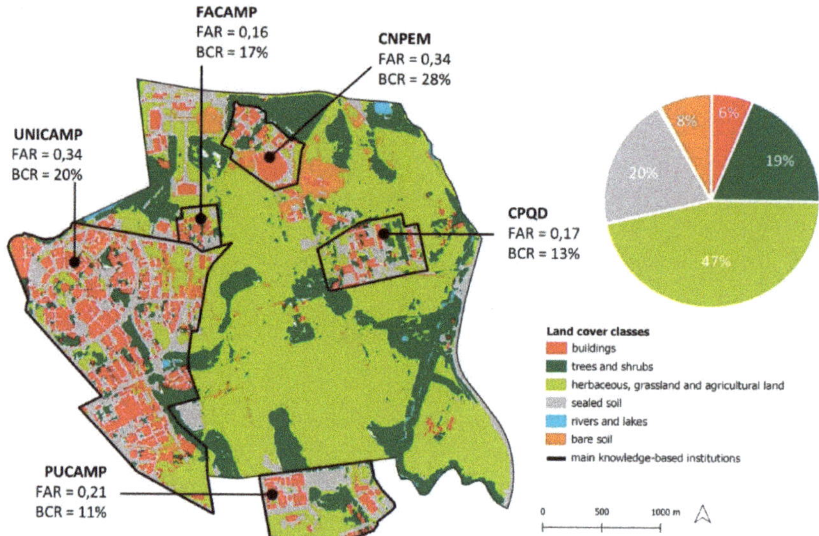

Figure 10. HIDS' land cover map and percentage of each class.

As expected, herbaceous and grass cover (including agricultural areas) is the predominant land cover (47%). However, the proportion of impermeable areas (20%), consisting of general pavements, parking lots and roads, exceeds the proportion of tree cover (19%). This is not surprising given the existing car culture, associated with precarious public transportation, making the automobile the primary means of mobility for many.

When we consider only the urbanised areas, the proportion of tree cover, herbaceous areas, and impermeable surfaces is very different from the HIDS as a whole, as shown in Figure 10. In this case, impermeable pavements are predominant, reaching 42% of the total, due to large areas destined for parking and paved roads. The buildings and tree cover reach similar values.

Low BCRs and FARs were observed, when compared to those practiced in the city of Campinas, and to the values established in the city's master plan. Analysed alongside the tree cover, it is evident that although buildings do not intensely occupy the ground, the open areas present a much more limited tree cover than the site could accommodate.

5.3. The Application of the Model

This first stage was carried out through an international workshop (July 2020) that asked participants to consider how greenness, compactness and density could be achieved simultaneously at the scale of the urban block, alongside other sustainability criteria, with the use of computational tools.

One of the urban design proposals developed during the workshop is presented in Figure 11. It shows the compatibility between concepts and parameters of compact and dense green blocks and other sustainability criteria. In this example, as set in the spatial model, the vegetation cover reached 35% of the block area with a minimum FAR of 1.5 (performed with a FAR of 4.3). Inner-block green spaces are evenly distributed and relate to one another across adjacent blocks. Housing represented 60% of the total land us. The density achieved was 286 dwellings/ha (which met the parameter of a minimum of 220 du/ha). The results show that the green space and building density parameters were not mutually exclusive, and sound proposals were achieved.

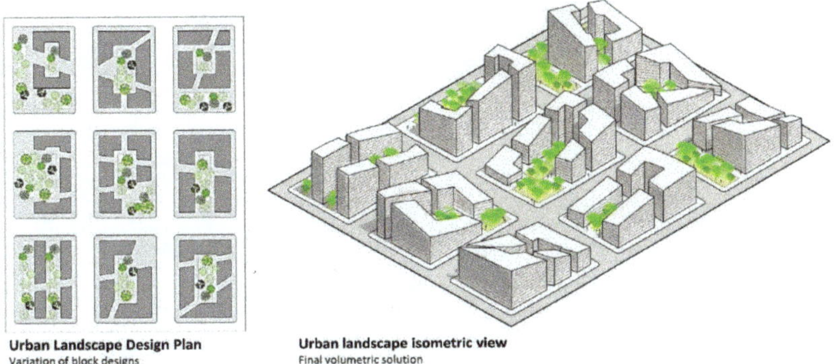

Figure 11. Urban design proposal developed during the international workshop in July 2020. Reproduced with permission of team participants.

Subsequently, the complete spatial concept model was applied to the full extent of the HIDS' area in two expert workshops with planning professionals and architects. Prior to the design explorations, a presentation of the model took place and a list of the indicators given to participants.

The second exercise consisted of applying the spatial concept model from the macro to micro scale (January–February 2021). Firstly, participants had to consider the site's related policies and characteristics (i.e., environmental regulations, water streams/bodies; natural forest patches, topography, existing buildings, road system, points of interest) and other requirements, such as sustainable mobility, at the macro scale. At the meso and micro scale, participants were challenged to develop a proposal for typical urban blocks for the HIDS' area.

The exercise's aims were to protect the current streams and natural vegetation, to promote high-density urban development, to connect the new urban blocks to current facilities in the HIDS and to the surrounding urban fabric. The spatial concept model was a tool offered to the participants to achieve these aims.

Nonetheless, each team was challenged with a different set of requirements: the first team had a pre-defined street network system as a constraint; the second team had to consider the current land ownership boundaries; and the third team did not have any constraints. This was intended as a way to explore the applicability of the model across different scenarios.

The urban designs presented some shared solutions: all teams worked with the model's green spaces typologies such as the ecological corridors and buffer strips along them, which predominantly host productive functions (i.e., urban agriculture, agroforestry, community gardens), and linear parks. The three teams also proposed a large park in the Anhumas Valley, located in the east part of the HIDS, to protect the Anhumas River and its embankments (Figures 12–14). Figure 12 shows one of such proposals. The team employed the urban density gradient logic set in the model as a starting strategy. As in the model, its greenways are independent of the street network, and linked to other green spaces categories, such as the Anhumas Park.

Figure 12. Urban design proposal by Team 1 (pre-defined street network requirement). Reproduced with permission of team participants.

Figure 13. Mobility network solution developed by Team 2 (fixed land ownership boundaries requirement). Reproduced with permission of team participants.

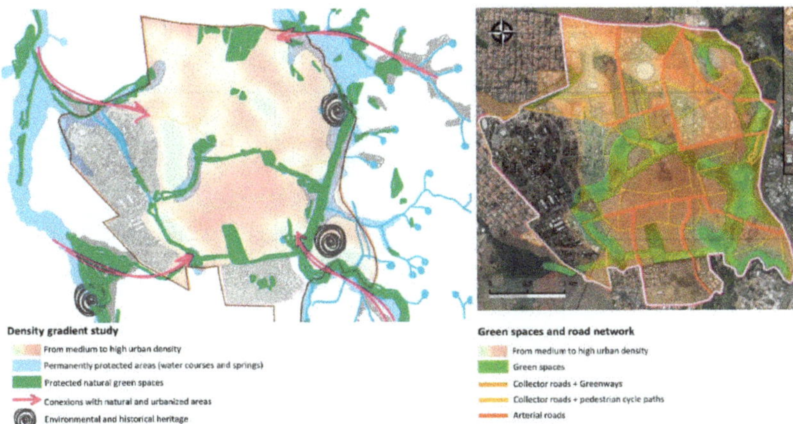

Figure 14. The gradient of urban density related to green spaces and arterial road network proposed by Team 3 (no constraints). Reproduced with permission of team participants.

Besides the green spaces categories suggested in the model, Team 2 proposed wetlands as a new green space category for water treatment purposes. They also included greenways along the boundary of Anhumas Park (Figure 13).

One of the main aspects of Team 3's proposal is the adoption of the model's density gradient - the closer to arterial roads, the higher the urban density; in contrast, the closer to protected natural green space (e.g., ecological corridors), the lower the urban density. Since the team was not bound by any design constraint, the model's gradient density helped the team design the road system avoiding placing arterial avenues close to ecologically significant green spaces (Figure 14). The greenways were also designed independently of the arterial road network and linked to other green spaces.

All the teams went forward in the design process, resulting also in a neighbourhood design proposal at the level of massing, and the design of urban blocks, as a first exploratory design which were improved in the third exercise to be shown as follows.

In the third and last exercise (March–June 2021) participants were asked to further develop a smaller area of HIDS, working in the meso and micro scales. Two teams were responsible to develop two different urban centralities, named Central Plateau and South Centre. These areas were chosen because the Central Plateau is HIDS' geographical centre, hence equidistant from existing facilities and institutions; and the South Centre because it presents the longest edge between public and private land within HIDS.

The teams were asked to develop mixed-use neighbourhoods (housing, retail, offices, public equipment/institutional use as schools, administration, etc., and science labs) and create two building density scenarios: FAR 2.5 and 3.5. Both design proposals adopted multifamily residential buildings as a solution to provide higher density. No buildings (residential, offices and retail) exceed seven storeys.

Although they pursued the same aim, each team adopted a distinctive urban morphological solution. The Central Plateau was developed considering the block as the minimum design unit, presenting the following morphological types: $\frac{1}{4}$ open block, half urban block and closed urban block [66]. These morphological solutions often generated a single larger green space courtyard in the block and assured aggregated green spaces, which helped achieve the maximum number of green patches and average green patch size set by the spatial model (Figure 15).

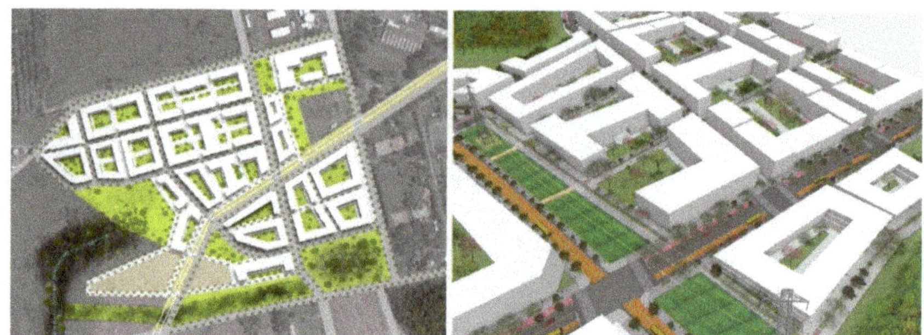

Figure 15. Urban design proposal for the Central Plateau. Reproduced with permission of team participants.

In contrast, the South Centre proposal was based on plots as the minimum design unit, presenting point-block buildings and slab buildings with plinths as the morphological typologies [66]. Nevertheless, the team sought to establish green space continuity throughout the blocks by employing privately-owned public open spaces (POPOS) as per the spatial model (Figure 16).

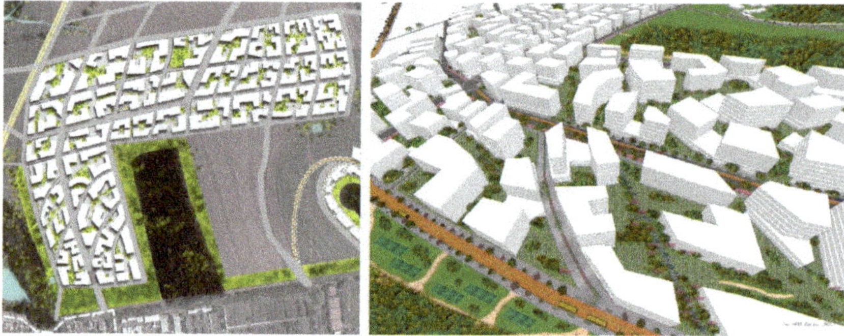

Figure 16. Urban design proposal for the South Centre. Reproduced with permission of team participants.

Quantitative data about the proposal's performance compared to the model's parameters is shown in Table 1. We can see that the green space area, housing density value, FAR, BCR, parking area and average dwelling size are either very close to the model's parameters or reaches even higher values and better performances.

Table 1. The design's performance compared to the model's parameters.

Parameters/Centrality	Central Plateau (per Block)	South Centre
Green space area	35%	20%
BCR	50%	55%
FAR	2,5/3,5	2,5/3,5
Housing density	136 du/ha/197du/ha	212 du/ha/285 du/ha
Parking area	No data	Up to 10%
Average dwelling size	100 m^2	100 m^2

5.4. Evaluation and Validation of the Model

One of the most relevant aspects of this research was validating the spatial planning model through its practical application in the design exercises discussed previously. Through the series of workshops, an iterative process of testing and evaluating results generated improved propositions that, in turn, took the designs in new directions [67]. The design science research approach undertaken meant the definition of evidence-based solutions and the evaluation of their results in real-time. This, in turn, can contribute towards reducing the gap between theory and practice. The ability to continuously assess both the projects and the model was one of the main benefits of the approach.

The survey carried out with the design teams after the third exercise confirmed that the quantitative and qualitative parameters provided at the beginning were constantly considered, even though some found it challenging to follow them through all stages of the design process. In the end, the participants considered the model helpful for the definition of an urban landscape with the desired qualities.

The green space categories at the meso scale were adopted by most participants as strategies for ecological connectivity, urban connectivity, to enhance biodiversity and to provide a range of ecosystem services. The density parameters were fully adopted or used as the starting point for the setting of other indicators.

As a model is a simplification of reality, it must be selective. In this research, the focus was on density and green spaces. Nevertheless, participants felt the need to integrate other aspects in the design process (e.g., urban climate, energy efficiency, mobility and TOD—Transit-Oriented Development, sustainable water management, vitality and liveability, etc.). In some cases, however, participants applied the model literally, laying down some spatial elements as schematically suggested, such as the linear parks and density gradient distribution. This misunderstanding of the model's application must be avoided, since it may lead to a limited and less creative proposal.

In all exercises, participants freely adopted an intermediate density level morphological solution influenced by the recommendation of recognized experts in urban planning and most recent literature [8,68,69] as a way to achieve a better-quality urban design. Thus, the model was not applied and tested in morphological solutions that are in the extremes of the BCR range, i.e., detached single family houses and high-rise buildings [70]. This is in line with the model's intention, since it focuses on knowledge-based developments in peri-urban areas, where these morphological types are not desirable.

Although the model has been applied in urban design proposals of intermediate density, we noticed how variable and diverse the morphological solutions could be for an existing area and a real context, demonstrating the feasibility to embed qualitative guidelines and reaching ideal green space and density parameters, thus enabling its validation.

6. Discussion

Low-density monofunctional sprawl on urban fringes remains one of the most unsustainable patterns of urbanisation. Although promising proximity to green areas or the countryside, their very proliferation threats both. They lead to the disaggregation of urban form and landscape fragmentation, relying on private modes of transportation and putting pressure on municipal governments to extend the reach of services [71]. This article proposes a mixed-use spatial planning model that balances compactness, densification, and the delivery of ecosystem services in peri-urban areas. It aims to support sustainable and resilient development at the urban fringes associated with the establishment of new knowledge-based areas, and to connect intra-urban areas and the hinterland [72].

Knowledge-based urban development requires space commonly found through urban regeneration projects or in peri-urban areas. It has been shown that such developments, besides spurring innovation and the dissemination of knowledge [73], can become new economic and social nodes in cities. Yet, they are often seen as enclaves with a limited range of land uses, and underdelivering on aspects of liveability and ecological connectivity. The results revealed insights on how to address these challenges concomitantly.

Firstly, as suggested by Artmann, Inostroza and Fan, there is a need for innovative concepts and comprehensive planning strategies across scales for compact and green cities [17]. This research confirms that a multi-scalar and systemic approach is necessary to bridge the gap between compactness, densification and the provision of green spaces in cities. Multifunctionality, argued by Hansen et al. [74] as a crucial strategy for green space in compact cities, is acknowledged as a vital principle. This was achieved by linking quantitative data with design science research, which allowed for the development of solutions in which evidence-based design and real-time testing are realised throughout the process. As models lack the complexity of reality itself, planning must ensure that challenges are holistically addressed. The model here proposed is flexible and must be tailored to context. Local policies and site-specific considerations ought to directly inform it.

Secondly, following the initial input of the model, design activities were crucial to assess its effectiveness to guide the development of proposals and finetune it. On the one hand, such an approach can enrich meaningful participatory processes [75], as a continuous loop [67] of testing, reaching results and assessing them occurs, and, on the other hand, addressing design quality [56]. In this process, after the model's evaluation, further considerations according to place-specific needs were identified. For instance, including additional building density parameters at the block level, such as maximum building height, to encourage compactness while providing morphological solutions at human-scale, could be helpful. Furthermore, mixed-use parameters, such as percentage of retail/offices or employment density, could contribute to walkability and liveability. The definition of "Ecological corridor and park" and "Ecological corridor and productive greens spaces" must minimise possible conflicts between vegetable growing and biodiversity conservation (e.g., wildlife risks due to increased predation, exotic species negatively affecting local ones, farming yields being affected by the consumption of vegetables by wild animals, etc). Furthermore, creating a minimum percentage for local food production could ensure that the benefits deriving from this activity are embedded into the model. There is scope for further research into computational approaches for implementing the block-level parameters and evolutionary algorithms to generate design options for the maximisation of specific attributes. In addition, future activities ought to further the use of the model as a tool for participatory design with local communities and stakeholders. Since the model is abstract, the agreement upon it can guarantee that ecological principles will be followed, but still leave room for the design of alternative urban forms, as shown above in the exercises developed.

Thirdly, studies about place quality in innovation districts highlight the importance of providing mixed uses and a variety of public open spaces designed and managed to spur interaction, learning and networking. In such contexts, public open spaces when "designed and programmed well (. . .) can be the connective tissue between people and firms, effectively serving as the heart of a healthy and vibrant innovation ecosystem" [76]. Furthermore, dense, walkable, and highly connected urban areas encourage a collaborative and open culture of innovation. They favour face-to-face encounters, an important aspect for innovation sectors that often demand the exchange of complex, tacit knowledge among their workers [77]. Recommendations have also been made for setting parameters for zoning-related parking to reduce it to a minimum and place it in specific sites to encourage multimodal transportation, including active mobility. The model addresses both issues. Linear parks, either as a category in itself or as buffer strips, establish a large perimeter of contact to mixed-use areas, enhancing connections to the urban fabric. The green spaces inside the blocks, whether squares or POPOS, enable "third places" which are "particularly desirable for young tech employees compelled to rent micro-units or share spaces to keep monthly rent low" [78]. The parameter of a maximum parking area at block scale also contributes to enlarging inner-block open spaces and encouraging active mobility when linked to greenways for slow mobility, as defined in the model.

Fourthly, addressing climate change through urban development must include principles of circular economy [62]. The question of the economic viability of sustainable solutions frequently challenges their consideration in favour of standard practices. This research shows that taking a positive approach towards policy "restrictions" and regulations and using them as guidelines for a configuration of hybrid and multifunctional spaces unlocks a range of economically viable benefits within the frame of sustainable urban development. Furthermore, such an approach enhances the sites' resilience, generating uses and functionalities that span various anthropic and natural needs. The research advances thinking in how the choices of direct ecosystem services to be delivered and understanding their socio-economic values can be catalysers for employing nature-based solutions. Direct ecosystem services such as food [79] and timber [80] that can be consumed and employed locally are just some of such viable solutions.

Finally, the research shows that a spectrum of residential and building density can be developed in tandem with ranges of green space provision. Furthermore, we argue that, especially in new development, the definition of green infrastructure and nature-based solutions must be established concomitantly to the built form, indexes of density and other land uses. The integration of GI and NBS into overarching planning processes maximises the possibility of a balanced environment where the benefits of urbanity and those from ecosystem services can be achieved.

7. Conclusions

Compactness, density and green space are not mutually exclusive. Green infrastructure and nature-based solutions need to be integrated into planning processes, and not considered as separate add-ons. The spatial structure suggested by the model provides the armature for a range of horizontal and vertical nature-based solutions of various scales to be included into the final plan.

Aligning density gradients with the provision of selected ecosystem services can enhance accessibility to green areas and minimize conflicts and trade-offs across potentially competing needs, such as the integrity of ecological corridors and high anthropic use of open spaces.

Peri-urban contexts are the very place where sprawl occurs. Such areas are normally object of low-density development, fragmentation due to infrastructural building and the force field where agricultural and natural land is taken. In turn, given the availability of land and consequential larger potential to offer ecosystem services, they must be a focus of attention. The model articulates development with the planning for enhanced ecosystem services, bringing higher densities to the urban fringe and establishing ecological connectivity between the consolidated urban areas and the hinterland.

Hybrid landscapes that articulate urban allotments and gardens and agroforestry with development areas can support circular economy principles. Access to zero-kilometre food and materials (i.e., timber) for building and furniture construction can strengthen well-being while boosting local jobs and economic benefits for residents. In addition, agroforestry can provide ecologically sound transitions from inner urban areas to rural landscapes, enhancing biodiversity and supporting agricultural production.

As such, the role of design is crucial. The model is a vehicle developed from quantitative and qualitative approaches to combine urban and environmental planning preoccupations and principles. It provides both urbanity and access to nature. Its application into plans must incorporate a comprehensive analysis of the territory and its specific needs. The transformation of the peri-urban areas as here proposed can allow for a reconfiguration of planning from the edges, supporting more sustainable, liveable, and resilient environments.

Author Contributions: Conceptualization, all authors; methodology, P.S. and G.C.; software, P.S.; validation, P.S. and G.C.; formal analysis, P.S. and G.C.; investigation, F.L.d.O.; data curation, P.S.; writing—original draft preparation, F.L.d.O. and P.S; writing—review and editing, F.L.d.O.; visualisation, P.S.; supervision, F.L.d.O. and G.C.; funding acquisition, all authors. All authors have read and agreed to the published version of the manuscript.

Funding: This research was funded by São Paulo Research Foundation (FAPESP), grant number 2019/25369-3, CNPq, grant number 304370/2020-7, and Politecnico di Milano. It was also financed in part by the Coordenação de Aperfeiçoamento de Pessoal de Nível Superior—Brasil (CAPES)—Finance Code 001.

Institutional Review Board Statement: This study was approved by the Ethics Committee of University of Campinas (protocol code: 50658921.0.0000.8142 and date of approval: 3 October 2021).

Informed Consent Statement: Informed consent was obtained from all subjects involved in the study.

Acknowledgments: We thank the reviewers for their comments, and all the participants of the design events. For the permission to reproduce their urban design images, we thank: Fabio Bellucci, Gabriela Bozzola, Juliana Milanez, Isabella Cavalcanti and Sandro Martinez (Figure 11); Nathalia da Mata Mazzonetto Pinto, Barbara de Holanda Maia Teixeira, Thais Michelão Martins, Helenylson Jesus Pereira and Tiffany Liu (Figure 12); Marcelo Meloni Montefusco, Juan Pietro Cucolo Marçula, Luiza de Bastos Gaillac, Rita de Cássia Gouveia Jácome and Valdiere Braga (Figure 13); Isabella Eloy Cavalcanti, Guilherme Pires Veiga Martins, Gabriela Rosa and Jessica Rabito Chaves (Figure 14); Helenylson Jesus Pereira, Barbara de Holanda Maia Teixeira, Jessica Rabito Chaves, Marcelo Meloni Montefusco and Juan Pietro Cucolo Marçula (Figure 15); and Luiza de Bastos Gaillac, Isabella Eloy Cavalcanti, Rita de Cássia Gouveia Jácome and Valdiere Custódio Braga (Figure 16).

Conflicts of Interest: The authors declare no conflict of interest.

References

1. United Nations. *The Sustainable Development Goals Report 2020*; Department of Economic and Social Affairs: New York, NY, USA, 2020.
2. European Commission. Land Take in Europe. Available online: https://ec.europa.eu/environment/strategy/biodiversity-strategy-2030_en (accessed on 10 November 2021).
3. Climate Action Tracker. *Climate Target Updates Slow as Science Ramps up Need for Action*; Climate Action Tracker: Berlin, Germany, 2021.
4. Lemes de Oliveira, F.; Bezerra, M.d.C.d.L.; Teba, T.; Oliveira, A. The environment-culture-technology nexus framework: An approach for assessing the challenges and opportunities for implementing nature-based solutions in Brazil. In *Nature-Based Solutions for Sustainable Urban Planning: Greening Cities, Shaping Cities*; Mahmoud, I.H., Morello, E., Lemes de Oliveira, F., Geneletti, D., Eds.; Springer: Berlin/Heidelberg, Germany, 2021. [CrossRef]
5. Jenks, M.; Burton, E.; Williams, K. *The Compact City: A Sustainable Urban Form?* E & FN Spon: London, UK, 1996; 250p.
6. Bibri, S.E.; Krogstie, J.; Kärrholm, M. Compact city planning and development: Emerging practices and strategies for achieving the goals of sustainability. *Dev. Built Environ.* **2020**, *4*, 100021. [CrossRef]
7. Andersson, E.; Barthel, S.; Borgstrom, S.; Colding, J.; Elmqvist, T.; Folke, C.; Gren, A. Reconnecting cities to the biosphere: Stewardship of green infrastructure and urban ecosystem services. *Ambio* **2014**, *43*, 445–453. [CrossRef] [PubMed]
8. Rogers, R.; Gumuchdjian, P. *Cities for a Small Planet*; Faber and Faber: London, UK, 1997.
9. Urban Task Force. *Towards An Urban Renaissance: The Report of The Urban Task Force Chaired by Lord Rogers of Riverside*; Urban Task Force: Sydney, Australia, 1999.
10. Williams, K.; Jenks, M.; Burton, E. *Achieving Sustainable Urban Form*; E & FN Spon: London, UK, 2000.
11. Breheny, M. Urban compaction: Feasible and acceptable? *Cities* **1997**, *14*, 209–217. [CrossRef]
12. Haaland, C.; van den Bosch, C.K. Challenges and strategies for urban green-space planning in cities undergoing densification: A review. *Urban For. Urban Green.* **2015**, *14*, 760–771. [CrossRef]
13. Littke, H. Planning the Green Walkable City: Conceptualizing Values and Conflicts for Urban Green Space Strategies in Stockholm. *Sustainability* **2015**, *7*, 11306–11320. [CrossRef]
14. Tappert, S.; Klöti, T.; Drilling, M. Contested urban green spaces in the compact city: The (re-)negotiation of urban gardening in Swiss cities. *Landsc. Urban Plan.* **2018**, *170*, 69–78. [CrossRef]
15. Brokking, P.; Mörtberg, U.; Balfors, B. Municipal Practices for Integrated Planning of Nature-Based Solutions in Urban Development in the Stockholm Region. *Sustainability* **2021**, *13*, 10389. [CrossRef]
16. Modarres, A.; Kirby, A. The suburban question: Notes for a research program. *Cities* **2010**, *27*, 114–121. [CrossRef]
17. Artmann, M.; Inostroza, L.; Fan, P. Urban sprawl, compact urban development and green cities. How much do we know, how much do we agree? *Ecol. Indic.* **2019**, *96*, 3–9. [CrossRef]
18. Wellmann, T.; Schug, F.; Haase, D.; Pflugmacher, D.; van der Linden, S. Green growth? On the relation between population density, land use and vegetation cover fractions in a city using a 30-years Landsat time series. *Landsc. Urban Plan.* **2020**, *202*, 103857. [CrossRef]
19. European Commission. *Green Infrastructure*; European Commission: Luxembourg, 2010.
20. European Commission. *Towards An EU Research and Innovation Policy Agenda for Nature-Based Solutions & Re-Naturing Cities—Final Report of the Horizon 2020 Expert Group on 'Nature Based Solutions and Re-Naturing Cities'*; European Commission: Luxembourg, 2015.
21. Lee, A.C.K.; Maheswaran, R. The health benefits of urban green spaces: A review of the evidence. *J. Public Health* **2011**, *33*, 212–222. [CrossRef]

22. Kabisch, N.; van den Bosch, M.; Lafortezza, R. The health benefits of nature-based solutions to urbanization challenges for children and the elderly—A systematic review. *Environ. Res.* **2017**, *159*, 362–373. [CrossRef] [PubMed]

23. Ward Thompson, C.; Roe, J.; Aspinall, P.; Mitchell, R.; Clow, A.; Miller, D. More green space is linked to less stress in deprived communities: Evidence from salivary cortisol patterns. *Landsc. Urban Plan.* **2012**, *105*, 221–229. [CrossRef]

24. Soga, M.; Gaston, K.J. Extinction of experience: The loss of human-nature interactions. *Front. Ecol. Environ.* **2016**, *14*, 94–101. [CrossRef]

25. da Schio, N.; Phillips, A.; Fransen, K.; Wolff, M.; Haase, D.; Ostoić, S.K.; Živojinović, I.; Vuletić, D.; Derks, J.; Davies, C.; et al. The impact of the COVID-19 pandemic on the use of and attitudes towards urban forests and green spaces: Exploring the instigators of change in Belgium. *Urban For. Urban Green.* **2021**, *65*, 127305. [CrossRef]

26. Uchiyama, Y.; Kohsaka, R. Access and use of green areas during the covid-19 pandemic: Green infrastructure management in the "new normal". *Sustainability* **2020**, *12*, 9842. [CrossRef]

27. Haase, D.; Larondelle, N.; Andersson, E.; Artmann, M.; Borgstrom, S.; Breuste, J.; Gomez-Baggethun, E.; Gren, A.; Hamstead, Z.; Hansen, R.; et al. A quantitative review of urban ecosystem service assessments: Concepts, models, and implementation. *Ambio* **2014**, *43*, 413–433. [CrossRef]

28. Kabisch, N.; Frantzeskaki, N.; Pauleit, S.; Naumann, S.; Davis, M.; Artmann, M.; Haase, D.; Knapp, S.; Korn, H.; Stadler, J.; et al. Nature-based solutions to climate change mitigation and adaptation in urban areas: Perspectives on indicators, knowledge gaps, barriers, and opportunities for action. *Ecol. Soc.* **2016**, *21*, 39. [CrossRef]

29. Kabisch, N.; Korn, H.; Stadler, J.; Bonn, A. *Nature-Based Solutions to Climate Change Adaptation in Urban Areas: Linkages between Science, Policy and Practice*; Springer: Amsterdam, The Netherland, 2017.

30. European Commission. EU Biodiversity Strategy for 2030. Available online: https://www.eea.europa.eu/policy-documents/eu-biodiversity-strategy-for-2030-1 (accessed on 10 November 2021).

31. Echenique, M.H.; Hargreaves, A.J.; Mitchell, G.; Namdeo, A. Growing Cities Sustainably: Does Urban Form Really Matter? *J. Am. Plan. Assoc.* **2012**, *78*, 121–137. [CrossRef]

32. Frey, H. *Designing The City: Towards a More Sustainable Urban Form*; E & FN Spon: London, UK, 1998.

33. Cortinovis, C.; Geneletti, D. Ecosystem services in urban plans: What is there, and what is still needed for better decisions. *Land Use Policy* **2018**, *70*, 298–312. [CrossRef]

34. Lemes de Oliveira, F. *Green Wedge Urbanism: History, Theory and Contemporary Practice*; Bloomsbury: London, UK, 2017.

35. Lemes de Oliveira, F. Towards a Spatial Planning Framework for the Re-naturing of Cities. In *Planning Cities with Nature: Theories, Strategies and Methods*; Lemes de Oliveira, F., Mell, I., Eds.; Springer: Amsterdam, The Netherland, 2019; pp. 81–95.

36. Artmann, M.; Kohler, M.; Meinel, G.; Gan, J.; Ioja, I.-C. How smart growth and green infrastructure can mutually support each other—A conceptual framework for compact and green cities. *Ecol. Indic.* **2019**, *96*, 10–22. [CrossRef]

37. Richter, B.; Behnisch, M. Integrated evaluation framework for environmental planning in the context of compact green cities. *Ecol. Indic.* **2019**, *96*, 38–53. [CrossRef]

38. D'Acci, L. A new type of cities for liveable futures. Isobenefit Urbanism morphogenesis. *J. Environ. Manag.* **2019**, *246*, 128–140. [CrossRef] [PubMed]

39. Orsi, F. Centrally located yet close to nature: A prescriptive agent-based model for urban design. *Comput. Environ. Urban Syst.* **2019**, *73*, 157–170. [CrossRef]

40. Hamdan, D.M.A.; Lemes De Oliveira, F. Urban planning and nature: Parametric modelling as a tool for responsive greening of cities. In *Planning for Transition: AESOP 2019 Conference—Book of Papers*; Association of European Schools of Planning: Portsmouth, UK, 2019; pp. 1143–1155.

41. Romero-Duque, L.P.; Trilleras, J.M.; Castellarini, F.; Quijas, S. Ecosystem services in urban ecological infrastructure of Latin America and the Caribbean: How do they contribute to urban planning? *Sci. Total Environ.* **2020**, *728*, 138780. [CrossRef]

42. Costa, S.M.F.D.; Forlin, L.G.; Carmo, M.B.S.; Silva, R.L.D. Study of gated communities in Brazil: New developments and typologies in the Paraíba Valley, SP. *Bol. De Geogr.* **2014**, *32*, 87. [CrossRef]

43. Villaça, F. São Paulo: Urban segregation and inequality. *Estud. Avançados* **2011**, *25*, 37–58. [CrossRef]

44. Polidoro, M.; de Lollo, J.A.; Barros, M.V.F. Environmental impacts of urban sprawl in Londrina, Paraná, Brazil. *J. Urban Environ. Eng.* **2011**, *5*, 73–83. [CrossRef]

45. Torres, H.; Alves, H.; Aparecida De Oliveira, M. São Paulo peri-urban dynamics: Some social causes and environmental consequences. *Environ. Urban.* **2007**, *19*, 207–223. [CrossRef]

46. Yigitcanlar, T.; Velibeyoglu, K.; Baum, S. (Eds.) *Knowledge-Based Urban Development: Planning and Applications in the Information Era*; IGI Global: Hershey, PA, USA, 2008; 23p, ISBN 978-1-59904-720-1.

47. Magdaniel, F.C. *Campuses, Cities and Innovation: 39 International Cases Accommodating Tech-Based Research*; TU Delft, Faculty of Architecture, Department of Management in the Built Environment: Delft, The Netherlands, 2018.

48. Esmaeilpoorarabi, N.; Yigitcanlar, T.; Guaralda, M.; Kamruzzaman, M. Evaluating place quality in innovation districts: A Delphic hierarchy process approach. *Land Use Policy* **2018**, *76*, 471–486. [CrossRef]

49. van Aken, J.E.; Romme, G. Reinventing the future: Adding design science to the repertoire of organization and management studies. *Organ. Manag. J.* **2009**, *6*, 5–12. [CrossRef]

50. Dresch, A.; Lacerda, D.P.; Valle Antunes, J.A., Jr. (Eds.) *Design Science Research: A Method for Science and Technology Advancement*; Springer: Berlin/Heidelberg, Germany, 2015.

51. March, S.T.; Smith, G.F. Design and natural science research on information technology. *Decis. Support Syst.* **1995**, *15*, 251–266. [CrossRef]
52. Lenzholzer, S.; Duchhart, I.; Koh, J. 'Research through designing' in landscape architecture. *Landsc. Urban Plan.* **2013**, *113*, 120–127. [CrossRef]
53. Roggema, R. Research by Design: Proposition for a Methodological Approach. *Urban Sci.* **2017**, *1*, 2. [CrossRef]
54. Nassauer, J.; Opdam, P. Design in science: Extending the landscape ecology paradigm. *Landsc. Ecol.* **2008**, *23*, 633–644. [CrossRef]
55. Sanches, P. *Cidades Compactas e Mais Verdes: Conciliando Densidade Urbana e Vegetação Por Meio do Desenho Urbano*; Universidade de São Paulo: Piracicaba, Brazil, 2020.
56. Madureira, H.; Monteiro, A. Going Green and Going Dense: A Systematic Review of Compatibilities and Conflicts in Urban Research. *Sustainability* **2021**, *13*, 10643. [CrossRef]
57. Kayden, J. *Privetly-Owned Public Space*; Wiley: New York, NY, USA, 2000.
58. Bentrup, G. *Conservation Buffers—Design Guidelines for Buffers, Corridors, and Greenways*; Department of Agriculture, Forest Service: Asheville, NC, USA, 2008.
59. Cook, E.A.; Van Lier, H.N. *Landscape Planning and Ecological Networks: Developments in Landscape Management and Urban Planning*; Elsevier: Amsterdam, The Netherland, 1994.
60. Fernández-juricic, E. Avian spatial segregation at edges and interiors of urban parks in Madrid, Spain. *Biodivers. Conserv.* **2001**, *10*, 1303–1316. [CrossRef]
61. MacGregor-Fors, I.; Schondube, J.E. Gray vs. green urbanization: Relative importance of urban features for urban bird communities. *Basic Appl. Ecol.* **2011**, *12*, 372–381. [CrossRef]
62. Korhonen, J.; Honkasalo, A.; Seppala, J. Circular Economy: The Concept and its Limitations. *Ecol. Econ.* **2018**, *143*, 37–46. [CrossRef]
63. Skar, S.L.G.; Pineda-Martos, R.; Timpe, A.; Polling, B.; Bohn, K.; Kulvik, M.; Delgado, C.; Pedras, C.M.G.; Paco, T.A.; Cujic, M.; et al. Urban agriculture as a keystone contribution towards securing sustainable and healthy development for cities in the future. *Blue-Green Syst.* **2020**, *2*, 1–27. [CrossRef]
64. Ortega-Álvarez, R.; MacGregor-Fors, I. Living in the big city: Effects of urban land-use on bird community structure, diversity, and composition. *Landsc. Urban Plan.* **2009**, *90*, 189–195. [CrossRef]
65. Gyurkovics, J.; Lukovics, M. Generations of Science Parks in the Light of Responsible Innovation. In *Responsible Innovation*; Buzás, N., Lukovics, M., Eds.; University of Szeged, Faculty of Economics and Business Administration: Szeged, Hungary, 2014; pp. 193–208.
66. A+T Research Group. *50 Urban Blocks*; A+T Architecture Publishers: Vitoria-Gasteiz, Spain, 2017.
67. Roggema, R. From Nature-Based to Nature-Driven: Landscape First for the Design of Moeder Zernike in Groningen. *Sustainability* **2021**, *13*, 2368. [CrossRef]
68. Farr, D. *Sustainable Urbanism: Urban Design with Nature*; John Wiley & Sons: Hoboken, NJ, USA, 2011.
69. Gehl, J. *Cities for People*; Island Press: Washington, DC, USA, 2013.
70. Pont, M.B.; Haupt, P.A. *Spacematrix. Space, Density and Urban Form*; NAi Publishers: Rotterdam, The Netherlands, 2010.
71. Hedblom, M.; Andersson, E.; Borgström, S. Flexible land-use and undefined governance: From threats to potentials in peri-urban landscape planning. *Land Use Policy* **2017**, *63*, 523–527. [CrossRef]
72. Salvati, L.; Ranalli, F.; Gitas, I. Landscape fragmentation and the agro-forest ecosystem along a rural-to-urban gradient: An exploratory study. *Int. J. Sust. Dev. World* **2014**, *21*, 160–167. [CrossRef]
73. Johnston, A. The roles of universities in knowledge-based urban development: A critical review. *Int. J. Knowl.-Based Dev.* **2019**, *10*, 213–231. [CrossRef]
74. Hansen, R.; Olafsson, A.S.; van der Jagt, A.P.N.; Rall, E.; Pauleit, S. Planning multifunctional green infrastructure for compact cities: What is the state of practice? *Ecol. Indic.* **2019**, *96*, 99–110. [CrossRef]
75. Mahmoud, I.H.; Morello, E.; Ludlow, D.; Salvia, G. Co-creation Pathways to Inform Shared Governance of Urban Living Labs in Practice: Lessons from Three European Projects. *Front. Sustain. Cities* **2021**, *3*, 690458. [CrossRef]
76. Wagner, J.; Davies, S.; Sorring, N.; Vey, J. *Advancing a New Wave of Urban Competitiveness: The Role of Mayors in the Rise of Innovation Districts*; Brookings: Washington, DC, USA, 2017.
77. Wagner, J.; Watch, D. *Innovation Spaces: The New Design of Work*; Anne, T., Robert, M., Eds.; Bass Initiative on Innovation and Placemaking at Brookings: Washington, DC, USA, 2017.
78. Vey, J.; Hachadorian, J.; Wagner, J.; Andes, S.; Storring, N. Assessing Your Innovation District: A How-to Guide. Brookings Institute. Available online: https://www.brookings.edu/research/assessing-your-innovation-district-a-how-to-guide (accessed on 11 November 2021).
79. Deksissa, T.; Trobman, H.; Zendehdel, K.; Azam, H. Integrating Urban Agriculture and Stormwater Management in a Circular Economy to Enhance Ecosystem Services: Connecting the Dots. *Sustainability* **2021**, *13*, 8293. [CrossRef]
80. Kampelmann, S. Wood works: How local value chains based on urban forests contribute to place-based circular economy. *Urban Geogr.* **2020**, *41*, 911–914. [CrossRef]

 *sustainability*

Article

Guidelines for Citizen Engagement and the Co-Creation of Nature-Based Solutions: Living Knowledge in the URBiNAT Project

Nathalie Nunes [1],*, Emma Björner [2],* and Knud Erik Hilding-Hamann [3]

[1] Centre for Social Studies, University of Coimbra (CES-UC), 3000-995 Coimbra, Portugal
[2] International Organisation for Knowledge Economy and Enterprise Development (IKED), SE-201 22 Malmö, Sweden
[3] Danish Technological Institute (DTI), DK-8000 Aarhus, Denmark; khi@teknologisk.dk
* Correspondence: nathalienunes@ces.uc.pt (N.N.); emma.bjorner@iked.org (E.B.)

Abstract: Participation and citizen engagement are fundamental elements in urban regeneration and in the deployment of nature-based solutions (NBS) to advance sustainable urban development. Various limitations inherent to participatory processes concerning NBS for inclusive urban regeneration have been addressed, and lessons have been learnt. This paper investigates participation and urban regeneration and focuses on the development of guidelines for citizen engagement and the co-creation of NBS in the H2020 URBiNAT project. The methodology first involves the collection of scientific and practical input on citizen engagement from a variety of stakeholders, such as researchers and practitioners, to constitute a corpus of qualitative data. This input is then systematized into guideline categories and serves as the basis for a deeper analysis with researchers, experts, and practitioners, both inside and outside URBiNAT, and in dialogue with other cases of participatory NBS implementation. The results highlight an 'ecology of knowledges' based on a 'living' framework, which aims to address the specific needs of various segments of citizens and to match citizen engagement to the participatory cultures of cities. Implications and further research are also discussed, with a special focus on the implementation of NBS. The conclusions broaden the research context to include the refinement of the NBS approach, with participation being seen as both a means and an end.

Keywords: guidelines; citizen engagement; co-creation; nature-based solutions; participation; urban regeneration; living knowledge; URBiNAT

Citation: Nunes, N.; Björner, E.; Hilding-Hamann, K.E. Guidelines for Citizen Engagement and the Co-Creation of Nature-Based Solutions: Living Knowledge in the URBiNAT Project. *Sustainability* **2021**, *13*, 13378. https://doi.org/10.3390/su132313378

Academic Editors: Israa H. Mahmoud, Eugenio Morello, Giuseppe Salvia and Emma Puerari

Received: 22 October 2021
Accepted: 28 November 2021
Published: 3 December 2021

Publisher's Note: MDPI stays neutral with regard to jurisdictional claims in published maps and institutional affiliations.

1. Introduction

The European Union (EU) has invested in research and innovation on nature-based solutions (NBS) in order to promote sustainable urban development and contribute to an evidence-based framework. The Horizon 2020 Programme on research and innovation (2014–2020) included an area dedicated to societal challenges, and financed, among other things, the implementation of innovative NBS in cities for inclusive urban regeneration, and, in particular, the regeneration of deprived urban districts [1]. These are districts which are often characterized by the presence of derelict infrastructure, environmental pollution, low employment rates, and high levels of urban poverty [2,3].

NBS is still a novel concept that is under development, and it is one that offers both challenges and opportunities [4]. As a new concept, it generates uncertainty because of a lack of operational and technical preparedness. Yet, it also brings with it possibilities for deploying new ways of addressing old problems, new and innovative approaches, and practices that are more inclusive [5]. NBS have been defined as solutions that make use of ecosystem services and nature to provide environmental, social, and economic benefits [6–8]. NBS have the potential to generate benefits for citizens and other stakeholders in urban areas in a multifunctional way and at many different levels, from providing services, regulating and maintaining ecological balance, and generating cultural, social, and economic benefits. NBS can generate positive outcomes, such as cooling, heat avoidance, opportunities for exercise, and gathering points for citizen interaction.

NBS may, however, also have negative implications, such as allergic reactions, a sense of insecurity, or spots where rubbish is offloaded or targeted for vandalism. Implementing and managing NBS is, therefore, a complex and difficult process [9]. A recent state-of-the-art publication focusing on NBS in EU-funded projects points to other critical issues regarding the transformative potential of NBS. This relates particularly to issues of social justice, as well as to a growing body of research that suggests the potential for the exacerbation of inequalities and results that are incompatible with the objectives of sustainable communities [10].

There is widespread consensus that the participation and involvement of citizens are necessary for the planning of nature-based adaptations [8]. Citizen involvement is said to increase fairness, relevance, acceptance, and sustainability [11–13]. Co-creation procedures and polycentric governance, with the inclusion of a variety of stakeholders, are also seen as more effective in the management of public assets. NBS projects are also said to benefit from collaborative governance models, something that the EU also greatly encourages [14].

Despite a generalized agreement that citizen involvement is both necessary and positive, empirical evidence showing that it supports NBS in ensuring a transformative and continuous change in cities is scarce [15–17]. Xiang, Yang, and Li [18] also argue that there is a dearth of research on what 'features' the regeneration projects that reflect inclusion should have, and how urban regeneration should be implemented. The authors are proposing a concept of 'inclusive urban regeneration', combining NBS with society-based solutions. Central to this concept is the notion of effective and sustainable forms of public participation in inclusive urban regeneration [19]. Thus, inclusive urban regeneration is a topic that has been deemed worthy of further investigation [18].

Frantzeskaki [20] discusses several points when it comes to participation and NBS in cities. Trust is emphasised as important in every participatory process that has to do with policy, planning, and experimentation. Clarity, transparency, and openness are also central to the trust-building process. Furthermore, diversity and learning from social innovation are emphasized as central to the co-creation of NBS, and Frantzeskaki [20] also states that an inclusive narrative can enable the integration of many urban agendas.

Furthermore, face-to-face communication has been identified as the greatest factor in increasing the likelihood of cooperation [21,22]. The Internet and modern communication technology have created the potential for blending the advantages of face-to-face interaction with online communication [21,23]. Social media also offers benefits by delivering synchronous and interactive communication between governments and citizens, bringing new impetus to citizen engagement [23–26]. To achieve successful citizen engagement, it is also important to be aware of the tools citizens use, including digital tools. In general, people marginalized by income and education more often depend on a smartphone than a PC for Internet access [27].

The European Commission report on the state of the EU-funded NBS projects also raises critical issues concerning participation and inclusion [10]. It warns of the limitations of co-design and co-production processes, such as in the case of participatory methods that are exploitative and that legitimize solutions that provide little contribution to the needs and ambitions of the communities [10].

To summarize the discussion so far, participation and citizen engagement are perceived as elements that are fundamental to urban regeneration and to the deployment of NBS for the advancement of sustainable urban development. When it comes to participation and NBS in cities, some ideas have been suggested (e.g., [20]). There is, however, a lack of evidence on how to arrive at successful citizen engagement, given the various limitations of the participatory processes using NBS for inclusive urban regeneration that have been looked at [10]. All in all, there is a scarcity of research on participation and citizen engagement concerning NBS and urban regeneration [8].

The approach taken in Horizon 2020 holds that solutions based on nature can regenerate disadvantaged neighbourhoods, for example, by reducing urban violence and social tensions through better social cohesion [1]. It also promotes the adoption of a project model in which cities are given a role in order to enable them to facilitate the rapid exploration, replication, and scaling up of solutions. Moreover, solutions could benefit from being co-designed, co-developed, and co-implemented in a transdisciplinary multistakeholder participatory context, involving a variety of stakeholders, such as residents, local authorities, community groups, companies, academics, and local communities [1].

Participation is, therefore, a critical and challenging research and policy agenda for the European Commission within the framework of EU-funded NBS projects [10], of which many have been launched since 2016. These projects have been contributing to the knowledge production around citizen engagement and the co-creation of NBS in different contexts, and have been both theoretical and practical, taking the form of research as well as the development of different tools, such as handbooks (e.g., [28]) and knowledge platforms (e.g., [29]). Moreover, the European Commission has been promoting clustering activities to maximize the impact of the Horizon 2020 and Horizon Europe programmes, bringing together EU-funded NBS projects by means of task forces in order to explore the development of joint guidelines on the co-creation of NBS [30].

This paper investigates the development of guidelines for citizen engagement and the co-creation of NBS in the URBiNAT project. It looks at the 'ecology of knowledges', as termed by Boaventura de Sousa Santos [31,32], that has been emerging inside and outside the project, and works towards the creation of a 'living' framework. This 'living' framework will be used to address citizen engagement and NBS co-creation on the basis of local and specific priorities, contexts, and challenges, thereby ensuring that the specific needs of the various segments of citizens are addressed, and matching citizen engagement to the participatory cultures of cities. In this regard, as developed by Ferreira in conceptualizing URBiNAT methodology to map local participatory cultures, it is not only about the formal participation of citizens in urban governance, but also about the participation of citizens in other kinds of collective initiatives, in a diversity of formats, both physical and digital [33,34]. It is important to understand how participation works locally, as well as to assess the challenges and opportunities involved in the engagement of citizens and stakeholders as a baseline for public liveability in neighbourhoods and the design of participatory processes [35]. Identification of the conditions needed for active, positive, and ethically sound participation paves the way for new experiments with different features and effects, as well as for finding specific anchors in the social fabric and its institutions.

URBiNAT is a project funded within the Horizon 2020 programme, centring on innovative NBS in cities for inclusive urban regeneration. Its acronym stands for 'URBan Inclusive and Innovative NATure'. The five-year project (2018–2023) focuses on the urban regeneration of deprived city neighbourhoods through the co-creation of healthy corridors made up of a combination of NBS. Central to the project is the co-creation of NBS within and between different neighbourhoods, working together with citizens and other stakeholders. Having the physical, mental, and social well-being of citizens as its main goal, URBiNAT aims to co-diagnose, co-design, co-implement, and co-evaluate healthy corridors in the form of innovative and flexible NBS, integrating several 'micro NBS'. Healthy corridors are being co-created in the European cities of Nantes (France), Porto (Portugal), Sofia (Bulgaria), Siena (Italy), Nova Gorica (Slovenia), Brussels (Belgium), and Høje-Taastrup (Denmark). Moreover, through its Community of Practice, which includes non-EU observers, work carried out in URBiNAT is also being followed in Iran, Brazil, China, Oman, Japan, and Cyprus [36].

The URBiNAT consortium is coordinated by the Centre for Social Studies of the University of Coimbra (CES-UC), based in Portugal. The CES-UC is a scientific institution focused on research and advanced training in the social sciences and humanities, through an inter- and transdisciplinary approach, and with a particular focus on the North–South and South–North dialogues [37]. This reinforces how the URBiNAT Community of Practice incorporates social sciences together with the diversity of its consortium partners to better understand and contextualize the development of solutions, given that the involvement of the social science and humanities disciplines in dialogue with other disciplines is required to adequately deal with complex societal challenges, thus fostering social innovation [1].

URBiNAT is of special relevance and interest in the context of the present study for three main reasons. The first reason is tied to the novel approach of URBiNAT to NBS, with its focus on four types of NBS: Territorial NBS, Technological NBS, Social and Solidarity Economy NBS, and Participatory NBS, with a particular focus on the inclusion of the last two. The inclusion of Participatory NBS as a category of NBS in its own right is based on the assumption made in URBiNAT that participatory activities create various benefits for citizens and other stakeholders, improve collaboration within communities, and empower individuals in the decision-making process. Participatory NBS are seen as both a means (to develop the co-creation process) and an end (helping to activate citizenship) [33].

Secondly, URBiNAT is of interest because extensive work has been carried out on exploring and identifying the categories of the significant factors impacting citizen engagement in urban regeneration and the co-creation of NBS, leading to the development of guidelines. Thirdly, URBiNAT is relevant and interesting in the context of the present study because of its focus on deprived urban districts, where the need for inclusive urban regeneration is especially acute [1], and where participation and co-creation come with certain hurdles. The NBS developed in URBiNAT address local issues in deprived urban areas, such as poverty and unemployment, problems with health, crime and vandalism, cultural differences and conflicts, and low-quality housing and infrastructure [38]. URBiNAT has been focusing on the specificities of these neighbourhoods, with reference to ethics requirements and a rights-based approach [33,39].

This paper is structured in the following way: the next section, 'Materials and Methods', describes how the guidelines for citizen engagement and NBS have been developed. The results are then presented as an overview of the guidelines around key categories, and how they add value as a 'living' framework, namely, concerning the emergence of learning points for the co-creation of NBS. Subsequently, in the 'Discussion' section, the results are evaluated in relation to previous research, and implications and further research are considered. A particular focus is given to the development of the understanding of the different dimensions for designing and implementing NBS, especially in times of COVID-19, and to addressing issues of exclusion. Lastly, in 'Conclusions', the research context of the NBS approach is widened, with participation being looked at as both a means and an end.

2. Materials and Methods

2.1. Initial Input on Citizen Engagement

URBiNAT sees inclusive urban regeneration as a process that goes beyond the usual practice of urban planning, working within a collaborative framework to bring together many different stakeholders, including citizens, municipal officials and public servants, researchers, practitioners, local partners, and other stakeholders. As a result, it comprises a diversity of sectors, both public and private, as well as a third sector that includes public services, local businesses, social enterprises, voluntary organisations, grassroots movements, collective initiatives, and local civil society [33]. To go beyond the 'usual suspects' in urban planning (i.e., those individuals, groups, and associations/organizations who always participate, and who are more engaged in terms of availability and professional/disciplinary skills) means to fundamentally focus on the engagement of the widest possible range of citizens, at all stages of the co-creation process, to identify assets, needs, challenges, opportunities, and ambitions. This is conducted in order to design and implement solutions provided by citizens, encouraging them to take ownership, and then monitoring the results and impacts of their actions.

URBiNAT promotes a type of regeneration in which participation is fundamentally valuable as a process in its own right. The aim is to activate citizenship, in the sense of empowering people within their own demo-diversity or within different democratic models and practices [40], enabling them to choose solutions more adjusted to their diverse interests, agendas, and needs [33]. Participation is regarded both as a means of achieving the objectives of co-creating solutions, and as an end in an ongoing process based on the development of the participants' capacities to engage in collective initiatives and expand their role as active citizens [33].

At the heart of inclusive and innovative urban regeneration, participation is included in the URBiNAT NBS catalogue in the form of Participatory NBS [41]. The URBiNAT catalogue challenges the conventional definitions of NBS with the integration of solutions inspired by nature, such as Territorial and Technological solutions, comprising products and infrastructures, together with Participatory solutions and Social and Solidarity Economy solutions, comprising processes and services [42]. Material and immaterial solutions are presented together in the URBiNAT NBS catalogue, with the aim of balancing their position in the public space and pushing forward the perception that material solutions do not produce urban regeneration without immaterial solutions [33,43]. It is also a symbolic statement and a reminder to attribute the same relevance to both in terms of investments in time, energy, and budget [33].

If Xiang, Yang, and Li are proposing a concept of 'inclusive urban regeneration', combining NBS with society-based solutions [18], URBiNAT, in turn, inserts participation and the social and solidarity economy into the framework of NBS in search of societal harmony with nature. Conceptualizing participation as NBS gives it visibility as a fundamental natural solution to the reintegration of nature in the public sphere and space by overcoming the artificial separation of humans and nature [44]. The social and solidarity economy is one of the pillars of the URBiNAT approach to urban regeneration and the identification of new partnerships and forms of financing, as well as to the introduction of the innovation cycle in order to generate new products in response to concrete social problems [45]. It also requires an understanding and uptake of the meanings and opportunities of solidarity networks as a new way of producing, consuming, and living in which solidarity is at the heart of life, promoting development that is ecologically sustainable, socially just, and economically viable [46].

Citizen engagement and the community-driven processes implemented within the project are aimed at both contributing to the enablement of solutions for the design of the urban fabric, as well as constituting, in themselves, solutions that aim for social cohesion, as defined by Manca [47]. Participatory NBS are solutions that contribute to the fostering of social cohesion, being related to connectedness and solidarity among groups in society. They are also social processes designed to enable social cohesion that, in turn, makes room for the plurality of citizenship. Moreover, co-creation in URBiNAT embodies a strategic participatory approach aimed at tackling inequality, socioeconomic disparities, and fractures in society. The same approach applies to Social and Solidarity Economy solutions.

The URBiNAT NBS catalogue is based on an initial collection of several NBS, in accordance with the knowledge, expertise, and research carried out by URBiNAT partners. The first version of the catalogue was compiled during the project proposal phase but has since evolved from the launch of the project and throughout the co-creation process. It constituted an initial input and inspiration for co-creation with citizens and stakeholders, and it currently constitutes a 'living' NBS catalogue that has grown, is growing, and will continue to grow during the implementation of the project [44]. Being 'living' is an essential characteristic of the URBiNAT NBS catalogue as it needs to be flexible and adaptable to the different features of local physical and sociocultural contexts, as well as to the needs and wishes of citizens, including the solutions proposed by them [44]. The URBiNAT NBS catalogue is, therefore, subject to ongoing review in order to take account of the developments in the field. The results of the co-creation process, from the engagement to the co-design, co-implementation, and co-monitoring are evaluated, as they depend on learning and feedback activities so that improvements can be implemented [44].

In the first few months of the project, URBiNAT partners brought together different perspectives, expertise, and experiences in the academic, technical, and political fields to establish the theoretical and methodological foundations of the project, including in the area of citizen engagement. The partners, members of URBiNAT advisory boards, as well as other researchers and practitioners, were invited to participate in their specific areas of expertise in a set of webinars centring on the topics that shape the main pillars of the URBiNAT approach to urban regeneration. The webinars were followed by written contributions from speakers, reflecting their views and taking into account the discussions following their presentations. They were also asked to include guidelines in their written contributions to support the finetuning of a reference and methodology framework to guide the community-driven processes in URBiNAT.

Contributions on citizen engagement focused on: the relationship between citizenship rights and inclusive, active, and culturally diverse participation in processes of urban regeneration; participation in the practice of cities, contextualized within the framework of their urban governance; co-creation processes, platforms, and tools to support them, and the role of co-creation versus co-production; the participation of the private and third sectors in the lifetime of NBS; and the monitoring and evaluation of the participatory and co-creative processes [48].

2.2. Systematization into Categories of Significant Factors Impacting Citizen Engagement

On the basis of these initial written contributions on citizen engagement, the members of the URBiNAT working group on participation have been further exploring the combinations of expertise of the consortium partners, which constitute an initial corpus of qualitative data, through a two-step process [33].

The first step consisted of extracting and organizing guidelines into categories for participation in urban regeneration processes as a qualitative data analysis in order to explore meanings, both manifest and latent [49]. The contents and contributors were referenced in each category. In methodological terms, within the framework of the Step 1 (extract and organize) codification processes [50,51], a process of continuous inductive reconfiguration took place upon analysis of the contributions on citizen engagement. The objective of the analysis performed was to map the patterns and characteristics common to various types of content produced by diverse actors (practitioners, researchers, and municipal staff), where participation is the subject of specific practices, projects, and policies.

This resulted in the identification of the strategic and operational dimensions of the guidelines, which were grouped into twenty initial categories: Citizenship rights; Innovation cycle; Regulation; Governance; Inclusion; Trust; Co-production; Cultural Mapping; Behavioural changes; Intensity and levels of participation; Communication and interaction; Facilitation; Transparency; Quality of deliberation; Where; When; Supportive methodologies and techniques; Integration of the results of participatory processes; Private sector; and Monitoring and evaluation. On the basis of the results of the inductive analysis, a review was carried out in the second step. It consisted of reviewing and aggregating the guidelines by identifying and elaborating on the overlaps and contradictions, as well as identifying missing elements and raising additional aspects to be further explored in some of the categories. It also resulted in the addition of the category: Risk assessment and mitigation measures.

Step 2 also reviewed the inclusion of ethics as a requirement of the research and participatory activities involving people, and the human rights and gender considerations as cross-cutting dimensions of the project. Both the ethics requirement and the human rights and gender considerations—cross-cutting dimensions in URBiNAT—underpin citizen engagement and the co-creation of NBS, which can contribute to tackling a complex combination of societal challenges in the context of urban regeneration [33,39]. Indeed, a multiplicity of cultural and socioeconomic aspects are present in the URBiNAT neighbourhoods, including vulnerable individuals and groups. These are identified throughout the co-creation process, so that the participatory activities can make use of strategies tailored to the specificities of people and groups, as referred to in the URBiNAT Code of Ethics and Conduct [33,52].

In its principle guidelines on ethics, URBiNAT defined specificities according to childhood, gender (including gender minorities/diversity), older adults, race and ethnicity, functional diversity, citizenship status (migrant/refugee/asylum-seeker conditions), and religious diversity [52]. URBiNAT has also established a framework for a rights-based approach on the basis of the preliminary guiding principles, with special attention to gender analysis, to be integrated into all phases of the project activities. These range from planning, formulation, and implementation all the way to assessment [39,53,54]. The URBiNAT framework for a rights-based approach comprises the following guiding principles: (i) People as citizens; (ii) Full citizenship; (iii) Applying all rights; (iv) Participation and access to the decision-making process; (v) Nondiscrimination and equal access; (vi) Inclusivity; (vii) Accountability; (viii) Transparency and access to information; and (ix) 'do no harm' [39,54]. The URBiNAT ethical code of conduct for the communication and dissemination of activities mirrors these principles [55].

In the systematization of the significant factors impacting citizen engagement and the corresponding development of guidelines, the URBiNAT working group on participation has not created specific categories with regard to ethical or human rights and gender issues. Instead, it has introduced references to ethics requirements and a rights-based approach throughout the guideline categories, in line with the URBiNAT cross-cutting approach to making these issues integral to the project. In practical terms, the cross-cutting dimensions are theoretical lenses, guiding principles, and methodological frameworks to be adopted by URBiNAT partners and stakeholders for all activities, and across all work packages [39].

2.3. Sharing and Learning with Practitioners from the Field, Inside and Outside URBiNAT, towards Living Knowledge

In addition to being based on the experiences of the diverse partners that make up its consortium, the methodological and practical developments achieved by URBiNAT are advanced by the sharing of best practices and knowledge, particularly relative to URBiNAT cities and their specific situations. The analysis, carried out as an internal exercise relative to Steps 1 and 2, was followed by a broadening dialogue in order to enrich the gathering of input and the feeding and reframing of the categories and the related guidelines. It served as a basis for a deeper analysis with researchers, experts, and practitioners in the field of participation, both inside and outside URBiNAT.

This dialogue began on the occasion of an external workshop conducted by the URBiNAT working group on participation during the Open Living Lab Days (OLLD) in September 2019, organized by the European Network of Living Labs (ENoLL) in Thessaloniki (Greece). The OLLD is an annual event that brings together the global Living Lab community, made up of public officials, companies, entrepreneurs, academics, living lab representatives, and innovators [56]. The participants of the workshop were asked through an online polling tool (Mentimeter) to rank the URBiNAT categories. This first prioritization enabled the URBiNAT working group on participation to direct the discussion and developments of the review to specific categories, as these were the most critical aspects of citizen engagement in different contexts.

Subsequently, the URBiNAT working group on participation shared the categories that are based on the guidelines and its ranking, which emerged from the workshop held at the OLLD in a series of three internal interactive sessions/online meetings by means of discussion, experience sharing, and feedback. The invited participants have deep knowledge and sound expertise gained from citizen participation in the intervention areas of the project, as they are either employees of the municipality, or of the agency in charge of the management of social housing, or they work for nonprofit organisations. The following questions framed the discussions: What is missing? What is most relevant to your practice? Why? For what kind of situation? For what kind of people? How is it useful? How does it relate to your city? What are the main challenges you experience concerning these categories? What best practice could you share? This exchange resulted in support for the refinement of categories, including the addition of four more categories: Ownership; Culture of participation; Why participation; Mediation.

Moreover, if the experiences of URBiNAT cities can inspire each other through networking and training, other means are also being explored for further inspiration through both large and small participatory cases. Since January 2020, the leader of the URBiNAT working group on participation has been promoting internal exchanges around interesting participatory cases, both from URBiNAT cities as well as from additional examples from other cities, where citizens take an active role in making their city or district a better place to live. The samples were chosen on the basis of their successfully achieving a broad and diverse engagement of citizens and stakeholders in NBS co-creation, and their relevance for deprived neighbourhoods in cities. Furthermore, cases were selected that fit into the four categories of NBS used within URBiNAT. For URBiNAT and professionals working with citizen engagement, it is an interesting study into what can be achieved when citizens and other stakeholders work together to create new opportunities and a better future for their community.

More specifically, the good practice study of citizen participation, based on already established NBS solutions, includes the mapping of more than 100 examples of best practices that all have the potential to create positive results in deprived neighbourhoods. These examples were presented on the URBiNAT online work platform in a blog that provides the URBiNAT community with the opportunity to comment and reflect. From the discussion promoted internally on the URBiNAT online work platform, several preliminary learning points are emerging that can be linked to the guideline categories for citizen engagement.

Finally, the URBiNAT project has joined the framework of the European Commission's clustering activities to maximize the impact of the Horizon 2020 and Horizon Europe programmes, and, since 2020, has been leading a task force on co-creation and co-governance (Task Force 6), engaging EU-funded NBS projects launched since 2016. With the support of NetworkNature, one of the aims of this task force is to co-develop joint guidelines on the co-creation and co-governance of NBS [30].

The collective and participatory pathway to knowledge production is described in the section of 'Materials and Methods' that accounts for the development of living knowledge, which emerges from an ecology of knowledges both inside and outside URBiNAT's consortium, supported by a qualitative focus [32]. The ecology of knowledges promotes the active coexistence of knowledges with the assumption that all of them, including scientific knowledge and the knowledge of other practitioners, can be enriched through dialogue. Consequently, this process is not only a means to, but also an end for engaging different actors, knowledges, and experiences.

The ecology of knowledges is tied to the intercultural translation of the diversity of knowledges that emerges from sharing and learning [32]. If the ecology of knowledges identifies the main bodies of knowledge that might highlight important dimensions, it must be completed by an intercultural translation, which is aimed at enhancing reciprocal intelligibility without dissolving identity. In practical terms, this means identifying complementarities and contradictions, common grounds and alternative visions, and developing new hybrid forms of cultural understanding and intercommunication [32,57], as living knowledge to be further developed with citizens and other stakeholders, including the NBS concept itself.

3. Results

The following subsections start with an overview of the 25 guideline categories that emerged from the systematization and review. This framework addresses participation in urban regeneration, particularly with regard to NBS co-creation projects, such as URBiNAT.

It highlights the significant factors impacting citizen engagement—the core leverages/key ingredients for successful citizen engagement.

Secondly, we elaborate on the relevance and added value of a 'living' framework of guidelines for citizen engagement and the co-creation of NBS, based on sharing and learning with practitioners from the field, which reveals a variety of priorities according to the situation and the diversity of practitioners. This process of sharing and learning also reveals the need to combine and tailor the categories of the guidelines to ensure that the specific needs of the various segments of citizens are addressed, and to match the activities of citizen engagement to the participatory cultures of cities.

Thirdly, we have analysed the participatory implementations of NBS with relevance to deprived areas and have integrated the core leverages for citizen engagement with the learning points that emerged. This reveals that sharing and learning from different contexts may inspire further developments of the categories and the corresponding guidelines, such as those concerning challenges, lessons learned, and best practices. This is also in line with the aim of the clustering activities with other EU-funded NBS projects, as the research, testing, and validation of URBiNAT guidelines will continue within the task force dedicated to co-creation and governance.

3.1. Overview of Guideline Categories for Citizen Engagement

Table 1 gives an overview of the initial URBiNAT framework, which addresses participation in the processes of urban regeneration. This framework highlights the significant factors impacting citizen engagement—the core leverages/key ingredients for successful citizen engagement—as well as the interrelations between them. Further details are given in Table A1 of Appendix A, which includes the combination of a strategic overview with operational details, and more information on practice-related impacts. The guideline categories that emerged from the systematization and two-stage review process are marked in light grey in the case of Step 1—extracting and organizing—and in medium grey for Step 2—reviewing and aggregating. The additional categories resulting from sharing and learning with practitioners from the field are in dark grey. The details presented in Table A1 of Appendix A arose from an analysis of the results of both the systematization and the two-stage review process, and the sharing and learning with practitioners from the field, both inside and outside URBiNAT.

Table 1. Overview of the initial URBiNAT framework on guideline categories addressing core leverages for successful citizen engagement in the co-creation of nature-based solutions (NBS).

Categories of Guideline [1]	Prioritization [2]	Overview of the Categories	Impact of Other Categories
Communication and interaction	1	Communicating specificities for interacting with citizens.	Trust
Behavioural changes	2	Instigating behavioural adjustments, or changes in behaviour, in some particular respect.	Communication and interaction
Trust	3	Improving or creating relationships of trust between citizens, and between citizens and city staff, politicians, and other agents.	Transparency, Inclusion, Communication and interaction, Governance
Co-production	4	Stimulating and improving the co-production of public services, participatory processes, and product development.	Trust and Behavioural change
Inclusion	5	Having specific guidelines to guarantee the inclusion of diversity.	Citizenship rights, Governance, Transparency, Regulation
Regulation	5	Clarifying rules and regulations for equal rights in the expression of visions and priorities.	Governance, Transparency, Trust

Table 1. *Cont.*

Categories of Guideline [1]	Prioritization [2]	Overview of the Categories	Impact of Other Categories
Governance	6	Balancing interactions among citizens, city staff, politicians, and other agents.	Trust, Transparency, Culture of participation
Innovation cycle	7	Adopting processes of rupture and searching for alternative solutions in order to address concrete social problems.	Citizenship rights
Transparency	8	Arguments for encouraging efforts to act in a transparent manner.	Trust, Governance, Why participation, Monitoring and evaluation
Intensity and levels of participation	8	Setting different approaches and levels of participation depending on the goals and real conditions for participation.	Culture of participation
Citizenship rights	9	Broadening the meaning of the appropriation of social, urban, political, and cultural rights, both internally in the collective imagination, and externally in rejuvenated relationships with local powers.	Inclusion
Cultural mapping	10	Articulating and making visible the multilayered cultural assets, aspects, and meanings of a place.	Inclusion, Innovation cycle, Supportive methodologies and techniques
Facilitation	-	Having specific guidelines to address facilitation that include other participatory guidelines.	Supportive methodologies and techniques
Quality of deliberation	-	Setting a meaningful deliberation process.	Regulation, Governance, Citizenship rights, Facilitation
Where	-	Having guidelines for the spaces in which the participatory events are held.	Communication and interaction, Facilitation
When	-	Identifying the best moment for the participatory events.	Communication and interaction, Facilitation
Supportive methodologies and techniques	-	Using specific methodologies and guidelines to support mobilization and inclusivity.	Communication and interaction, Facilitation, Cultural mapping
Integration of the results of participatory processes	-	Enlarging the scope of co-creation to validate the ideas developed.	Communication and interaction, Facilitation, Supportive methodologies and techniques
Private sector	-	Mapping the relevant private sector actors with interests in, and input to, the NBS targeted area.	Co-production, Innovation cycle
Monitoring and evaluation	-	Addressing the monitoring and evaluation of the participatory processes.	Transparency, Ownership
Risk assessment and mitigation measures	-	Identifying the factors influencing the co-creation processes, as well as those leading to the failure of co-creation and co-production.	Monitoring and evaluation, Transparency
Ownership	-	Citizens having ownership of both problems and solutions.	Trust, Communication and interaction
Culture of participation	-	Enabling regular interaction with citizens, and increasing the culture of participation.	Governance, Intensity and levels of participation
Why	-	Being clear as to why we need to engage citizens and support participatory processes.	Transparency, Intensity and levels of participation
Mediation	-	Dialogue and collaboration.	Communication and interaction, Trust, Facilitation

[1] Categories that emerged from both the systematization and two-stage review process of URBiNAT guidelines for citizen engagement, and the sharing and learning with practitioners from the field in URBiNAT cities: categories that emerged from Step 1 of systematization (extract and organize) are in light grey; those from Step 2 (review and aggregate) are in medium grey; additional categories that resulted from sharing and learning are in dark grey. [2] Prioritization of guideline categories by members of the global Living Lab community, in which the 10 most critical aspects of citizen engagement are ranked. This was the result of an external workshop conducted by the URBiNAT working group on participation during the Open Living Lab Days in September 2019, organized by the European Network of Open Living Labs (ENoLL) in Thessaloniki (Greece).

Moreover, the right-hand columns in Table 1 below, and Table A1 of Appendix A, account for the practice-related impacts in the sense of interdependent categories. They highlight a greater focus on a specific category or specific categories while not excluding the others, whose relevance and connection depend on the practical context and stage of citizen engagement. Further practical-related impacts can emerge and complement this initial framework as living knowledge.

Prior to conducting an interactive review with practitioners from the field in URBiNAT cities, a workshop was held with members of the global Living Lab community—practitioners from outside the project. This resulted in the 10 most critical aspects of citizen engagement, in the form of guideline categories and ranked in order of priority, as shown in the second column of Table 1:

- At the top of the ranking are: (1) Communication and interaction (16% of participants suggested this category), (2) Behavioural changes (14%), and (3) Trust (12%);
- In the intermediate position are: (4) Co-production (10%), (5) Inclusion (9%), as well as visions and priorities, i.e., Regulation (9%), and (6) Governance (8%);
- The lowest ranking categories include: (7) Innovation cycle (6%), (8) Transparency (5%), as well as the levels and conditions of participation, i.e., Intensity and levels of participation (5%), (9) Citizenship rights (4%), and (10) Cultural mapping (2%);
- Categories not scored/prioritized by participants include: Facilitation, Quality of deliberation, Where, When, Supportive methodologies and techniques, Integration of the results of participatory processes, Private sector, and Monitoring and evaluation.

This initial framework of 25 guideline categories continues to be used as a basis for deeper analysis within the project, alongside researchers, experts, and practitioners in the field of participation, both inside and outside URBiNAT, such as in the cases of clustering activities with other EU-funded NBS projects and other participatory cases.

3.2. Relevance and Added Value of 'Living' Guidelines for Citizen Engagement and NBS Co-Creation

Beyond the addition of more categories (Ownership, Culture of participation, Why participation, Mediation), sharing and learning with practitioners from the field in URBiNAT intervention areas resulted in a feeding into and a reframing of the categories and their guidelines on the basis of the following questions: (i) What—what is missing and what is most relevant, including challenges and best practices? (ii) Why—the reasons behind what, related to types of situations and people; (iii) How—how it is useful and related to the specific context of cities?

During the internal interactive sessions, these practitioners confirmed the importance of the categories at the top of the ranking made by members of the Living Lab community, and they also highlighted that some are more critical to their specific local context. This is the case of Communication and interaction, as even though all interaction is digital now, what works is very local and hands-on (e.g., circulating sheets of paper and putting up posters is more appropriate than digital tools in some contexts, or the use of digital tools/social media, such as Facebook and WhatsApp, is limited to the incentivizing of doing stuff together) [23]. Moreover, beyond confirming that all categories are interconnected in the engagement of citizens, the practitioners from the field in the URBiNAT cities also indicated that some of the categories are more specifically intertwined, which may also correspond to local specificities, such as in the case of Transparency and Trust.

Some of the categories may not have been prioritised because they can be characterised as subsets of other more comprehensive guideline categories. As such, Facilitation, and Quality of deliberation can be seen as subsets of the more comprehensive Communication and interaction. This is also perhaps the case for Supportive methodologies and techniques. The Where and When are relevant for certain specific NBS projects, where the timing and location of the engagement process are particularly important. The Integration of participatory process results and Monitoring and evaluation concern the reflections on what needed to improve. Perhaps practitioners, in particular, are focusing on guideline categories that activate citizens and less on the overall impact. Furthermore, NBS development is primarily seen, rightly or wrongly, as a co-creation task between city administrators and citizens. However, many examples have shown that, when these two groups are joined by the private sector and associations, significant value can be added to the process and outcomes.

Another important contribution from the interactive review with practitioners in the URBiNAT cities relates to how guideline categories can be combined and tailored to address the specific needs of the various segments of citizens, with the aim of bringing people together. While certain categories are at the core of citizen engagement, others represent methods/tools/ways of improving citizen engagement, and still others take the form of the preconditions/elementary conditions that enable citizen engagement. This is illustrated in Figure 1 below, which shows the interconnections and ties between the categories of guidelines. However, this also means that general guidelines cannot be applied to citizen groups with different needs, for whom certain elementary conditions are not met.

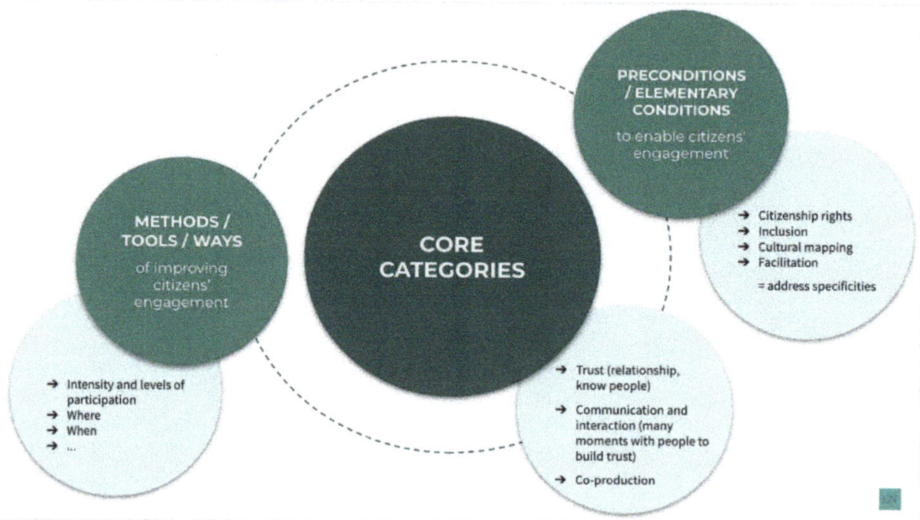

Figure 1. Interconnecting and organizing categories of guidelines to address the specific needs of the various segments of citizens.

As pointed out by practitioners in the URBiNAT cities, the local application of guidelines needs to be organized in accordance with specific challenges, namely, with reaching specific segments of citizens and keeping engagement simple and close to people's daily lives. This is all the more relevant in the case of NBS co-creation for deciding where and how to conduct specific interventions. Examples include the diagnosis and identification of problems, going beyond the sole intervention of the public sector or the focus on bureaucratic aspects.

Further research is therefore needed to deepen the categorization of the guidelines and to understand the timing and contexts involved in their application. This is particularly pertinent relative to the participatory cultures of cities and their specific challenges. The tailoring of participatory methods is informed by the local culture of participation, i.e., how participation works locally, by understanding and identifying the potential participants in co-creation, and by being able to assess the challenges and, especially, the opportunities for the engagement of citizens and stakeholders [33,35]. These specificities cannot be captured in a generic consultation, such as the one conducted with members of the Living Lab community through the ranking of the 10 most critical aspects of citizen engagement. It would need to be informed by research work on the local culture of participation, not only to look at the formal participation of citizens in urban governance, but also to consider the participation of citizens in other kinds of collective initiatives—ones that contribute to finding specific anchors in social fabrics and institutions, paving the way to new experiments with different features and effects, and which must also be included. For example, URBiNAT has mapped the local participatory culture in order to devise community-driven processes, which are the results of collectively designed strategies [33–35]. The research work consists of identifying and collecting a wide variety of data and documentation, organizing workshops, holding formal and informal meetings, and conducting semistructured interviews [33–35].

URBiNAT guideline categories offer an initial framework with which to address core leverages for successful citizen engagement in the co-creation of NBS, in accordance with local and specific priorities, contexts, and challenges. It is a 'living' framework in the sense that it makes room for co-creation in terms of combining participatory approaches and methods. It also constitutes a 'living' framework, with its categories being reviewed, discussed, and enriched through an ongoing process of sharing and learning. Advances will also be achieved with the help of the perspectives of citizens and other stakeholders engaged in the URBiNAT intervention areas. During the implementation of NBS projects based on co-creation, the engagement of citizens and stakeholders can further inform and inspire ways and tools to trigger engagement, as these groups have direct knowledge of the local participatory culture. In community-driven approaches, the inhabitants and stakeholders of the intervention areas can reveal strengths, weaknesses, and gaps in interaction that will guide the design of participatory processes, that is, the decisions on where to invest time, energy, and resources, which would feed a sustainable co-creation process.

Applying the lenses of the ethical requirements and human rights and gender, which have been mainstreamed in the URBiNAT guideline categories, is of particular relevance in this respect as it contributes to the unveiling of deep-seated inequalities that need to be overcome. It is also helpful to the design of strategies that promote respect for diversity, the acceptance of complexity, and improved conditions for participation [39], which corresponds to addressing the preconditions/elementary conditions to enable citizen engagement.

Given the research into participation and urban regeneration taking place in URBiNAT, learning points have been emerging in dialogue with participatory cases, both inside and outside the project, paving the way for further discussion and the development of a 'living' framework with which to address citizen engagement and NBS co-creation, as well as contributing to an evidence-based framework.

3.3. Emerging Learning Points in Relation to NBS Co-Creation

Now that more than 100 participatory implementations of NBS with relevance to deprived areas have been mapped, and the existing designs and implementations in the URBiNAT frontrunner and follower cities have been studied, several preliminary learning points have emerged that can be linked to the guideline categories for citizen engagement [33,35,58] and adaptation [20]. Some examples are provided in Table A2 of Appendix B, which reports that citizen engagement in the co-creation of NBS results in:

- NBS that are aesthetically, socially, economically, and charitably appealing to citizens and stakeholders;
- New urban spaces where people with common interests can regularly gather and engage;
- NBS diagnostics, design, and implementation relying on a community of stakeholders;
- Strong common projects between actors with different organisational goals as the propellers of social innovation;
- Inclusive and multistakeholder governance as a result of a collaborative approach;
- Bridging differences through an inclusive and highly attractive narrative;

- Effectiveness, achievements, scaling up, and replication as a result of monitoring and evaluation.

We can identify links to a series of guideline categories in these learning points, which make up a possible pathway to successful citizen engagement and the co-creation of NBS. The existing designs and implementations in the URBiNAT frontrunner and follower cities offer interesting and different participatory cases. As an example, four of these are analysed in relation to the guideline categories:

(i) In the pre-established community garden of Gadehavegård, in Høje-Taastrup (Denmark), a workshop was conducted as part of the URBiNAT local kick-off event, in which huge planter boxes with flowers and berry bushes were co-created in order to provide an instant reward to co-creators and inhabitants of the neighbourhood (Figure 2). This NBS example is even more appealing to citizens and stakeholders, as the harvesting of flowers, berries, and herbs at a greater scale will be made possible in the near future. The chosen place and the setting that framed the engagement of the neighbourhood's inhabitants (Where) around the existing meaningful endeavour of the community garden (Why), as a purpose for a participatory activity that integrates collective and individual gardening knowledge (Integration of participatory process results), constituted the key ingredients for successful citizen engagement in the NBS co-creation.

Figure 2. Planting box workshop during the URBiNAT kick-off event in the pre-established community garden of a Gadehavegård neighbourhood in Høje-Taastrup (Denmark), on 14 June 2020. Harvesting flowers, berries, and herbs at a greater scale will be made possible in the near future. Picture by Knud Erik Hilding-Hamann.

(ii) In Porto (Portugal), a task force made up of key stakeholders was formed to coordinate and provide project governance for many experiments now being designed in the designated social neighbourhood of Campanhã (Figure 3). The local task force initially brought together municipal technicians, experts, and researchers, and has been opening up to the involvement of citizens and other stakeholders throughout the engagement and co-creation process. For deprived neighbourhoods, NBS diagnostics, design (experiments), and implementation processes rely on and feed into Trust, co-Ownership, Governance, and

Regulation between the members of a community of stakeholders, identified as "participants", as do the experiments themselves. Several other factors also play a role, depending on the characteristics of the neighbourhood and the NBS.

Figure 3. Workshop organized by the URBiNAT local task force, on 14 July 2020, to share results and design actions of co-created ideas for the healthy corridor of Campanhã in Porto (Portugal). The local task force initially brought together municipal technicians, experts, and researchers, opening up to citizens and stakeholders throughout the engagement process. Picture by Nathalie Nunes.

(iii) In Nantes (France), the co-creation of a green loop as part of the URBiNAT healthy corridor is bringing the goals of municipal technicians more in line with those of the inhabitants of the Nantes Nord neighbourhood (Mediation). The different participatory activities promoted by the municipality and the local scientific partner (Intensity and levels of participation) made use of a communication campaign (Communication and interaction), raising interest among the inhabitants with the use of a subjective map, inviting them to use the green loop on their own (Ownership), and mobilizing them for a walkthrough and a co-selection workshop (Figure 4). Two walks were also organized by the municipality around the topic of food, together with a group of hikers involved in a working group on healthy food led by the communal centre for social action and the municipal public health division (Behavioural change). Facilities and activities are emerging through social innovation along a green path that connects the deprived neighbourhood with the rest of the city (Co-production). Pre-existing citizen initiatives plug into the work and help make it a reality (Culture of participation).

(iv) In Sofia (Bulgaria), a nonprofit organisation has successfully implemented a solution that was included in the URBiNAT NBS catalogue as a Social and Solidarity Economy NBS: the Bread Houses (Figure 5) [59]. The Bread Houses Network is an initiative of the International Council for Cultural Centers Association, which creates and unites centres for community-building, creativity, and social entrepreneurship [60]. The mission of the Bread Houses Network is to inspire individuals and communities to develop their creative potential and cooperate across all ages, professions, genders, special needs, and ethnic backgrounds (Cultural mapping) through collective bread-making, accompanying art forms and education in ecological sustainability (Co-production) [60]. The Bread Houses Network is supported by other actors and stakeholders (Private sector) to co-deliver the benefits of participation (Governance) to citizens in the district of Nadezhda (Communication and interaction).

Figure 4. Communication material to invite reflection and action among inhabitants about a green loop in Nantes Nord. A workshop was organized on 14 October 2020 with local inhabitants and associations to discuss plans and hopes for the URBiNAT healthy corridor in Nantes (France). Picture by Tom Mackenzie.

This brief analysis of the emerging learning points reinforces the relevance of sharing and learning from different contexts on the basis of the 'living' guidelines for citizen engagement and NBS co-creation. It is particularly relevant in relation to the challenges, lessons learned, and best practices, which may inspire further developments of the categories and corresponding guidelines. These aspects constitute starting points for the inclusion of newly developed categories of guidelines or the enhancement of these aspects in the existing categories.

URBiNAT has been particularly involved in sharing and learning with H2020 NBS sister projects (e.g., the online workshop organized at the Nantes Innovation Forum in October 2020 [61,62]) and other EU-funded NBS projects within the framework of clustering activities [30]. These exchanges around the plurality of co-creation models and strategies adopted by EU-funded NBS projects echo the approaches of the emerging learning points.

As an illustration of an approach following a collective and participatory pathway to knowledge production, Figure 6 presents the methodology proposed and applied in the online workshop organized together with URBiNAT's sister projects on "Co-creating solutions with local citizens and stakeholders within European projects". This co-organized workshop was aimed at promoting co-creation for inclusive urban regeneration by showcasing the plurality of the models and strategies of the NBS sister projects, as well as with the purpose of fostering advances among projects by addressing their challenges, lessons learned, and best practices. The methodology of such exchanges promotes sharing, learning, and interaction around the examples, issues, and solutions from different projects. These are at the core of living knowledge, which results in prioritizing and analysing critical issues, key findings, and recommendations. These are also at the core of the development of URBiNAT's 'living' guidelines for citizen engagement and NBS co-creation.

Figure 5. The URBiNAT project participated in a workshop at the Bread Houses Network in Sofia (Bulgaria), on 26 January 2019, during a meeting of its partners. Bread-making fosters cooperation and collective experience across cultures, professions, and ages. Picture by Rune Strunge.

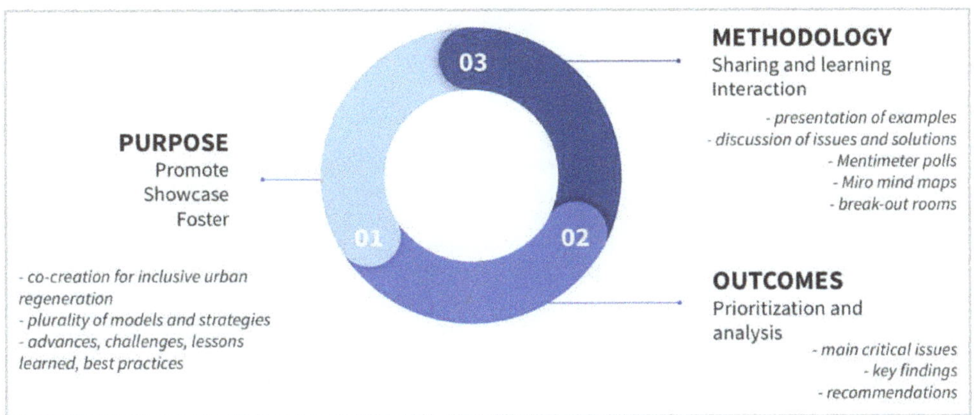

Figure 6. Methodology proposed and applied in the online workshop organized at the Nantes Innovation Forum on 8 October 2020, "Co-creating solutions with local citizens and stakeholders within European projects", together with URBiNAT's sister projects, CLEVER Cities, EdiCitNet, and proGIreg. Source: Presentation and results of the workshop [61].

The research, testing, and validation of the URBiNAT guidelines will continue to be overseen by a task force dedicated to co-creation and governance, together with experts and representatives from European projects working on improving and promoting citizen engagement for the co-creation of NBS. The task force works in five workstreams covering: (i) Why co-create and what may stand in the way; (ii) Who to involve in the process;

(iii) How to integrate co-creation into policies and the co-governance of NBS; (iv) How to co-create NBS, showing different pathways for the co-creation of NBS; and (v) Monitoring and the evaluation of the co-creation of NBS.

The guidelines developed within URBiNAT as one of the European projects will be discussed during workstream meetings and worked on further, which will enable the advancement of a 'living' framework of guidelines for co-creation and citizen engagement. Such a 'living' framework can take the form of a wiki-project, as an open knowledge-building process, as well as promoting the concept of NBS co-creation to the wider audience of stakeholders engaging in NBS development.

We have seen that learning points concerning NBS co-creation can emerge from co-creation in the field of NBS project intervention areas, as well as inside and outside such projects, by means of interdisciplinary and intercultural sharing and learning. This diversity paves the way for an ecology of knowledges, based on the diversity of cultures and knowledges and the recognition of difference, as put forward by Boaventura de Sousa Santos [63].

4. Discussion

The guidelines, categories, and learning points that emerge from a diversity of participatory cases can evidence pathways to successful citizen engagement and the co-creation of NBS, thereby contributing to an evidence-based framework that also includes the prioritization and analysis of critical issues. Although great efforts have been made as part of the URBiNAT project to make room for citizen participation in the NBS co-design process, URBiNAT is, however, still challenged by the efficacy of its co-creation process, as much as other EU-funded projects dedicated to NBS, particularly in relation to participation and inclusion [64]. At the same time, these challenges constitute research opportunities to be further explored. To this end, we have organized three main directions for discussion: an interdisciplinary and intercultural approach to the development of NBS; rethinking engagement, especially in times of COVID-19; and sharpening participation for inclusive and innovative urban regeneration with NBS.

4.1. Working Interdisciplinary and Interculturally in Developing NBS

The European Commission report on the state of EU-funded NBS projects raises a range of critical issues in relation to participatory methods, such as those mentioned in the introduction to this paper [10]. Moreover, the analysis stresses that fundamental questions of politics, arising with the involvement of highly diverse urban communities in the development of NBS, are often neglected. When efforts are made to increase participation and inclusion in the co-design and implementation of NBS, emphasis is often on minimising conflict and reaching consensus [10,65,66]. Consequently, the limitations of the co-design and co-production processes must be taken into consideration when it comes to fostering new processes for participation and inclusion [10].

These do not only provide a warning of the critical issues, but also of important considerations highlighting the needs for future research and innovation in the field of NBS. If sufficient technical knowledge on the design of specific NBS types already exists, there is still a need to develop the understanding of the economic, social, political, and cultural dimensions of designing and implementing NBS, moving beyond seeing the implementation challenge as primarily a 'technical' issue [67,68]. This includes revealing the structural aspects or aspects that have been revealed in the development of NBS, such as conflicts that arise but that are not taken on board and dealt with when a purely technical approach to participation is used [10,65,66], as well as power imbalances relative to marginalised voices [28], and power manifestations in multistakeholder collaborations, including academic disciplines, experts, and social circles in civil society [69].

In the case of URBiNAT, revealing and addressing the dimensions of the design and implementation of NBS was introduced relative to the ethics requirements and the rights-based approach, for example, in line with a cross-cutting approach, in order to make these issues integral dimensions of the project [33,39]. This also consists of mapping and analysing local participatory cultures in order to pave the way for testing a strategy for a municipal roadmap for the healthy corridor, adjusted to the local needs, cultures, and ambitions of each city [34,35]. The strategy of the municipal roadmap addresses the commitment of advancing innovation in the decision-making process of each city and is aimed at improving the quality of participation as both a means and an end [34,35].

The participatory cultures that URBiNAT builds on in each of the seven city districts are characterised by a population of citizens, the vast majority of whom have a good grasp of all the things that need to be improved in their neighbourhood. However, as the literature shows, when the majority of citizens do not see hope, do not trust the authorities, and have social and or health issues, engaging them in co-creation for the benefit of all requires an extraordinary level of mobilisation in order to bring about an individual and community readiness to engage in, and lead, change in the community [70]. As highlighted by practitioners from the field in URBiNAT cities, when reviewing the URBiNAT guideline categories, some significant factors, such as Trust and Transparency, may impact citizen engagement to a greater extent depending on the local context. This is particularly the case of contexts marked by distrust or a history of failure and disappointment, which require that different mechanisms be explored [71].

Therefore, building a 'living' framework that addresses the core leverages for successful citizen engagement in the co-creation of NBS requires further exploration of the interdisciplinary and intercultural work involved. In the case of the guideline category, Trust, it is the guideline itself that needs to be analysed, both as a means to enabling citizen engagement (building trust and motivating despondent citizens), and as an end, resulting in empowered participation in accordance with the local context. This involves differentiating how groups of citizens, including administrators, organisational stakeholders, and 'leaders' are motivated and engaged, what role they can play, the level of resources they can commit to a co-creation process, and what it will take to achieve sustained participation.

Researchers can also contribute with reflexivity to bring blind spots to the surface in the framework of action research on the basis of, for example, the conceptual framework of the sociology of absences and the sociology of emergences developed by Santos [72]. On the one hand, the sociology of absences contributes to identifying what has been made invisible, devalued, or rendered nonexistent; on the other hand, the sociology of emergences contributes to valuing the resistance of social groups and identifies in this resistance principles and practices of governance that point to other experiences [73]. The ecology of knowledges and intercultural translation intervene as tools for the diversity of knowledges made visible by the sociology of absences and the sociology of emergences, thereby reinforcing a research agenda that promotes interdisciplinary and intercultural approaches in the development of NBS. These are, indeed, key to further developing the guidelines for citizen engagement and NBS co-creation as living knowledge.

4.2. Rethinking Engagement, Especially in Times of COVID-19

The impact of COVID-19 on public space and urban planning and design is a pressing subject of research, particularly for addressing how the needs of vulnerable groups (e.g., racial minorities, immigrants, women, older adults, children, people with functional diversity, and the homeless) will be accounted for in the future designs, practices, and rules for public spaces [74]. In this respect, the measurement of changes in use and the perceptions of public spaces will be critical, especially with regard to the possibilities they offer for socialization, recreation, claim-making, community building, and identity formation [74].

Meanwhile, cities are coping with limitations on interaction because of the pandemic, which affects both in-person and virtual interactions. Specific vulnerabilities have to be taken into account in the case of social neighbourhoods and deprived areas, and these are becoming more evident than in the past because of the COVID-19 pandemic [75,76]. These specific vulnerabilities are the pre-existing unequal conditions of structurally vulnerable neighbourhoods, where morbidity and mortality may be hardest felt, as a consequence, for example, of racial/ethnic health inequities [77]. If significant factors impacting citizen engagement remain key, some of the issues at stake have become more critical in the aftermath of the pandemic, in particular and most importantly, how people feel about their current situation and their perspectives on the future, and, subsequently, how people want to engage now, by what means, and through which channels, as well as how to rethink the methods and tools.

In this context, URBiNAT cities gathered information and came up with a picture of some of the aspects related to the repercussions, challenges, responses, and alternatives that emerged from the pandemic [75]. URBiNAT cities are committed to sharing with and learning from each other, and this effort has produced different but complementary pictures, analyses, and perspectives for the tackling of increased and new social challenges, which may possibly lead to common strategies embedded in solidarity.

This consultation process, in which the cities involved shared the impact of COVID-19 with URBiNAT, confirmed that the crisis primarily highlights and amplifies existing inequalities and increases the vulnerability of large sections of the populations of social neighbourhoods, such as in the case of Brussels (Belgium) [78], Nantes (France) [79], and Porto (Portugal) [80]. These increased and new challenges faced by the populations in the URBiNAT intervention areas include economic shortages in some households, food emergencies, the digital divide, psychological distress, and psychic suffering, to name some of the points.

Building on lessons learned is about rethinking many aspects of life in the city. This is a time and opportunity for cities to rethink the use and development of housing, transport, and public spaces, particularly in relation to citizens in the most vulnerable conditions [81], as highlighted by representatives from the URBiNAT cities of Sofia (Bulgaria), Høje-Taastrup (Denmark), Siena (Italy), and Nova Gorica (Slovenia). It involves, for example, working to preserve physical and mental health [82], developing a resilience to dealing with uncertainty, and similar challenges in the future, which will differ across different groups [83]. The renaturing of cities and the provision of healthy spaces for leisure [84] must also be prioritized.

Most importantly, building an 'alternative future' begins, crucially, with the communities that control the management, care, and regeneration of green areas and other common spaces. These could be delegated to citizens by creating collaboration agreements between them and the administrative bodies [85]. Participatory spaces are, therefore, important tools for asking people about their current situation and their perspectives on the future, for sharing experiences of pain and suffering in a social setting, and for shifting from the sphere of individual experience to that of collective processing [86]. This is increasingly important to the understanding of the economic, social, political, and cultural dimensions of designing and implementing NBS, both during and after the pandemic.

The limitations on interactions with citizens and stakeholders because of the current COVID-19 crisis also require us to rethink participation by developing sustainable models for keeping citizens engaged. The guidelines for citizen engagement and the co-creation of NBS, based on a 'living framework', need to include questions on the 'after', how much the participatory processes of the 'new normal' will have to change, and how to rethink the participatory devices for involving inhabitants in public policies [87]. More broadly, beyond methodologies and outreach, it also raises the need to rethink formal citizenship and to adopt affirmative interventions as a way of reducing democratic gaps and disseminating new forms of participation. Particular attention must be paid to supporting the most vulnerable groups and territories [87], as these were especially susceptible to the increased and new challenges.

4.3. Sharpening Participation for an Inclusive and Innovative Urban Regeneration with NBS

The publication on the state of the EU-funded NBS projects points to the need for NBS initiatives to be designed and implemented with the explicit intention of addressing the inequalities and tensions underlying urban development, so that they can potentially contribute to the realization of sustainable communities [10]. In the case of URBiNAT, the challenges and responses devised in the field of the social and solidarity economy, as a pillar of its approach to urban regeneration and inserted in the NBS framework, address several related aspects: the problematization of the multidimensional and intersectional causes of inequalities in the urban space; the realization of the social well-being of vulnerable individuals and groups through opportunities for strengthening social relations, autonomy, and economic conditions; new models of governance aimed at community development by influencing public policies and through the empowerment of people for social change [34,45].

Another aim of the project is equality and equity for all, the latter being related to the intersectional approach, which is based on the notion that specific modalities of subordination and discrimination act in an integrated manner, as experienced by racialized and minoritized peoples and communities [33]. In this respect, the application of ethical requirements and a rights-based approach combine a series of principles that are complex and challenging to implement in both theory and practice [54].

Firstly, to make these themes integral dimensions of URBiNAT involves always having them present in the planning and development of activities. It also involves as many partners of the consortium as possible adopting these lenses in their internal and organizational agendas, and in their analyses and perspectives on the project's progress and results. However, it may also require changes in the established procedures and cultures of partners and stakeholders and their values and practices inside the deep core of hierarchies and organizational cultures and practices [39], especially in relation to gender mainstreaming and intersectionality [88]. Putting into effect a rights-based framework requires an awareness of, and an ability to manage, controversies, including complaints, and must also take into consideration that the behaviour of project researchers, technicians, and experts in the field, such as being accessible, responsive, and transparent, is at the heart of these interactions [39,89,90].

Furthermore, reaching and engaging marginalized voices requires not only strategies and methods for the inclusion of their visions and perceptions in the development of NBS, but also the investment of time, energy, and resources to enable consistent improvement in the quality of participation as both a means and an end. It is about inclusiveness beyond the term of the project and looking at the deep-seated inequalities that are present when inclusive and innovative urban regeneration projects handle a complex combination of societal challenges that aim to contribute to the right to the city. This again has to do with developing an understanding of the economic, social, political, and cultural dimensions of designing and implementing NBS, and moving beyond seeing the implementation challenge as primarily an immediate 'technical' issue [67,68].

5. Conclusions

This paper systematizes the efforts and results involved in the development of guidelines for citizen engagement in the first three years of the URBiNAT project, which have also contributed to establishing the URBiNAT Community of Practice [36]. This includes sharing and building on differences and highlighting core leverages for successful participation, as well as sharing visions about the results of community-driven processes. This approach is at the core of co-creation and encourages researchers and practitioners to build a mixed knowledge base with key stakeholders [91]. It also takes into consideration local participatory cultures and the specificities of segments of individuals and citizens [92], hence stimulating the scaling up of local sustainable citizen-driven initiatives.

Further research is needed to deepen the categorization of guidelines for citizen engagement and to understand the timings and contexts for their application. This would also contribute to improving NBS as an approach, another area in which research is required [2,93]. Participation is considered to be both a means and an end to the shaping of the urban environment and promoting active citizenship and social cohesion by the means of strategies that are collectively designed.

The need to improve NBS as an approach has emerged, enabling them to become more comprehensive and holistic. This is particularly true in terms of social embedding and the impacts that need to be considered so that NBS can become more than just tools, technologies, and instruments [93]. Research on community-based and policy-based initiatives aimed at improving sustainability and liveability, and that fosters inclusivity and social justice, has also evidenced the transformative social impact of NBS; new social relationships and configurations are mediated, contributing to social innovation in cities and changing perceptions of nature and human–nature relationships in urban contexts [2]. URBiNAT, in turn, aims to enrich and complement the NBS concept with new perspectives, such as the ones introduced with Participatory NBS and Social and Solidarity NBS, and more ingredients, such as the ones that compose its 'living' framework of guidelines for successful citizen engagement in the co-creation of NBS.

The COVID-19 pandemic has reinforced the relevance of a systemic approach with mixed methods and multichannel systems of engagement [94], which can be important in challenging traditional NBS by confronting them with new visions. This is not only about targeting specific segments of the population, but also about building upon complementary participatory processes that provide both immediate results and medium- and long-term visions, towards complexifying the visions of all the intervening actors and going beyond immediatism and self-referentiality [95]. For instance, the practice of participatory budgets, which tends to encompass immediate investments and short-term perspectives [95], has been transforming and improving with other instruments of planning and visioning [96]. This then permits a dialogue that contains strategic visions brought about through long-term visioning exercises, a process in which a community envisions the future it wants and makes plans for how to achieve it [97]. In rethinking many aspects of life in the city, citizen engagement and the co-creation of NBS may contribute to a paradigm shift in society's relationship with nature, in line with the promotion of multisectoral and multidimensional approaches towards healthier cities [98–100]. In the context of a 'living' framework, citizen engagement constitutes a critical aspect of the development of NBS so that NBS reach their full potential, as well as the advancement of the science and practice of NBS not only as a scientific rethinking [101,102], but also as part of an ecology of knowledges where participation is seen as both a means and an end.

Author Contributions: Conceptualization, N.N., E.B. and K.E.H.-H.; Investigation, N.N., E.B. and K.E.H.-H.; Methodology, N.N., E.B. and K.E.H.-H.; Writing—original draft, N.N., E.B. and K.E.H.-H.; Writing—review & editing, N.N., E.B. and K.E.H.-H. All authors have read and agreed to the published version of the manuscript.

Funding: This research was funded within the framework of project 'URBiNAT—Healthy corridors as drivers of social housing neighbourhoods for the co-creation of social, environmental and marketable NBS'. This project has received funding from the European Union's Horizon 2020 research and innovation program under grant agreement No. 776783.

Institutional Review Board Statement: Not applicable.

Informed Consent Statement: Not applicable.

Data Availability Statement: The initial URBiNAT guideline categories for citizen engagement emerged from the systematization of guidelines contained in written contributions as part of an URBiNAT-deliverable D1.2—Handbook on the theoretical and methodological foundations of the project [48]. These written contributions are not publicly available because of the internal character of the deliverable. URBiNAT deliverables are referenced on the project website at https://urbinat.eu/resources/ (accessed on 20 October 2021), where permission to access these documents can be obtained upon request. The preliminary learning points emerging from the participatory implementation of NBS with relevance to deprived areas are derived from good practice that has been presented and discussed on the URBiNAT online work platform, in a nonpublic blog that allows project members to comment and reflect on the issues. It is available upon request to the corresponding author [58].

Acknowledgments: The methodological approach and its application, as described in this paper, was developed with the contribution of Isabel Ferreira, in particular, for the framing of the methodology and the extracting of the guidelines from Chapter 1 of the URBiNAT-deliverable D1.2 dedicated to citizen engagement, as well as for the organizing of the guidelines into categories. This contribution was made as part of Isabel Ferreira's PhD research, under the topic: "Governance, citizenship and participation in small and medium-sized cities: a comparative study between Portuguese and Canadian cities", funded by Fundação para a Ciência e Tecnologia, Fundacção Calouste Gulbenkian, and the International Council for Canadian Studies. The results presented in this paper benefited from the work developed within the URBiNAT working group on participation, together with Ingrid Andersson, Américo Mateus, and Sofia Martins. We are particularly grateful for the contributions made by our URBiNAT colleagues, Giovanni Allegretti and Isabel Ferreira, with their valuable suggestions and discussions. Finally, the authors would like to thank the editors of this Special Issue, as well as the reviewers for the time and effort dedicated to providing insightful feedback on ways to strengthen our paper, which gave us the opportunity to introduce corrections and clarifications and to improve the quality of our manuscript.

Conflicts of Interest: The authors declare no conflict of interest. The funders had no role in the design of the study; in the collection, analysis, or interpretation of the data; in the writing of the manuscript; or in the decision to publish the results.

Appendix A

Table A1. URBiNAT-detailed initial framework on guideline categories addressing core leverages for successful citizen engagement in the co-creation of nature-based solutions (NBS).

Category of Guidelines [1]	Prioritization [2]	Strategic Overview	Operational Details	Impact of Other Categories
Communication and interaction	1	Communicating specificities for interacting with citizens.	Covers, in operational terms: (1) Communication strategies; (2) Communication materials and channels; (3) Multichannel interaction; (4) Codes of conduct related to communication and ethics.	This category ties in with the category of Trust, especially with regard to how communication and interaction with residents happens. For instance, depending on the city, building trust may be based on meetings and face-to-face encounters, whereas the use of digital tools is limited and aimed at incentivising being together. Social circles of residents may also be limited to close relatives, which enhances the importance of providing a space where people can communicate and not be frightened of being together. Furthermore, organisations working with specificities constitute important partners in establishing communication and interaction with particular groups and individuals.
Behavioural changes	2	Instigating behavioural adjustments, or changes in behaviour, in some particular respect.	Namely, by: (1) Challenging traditional models of governance, expert advice, and implementation; and (2) Instigating adjustments to attitudes, mindsets, and behaviours in support of participation and collaboration.	Ties in with the category of Communication and interaction in many ways, including: how residents are shown that their inputs are valuable and can be applied for the creation of change; identifying and engaging agents of change; promoting participatory and creative activities to address specific behaviours (e.g., aggression, intolerance, lack of openness, and looking at the different cultures and existing boundaries built on the differences).
Trust	3	Improving or creating relationships of trust between citizens, and between citizens and city staff, politicians, and other agents.	With particular attention to: (1) Confidence and team dynamics; and (2) Language.	Ties in with the categories of Transparency, Inclusion, Communication and interaction, and Governance by: ensuring that everyone is part of the conversation and deliberations; documenting the activities to promote ownership; qualifying local ideas instead of bringing many ideas from practitioners/experts; properly communicating and translating what the residents feel, as well as repeating people's opinions. It also ties in with the categories of Culture of participation and Cultural mapping since, as highlighted by practitioners from the field in URBiNAT cities, Trust and Transparency may impact citizen engagement to a greater extent according to the local context. This is the case of contexts marked by distrust or a history of failure and disappointment, which require the exploration of different mechanisms [71].

Table A1. *Cont.*

Category of Guidelines [1]	Prioritization [2]	Strategic Overview	Operational Details	Impact of Other Categories
Co-production	4	Stimulating and improving the co-production of public services, participatory processes, and product development.	Focusing in particular on: (1) The process by which citizens participate in the implementation and delivery phase; (2) An open process of participation that includes a wide range of key actors, namely, end-users.	Ties in with the categories of Trust and Behavioural change, in particular relative to team dynamics at the different stages of the NBS co-creation process, and by challenging traditional models of implementation.
Inclusion	5	Having specific guidelines to guarantee the inclusion of diversity.	Concerning: (1) The different modalities of the participatory process; (2) Not only pursuing the 'usual suspects' who always participate and are more engaged because of their availability, resources, and professional/disciplinary skills, and may also constitute an exclusive group; (3) Capacity and tools to address and welcome diversity.	Ties in with the categories of Citizenship rights, Governance, Transparency, and Regulation, given that inclusion must be shown through bonds and by sticking together (e.g., the scene is the house for volunteer associations where they can bring more people). It also ties in with the category of Supportive methodologies and techniques in terms of the modalities for mobilization and inclusivity in the participatory process.
Regulation	5	Clarifying rules and regulations for equal rights in the expression of visions and priorities.	It means not only to: (1) Establish rules and regulations for the participatory process; but also (2) Promote co-decisional processes.	Tying in with the categories of Governance, Transparency, and Trust, the local contexts may bring additional critical issues, such as when rules are not followed.
Governance	6	Balancing interactions among citizens, city staff, politicians, and other agents.	The Governance category focuses on: (1) Opening doors in the public sphere and balancing power relations; (2) More liveable and balanced interactions; (3) Organization of participation in an integrated manner, going beyond the institutional division of municipal departments.	It ties in with the categories of Trust, Transparency, and Culture of participation. The use of Participatory NBS is particularly relevant in this respect, by promoting participatory activities that improve collaboration and empower individuals in the decision-making process.
Innovation cycle	7	Adopting processes of rupture and searching for alternative solutions to address concrete social problems.	By: (1) Breaking the crystallized image of a problematic neighbourhood, including observing a code of conduct for the communication and dissemination of activities; (2) Connecting people, introducing creativity, and mobilizing energy.	The use of Participatory NBS is relevant in promoting such an innovation cycle when they focus on the available resources, assets, and relationships of solidarity in the community. Ties in with the category of Citizenship rights relative to strengthening the capabilities and empowerment of the population as well as the satisfaction of needs and the corresponding access to rights.

Table A1. *Cont.*

Category of Guidelines [1]	Prioritization [2]	Strategic Overview	Operational Details	Impact of Other Categories
Transparency	8	Arguments for encouraging efforts to act in a transparent manner.	With an emphasis on: (1) Reflecting on why people should participate in the process, and being clear about purposes and rules; (2) Avoiding hidden agendas or information.	Ties in, to a great extent, with the categories of Trust and Governance. When there is a lack of trust between citizens and politicians, transparency is even more necessary from the municipality as a governance issue to build trust. It also ties in with the categories Why participation and Monitoring and evaluation, given the need to provide information enable discussion of the results of each stage, and to include a systematic follow-up. This means being able to speak about expected results that are both positive and negative, and to give feedback about what is going well and what is not, which will impact expectations and trust.
Intensity and levels of participation	8	Setting different approaches and levels of participation depending on the goals and real conditions for participation.	Through: (1) Systematic awareness of the conditions under which citizens are prepared to engage in actions of social innovation; and (2) Thinking about different steps to citizen engagement (e.g., communication, information, consultation, participation, co-building).	Ties in with the Culture of participation category, which can reveal strengths, weaknesses, and gaps in interaction and will guide the design of participatory processes. To be community-driven, the design must focus on raising the intensity of interactions among citizens, stakeholders, organizations, and institutions.
Citizenship rights	9	Broadening the meaning of the appropriation of social, urban, political, and cultural rights, both internally in the collective imagination, and externally in rejuvenated relationships with local powers.	In operational terms, the emphasis would be on addressing: (1) Access to and implementation of rights; (2) Engaging/empowering; (3) Strategies designed to promote participation according to specificities; (4) Codes of conduct related to research and ethics.	Ties in with the category of Inclusion, namely, concerning the modalities of the participatory process which addresses, welcomes, and promotes diversity. The use of Participatory NBS, such as Forum Theatre, is relevant in this respect in that it involves the community in the analysis and discussion of problems, thereby raising awareness and encouraging citizen participation.
Cultural mapping	10	Articulating and making visible the multilayered cultural assets, aspects and meanings of a place.	It: (1) Encourages the attachment of citizens to a location; and (2) Acts as a catalyst to the process.	Tying in with the category of Inclusion and more than simply just another tool, cultural mapping is relevant if used as an approach to get to know people, address their specificities, what they like to do, and what they want to do, as used in the co-diagnostic phase of the URBiNAT co-creation process. It also ties in with the categories of Innovation cycle and Supportive methodologies and techniques as a Participatory NBS, since cultural mapping emphasizes processes that enable projects to be platforms for discussion, engagement, and empowerment.

Table A1. *Cont.*

Category of Guidelines [1]	Prioritization [2]	Strategic Overview	Operational Details	Impact of Other Categories
Facilitation	-	Having specific guidelines to address facilitation that include other participatory guidelines.	Defining: (1) The main attributes of facilitation in understanding the role that is expected; (2) The different steps of the co-creation process, including information about NBS; (3) How to hold successful public meetings based, for example, on successful participatory cases; (4) The principles and requirements of ethics.	These aspects need to be covered in guidelines, training of local facilitators, and corresponding support materials. Participatory NBS, such as Learning for Life or community workshops, constitute useful resources to help organize these aspects and put them into practice, by providing protocols and approaches that will support facilitation. This category, therefore, ties in with the category of Supportive methodologies and techniques.
Quality of deliberation	-	Setting a meaningful deliberation process.	Through: (1) Authentic deliberation; (2) A clear decision-making process; and (3) Ensuring equal rights of expression.	More than simply voting, the focus is on interaction, democratic decisions, and expression, which ties this category in with the categories of Regulation, Governance, Citizenship rights, and Facilitation.
Where	-	Having guidelines for the spaces in which the participatory events are held.	Addressing: (1) Place/setting: as well as (2) Form and quality.	Ties in with the categories of Communication and interaction and Facilitation since the definition of these aspects is all the more relevant when dealing with a lack of space in which to speak and do things together. Spaces to not ony share visions, values, roles, dialogue with people, but also to create a dialogue between people. For example, older adults and victims of violence need to be heard in a space where they can voice what they want to do and what they need. Determination of how to devise a model for a space that incentivises people to work constructively together is key.
When	-	Identifying the best moment for the participatory events.	Including: (1) Time/day; (2) Date; and (3) Phase.	In practical terms, this category implies meeting the community and knowing as much as possible about the needs of the people who live in the area of intervention, as well as their habits and traditions, so that the participatory activities can be tailored to fit. Additionally, to be relevant, participation cannot happen at the end of the process of planning a project. Ideally, we should all begin together, with an empty page or question to be addressed, but it depends on the project, on whether it has already started, and on its technical level. What needs to be assessed is the decision on the right time/phase to engage. Therefore, this ties in with the categories of Communication and interaction and Facilitation.

Table A1. *Cont.*

Category of Guidelines [1]	Prioritization [2]	Strategic Overview	Operational Details	Impact of Other Categories
Supportive methodologies and techniques	-	Using specific methodologies and guidelines to support mobilization and inclusivity.	Considering: (1) Culture as a platform; (2) Lower degrees of formalization; and (3) Articulation of knowledge.	Ties in particularly with the categories of Communication and interaction, Facilitation, and Cultural mapping. Generally speaking, in practical terms, arts and community events can facilitate creativity. We can also consider here the appropriation of complex languages by including people's knowledge in dialogue with technical and scientific knowledge. Going beyond Cultural Mapping, Forum Theatre, or community-based art, which makes specific use of the arts, Participatory NBS comprise protocols, approaches, and methods aimed at engaging citizens at different stages of the co-creation process.
Integration of the results of participatory processes	-	Enlarging the scope of co-creation to validate the ideas developed.	Through: (1) Cross-pollination; (2) Validation; (3) Systematization; and (4) Definition of purpose.	This ties in particularly well with the categories of Communication and interaction, Facilitation, and Supportive methodologies and techniques, concerning the use of communication materials and channels, the definition of the different steps of the co-creation process, and with the articulation of knowledge from people, technicians, experts, and researchers.
Private sector	-	Mapping the relevant private sector actors with interests in, and input to, the NBS targeted area.	Requires: (1) Mapping who has links and can facilitate contacts with private actors (e.g., business associations, local companies, private owners), as well as their eventual roles in the co-creation of NBS; (2) Conducting meetings and workshops with specific groups to understand visions, priorities, and interests, as well as bringing all participating groups together to devise a common vision and project, as well as to seek formal commitment.	Highlighted here is the definition of the relevance of the private sector, not limited to actors, among others. It ties in with the categories of Co-production and Innovation cycle relative to the development of products and services, the involvement of a wide range of key actors, and connecting them based on creativity and the mobilization of energy.

Table A1. *Cont.*

Category of Guidelines [1]	Prioritization [2]	Strategic Overview	Operational Details	Impact of Other Categories
Monitoring and evaluation	-	Addresses monitoring and evaluation of the participatory process.	The aspects related to monitoring and evaluation of the participatory process cover: (1) The process itself; (2) Results and impact of participation; (3) The different aspects of evaluation guiding the selection of methods for impact assessment, taking into consideration: (4) Participatory monitoring and evaluation; as well as (5) Participatory impact assessment.	Ties in with the categories of Transparency in terms of information and follow-up, and Ownership of the co-creation process and the corresponding results.
Risk assessment and mitigation measures	-	Identifying the factors influencing co-creation processes, as well as those leading to the failure of co-creation and co-production.	In operational terms, it covers the identification of both: (1) Basic requirements in the risk assessment of co-creation processes; (2) Mitigation measures corresponding to risk factors related to the process of engaging citizens in co-creation and their participation in the implementation and delivery phase.	It ties in with the categories of Monitoring and evaluation and Transparency regarding clarity of the participatory process, its assessment, and improvement.
Ownership	-	Citizens having ownership of both problems and solutions.	This depends on: (1) The assumption that practitioners can only bring knowledge if people own the process, by providing the framework but not taking the lead; (2) Enabling inputs from people by showing that contribution is possible and providing safe spaces, as well as implementing a diversity of appealing activities.	This category, therefore, ties in with the categories of Trust and Communication and interaction.
Culture of participation	-	Enabling regular interaction with citizens, and increasing the culture of participation.	Requirements: (1) Transversally increasing the culture of participation in all departments of the municipality by introducing new models that involve all people and services, as well as building bridges between the public sector, the private sector, and citizens; (2) Enabling initiatives by citizens, with consideration given as to how to encourage, receive, and adapt to spontaneous initiatives that they make, and how to listen to and receive these initiatives, some of which will be off the municipal radar.	Ties in with the categories of Governance and Intensity and levels of participation, regarding interactions between city staff and citizens.

Table A1. *Cont.*

Category of Guidelines [1]	Prioritization [2]	Strategic Overview	Operational Details	Impact of Other Categories
Why participation	-	Being clear as to why we need to engage citizens and support participatory processes.	Includes: (1) The object of participation, and the things we want to discuss and do with people; (2) The purpose of participation, why participation is important to the project in question, and what motivates people to participate; (3) Ways of carrying out participation, and why we use specific methodologies; (4) The relevance of participation, since not everything needs to be in the form of dialogue/discussion. Participation is not always the solution, and sometimes inputs can be received in other ways.	Ties in with the categories of Transparency and Intensity and levels of participation relative to the clarity of purpose and rules, as well as to the consideration of different approaches in accordance with the goals and real conditions of participation.
Mediation	-	Dialogue and collaboration.	Covering both: (1) The resolution of conflicts; and (2) The use of dialogue to foster collaboration between people who do not have much experience in this type of problem solving.	Ties in with the categories of Communication and interaction, Trust and Facilitation, concerning strategies that are sensitive to local history and existing relationships, to build trust and foster being/working together, as well as the specific attributes and expected role of the mediator.

[1] Categories that emerged from both the systematization and two-stage review process of URBiNAT guidelines for citizen engagement, and the sharing and learning with practitioners from the field in URBiNAT cities: categories that emerged from Step 1 of systematization (extract and organize) are in light grey; those from Step 2 (review and aggregate) are in medium grey; additional categories that resulted from sharing and learning are in dark grey. [2] Prioritization of guideline categories by members of the global Living Lab community in which the 10 most critical aspects of citizen engagement are ranked. This was the result of an external workshop conducted by the URBiNAT working group on participation during the Open Living Lab Days in September 2019, organized by the European Network of Open Living Labs (ENoLL) in Thessaloniki (Greece).

Appendix B

Table A2. Preliminary learning points emerging from citizen engagement in the co-creation of nature-based solutions (NBS).

Learning Points	Overview and Discussion	Examples of NBS Participatory Cases	Links of Learning Points/Categories
The appeal of NBS to citizens and stakeholders aesthetically, socially, economically, and charitably.	NBS need to be aesthetically, socially, economically, and charitably appealing to citizens and stakeholders so that they are encouraged to engage with, further develop, maintain, and protect them.	A workshop was conducted at Høje-Taastrup (Denmark) in which huge planter boxes with flowers and berry bushes were co-created to provide an instant reward to co-creators and the inhabitants of the neighbourhood.	Why participation, Where and Integration of participatory process results.
NBS and new urban spaces where people with common interests can regularly gather and engage.	NBS create new urban collective spaces at the neighbourhood level and across the city. This is achieved both virtually, through online platforms, and physically, by creating attractive physical spaces where people with common interests can gather and engage on a regular basis, as part of their daily lives or for special events.	In Milan (Italy), a previously deprived neighbourhood was turned into an attractive neighbourhood through the online coordination of many different events and facilities, including local radio, breakfast meetings on street corners, and support for the opening of small shops and services.	Where, Innovation cycle, and Ownership.
Diagnostics, design, and implementation of NBS rely on a community of stakeholders.	For deprived neighbourhoods, NBS diagnostics, design (experiments), and implementation processes rely on and feed into trust, co-ownership, governance, and regulation between the five main stakeholders, identified as "participants" in the URBiNAT model [33]: (1) The municipality (political representatives and technicians of different departments); (2) Housing administrators (responsible for the management of social housing); (3) NGOs, businesses and other private and public organisations working in the intervention areas; (4) Champions/ambassadors and facilitators; and (5) Citizens, as well as in the experiments themselves. Several other factors also play a role depending on the characteristics of the neighbourhood and the NBS.	In Porto (Portugal), a task force made up of these five actors has been formed to coordinate and provide project governance to many experiments now being designed in the designated social neighbourhood.	Many of the guideline categories and especially Trust, Ownership, Governance, and Regulation.
Strong common projects between actors with different organisational goals as propellers of social innovation.	Sustainable NBS emerge strongly from the fabric of social innovation at the neighbourhood and urban levels, building success on the back of strong common projects between actors with different organisational goals (link to a sense of co-ownership). Social innovation is referred to here as a process, implying changes in social relations and power relations, or as a product, by means of the construction of methodologies, artifacts, and/or services, especially those aimed at strengthening the capabilities of the population, the satisfaction of needs, and the access to rights [103–106].	In Nantes (France), facilities and activities are emerging through social innovation along a green path that connects the deprived neighbourhood with the rest of the city. Pre-existing citizen initiatives plug into the work and help make it a reality.	Communication and interaction, Ownership, Intensity and levels of participation, Culture of participation, Mediation, Behavioural change, and Co-production.

Table A2. *Cont.*

Learning Points	Overview and Discussion	Examples of NBS Participatory Cases	Links of Learning Points/Categories
Inclusive and multistakeholder governance as a result of a collaborative approach.	NBS require a collaborative approach that involves and engages all five stakeholders, developing governance that both influences and sets limits for all five, while ensuring that the strengths and weaknesses of each actor are integrated into the division of roles and responsibilities.	In Sofia (Bulgaria), the Bread Houses Network is supported by the other actors and stakeholders to co-deliver the benefits of participation to citizens in the district of Nadezhda.	Communication and interaction, Private sector, Co-production, Cultural mapping, and Governance.
Bridging differences through an inclusive and highly attractive narrative.	An inclusive and highly attractive narrative (vision and mission) for the implementation of NBS can help bridge differences between municipal departments, participating businesses, and organisations, and between the different housing blocks in neighbourhoods.	The vision for Helsingborg (Sweden) is that, by 2035, the city will be creative, pulsating, inclusive, and balanced for people and businesses. The city by then will be exciting, attractive, and sustainable. This common vision, which is defined in more detail locally, informs and governs all NBS and other regenerative activities in the city.	Inclusion, Trust, Governance, and Communication and interaction.
Effectiveness, achievements, scaling up, and replication as a result of monitoring and evaluation.	The design, monitoring, and evaluation of NBS need to be arranged so that the effectiveness and achievements can be measured and analysed, enabling successful NBS to be scaled up and repeated in other similarly deprived areas.	In the municipality of Høje-Taastrup (Denmark), the stakeholders are transferring documented models using URBiNAT co-creation methods from one deprived neighbourhood, Charlottekvarteret, where the quality of life has improved, to another neighbourhood in the city, where there is the potential to achieve similar results.	Monitoring and evaluation, Private sector, Behavioural change, and Co-production.

References

1. European Commission. Horizon 2020 Work Programme 2016–2017: 17. Cross-Cutting Activities (Focus Areas). 2016. Available online: https://ec.europa.eu/research/participants/data/ref/h2020/wp/2016_2017/main/h2020-wp1617-focus_en.pdf (accessed on 20 October 2021).
2. Frantzeskaki, N.; Borgström, S.; Gorissen, L.; Egermann, M.; Ehnert, F. Nature-Based Solutions Accelerating Urban Sustainability Transitions in Cities: Lessons from Dresden, Genk and Stockholm Cities. In *Nature-Based Solutions to Climate Change Adaptation in Urban Areas: Linkages between Science, Policy and Practice*; Kabisch, N., Korn, H., Stadler, J., Bonn, A., Eds.; Theory and Practice of Urban Sustainability Transitions; Springer International Publishing: Cham, Switzerland, 2017; pp. 65–88. [CrossRef]
3. Vaz, E.; Anthony, A.; McHenry, M. The geography of environmental injustice. *Habitat Int.* **2017**, *59*, 118–125. [CrossRef]
4. Seddon, N.; Chausson, A.; Berry, P.; Girardin, C.A.J.; Smith, A.; Turner, B. Understanding the value and limits of nature-based solutions to climate change and other global challenges. *Philos. Trans. R. Soc. B Biol. Sci.* **2020**, *375*, 20190120. [CrossRef] [PubMed]
5. Egusquiza, A.; Cortese, M.; Perfido, D. Mapping of innovative governance models to overcome barriers for nature based urban regeneration. *IOP Conf. Ser. Earth Environ. Sci.* **2019**, *323*, 012081. [CrossRef]
6. Directorate General for Research and Innovation (European Commission). *Towards an EU Research and Innovation Policy Agenda for Nature-Based Solutions & Re-Naturing Cities: Final Report of the Horizon 2020 Expert Group on 'Nature-Based Solutions and Re-Naturing Cities' (Full Version)*; Publications Office of the European Union: Luxembourg, 2020.
7. Maes, J.; Jacobs, S. Nature-Based Solutions for Europe's Sustainable Development. *Conserv. Lett.* **2017**, *10*, 121–124. [CrossRef]
8. Wamsler, C.; Alkan-Olsson, J.; Björn, H.; Falck, H.; Hanson, H.; Oskarsson, T.; Simonsson, E.; Zelmerlow, F. Beyond participation: When citizen engagement leads to undesirable outcomes for nature-based solutions and climate change adaptation. *Clim. Chang.* **2020**, *158*, 235–254. [CrossRef]
9. Somarakis, G.; Stagakis, S.; Chrysoulakis, N. (Eds.) *ThinkNature Nature-Based Solutions Handbook, Project Funded by the EU Horizon 2020 Research and Innovation Programme under Grant Agreement No. 730338*; Foundation for Research and Technology—Hellas: Heraklion, Greece, 2019. [CrossRef]
10. Bulkeley, H. Nature-based Solutions: Towards sustainable communities—Analysis of EU-funded projects. In *Nature-Based Solutions—State of the Art of EU-Funded Projects*; Directorate-General for Research & Innovation (European, Commission); Wild, T., Freitas, T., Vandewoestijne, S., Eds.; Publications Office of the European Union: Luxembourg, 2020; pp. 156–180.
11. Burton, P.; Mustelin, J. Planning for Climate Change: Is Greater Public Participation the Key to Success? *Urban Policy Res.* **2013**, *31*, 399–415. [CrossRef]
12. Renn, O.; Schweizer, P.-J. Inclusive risk governance: Concepts and application to environmental policy making. *Environ. Policy Gov.* **2009**, *19*, 174–185. [CrossRef]
13. Mees, H.L.P.; Driessen, P.P.J.; Runhaar, H.A.C. "Cool" governance of a "hot" climate issue: Public and private responsibilities for the protection of vulnerable citizens against extreme heat. *Reg. Environ. Chang.* **2015**, *15*, 1065–1079. [CrossRef]
14. Zingraff-Hamed, A.; Hüesker, F.; Lupp, G.; Begg, C.; Huang, J.; Oen, A.; Vojinovic, Z.; Kuhlicke, C.; Pauleit, S. Stakeholder Mapping to Co-Create Nature-Based Solutions: Who Is on Board? *Sustainability* **2020**, *12*, 8625. [CrossRef]
15. Glaas, E.; Neset, T.-S.; Kjellström, E.; Almås, A.-J. Increasing house owners adaptive capacity: Compliance between climate change risks and adaptation guidelines in Scandinavia. *Urban Clim.* **2015**, *14*, 41–51. [CrossRef]
16. Hegger, D.L.T.; Mees, H.L.P.; Driessen, P.P.J.; Runhaar, H.A.C. The Roles of Residents in Climate Adaptation: A systematic review in the case of the Netherlands. *Environ. Policy Gov.* **2017**, *27*, 336–350. [CrossRef]
17. Mees, H.L.P.; Uittenbroek, C.J.; Hegger, D.L.T.; Driessen, P.P.J. From citizen participation to government participation: An exploration of the roles of local governments in community initiatives for climate change adaptation in the Netherlands. *Environ. Policy Gov.* **2019**, *29*, 198–208. [CrossRef]
18. Xiang, P.; Yang, Y.; Li, Z. Theoretical Framework of Inclusive Urban Regeneration Combining Nature-Based Solutions with Society-Based Solutions. *J. Urban Plann. Dev.* **2020**, *146*, 04020009. [CrossRef]
19. Ferilli, G.; Sacco, P.L.; Tavano Blessi, G. Beyond the rhetoric of participation: New challenges and prospects for inclusive urban regeneration. *City Cult. Soc.* **2016**, *7*, 95–100. [CrossRef]
20. Frantzeskaki, N. Seven lessons for planning nature-based solutions in cities. *Environ. Sci. Policy* **2019**, *93*, 101–111. [CrossRef]
21. Carpini, M.X.D.; Cook, F.L.; Jacobs, L.R. Public Deliberation, Discursive Participation, and Citizen Engagement: A Review of the Empirical Literature. *Annu. Rev. Political Sci.* **2004**, *7*, 315–344. [CrossRef]
22. Ostrom, E. A Behavioral Approach to the Rational Choice Theory of Collective Action: Presidential Address, American Political Science Association, 1997. *Am. Political Sci. Rev.* **1998**, *92*, 1–22. [CrossRef]
23. URBiNAT. *Deliverable D3.3—Portfolio of Purposes, Methods, Tools and Content: Forming Digital Enablers of NBS. Project URBiNAT, Funded by the European Union's Horizon 2020 Research and Innovation Program under Grant Agreement No 776783*; IKED: Malmö, Sweden, 2020.
24. Agostino, D.; Arnaboldi, M. A Measurement Framework for Assessing the Contribution of Social Media to Public Engagement: An empirical analysis on Facebook. *Public Manag. Rev.* **2016**, *18*, 1289–1307. [CrossRef]
25. Bonsón, E.; Perea, D.; Bednárová, M. Twitter as a tool for citizen engagement: An empirical study of the Andalusian municipalities. *Gov. Inf. Q.* **2019**, *36*, 480–489. [CrossRef]
26. Chen, Q.; Min, C.; Zhang, W.; Wang, G.; Ma, X.; Evans, R. Unpacking the black box: How to promote citizen engagement through government social media during the COVID-19 crisis. *Comput. Human Behav.* **2020**, *110*, 106380. [CrossRef]

27. Pearce, K.E.; Rice, R.E. Digital Divides From Access to Activities: Comparing Mobile and Personal Computer Internet Users. *J. Commun.* **2013**, *63*, 721–744. [CrossRef]

28. Hörschelmann, K.; Werner, A.; Bogacki, M.; Lazova, Y. Taking Action for Urban Nature: Citizen Engagement Handbook, NATURVATION Guide. Project Funded by the European Union's Horizon 2020 Research and Innovation Programme under Grant Agreement No. 730243. 2019. Available online: https://naturvation.eu/result/taking-action-urban-nature-citizen-engagement (accessed on 20 October 2020).

29. Nature4Cities. Knowledge and Assessment Platform for Nature Based Solutions. Project Funded by the European Union's Horizon 2020 Research and Innovation Programme under Grant Agreement No 730468. 2017. Available online: https://nature4 cities-platform.eu/#/ (accessed on 20 October 2021).

30. NetworkNature. Nature-based Solutions Task Forces. Project Funded by the European Union's Horizon 2020 Research and Innovation Programme under Grant Agreement No. 887396. 2021. Available online: https://networknature.eu/networknature/ nature-based-solutions-task-forces (accessed on 20 October 2021).

31. Santos, B.d.S. (Ed.) *Cognitive Justice in a Global World: Prudent Knowledges for a Decent Life*; Lexington Books: Lanham, MD, USA, 2007; p. 462.

32. Santos, B.d.S. *The End of the Cognitive Empire: The Coming of Age of Epistemologies of the South*; Duke University Press Books: Durham, NC, USA, 2018; p. 392.

33. URBiNAT. *Deliverable D3.1—Strategic Design and Usage of Participatory Solutions and Relevant Digital Tools in Support of NBS Uptake. Project URBiNAT, Funded by the European Union's Horizon 2020 Research and Innovation Program under Grant Agreement No 776783*; Danish Technological Institute (DTI): Aarhus, Denmark, 2019.

34. Ferreira, I.; Caitana, B.; Nunes, N. *Policy Brief: Municipal Committees Experimenting in Innovative Urban Governance and Nature-Based Projects, Aimed at Inclusive Urban Regeneration*; ICLD—Swedish International Centre for Local Democracy: Visby, Sweden, in press.

35. URBiNAT. *Deliverable D3.2—Community-Driven Processes to Co-Design and Co-Implement NBS. Project URBiNAT, Funded by the European Union's Horizon 2020 Research and Innovation Program under Grant Agreement No 776783*; Centre for Social Studies of the University of Coimbra (CES-UC): Coimbra, Portugal, 2019.

36. URBiNAT. *Deliverable D2.3—On the Establishment of URBiNAT's Community of Practice (CoP). Project URBiNAT, Funded by the European Union's Horizon 2020 Research and Innovation Program under Grant Agreement No 776783*; IKED: Malmö, Sweden, 2020.

37. CES-UC. About CES—Overview. 2021. Available online: https://ces.uc.pt/en/ces/sobre-o-ces (accessed on 20 October 2021).

38. Madzak, M.V.; Hougaard, K.F.; Hilding-Hamann, K.E. *Kortlægning af Interne og Eksterne Initiativer Med Fokus på Entreprenørskab Inden for Udsatte Boligområder*; Danish Technological Institute (DTI): Aarhus, Denmark, 2020.

39. URBiNAT. *Deliverable D1.5—Compilation and Analysis of Human Rights and Gender Issues (Year 1). Project URBiNAT, Funded by the European Union's Horizon 2020 Research and Innovation Program under Grant Agreement No 776783*; Centre for Social Studies of the University of Coimbra (CES-UC): Coimbra, Portugal, 2019.

40. Santos, B.d.S.; Avritzer, L. Introduction: Opening up the Canon of Democracy. In *Democratizing Democracy: Beyond the Liberal Democratic Canon*; Santos, B.d.S., Ed.; Verso: London, UK; New York, NY, USA, 2005; pp. xxxiv–lxxiv.

41. URBiNAT. URBiNAT NBS Catalogue. 2021. Available online: https://urbinat.eu/nbs-catalogue/ (accessed on 20 October 2021).

42. Moniz, G.C.; Ferreira, I. Healthy Corridors for Inclusive Urban Regeneration. In *Rassegna di Architettura e Urbanistica—How Many Roads, 158*; Quodlibet: Rome, Italy, 2019; pp. 51–59.

43. Ferreira, I.A.R.G. Desenvolvimento Sustentável e Corredores Verdes. Barcelos Como Cidade Ecológica. Master's Thesis, University of Nova de Lisboa, Lisbon, Portugal, 2005.

44. URBiNAT. *Deliverable D4.1—New NBS. Co-Creation of URBiNAT NBS (Live) Catalogue and Toolkit for Healthy Corridor of the URBiNAT Project, Funded by the European Union's Horizon 2020 Research and Innovation Program under Grant Agreement No 776783*; IAAC: Barcelona, Spain, 2021.

45. Caitana, B.; Nunes, N. Social and solidarity economy and rights-based approach in URBiNAT. In *Deliverable D1.8/1.9—Compilation and Analysis of Human Rights and Gender Issues (Years 2 & 3) of the Project URBiNAT, Funded by the European Union's Horizon 2020 Research and Innovation Program under Grant Agreement No 776783*; URBiNAT, Ed.; Centre for Social Studies of the University of Coimbra (CES-UC): Coimbra, Portugal, 2021; pp. 20–23.

46. Mance, E.A. Redes de Colaboração Solidária. In *Dicionário Internacional da Outra Economia*; Cattani, A.D., Laville, J.-L., Gaiger, L.I., Hespanha, P., Eds.; Almedina: Coimbra, Portugal, 2009; pp. 278–283.

47. Manca, A.R. *Social Cohesion. Encyclopedia of Quality of Life and Well-Being Research*; Springer: Dordrecht, The Netherlands, 2014; pp. 6026–6028. [CrossRef]

48. URBiNAT. Chapter 1—Citizens engagement. In *Deliverable D1.2—Handbook on the Theoretical and Methodological Foundations of the Project URBiNAT, Funded by the European Union's Horizon 2020 Research and Innovation Program under Grant Agreement No 776783*; URBiNAT, Ed.; Centre for Social Studies of the University of Coimbra (CES-UC): Coimbra, Portugal, 2018; pp. 13–91.

49. Bernard, R.H.; Wutich, A.; Ryan, G.W. *Analyzing Qualitative Data: Systematic Approaches*, 2nd ed.; SAGE Publications: Los Angeles, CA, USA, 2017; p. 552.

50. Strauss, A.L.; Corbin, J.M. *Basics of Qualitative Research: Techniques and Procedures for Developing Grounded Theory*, 2nd ed.; Sage Publications: Thousand Oaks, CA, USA, 1998; p. 312.

51. Bernard, H.R.; Ryan, G.W. *Analyzing Qualitative Data: Systematic Approaches*; Sage: Thousand Oaks, CA, USA, 2010; p. 451.

52. URBiNAT. URBiNAT Ethical Principles. 2021. Available online: https://urbinat.eu/ethics/ (accessed on 20 October 2021).

53. URBiNAT. [Human Rights & Gender] Co-Designing Strategies for Inclusion, Based on Gender and Intersectional Approaches. 3 May 2021. Available online: https://urbinat.eu/articles/co-designing-strategies-for-inclusion-based-on-gender-and-intersectional-approaches/ (accessed on 20 October 2021).

54. URBiNAT. [Human Rights & Gender] Towards a Rights-Based Approach for an Inclusive Urban Regeneration with Nature-Based Solutions. 25 August 2021. Available online: https://urbinat.eu/articles/human-rights-gender-towards-a-rights-based-approach-for-an-inclusive-urban-regeneration-with-nature-based-solutions/ (accessed on 20 October 2021).

55. URBiNAT. *Deliverable D6.1—Communication and Dissemination Plan (V2). Project URBiNAT, Funded by the European Union's Horizon 2020 Research and Innovation Program under Grant Agreement No 776783*; ITEMS: Paris, France, 2020.

56. European Network of Living Labs (ENoLL). About—Open and Digital Living Lab Days. 2021. Available online: https://openlivinglabdays.com/about-us/ (accessed on 20 October 2021).

57. Santos, B.d.S. *Epistemologies of the South: Justice against Epistemicide*; Routledge: London, UK; New York, NY, USA, 2016; p. 282.

58. Hilding-Hamann, K.E. Interesting participatory cases. Non-public blog on URBiNAT's Internal Collaborative Platform. Work Package 3 on Citizen Engagement in Support of NBS (Available upon Request to the Corresponding Author). 2020.

59. URBiNAT. Bread Houses: Social and Solidarity Economy Nature-Based Solution. 2021. Available online: https://urbinat.eu/nbs_catalogue/bread-houses/ (accessed on 20 October 2021).

60. Savova-Grigorova, N. About Us | Bread Houses Network. 2021. Available online: https://www.breadhousesnetwork.org/about-us/ (accessed on 20 October 2021).

61. URBiNAT. Presentation and Results of the Workshop "Co-Creating Solutions with Local Citizens and Stakeholders within European Projects". 8 October 2020.

62. URBiNAT. Deliverable of the Workshop "Co-Creating Solutions with Local Citizens and Stakeholders within European Projects". 8 October 2020.

63. Santos, B.d.S.; Nunes, J.A.; Meneses, M.P. Introduction: Opening Up the Canon of Knowledge and Recognition of Difference. In *Another Knowledge Is Possible: Beyond Northern Epistemologies*; Santos, B.d.S., Ed.; Reinventing Social Emancipation toward New Manifestos; Verso: London, UK; New York, NY, USA, 2008; pp. xix–lxii.

64. Wild, T.; Freitas, T.; Vandewoestijne, S. (Eds.) *Directorate-General for Research and Innovation (European Commission). Nature-Based Solutions—State of the Art in EU-Funded Projects*; Publications Office of the European Union: Luxembourg, 2020.

65. Connolly, J.J.T.; Trebic, T.; Anguelovski, I.; Wood, E.; Thery, E. (Eds.) *Green Trajectories: Municipal Policy Trends and Strategies for Greening in Europe, Canada and United States (1990–2016)*; BCNUEJ: Barcelona, Spain, 2018.

66. Turnhout, E.; Metze, T.; Wyborn, C.; Klenk, N.; Louder, E. The politics of co-production: Participation, power, and transformation. *Curr. Opin. Environ. Sustain.* **2020**, *42*, 15–21. [CrossRef]

67. Wild, T. Research & innovation priorities in Horizon Europe and beyond. In *Nature-Based Solutions—State of the Art of EU-Funded Projects*; Directorate-General for Research and Innovation (European Commission); Wild, T., Freitas, T., Vandewoestijne, S., Eds.; Publications Office of the European Union: Luxembourg, 2020; pp. 223–233.

68. Bulkeley, H. Governing NBS: Towards transformative action. In *Nature-Based Solutions—State of the Art of EU-Funded Projects*; Directorate-General for Research & Innovation (European Commission); Wild, T., Freitas, T., Vandewoestijne, S., Eds.; Publications Office of the European Union: Luxembourg, 2020; pp. 181–202.

69. Nature4Cities. Deliverable D5.2—Citizen and Stakeholder Engagement Strategies and Tools for NBS Implementation. Project Funded by the European Union's Horizon 2020 Research and Innovation Programme under Grant Agreement No 730468. 2018. Available online: https://www.nature4cities.eu/results (accessed on 20 October 2021).

70. Kaiser, M.L.; Hand, M.D.; Pence, E.K. Individual and Community Engagement in Response to Environmental Challenges Experienced in Four Low-Income Urban Neighborhoods. *Int. J. Environ. Res. Public Health* **2020**, *17*, 1831. [CrossRef] [PubMed]

71. Fung, A. *Empowered Participation: Reinventing Urban Democracy*; Princeton University Press: Princeton, NJ, USA, 2004; p. 278.

72. Santos, B.d.S. A Critique of Lazy Reason: Against the Waste of Experience and Toward the Sociology of Absences and the Sociology of Emergences. In *Epistemologies of the South*; Routledge: London, UK; New York, NY, USA, 2016; pp. 164–187.

73. Santos, B.d.S.; Mendes, J.M. (Eds.) *Demodiversity: Towards Post-Abyssal Democracies*, 1st ed.; Routledge: New York, NY, USA, 2020; p. 260.

74. Honey-Rosés, J.; Anguelovski, I.; Chireh, V.K.; Daher, C.; Bosch, C.K.v.d.; Litt, J.S.; Mawani, V.; McCall, M.K.; Orellana, A.; Oscilowicz, E.; et al. The impact of COVID-19 on public space: An early review of the emerging questions—design, perceptions and inequities. *Cities Health* **2020**, 1–17. [CrossRef]

75. URBiNAT. In times of pandemic: Mapping backlashes, social challenges and solidarity responses with URBiNAT cities. In *Deliverable D1.8/1.9—Compilation and Analysis of Human Rights and Gender Issues (Years 2 & 3) of the Project URBiNAT, Funded by the European Union's Horizon 2020 Research and Innovation Program under Grant Agreement No 776783*; URBiNAT, Ed.; Centre for Social Studies of the University of Coimbra (CES-UC): Coimbra, Portugal, 2021; pp. 34–77.

76. Megglé, C. Coronavirus, Confinement et Quartiers Populaires: Des Vulnérabilités Particulières à Prendre en Compte. Localtis, un Média Banque des Territoires. 25 March 2020. Available online: https://www.banquedesterritoires.fr/coronavirus-confinement-et-quartiers-populaires-des-vulnerabilites-particulieres-prendre-en-compte (accessed on 20 October 2021).

77. Berkowitz, R.L.; Gao, X.; Michaels, E.K.; Mujahid, M.S. Structurally vulnerable neighbourhood environments and racial/ethnic COVID-19 inequities. *Cities Health* **2020**, 1–4. [CrossRef]

78. Campos, R. [COVID-19] Challenges, Responses and Solidarity in Brussels. URBiNAT—News. 4 June 2021. Available online: https://urbinat.eu/articles/covid-19-challenges-responses-and-solidarity-in-brussels/ (accessed on 20 October 2021).
79. Campos, R. [COVID-19] Challenges, Responses and Solidarity in Nantes. URBiNAT—News. 18 May 2021. Available online: https://urbinat.eu/articles/covid-19-challenges-responses-and-solidarity-in-nantes/ (accessed on 20 October 2021).
80. Campos, R. [COVID-19] Challenges, Responses and Solidarity in Porto. URBiNAT—News. 11 May 2021. Available online: https://urbinat.eu/articles/covid-19-reflections-on-the-impact-of-the-pandemic-part-i-porto/ (accessed on 20 October 2021).
81. Anguelovski, I. COVID-19 Highlights Three Pathways to Achieve Urban Health and Environmental Justice. International Institute for Environment and Development—Urban. 27 August 2020. Available online: https://www.iied.org/covid-19-highlights-three-pathways-achieve-urban-health-environmental-justice (accessed on 20 October 2021).
82. Campos, R. [COVID-19] Challenges, Responses and Solidarity in Sofia. URBiNAT—News. 19 May 2021. Available online: https://urbinat.eu/articles/covid-19-challenges-responses-and-solidarity-in-sofia/ (accessed on 20 October 2021).
83. Campos, R. [COVID-19] Challenges, Responses and Solidarity in Høje-Taastrup. URBiNAT—News. 4 June 2021. Available online: https://urbinat.eu/articles/covid-19-challenges-responses-and-solidarity-in-hoje-taastrup/ (accessed on 20 October 2021).
84. Campos, R. [COVID-19] Challenges, Responses and Solidarity in Nova Gorica. URBiNAT—News. 4 June 2021. Available online: https://urbinat.eu/articles/covid-19-challenges-responses-and-solidarity-in-nova-gorica/ (accessed on 20 October 2021).
85. Campos, R. [COVID-19] Challenges, Responses and Solidarity in Siena. URBiNAT—News. 9 June 2021. Available online: https://urbinat.eu/articles/covid-19-challenges-responses-and-solidarity-in-siena/ (accessed on 20 October 2021).
86. Gelli, F. Partecipazione e Fragilità. Osservatorio su Città e Trasformazioni Urbane. Fondazione Giangiacomo Feltrinelli. 19 October 2020. Available online: https://fondazionefeltrinelli.it/partecipazione-e-fragilita/ (accessed on 20 October 2021).
87. Allegretti, G. Ricostruire la Partecipazione Civica Nella Nuova Normalità. Alcuni Indirizzi Per Una Possibile Rifondazione. *Contesti Città Territ. Progett.* **2020**, 177–194. [CrossRef]
88. Dorronsoro, B. Gender mainstreaming. In *Deliverable D1.2—Handbook on the Theoretical and Methodological Foundations of the Project URBiNAT, Funded by the European Union's Horizon 2020 Research and Innovation Program under Grant Agreement No 776783*; URBiNAT, Ed.; Centre for Social Studies of the University of Coimbra (CES-UC): Coimbra, Portugal, 2018; pp. 215–218.
89. Lettoun, S. Human rights-based approach in urban regeneration. In *Deliverable D1.2—Handbook on the Theoretical and Methodological Foundations of the Project URBiNAT, Funded by the European Union's Horizon 2020 Research and Innovation Program under Grant Agreement No 776783*; URBiNAT, Ed.; Centre for Social Studies of the University of Coimbra (CES-UC): Coimbra, Portugal, 2018; pp. 211–214.
90. Lettoun, S. Applying SDGs framework. In *Deliverable D1.2—Handbook on the Theoretical and Methodological Foundations of the Project URBiNAT, Funded by the European Union's Horizon 2020 Research and Innovation Program under Grant Agreement No 776783*; URBiNAT, Ed.; Centre for Social Studies of the University of Coimbra (CES-UC): Coimbra, Portugal, 2018; pp. 236–239.
91. Voorberg, W.H.; Bekkers, V.J.J.M.; Tummers, L.G. A Systematic Review of Co-Creation and Co-Production: Embarking on the social innovation journey. *Public Manag. Rev.* **2015**, *17*, 1333–1357. [CrossRef]
92. Ferreira, I.; Duxbury, N. Cultural projects, public participation, and small city sustainability. In *Culture in Sustainability: Towards a Transdisciplinary Approach*; Asikainen, S., Brites, C., Plebańczyk, K., Rogač Mijatović, L., Soini, K., Eds.; SoPhi, University of Jyväskylä: Jyväskylä, Finland, 2017; pp. 46–61.
93. Haase, A. The Contribution of Nature-Based Solutions to Socially Inclusive Urban Development—Some Reflections from a Social-environmental Perspective. In *Nature-Based Solutions to Climate Change Adaptation in Urban Areas: Linkages between Science, Policy and Practice*; Kabisch, N., Korn, H., Stadler, J., Bonn, A., Eds.; Theory and Practice of Urban Sustainability Transitions; Springer International Publishing: Cham, Switzerland, 2017; pp. 221–236. [CrossRef]
94. Spada, P.; Allegretti, G. When Democratic Innovations Integrate Multiple and Diverse Channels of Social Dialogue: Opportunities and Challenges. In *Using New Media for Citizen Engagement and Participation*; Adria, M., Ed.; IGI Global: Hershey, PA, USA, 2020; pp. 35–59.
95. Allegretti, G. Participatory democracies: A slow march toward new paradigms from Brazil to Europe? In *Cities into the Future*; Lieberherr-Gardiol, F., Solinís, G., Eds.; Les Classiques des Sciences Sociales: Chicoutimi, QC, Canada, 2014; pp. 141–177.
96. Allegretti, G.; Antunes, S. The Lisbon Participatory Budget: Results and perspectives on an experience in slow but continuous transformation. *Field Actions Sci. Rep. J. Field Actions* **2014**. Available online: https://journals.openedition.org/factsreports/3363 (accessed on 20 October 2021).
97. D'Hont, F. *Visioning as Participatory Planning Tool: Learning from Kosovo Practices*; UN-Habitat, SIDA: Nairobi, Kenya, 2012; p. 100.
98. Wilding, H.; Gould, R.; Taylor, J.; Sabouraud, A.; Saraux-Salaün, P.; Papathanasopoulou, D.; de Blasio, A.; Nagy, Z.; Simos, J. Healthy Cities in Europe: Structured, Unique, and Thoughtful. In *Healthy Cities: The Theory, Policy, and Practice of Value-Based Urban Planning*; de Leeuw, E., Simos, J., Eds.; Springer: New York, NY, USA, 2017; pp. 241–292.
99. World Health Organization. Copenhagen Consensus of Mayors. Healthier and Happier Cities for All. 13 February 2018. Available online: https://www.euro.who.int/en/health-topics/environment-and-health/urban-health/publications/2018/copenhagen-consensus-of-mayors.-healthier-and-happier-cities-for-all-2018 (accessed on 20 October 2021).
100. De Leeuw, E. One Health(y) Cities. *Cities Health* **2020**, 1–6. [CrossRef]
101. Igoe, M. Can 'Nature-Based Solutions' be More than a Buzzword? Devex News. 13 December 2019. Available online: https://www.devex.com/news/sponsored/can-nature-based-solutions-be-more-than-a-buzzword-96216 (accessed on 20 October 2021).

102. Nelson, D.R.; Bledsoe, B.P.; Ferreira, S.; Nibbelink, N.P. Challenges to realizing the potential of nature-based solutions. *Curr. Opin. Environ. Sustain.* **2020**, *45*, 49–55. [CrossRef]
103. Moulaert, F.; MacCallum, D.; Mehmood, A.; Hamdouch, A. *The International Handbook on Social Innovation: Collective Action, Social Learning and Transdisciplinary Research*; Edward Elgar Pub: Cheltenham, UK; Northampton, MA, USA, 2014.
104. André, I.; Abreu, A. Dimensões e espaços da inovação social. *Finisterra* **2006**, *41*. [CrossRef]
105. Murray, R.; Caulier-Grice, J.; Mulgan, G. *The Open Book of Social Innovation: Ways to Design, Develop and Grow Social Innovation*; Young Foundation and NESTA: London, UK, 2010; p. 222.
106. Nunes, N.; Caitana, B. The appropriation of citizenship rights in the promotion of social cohesion and urban social innovation. In *Deliverable D1.2—Handbook on the Theoretical and Methodological Foundations of the Project URBiNAT, Funded by the European Union's Horizon 2020 Research and Innovation Program under Grant Agreement No 776783*; URBiNAT, Ed.; Centre for Social Studies of the University of Coimbra (CES-UC): Coimbra, Portugal, 2018; pp. 18–24.

Article

How Do Nature-Based Solutions' Color Tones Influence People's Emotional Reaction? An Assessment via Virtual and Augmented Reality in a Participatory Process

Barbara Ester Adele Piga [1,*], **Gabriele Stancato** [1], **Nicola Rainisio** [2] and **Marco Boffi** [2]

[1] Laboratorio di Simulazione Urbana Fausto Curti, Department of Architecture and Urban Studies (DAStU), Politecnico di Milano, 20133 Milan, Italy; gabriele.stancato@polimi.it
[2] Department of Cultural Heritage and Environment, University of Milan, 20122 Milan, Italy; nicola.rainisio@unimi.it (N.R.); marco.boffi@unimi.it (M.B.)
* Correspondence: barbara.piga@polimi.it

Abstract: Simulations of urban transformations are an effective tool for engaging citizens and enhancing their understanding of urban design outcomes. Citizens' involvement can positively contribute to foster resilience for mitigating the impact of climate change. Successful integration of Nature-Based Solutions (NBS) into the urban fabric enables both the mitigation of climate hazards and positive reactions of citizens. This paper presents two case studies in a southern district of Milan (Italy), investigating the emotional reaction of citizens to existing urban greenery and designed NBS. During the events, the participants explored in Virtual Reality (VR) ($n = 48$) and Augmented Reality (AR) ($n = 63$) (i) the district in its current condition and (ii) the design project of a future transformation including NBS. The environmental exploration and the data collection took place through the exp-EIA© method, integrated into the mobile app City Sense. The correlations between the color features of the viewed landscape and the emotional reaction of participants showed that weighted saturation of green and lime colors reduced the unpleasantness both in VR and AR, while the lime pixel area (%) reduced the unpleasantness only in VR. No effects were observed on the Arousal and Sleepiness factors. The effects show high reliability between VR and AR for some of the variables. Implications of the method and the benefits for urban simulation and participatory processes are discussed.

Keywords: urban design; Augmented Reality; Virtual Reality; emotions; co-design; computer vision; simulation; Environmental Psychology; colors; Nature-Based Solutions

Citation: Piga, B.E.A.; Stancato, G.; Rainisio, N.; Boffi, M. How Do Nature-Based Solutions' Color Tones Influence People's Emotional Reaction? An Assessment via Virtual and Augmented Reality in a Participatory Process. *Sustainability* **2021**, *13*, 13388. https://doi.org/10.3390/su132313388

Academic Editor: Ivo Machar

Received: 3 November 2021
Accepted: 29 November 2021
Published: 3 December 2021

Publisher's Note: MDPI stays neutral with regard to jurisdictional claims in published maps and institutional affiliations.

1. Introduction

Nature-Based Solutions (NBS) are increasingly adopted in a logic of risk management, resilience, mitigation, and adaptation to face urgent socio-economical-environmental issues such as climate change, natural disasters, food and water security, biodiversity loss, social cohesion, health [1]. NBS encompasses several eco-system-based approaches, such as ecosystem-based adaptation, ecosystem-based disaster-risk reduction, ecosystem-based mitigation [2,3]. With other initiatives related to the 2030 Agenda for Sustainable Development, the systematic approach linking human wellbeing and natural systems emerge as crucial for proper sustainable growth. Governments, businesses, and civil society show a growing interest in such a perspective. In the NBS panorama framed by the European Commission, interdisciplinary and systematic approaches and solutions are relevant and should lead to a mutual and "balanced benefit for nature and society" [4] (p. 1217). This approach towards more sustainable development can benefit from rapid technological advancements. As highlighted by Bishop [5], in the field of landscape planning and particularly concerning environmental information, already in the past and even more today, "computer developments have created new opportunities for the landscape researcher in all areas of data storage, modeling and visualization" [5] (p. 112). The author continues stressing that it is highly probable that Augmented Reality (AR) and Virtual Reality (VR), in conjunction with immersive modes of visualizations, will play a significant role in the field in the next 10 years.

In this article, we explore a specific application of computer vision to deepen the relationship between NBS and people's emotions, with a dual aim = On the one hand, to investigate how color tones of NBS influence the subjective emotional experience in urban spaces, a topic that is poorly addressed in the literature so far. On the other hand, to develop a reliability analysis on two emerging technologies (AR and VR) regarding the aforementioned relationship. The results will therefore provide new insights both in the design of NBS as emotionally supportive environments, and in the field of urban simulation.

In detail, the article presents: (i) a literature review that relates Nature-Based Solutions (NBS) and citizens engagement, (ii) a brief framework of AR and VR solutions for citizens' involvement, (iii) the relationship between green solutions and their emotional effects, (iv) the objective and methodology of the study, (v) the analysis of results, and (vi) the outcomes discussion, the limitations of the current research, and the future work to develop. The analytical method is part of a research project that led to developing the AR4CUP mobile application (distributed as City Sense).

2. Literature Review

2.1. NBS and the Relevance of Citizens Engagement

The many positive effects of NBS, often highlighted in the literature, should not be confused with an indistinct process of 'biophilic washing' that applies standardized solutions to different urban contexts and does not consider the social, emotional, and community dimension of the transformation processes [4,6,7]. Different biophilic design strategies have various cost-benefit ratios and different acceptance levels [8–10]. According to the review of 42 different design strategies carried out by Xue in collaboration with 30 experts [11], the solutions with the best Benefit-Cost Ratio (BCR) include: 'biophilic infrastructure' (i.e., green space coverage ratio, plants canopy configuration), 'sensory design', (i.e., visual connections with nature, green walls, and others), and 'natural landscape promotion with minimal management' (i.e., green roof). Among those strategies, one of the most preferred for investments is the 'green space coverage ratio', which focuses on the correct ratio between green elements and artificial ones [12] to obtain a positive response by observers. Indeed, according to Jiang and colleagues' observations, the appreciation curve tends to have an asymptotic trend, reaching the plateau around 41% density of the tree canopy, which is consistent with the notion of balance between understanding and exploration supported by the preference matrix theory [13]. This variation in response to green distribution shows that it is not possible to take for granted that an NBS intervention is functional and appreciable in itself. However, positive effects of buildings, including greenery on their façades, were observed on aesthetic, restorative, and affective dimensions [14].

Nevertheless, Wolch [15] suggests that urban green projects may create a paradox. On the one hand, they make the city healthier both physically and mentally (see for instance [10,16]). On the other hand, the most effective NBS interventions are usually applied to urban degraded areas, where they often induce the increase of the real estate value fostering gentrification; as a result, these renovated areas become economically unsustainable for the population living there [17–19]. To avoid inducing a phenomenon of social injustice and the related conflicts resulting from an urban intervention, some authors propose finding a 'just green enough' balance [20] intended as an optimal and balanced solution between the community's needs and the developers' tradeoff. Wolch and colleagues [15] argue that this approach implies the development of urban planning strategies based on the wishes and needs of the communities involved, rather than grounding design projects on conventional solutions focused exclusively on ecological issues. As urban greenspace has a significant impact also on real estate values [21,22], its implementation plays a crucial role in increasing or reducing social justice and equity [23]. Equity factors are closely related to the urban greenery accessibility and proximity to the public [24], to the point that the spatial distribution of greenery can even draw social geography of inequality [24–26]. Therefore, in NBS interventions, it is necessary to consider both the physical environmental effects and related long-term social impacts.

In this regard, participatory processes in green areas are fundamental to prevent and reduce conflicts between stakeholders [27]; moreover, the experience of being involved in the decision-making process increases end-users awareness about the importance of implementing and preserving the green areas [28]. Furthermore, some critical issues may emerge in NBS processes if the citizen's perceptual perspective is not adequately considered [29]. Indeed, various actors perceive vegetation differently [30]; in some cases, social groups may object to tree planting since they perceive it as a potential source of indirect disservices [31]. Citizens' involvement in transformation processes generally mitigates these types of disagreements [32]. However, traditional participatory processes may encounter difficulties in engaging the weaker segments of the population [24,26], which implies the need for a contextual design of the engagement strategy to favor sustainable participation for citizens [33].

Information technologies might play an important role in such perspective, extending the possibilities of participation [34] by overcoming some limitations of traditional methods through digital inclusion [35], such as the difficulty of engaging many people simultaneously or the availability of schedules for specific categories of workers [36]. The widespread use of mobile devices and the continuous flux of information exchange led to the idea that it is possible to describe the relationship between citizen and city as a spatialized intelligence [37]. Nevertheless, these devices should be considered as an integrative tool for more comprehensive participatory processes and not as a stand-alone solution; this is particularly relevant when dealing with specific populations affected by low digital literacy, such as older people, which may benefit from more traditional activities (see for instance [38,39]).

2.2. Augmented and Virtual Reality as Citizens' Engagement Solutions

Although the forms of smart participation are relatively recent, two main approaches emerge when considering the type of information collected involving citizens: (i) the environmental-centered approach, which uses objective data for studying the environment, e.g., by evaluating environmental parameters through cell phone sensors; (ii) the people-centered approach, which studies the human perceptual dimension exploring subjective data [40]. Often these solutions make use of mobile applications: the first approach encompasses APPs dealing with the urban environment under different meanings such as 'environmental risk and adaptation' [41] and 'urban transformation modeling' [42,43]; the second approach encompasses APPs dealing with citizens' perceptions, through sensory assessment (e.g., physical comfort) [40], attitudes (e.g., safety and security) [44], or emotional assessment (e.g., environmental affective quality) [45].

After the early 2000s, VR has enriched e-participation (participation through ICT) [46]; more recently, AR has also become relevant in participatory design [47–49]. These solutions can be exploited to visualize the design projects or their alternatives or even allow the direct modification by the user of the 3D model components [50,51]. The three-dimensional model visualization can directly occur on the construction site through Augmented Reality (e.g., APPs such as City Sense; Urban CoBuilder; AR Sketchwalk; ARki) or off-site in Virtual Reality (e.g., APPs such as City Planner Online—KALASATAMA; Virtual Singapore; 3D Rotterdam 2.0; TYGRON). The visualization via VR and AR often happens in a subjective perspective using photorealistic scenarios. Such representations are named 'experiential simulations' since they mimic reality with an eye-level point of view. When such solutions follow parameters ensuring representation realism and fidelity [52–55], their use in participatory urban processes ensures that citizens' reactions to the simulated environments are comparable to those they would have experienced by looking at the actual environment [56]. Several studies demonstrated that observing real natural landscapes or accurate experiential simulations of the same environment generates comparable physiological and psychological reactions [57–61]. Among the different ways to visualize urban scenarios, VR and AR open a crucial possibility: they can anticipate future urban transformation easy-to-understand, often enabling a 'naturalistic interaction' [62] with the environment. Indeed, 3D visualization methods are considered crucial for properly conveying spatial features of places to laypeople, both in their current and future conditions, enabling new forms of citizens engagement [63]. According to the case study presented by Edler et al. [64], one of the main advantages of VR lies in its interactive nature, which offers the possibility of actively exploring the modelled landscapes. Combining the possibility of freely positioning the virtual camera, which overcomes the limitations of physical environments, and the support offered by navigation aids (e.g., mini-maps, signifying footprints, pointer teleportation, teleport stations), participants can access a more detailed experience of the simulated environment in a short time than with traditional tools. Likewise, Loyola et al. [65] exploited the natural sense of presence of immersive VR simulations to present a project of an urban park. The authors reported a higher level of comprehension of the physical features of the design proposal (e.g., presence of various functional areas, characteristics of urban furniture and vegetation, relationship with the surroundings) for VR users compared to those involved with traditional means. Similar considerations are drawn from case studies applying AR for citizens engagement, which is considered particularly effective as its novelty increases the willingness to participate [66] and can be fruitfully integrated in existing participatory practices [67]. In particular, AR offers two fundamental advantages: (i) the in-situ immersive experiences favor a suspension of disbelief and, therefore, ease spontaneous reactions to visualizations [63]; (ii) the superimposition of the design project onto the physical reality fosters the comparison between current and potential condition. Despite this, it is essential to note that in AR the existing context is perceived in its actual conditions and not according to the modifications induced by the urban transformation on its surroundings. Thus, the 'semi-permanent' urban elements (e.g., urban furniture), the 'recursive' ones (e.g., seasonal or hourly cycles), and the 'temporary conditions' (e.g., people, cars) [54] are consistent with the current conditions and are not affected by the designed transformation, as it can instead happen in VR. Despite their differences, VR and AR are considered among the key technologies to enable a smart urban greenery management, which is conceived as "the design, establishment, monitoring, and management of urban trees and vegetation through the use of digital technologies, for the joint purpose of improving the urban environment and engaging all relevant stakeholders in its governance" [68] (p. 8).

2.3. Green Effects: Natural Elements and Color Clues

According to the psychological literature, natural and/or green elements are strongly associated with positive effects [69,70]. Momentary or prolonged exposure to natural environments was found to be related to a broad spectrum of positive psychological states, namely stress reduction [71,72], restoration of optimal attention span [73], flow or peak experiences [7], positive emotions increase [74,75]. Furthermore, these elements were identified as antecedents of broader experiential, social, and performance outcomes, including pain reduction [76,77], faster post-surgery recovery [72], better results in logic tasks [78], decreased aggressivity [79], and increased proximity sociality [80]. Referring to an epidemiological framework, more frequent exposure to natural environments was also associated with better children development [81] and a lower prevalence of psychiatric pathologies (schizophrenia and anxiety) [82,83]. These results were mainly explained by the ability of the natural environment to attract involuntary attention [13,84] and/or automatically reduce stress [16] due to the evolutionary bond between humans and the natural environment as a primary source of food and shelter. According to Korpela and colleagues [85,86], this evolutionary link is well recognized and present in people's mental life, as natural environments are consciously used as tools for emotional self-regulation (environmental self-regulation hypothesis).

These beneficial effects occur in the direct presence of natural environments and also when such environments are presented through photographs, videos, or VR [57,87–89]. Many studies in this field are conducted in laboratories to control the intervening variables (see [70] for a methodological summary). On the one side, the virtual scenario reduces variables and allows scholars to manipulate them according to the experimental goals, e.g., studying the influence of light or weather conditions on emotional states [90]. On the other side, researches may be focused on the simulation tools themselves; for instance, de Kort [91] showed that immersive simulations increase the restorative effect of projected natural environments, even though such effect is recorded only for physiological measures (HR and skin conductance) and not for self-reported affect. The use of immersive devices, e.g., Head-Mounted Displays (HMD), showed that exposure to natural scenarios in VR induces the same anxiety-reducing effects as exposure to real natural environments [92], but without inducing the expected positive attentional restoration effects [93]. Despite this, VR scenarios allowed scholars to show that environments partially covered by vegetation are preferred to environments wholly covered with greenery or completely open green spaces, both considering physiological sensors [94] and EEG [95].

Positive green effects are also found in the absence of natural elements, as mere exposure to color clues (see [96] for a review). Exposure to green color was indeed correlated to better performances in logic [97] and creativity tasks [98], reduced perception of physical fatigue [99], and was associated with a general feeling of calm [100], in line with the results presented above. Green is often contrasted with red, as these are two antagonistic additive primary colors, which is instead related to greater aggressivity, sexual attraction, and better sports performance [101,102]. The influence of the green color was also tested by showing a video of a route in a natural environment represented in three variants (unedited, achromatic, and red filter) to three groups of volunteers undergoing physical exertion; only those who observed nature without alteration obtained a benefit in terms of performance [99]. Considering the colors' mutual influence, Bartram and colleagues [103] used network analysis to represent colors as a network of assessed psychological relationships, arguing that the brightest colors do not transmit negative sensations and green tones represent a significant cluster related to positive sensations. The lightness of neutral tones, often related to artificial elements, seems to reduce the arousal values [100,104,105]. Applying such an approach to architectural settings requires considering many factors, as the chromatic composition of an urban environment is articulated and complex [106]. Thus, Manav [107] applied a segmentation and dominant color extraction method to discretize the volunteers' emotional answers to panoramic photos, relating color tones to emotions. It emerged that the urban context with the more massive presence of green vegetation was identified as the most restful. Different theoretical explanations for such green color effects appear to be complementary. Elliot and Maier [96] hypothesized that colors might have a signal function to maximize animal fitness, eliciting automatic psycho-physiological reactions and orienting the individual behavior. Moreover, the green color is historically associated with positive meanings in popular culture, particularly with fertility, hope, and renewal [98]. Consistently with such studies, Palmer and Schloss [108] argued that human preference for specific tones is both an evolutionary effect and an association of ideas between colors and known objects.

The studies presented in this article investigate how different color tones of NBS can influence people's emotional reactions towards urban areas and whether this influence relationship is equivalent in AR and VR. To this end, a two-step analysis was carried out:

(1)   Analysis of the correlations between different color tones and emotions.
(2)   Reliability analysis of the detected correlations, comparing the results obtained in AR and VR.

3. Methods

3.1. Materials

Data collection was carried out in two case studies in the Porta-Romana district (south Milan, Italy) with a quasi-experimental design. From the eighties the city of Milan passed from an industry-based to a service-based city; this long and relevant process of urban renewal of unused areas, mainly former industries and railway yards, is still undergoing [109]. The southern part of the Porta-Romana district, where the two studies are located, is one of these areas, and it is still under an important renovation process that is changing the district's identity from industrial to business-oriented. In general, Milan, a city of the flat Po Valley, suffers from severe heatwaves in summer and flooding in winter [110]. Despite this, no data seem to suggest a relevant effect of weather on daily mood in such a context [111].

The two studies were conducted under different climatic conditions. Study 1 was conducted indoors, observing panoramic photographs taken in June 2018; Study 2 was conducted outdoors in December 2019. In Milan. According to ARPA (*Richiesta dati misurati—Meteorologia | ARPA Lombardia* (2021). Available at: https://www.arpalombardia.it/Pages/Meteorologia/Richiesta-dati-misurati.aspx, accessed on 19 November 2021) (Agenzia Regionale per la Protezione Ambientale) data, the environmental conditions in June on average are: temperature 24.02 °C, relative humidity 82.69%, rainfall 5.53 mm, wind speed 1.45 m/s, daylight hours 480. In December, the environmental conditions on average are: temperature 6.24 °C, relative humidity 83.90%, rainfall 40.32 mm, wind speed 1.39 m/s, daylight hours 266. According to the criteria of the UTCI (Universal Thermal Climate Index) [112,113], a potential discomfort condition in June can be indicated for 12% of the hours of the month; in December, there a potential discomfort can be experienced for 88% of the hours of the month. The average NDVI (Normalized Difference Vegetation Index) [114], evaluated on the base of Sentinel2 data, in June 2018 is M:0.25 s.d.:0.00813, and in December 2019 is M: 0.11 s.d.: 0.00740; comparing the two months NDVI decreases by 44%. Study 1 uses VR (i.e., panoramic photographs) of four representative points of view surrounding the Fondazione Prada and piazza Olivetti, recently renovated. Study 2 uses AR to show the urban design project VITAE by Covivio, Carlo Ratti Associati, and Partners (via Serio). Study 1 presents the actual urban area with the existing vegetation, whereas Study 2 is a biophilic design project with NBS solutions, including a walkable green spiral with terraces running from the ground to the rooftop of the building.

In both case studies, participants observed experiential simulations with vegetation and artificial elements. Study 1 presented pre-selected StreetView™ pannable panoramas from four fixed points of view. Study 2 presented the photorealistic render of the VITAE design project superimposed on-site to the actual environment through AR using the City Sense app; the rendered photorealistic 3D model of the urban transformation is automatically located in the right place and consistently anchored to the actual context by the app.

The parts of the simulations belonging to the chromatic range of lime and green tones (Hue: 38–67°) were exclusively vegetal elements, both in VR (Study 1) and AR (Study 2); the chromatic preference for these tones is therefore connected to the existing or designed vegetation. Neutral tones (Saturation <10%) are associated with built elements (mainly buildings, sidewalks, streets).

The emotions experienced by the participants in both simulated urban environments were assessed through a questionnaire consisting of 20 items rated on a 5-points Likert scale [115]. The questionnaire's answers allow describing the emotions through four factors, namely Pleasantness, Unpleasantness, Arousal, and Sleepiness. Those factors are conceived as two pairs of oppositional values on the Unpleasant-Pleasant continuum, which indicates the level of pleasure of the emotions, and the Sleepy-Arousing continuum, which indicates the level of activation of the emotions. According to such a theoretical model, considering the values of the four scales provides a holistic description of a person's affective state. Moreover, the values obtained on each continuum can be used as coordinates to place the resulting emotions in the circumplex model, which describes a cartesian plane where affective states have a univocal label.

3.2. Procedure and Participants

In this case, 48 students (age M = 26; s.d. = 12.12) from Università degli Studi di Milano attended Study 1. The four stimuli, i.e., spherical panoramas of the existing condition projected on a widescreen, were presented to participants. After a short visual exploration of the urban context by panning the panorama, the virtual camera was brought back to the initial target point, i.e., the urban perspective to assess. At the end of the virtual experience of each point of view, students had to fill in the questionnaire.

Here, 63 citizens (age M = 41; s.d. = 12.81) attended Study 2. During the first public event for presenting the VITAE design project (*Experiencing VITAE—LABSIMURB* (2021). Available at: http://www.labsimurb.polimi.it/research/ar4cup/experiencing_vitae/, accessed on 19 November 2021), participants made a semi-guided exploration of the project area using the City Sense app in AR mode. The app showed the photorealistic model of the urban transformation superimposed to the actual context and applied the exp-EIA© method for assessing the experience in the environment, including the psychological questionnaire formerly described. The organizers identified three main relevant perspectives (two in front of the designed project and one on its back) for stopping the walk, looking around and towards the VITAE project, and assessing the urban scenario via the app.

3.3. Analyses

Data collected through the emotions' questionnaire were treated in three ways. Firstly, descriptive statistics were used to locate on a cartesian plane the emotional state experienced from each point of view; different colors are assigned to the emotions distributed on the cartesian plane. Secondly, emotions were integrated with geographic information of the users' position and their visual target, producing a semantic isovist map; according to the exp-EIA method©, colors corresponding to the experiences on the cartesian plane were applied to the related partial isovist, i.e., the portion of space visible from a specific point of view and with a single target [116,117]. Thirdly, inferential statistics were used to detect significant correlations between the colors of the urban landscape and the emotional factors.

The participants' answers gathered in the two case studies were clustered according to the StreetView™ camera location (Study 1) and the GPS observers' locations (Study 2) using the DBSCAN method [118] with Scikit-learn 0.22 and Python 3.8 libraries. This procedure allowed us to identify different clusters of participants based on their spatial exploration. For each cluster, the average value of the answers to the emotions' questionnaire was calculated. In Study 1, four main clusters were identified, i.e., the four target points assigned for the experimental task. In Study 2, three main clusters were identified, distributed around the building simulated in AR. Each cluster was associated with the specific image representing its view and the related answers, which served for building the color-emotion correlation matrix. Through color segmentation [119], we extracted and quantified, via Computer Vision processing with OpenCV 3 library and Python 3.8, 45 chromatic features mined from Lightness, a* and b* (CIELAB) color-space, and Hue Saturation Value (HSV) color-space. The 45 color features identified and evaluated are organized in: (i) Lightness; (ii) oppositive channels a* b*; (iii) Hue; (iv) Saturation. More in detail:

- LIGHTNESS: "The brightness of an area judged relative to the brightness of a similarly illuminated area that appears to be white or highly transmitting" [120] (p. 88), analyzed in the following ways: (1) average brightness of the image; (2) percentage of low-brightness pixels (L < 15%); (3) percentage of high-brightness pixels (L > 85%); (4) average brightness of low-saturation pixels only.
- OPPOSITIVE CHANNELS: "The a* and b* dimensions approximately correlated with red-green and yellow-blue chroma perceptions" [120] (p. 202), analyzed in the following ways: (1) average oppositional a* green-red; (2) average oppositional b* blue-yellow.
- HUE: "Attribute of a visual perception according to which an area appears to be similar to one of the colors—red, yellow, green, and blue—or to a combination of adjacent pairs of these colors considered in a closed ring" [120] (p. 88), calculated in a range [0°:180°], it is analyzed in the following ways: (1) percentage of orange tones pixels ($8° \leq H < 23°$); (2) percentage of yellow tones pixels ($23° \leq H < 38°$); (3) percentage of lime tones pixels ($38° \leq H < 53°$); (4) percentage of green tones pixels ($53° \leq H < 68°$); (5) percentage of turquoise tones pixels ($68° \leq H < 83°$); (6) percentage of cyan tones pixels ($83° \leq H < 98°$); (7) percentage of cobalt tones pixels ($98° \leq H < 113°$); (8) percentage of blues tones pixels ($113° \leq H < 143°$); (9) percentage of violet tones pixels ($128° \leq H < 143°$); (10) percent-

age of magenta tones pixels ($143° \leq H < 158°$); (11) percentage crimson tones pixels ($158° \leq H < 173°$); (12) percentage of red tones pixels ($173° \leq H < 179°$ and $0° \leq H < 8°$).

- SATURATION: "Colorfulness of an area judged in proportion to its brightness" [120] (p. 91), it is calculated for each image as: (1) mean saturation of the entire image; (2) the percentage of the pixels' area belonging to the same hue, or more simply, the image surface with the same color tones, (3) the 'mean saturation' of a specific hue, that is the average saturation of a color's tone range, and (4) the 'weighted saturation', i.e., the combination of the previous two, that is the ratio between the mean saturation and the pixels' area of a specific hue. Furthermore, each image was filtered on the base of the L channel (CIELAB) to analyze: (i) saturation of low lightness pixels; (ii) saturation of high lightness pixels.

The datasets containing the chromatic features and the emotional response values (structured separately for the two different studies) were normalized using the ScikitLearn MinMaxScaler method (range [0:1]) to make the variables comparable. A correlation matrix was then generated by checking the chromatic feature/emotional response pairs.

Due to the different simulation solutions of the two studies (VR in Study 1, AR in Study 2), the correlation values between colors and emotions in one system and the other may differ. For this reason, we first considered the correlation's statistical significance and then applied the Bland-Altman evaluation [121,122] to establish the concordance between the correlations found in the two case studies. In Bland-Altman's graphs, the mean of the correlations found in the two cases is shown on the abscissas and the difference between the two correlations values on the ordinates. According to this representation, the more the pairs of correlations agree, the closer they are to the indifference line (delta = 0.00) on the y-axis. The closer they are to the indifference line, the closer they are to the probable real value. Furthermore, the more the correlation values are in agreement and the higher the correlation value is, the more they are placed at the two extremes of the x-axis (mean). In identifying the most relevant correlations, we classified the concordance based on the following criteria: (i) to be High level, the correlations must be contained within the confidence range of the difference values (mean-t_test_confidence) and have an absolute value r > 0.75 (in charts the points inside the azure stripe); (ii) to be Medium level the correlations must be in the intermediate bands between the confidence area of the difference values and the t-confidence boundaries (±1.96 std) of the data (the charts' area outside the azure stripe and inside the red dashed lines); (iii) all the other correlations are classified as a Low level of concordance.

Using the Bland-Altman chart coordinates, we applied a DBSCAN clustering to check possible groups of features to deeper interpret the correlations as agreement patterns; the cluster analysis is conducted on the emotional characteristics that show higher agreement and higher statistical significance in correlations.

4. Results

4.1. Emotional Reactions to the Simulated Urban Environment

The cartesian plane described by the circumplex model presents pleasant emotions on the right and unpleasant emotions on the left, arousing emotions on the top and sleepy emotions on the bottom. The cartesian plane is divided into sections labeled with basic emotions. In Figure 1, each dot represents the average value of the emotional reaction from a single point of view (PoV). PoVs A and B of Study 1 were categorized as depressed. PoV C of Study 1 was categorized as alert-excited. PoV D of Study 1 was categorized as tense. PoVs 1 and 3 of Study 2 were categorized as fatigued. PoV 2 of Study 2 was categorized as calm. In Figure 2, the isovists corresponding to the different PoVs (Figure 3) rated by participants are shown on a map. The color of each isovist corresponds to the color of the position the PoV has on the cartesian plane.

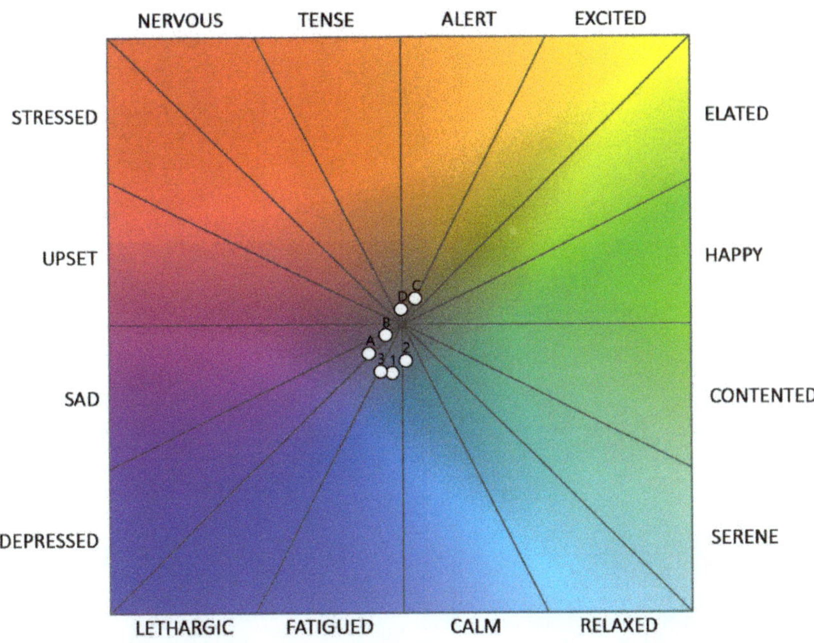

Figure 1. Mean values on the cartesian plane described by the circumplex model for the PoVs assessed in Study 1 (PoVs A, B, C, and D) and Study 2 (PoVs 1, 2, and 3). Source: chart based on Russell's circumplex model, elaboration by the authors.

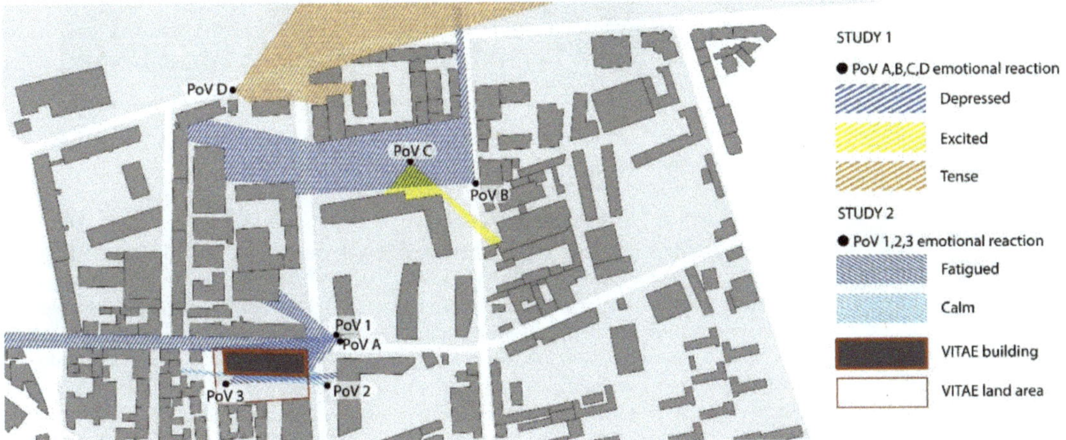

Figure 2. The Porta-Romana district (Milan, Italy), with the isovists of the PoVs assessed in Study 1 (PoVs A, B, C, and D) and Study 2 (PoVs 1, 2, and 3), colored with the respective colors resulting from the affective state described by the circumplex model. Source: the authors.

Figure 3. Points of view: Study 1, PoV A-B-C-D (without the VITAE project); Study 2 PoV 1-2-3. Sources: PoV A-B-C-D, Google StreetView™; PoV 1-2-3, City Sense™ (in Augmented Reality mode) screenshots.

4.2. Color Features and Emotional Reactions

Figure 4 (PoV D, case Study 1), Figure 5 (PoV 1, case Study 2), and Figure 6 (PoV 2, case Study 2) provide an example of the viewed urban landscape and the related green and lime elements identification, including the pie-charts of the hues proportion and the eight main colors of the scene. For the present paper and based on the literature suggesting the key role of different types of green in assessing vegetation effects on people [123,124], we only present here the correlations of lime and green tones with the emotions' factors (Table 1). The greenery in the pictures was mainly represented by lime pixels: in Study 1, Lime M = 7.99% s.d. = 9.99%, Green M = 2.04% s.d. = 1.73%; in Study 2, Lime M = 5.71% s.d. = 1.80%, Green M = 0.85% s.d. = 1.13%. As a first step, a correlational inquiry was performed. In Study 1 (VR, Table 1), results suggested the existence of significant correlations ($p < 0.05$) of the Unpleasant factor with two variables concerning the lime color and one variable concerning the green color. No other variables correlated significantly with any other detected emotion. Considering the lime color, the variables significantly correlated (negative correlation) to the Unpleasant factor were the pixel area (%) (r = −0.98, $p < 0.05$), namely the amount of image surface covered by lime pixels, and the "weighted saturation" (r = −0.96, $p < 0.05$), calculated as the pixels' average saturation value in the lime tone range ($38° \leq$ Hue $< 53°$). Regarding the green color, only this latter variable correlated significantly with the Unpleasant factor (r = −0.95, $p < 0.05$). In Study 2 (AR, Table 1), only one significant (negative) correlation between color variables (lime pixel area %) and emotions (Pleasant factor) was detected ($r = -0.98$, $p < 0.05$). Moreover, in Study 2 significant (positive) correlations were detected between the percentage of area covered by high lightness pixels and the Unpleasant/Pleasant continuum ($r = 0.99$, $p < 0.05$), and between the mean lightness of neutral hues areas and the Deactivation/Activation continuum ($r = 0.99$, $p < 0.05$).

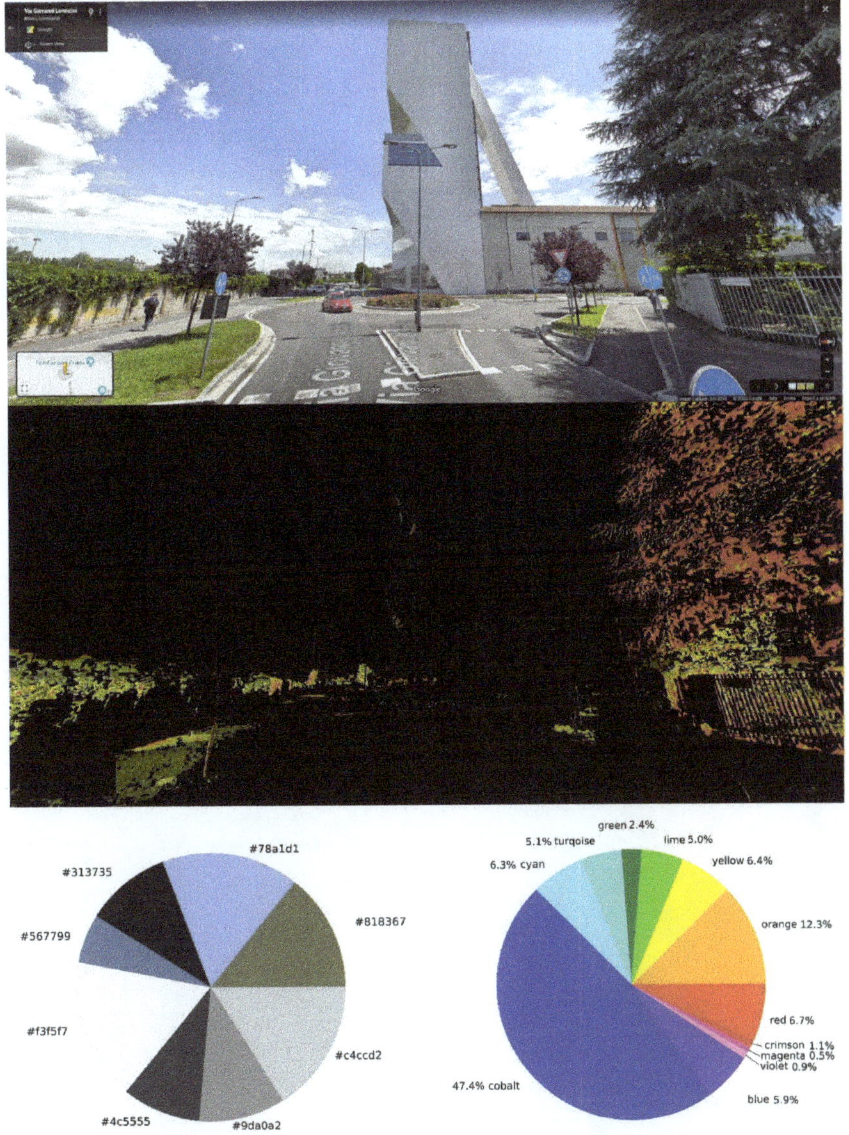

Figure 4. Analysis of the image of StreetView™ from via Giovanni Lorenzini (Milan)—PoV D— towards Fondazione Prada (existing condition: panoramic photo). Top-left, StreetView™ screenshot; bottom-left, lime areas identification ($38° \leq$ Hue $< 53°$, not depending on the saturation); top-right, the eight principal colors proportions; bottom-right, the proportions of the hues [Credits: the authors]. Sources: photo by Google StreetView™; color segmentation and charts by the authors.

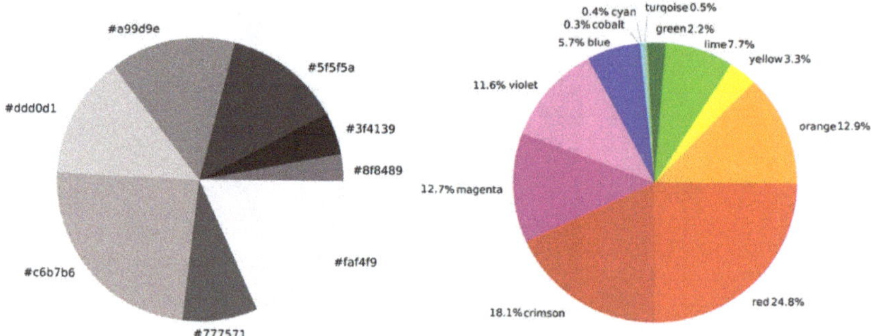

Figure 5. Analysis of the AR view from via Vezza d'Oglio (Milan)—PoV 1—towards the VITAE project. From left to right: future condition: simulation in AR; lime areas identification (38° ≤ Hue < 53°, not depending on the saturation); the eight principal colors proportions; the proportions of the hues [Credits: the authors]. Sources: photo by Google StreetView™; color segmentation and charts by the authors.

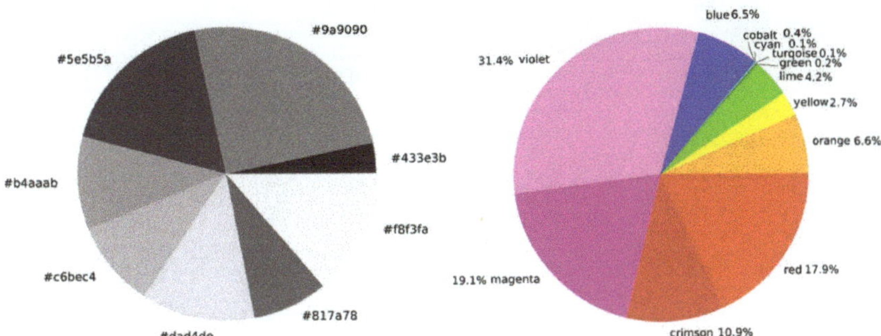

Figure 6. Analysis of the AR view from via Condino (Milan)—PoV 2—towards the VITAE project (future condition: simulation in AR). From left to right: future condition: simulation in AR; lime areas identification ($38° \leq$ Hue $< 53°$, not depending on the saturation); the eight principal colors proportions; the proportions of the hues [Credits: the authors]. Sources: photo by Google StreetView™; color segmentation and charts by the authors.

As a second step, Bland-Altman analysis was applied to measure the level of agreement between the correlations obtained in the two studies (VR and AR).

The results from the first and second steps were used to identify the variables that met the following restrictive criteria of significance:

1. To show a significant correlation in at least one of the two studies: $p < 0.05$.
2. To be included in the Bland-Altman interval of confidence: Difference in the range [(mean distance from equality—t_confidence) : (mean distance from equality + t_confidence)], where the mean distance from equality is the mean of all difference values related to an emotional parameter.
3. To show a high level of agreement in the two studies comparisons: |mean correlation| > 0.75.

Table 1. Study 1 (VR) and Study 2 (AR) correlations' matrix of chromatic features and emotions. The first four columns relate to single parameters, the last two columns relate to the two axes of Russell's chart. Cells values are: orange r < −0.80 and $p < 0.05$; light orange r < −0.80 and $p > 0.05$; green r > 0.80 and $p < 0.05$; light green r > 0.80 and $p > 0.05$. Source: the authors.

Image Features	Unpleasant	Pleasant	Sleepiness	Arousal	Unpleasant/Pleasant Continuum	Deactivation/Activation Continuum	Bland-Altman Chart Annotation
Study 1 (VR)							
CIELAB color space							
Low Lightness area (%)	−0.65	0.11	−0.55	−0.21	−0.32	−0.37	[1]
High Lightness area (%)	0.41	0.36	0.61	0.46	0.36	0.52	[2]
Lightness mean (entire pic)	0.65	−0.06	0.58	0.24	0.32	0.39	[3]
Mean lightness of neutral hues areas	0.19	−0.04	0.33	0.09	0.02	0.18	[4]
Mean green-red	0.86	0.46	0.68	0.65	0.79	0.69	[5]
HSV color space							
Mean saturation area (%)	−0.18	0.62	−0.02	0.35	0.23	0.19	[6]
Low light pixel saturation	−0.01	−0.74	−0.16	−0.51	−0.41	−0.36	[7]
High light pixel saturation	−0.79	−0.05	−0.68	−0.37	−0.49	−0.52	[8]
LIME (HSV)							
Lime pixel area (%)	**−0.98**	−0.45	−0.87	−0.71	−0.83	−0.80	[9]
Mean saturation lime	−0.84	−0.15	−0.62	−0.44	−0.62	−0.54	[10]
Weighted saturation lime	**−0.96**	−0.36	−0.84	−0.65	−0.77	−0.75	[11]
GREEN (HSV)							
Green pixel area (%)	−0.89	−0.36	−0.70	−0.59	−0.75	−0.67	[12]
Mean saturation green	−0.83	−0.07	−0.68	−0.39	−0.54	−0.53	[13]
Weighted saturation green	**−0.95**	−0.33	−0.81	−0.61	−0.75	−0.72	[14]
Study 2 (AR)							
CIELAB color space							
Low Lightness area (%)	−0.94	0.34	0.45	−0.80	−0.77	−0.04	[1]
High Lightness area (%)	0.95	0.29	0.19	0.28	**0.99**	0.64	[2]
Lightness mean (entire pic)	0.99	0.09	−0.02	0.47	0.97	0.46	[3]
Mean lightness of neutral hues areas	0.33	0.94	0.90	−0.61	0.62	**0.99**	[4]
Mean green-red	0.97	−0.24	−0.35	0.73	0.84	0.15	[5]
HSV color space							
Mean saturation area (%)	−0.99	0.11	0.22	−0.64	−0.90	−0.28	[6]
Low light pixel saturation	0.79	0.61	0.52	−0.08	0.95	0.87	[7]
High light pixel saturation	−0.97	0.27	0.37	−0.75	−0.82	−0.12	[8]
LIME (HSV)							
Lime pixel area (%)	−0.06	**−0.99**	−0.98	0.80	−0.39	−0.95	[9]
Mean saturation lime	−0.85	0.53	0.62	−0.91	−0.62	0.17	[10]
Weighted saturation lime	−0.99	−0.10	0.01	−0.46	−0.97	−0.47	[11]
GREEN (HSV)							
Green pixel area (%)	0.30	−0.96	−0.98	0.96	−0.04	−0.77	[12]
Mean saturation green	−0.86	0.52	0.61	−0.90	−0.63	0.16	[13]
Weighted saturation green	0.26	−0.97	−0.99	0.95	−0.08	−0.80	[14]

Table 2 shows the difference values between Study 1 and 2 correlations; Tables 3–5 shows mean values of Study 1 and 2 correlations; Tables 4 and 5 show Bland-Altman analysis results, representing the presence within the confidence interval and the level of agreement. The correlation between weighted saturation lime and the Unpleasant factor was the only one respecting the criteria of significance. No other color variables showed statistical significance and high agreement strength in their interaction with emotional factors. We computed a post-hoc power analysis [125] for the Unpleasant variable, resulting $p = 94.90\%$: $\alpha = 0.05$, $\Delta = 0.70$, n1 = 48, n2 = 63, s1 = 0.74, s2 = 1.29.

Table 2. Correlations' difference between the two case studies. In light green: differences between −0.10 and 0.10 are considered strongly converging. Source: the authors.

Image Features	Unpleasant	Pleasant	Sleepiness	Arousal	Unpleasant/Pleasant Continuum	Deactivation/Activation Continuum
CIELAB color space						
Low Lightness area (%)	0.29	−0.23	−1.00	0.59	0.45	−0.33
High Lightness area (%)	−0.54	0.07	0.42	0.18	−0.63	−0.12
Mean Lightness area (%)	−0.34	−0.15	0.60	−0.23	−0.65	−0.07
Mean lightness of neutral hues areas	−0.14	−0.98	−0.57	0.70	−0.60	−0.81
Mean green-red	−0.11	0.70	1.03	−0.08	−0.05	0.54
HSV color space						
Mean saturation area (%)	0.81	0.51	−0.24	0.99	1.13	0.47
Low light pixel saturation	−0.80	−1.35	−0.68	−0.43	−1.36	−1.23
High light pixel saturation	0.18	−0.32	−1.05	0.38	0.33	−0.40
LIME (HSV)						
Lime pixel area (%)	−0.92	0.54	0.11	−1.51	−0.44	0.15
Mean saturation lime	0.01	−0.68	−1.24	0.47	0.00	−0.71
Weighted saturation lime	0.03	−0.26	−0.85	−0.19	0.20	−0.28
GREEN (HSV)						
Green pixel area (%)	−1.19	0.60	0.28	−1.55	−0.71	0.10
Mean saturation green	0.03	−0.59	−1.29	0.51	0.09	−0.69
Weighted saturation green	−1.21	0.64	0.18	−1.56	−0.67	0.08
Standard deviation	0.59	0.65	0.75	0.86	0.63	0.50
Mean (distance from equality)	−0.28	−0.11	−0.31	−0.12	−0.21	−0.24
Standard error	0.16	0.17	0.20	0.23	0.17	0.13
Confidence	0.34	0.38	0.43	0.50	0.37	0.29
Confidence—lower limit mean	−0.62	−0.48	−0.74	−0.62	−0.57	−0.53
Confidence—upper limit mean	0.06	0.27	0.12	0.37	0.16	0.05

Table 3. Correlations mean of the two case studies. In light green: means lower than −0.75 or higher than 0.75 are considered strongly converging. Source: the authors.

Image Features	Unpleasant	Pleasant	Sleepiness	Arousal	Unpleasant/Pleasant Continuum	Deactivation/Activation Continuum
CIELAB color space						
Low Lightness area (%)	−0.80	0.23	−0.05	−0.51	−0.55	−0.21
High Lightness area (%)	0.68	0.33	0.40	0.37	0.68	0.58
Mean Lightness area (%)	0.82	0.02	0.28	0.36	0.65	0.43
Mean lightness of neutral hues areas	0.26	0.45	0.62	−0.26	0.32	0.59
Mean green-red	0.92	0.11	0.17	0.69	0.82	0.42
HSV color space						
Mean saturation area (%)	−0.59	0.37	0.10	−0.15	−0.34	−0.05
Low light pixel saturation	0.39	−0.07	0.18	−0.30	0.27	0.26
High light pixel saturation	−0.88	0.11	−0.16	−0.56	−0.66	−0.32
LIME (HSV)						
Lime pixel area (%)	−0.52	−0.72	−0.93	0.05	−0.61	−0.88
Mean saturation lime	−0.85	0.19	0.00	−0.68	−0.62	−0.19
Weighted saturation lime	−0.98	−0.23	−0.42	−0.56	−0.87	−0.61
GREEN (HSV)						
Green pixel area (%)	−0.30	−0.66	−0.84	0.19	−0.40	−0.72
Mean saturation green	−0.85	0.23	−0.04	−0.65	−0.59	−0.19
Weighted saturation green	−0.35	−0.65	−0.90	0.17	−0.42	−0.76

Table 4. The inclusion of the agreement values of Study 1 and Study 2 within the mean confidence interval of the Bland-Altman chart. Cells with bold borders relate to a significant correlation in at least one case Study; correlations outside the confidence interval are not considered for significance. Source: the authors.

Image Features	Study 1 and Study 2 Correlations within the Mean Confidence Interval					
	Unpleasant	Pleasant	Sleepiness	Arousal	Unpleasant/Pleasant Continuum	Deactivation/Activation Continuum
CIELAB color space						
Low Lightness area (%)	OUT	IN	OUT	OUT	OUT	IN
High Lightness area (%)	IN	IN	OUT	IN	OUT	IN
Mean Lightness area (%)	IN	IN	OUT	IN	OUT	IN
Mean lightness of neutral hues areas	IN	OUT	IN	OUT	OUT	OUT
Mean green-red	IN	IN	OUT	IN	IN	OUT
HSV color space						
Mean saturation area (%)	OUT	IN	IN	OUT	OUT	OUT
Low light pixel saturation	OUT	OUT	IN	IN	OUT	OUT
High light pixel saturation	OUT	IN	OUT	OUT	OUT	IN
LIME (HSV)						
Lime pixel area (%)	OUT	IN	IN	OUT	IN	OUT
Mean saturation lime	IN	OUT	OUT	OUT	IN	OUT
Weighted saturation lime	IN	IN	OUT	IN	OUT	IN
GREEN (HSV)						
Green pixel area (%)	OUT	IN	OUT	OUT	OUT	OUT
Mean saturation green	IN	OUT	OUT	OUT	IN	OUT
Weighted saturation green	OUT	IN	OUT	OUT	OUT	OUT

Table 5. Level of agreement of Study 1 and Study 2 based on mean correlation and inclusion within the mean confidence interval. Cells with bold borders relate to significant correlation in at least one case study; correlation with low or medium agreement is not considered for significance. Source: the authors.

Image Features	Level of Agreement Between Study 1 (VR) And Study 2 (AR)					
	Unpleasant	Pleasant	Sleepiness	Arousal	Unpleasant/Pleasant Continuum	Deactivation/Activation Continuum
CIELAB color space						
Low Lightness area (%)	MED.	LOW	LOW	LOW	LOW	LOW
High Lightness area (%)	LOW	LOW	LOW	LOW	LOW	LOW
Mean Lightness area (%)	HIGH	LOW	LOW	LOW	LOW	LOW
Mean lightness of neutral hues areas	LOW	LOW	LOW	LOW	LOW	LOW
Mean green-red	HIGH	LOW	LOW	LOW	HIGH	LOW
HSV color space						
Mean saturation area (%)	LOW	LOW	LOW	LOW	LOW	LOW
Low light pixel saturation	LOW	LOW	LOW	LOW	LOW	LOW
High light pixel saturation	MED.	LOW	LOW	LOW	LOW	LOW
LIME (HSV)						
Lime pixel area (%)	LOW	LOW	HIGH	LOW	LOW	MED.
Mean saturation lime	HIGH	LOW	LOW	LOW	LOW	LOW
Weighted saturation lime	HIGH	LOW	LOW	LOW	MED.	LOW
GREEN (HSV)						
Green pixel area (%)	LOW	LOW	MED.	LOW	LOW	LOW
Mean saturation green	HIGH	LOW	LOW	LOW	LOW	LOW
Weighted saturation green	LOW	LOW	MED.	LOW	LOW	MED.

4.3. Agreements Cluster Analysis

The emotional factor resulting as significant from previous analyses was the Unpleasant (weighted saturation lime difference = 0.03, mean = −0.98). Furthermore, the Unpleasant parameter presents most of the agreement on the emotional effect of chromatic features: 36% of high agreements, 14% of medium agreements. Cluster analysis run on Unpleasant Bland-Altman chart for the correlations' agreement generated two clusters related to negative (cluster A) and positive correlations (cluster B). In the Unpleasant graph agreement (Figure 7), cluster A groups percentage area of low lightness, saturation of high lightness areas, mean saturation of lime areas, weighted saturation of lime area, mean saturation green; cluster B groups mean green-red value, percentage of area covered in mean lightness pixel, high lightness area (%).

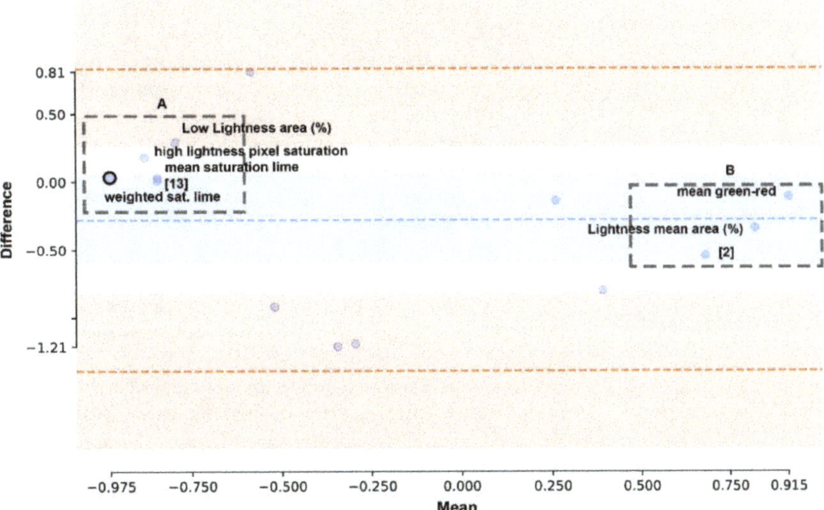

Figure 7. Bland-Altman chart of Unpleasant and chromatic features resulting from the comparison of Study 1 and Study 2; dots with black border show correlations with high agreement and statistical significance, with grey border correlations with medium-high agreement but no sufficient statistical significance. Dashed lines are correlations clusters (DBSCAN): cluster A groups percentage area of low Lightness, saturation of high lightness areas, mean saturation of lime areas, weighted saturation of lime area, mean saturation green; cluster B groups mean green-red value, percentage of area covered in mean lightness pixel, high Lightness area (%). Source: the authors.

5. Discussion

Our research investigated the relationship between colors and emotions in actual urban areas, including existing vegetation in VR (Study 1) and a designed project with NBS in AR (Study 2) through a two steps process. Firstly, we analyzed the general effects of urban scenes colors on emotions, focusing on lime and green colors traditionally associated with natural elements [123]. Secondly, we tested the level of agreement between two different simulation solutions by comparing VR and AR.

Compared with the generally positive effects of natural elements reported in the literature, the relationship between lime and green color tones and affective states was not straightforward in our studies. In the first place, green tones show a significant correlation with emotions only in one case (green weighted saturation reduces the Unpleasant factor) and only in the VR experimental condition. Lime tones show two significant correlations: both lime pixel area (%) and lime weighted saturation reduce the Unpleasant factor, and the latter effect has a high agreement between VR and AR. The effects of lime are consistent with previous studies on yellowish-green plants associated with positive emotions and happiness [123,124]. Despite this, it is worth noting that lime pixel area (%) also has a negative correlation with the Pleasant factor in AR, which is anyway in disagreement with VR. Most of the positive effects observed with lime and green tones are consistent with previous literature, whereas further studies are needed to understand better this negative effect of the lime pixel area in AR.

More in general, the results suggest that the presence of green and lime tones via AR and VR reduce the unpleasantness rather than increasing the pleasantness in the urban environment. This finding is coherent with psychological models, which stressed the independence between positive and negative affects [126]. It is possible to argue that, in the examined urban conditions, the perception of urban greenery (lime tones in AR and VR, green tones in AR only) triggers less intense sensations of dissatisfaction and repulsion but does not significantly stimulate the individual perception of beauty and pleasantness. Furthermore, the results show no significant relationships between green tones and emotions concerning positive (i.e., calm, relaxation) or negative (i.e., boredom) deactivation. This encourages a non-mechanistic view of the relationship between greenery and pleasantness or relaxation in urban environments. Indeed, despite the well-established positive effect of greenery presented in the literature, it is necessary to contextualize each case study. For example, referring to two classic psychological frameworks, we can hypothesize that in our case studies greenery is not able by itself to generate affordances [127] eliciting emotional states of activation/deactivation [128] or to trigger a restorative experience increasing people's perception of fascination, being away, extent and compatibility [84].

Finally, the results offer a remarkable suggestion regarding the reliability between AR and VR. Indeed, data suggest that lime and green tones' influence on some emotional variables is partially consistent with VR and AR scenarios, especially regarding the Unpleasant factor. Considering high-level agreements between VR and AR, including both significant and non-significant correlations, the number of assessments in agreement increases from one to seven. It is worth noting that the majority of such agreements include the Unpleasant factor, which appears as the most stable variable that can be assessed comparatively with AR and VR, at least regarding green tones. These results pave the way for future analyses comparing VR and AR.

The results suggest that the positive effects of natural elements [57,87] should be explored more in detail in the future. The different reactions to green and lime colors, as well as the varying effect that the considered lime variables had on pleasantness, call for a deeper understanding of the role played by several natural factors. As suggested by Han and Ruan [129], future research should tackle issues including plants' amount, size, color, scent, and type (including flowers, foliage, shape). In such a perspective, it is also relevant to consider time (e.g., seasonal conditions) in relation to the geographical location. In developing such researches, it is important considering that the relationship between human and environmental factors is an interdisciplinary topic, investigated in various disciplines with many different theoretical and methodological traditions such as psychology, architecture, landscape design, or agriculture. Hence, the research conducted by monodisciplinary teams are more prone to be methodologically sounder on environmental factors or on human ones but not on both of them. As Bringslimark, Hartig and Patil [130] suggested in their review on indoor plants, more collaboration between environmental psychologists and horticulturists would be beneficial. Similarly, such reflections can be extended to the greenery in public spaces, which calls for even broader collaboration [131].

Results also stress once more that an ethic for simulation usage is needed and urgent [132–134] since the improper alteration of the simulation elements, including colors, can create distortions in the proper understanding of the depicted environment; in the worst case, this can lead to poor urban/landscape planning and design decision that impact on society. Showing biased representations that do not trustfully anticipate the urban transformation is also a powerful—and thus hazardous—tool for manipulating public opinion and directly impacting citizens' emotional states. This consideration is relevant both when presenting urban design projects to a pre-selected private audience or when using it in public participatory processes. In the same perspective, it is also essential that the scientific research continues investigating such a topic for leading the fair preparation and application of simulation in relation to the specific knowledge goals and the available resources (e.g., available time and abilities, economic and human resources). This proper cost/benefit balance is crucial to positively impact cities and societies. Indeed, the recent advancement of solutions such as VR and AR suggests their incremental usage in urban processes, but the impact of costs for producing reliable realistic simulations should not be underestimated. Indeed, the proper production of simulations can be affected by this economic aspect; thus, correctly identifying the needed fidelity level of the simulation can potentially contribute to its wider unbiased application.

6. Conclusions

Our results bear some limitations to be carefully considered. Our analysis focused on color features extracted from the images, which included a high amount of lime pixels compared to green ones, both in VR and AR. In addition, we did not consider the semantic value of the green elements included in the images; hence all types of vegetation are considered equal, including trees, bushes, grass, and flowers. These elements were present in varying proportions in the actual and in the designed scenario. Similarly, no variables were considered to distinguish lush and cultivated vegetation from unkempt and spontaneous greenery, which varied across the scenarios. Moreover, it is worth noting that the weather conditions, another environmental feature that influences affective states, varied across the scenarios. VR presented a sunny environment, whereas AR superimposed the building with NBS on the urban landscape on a cloudy winter day, which may have affected participants' emotions. The limitations of the quasi-experimental design, which hinders the full comparability of the two scenarios, and the choice of environments unbalanced towards lime tones are due to the data collection during a real case study. The needs of the participatory process limited the choice of the environment to explore, prevented us from manipulating the designed NBS, narrowed the choice of tools for data collection, and influenced the choice of participants. Such considerations suggest two main fields of research for the future. The first concerns theoretical research, which implies experimental design, the manipulation of single environmental variables (e.g., types of vegetation, weather and seasonal conditions, architectural solutions), and broader categories of participants. Laboratory research would also allow scholars to examine more in depth the psycho-physiological effects of such technologies, offering a complementary perspective to the emotional description obtained through psychometric scales; one of the main obstacles for the inclusion of such physiological tools in actual participatory processes lies in the difficulty of having reliable measures with low-cost and non-intrusive instruments. The second field is related to applied research, investigating the most effective procedures for integrating these technologies and the related devices into concrete case studies, devoting specific attention to sensitive populations (e.g., with low digital literacy, color vision or other sight deficiencies). In addition to the citizens' perspective, it is also worth considering other stakeholders' perception: the opinion of experts working in public institutions and private companies involved in relevant urban transformations are a key factor for designing successful engagement strategies.

Despite these limitations, our studies showed that it is possible to obtain a reliable assessment of the emotional reaction to NBS, even when comparing data gathered with different technologies such as VR and AR. In light of such results, we conceive participatory apps with VR/AR solutions as valuable tools for participatory urban processes; they can represent a quick and affordable assessment tool for investigating the relevant issue of alignment with the community needs [15] and monitoring the differences among sub-populations [30]. Indeed, the gained results are relevant sources for checking the congruence between users' perceptions and design desiderata. Moreover, the application of AR and VR solutions generally engages citizens and fosters inclusiveness thanks to the ease of understanding of the design outcomes. Our studies show that integrating these apps in a participatory process enables the collection of specific contextual information about the people-environment relationship focusing on NBS. Indeed, with such an approach, it is possible to combine objective (e.g., environmental features such as chromatic elements) and subjective data analysis (e.g., emotional and/or cognitive reaction). The explanatory capability of subjective data is increased when associated with socio-demographic variables, allowing a more detailed explanation or a targeted analysis of results.

In this perspective, the proposed approach supports improving collective wellbeing by favoring the creation of places capable of fulfilling the community goals and inclinations [135]. Such an approach would allow different stakeholders to assess the wellbeing experienced by different population segments and thus consistently inform the design or decision-making phase. Nevertheless, participatory apps are not conceived as autonomous tools for guiding design solutions; rather, they can be seen as a tool for fostering people's perspectives in urban processes (i.e., human-centered design) by opening a debate among the stakeholders involved in the transformation process. The interpretation of the results gained via such smart participatory solutions is assigned to professionals with a background in social sciences and architecture/urban planning, and experts of the local context capable of including cultural variables regarding artificial and natural elements and expected behaviors of local/global communities. The translation of such information into meaningful physical features is entirely in charge of architects and planners.

7. Patents

The AR4CUP APP, distributed as City Sense by Artefacto, provides realistic and immersive environment replicas via AR or VR. Through an architectural\psychological integrated framework, the interaction with the simulated environment triggers an experience that can be reliably assessed using established psychological constructs (exp-EIA©—Experiential Environmental Impact Assessment—Copyright BOIP N. 123453—6 May 2020 and N. 130516—25 February 2021; Patent for Invention application N. 102021000017168—30 June 2021).

Author Contributions: The overall effort on the article and the writing is equally distributed among the authors, whose contribution according to CRediT (Contributor Roles Taxonomy) is described as follows. Conceptualization, B.E.A.P. and M.B.; methodology, B.E.A.P., G.S., N.R. and M.B.; formal analysis, G.S.; investigation, B.E.A.P., G.S., N.R. and M.B.; data curation, G.S.; writing—original draft preparation, G.S. and N.R.; writing—review and editing, B.E.A.P. and M.B.; visualization, B.E.A.P. and G.S.; supervision, B.E.A.P. and M.B.; funding acquisition, B.E.A.P. All authors have read and agreed to the published version of the manuscript.

Funding: This project has received funding from the European Union's Horizon 2020 research and innovation program EIT Digital 2019 and 2020 "AR4CUP: Augmented Reality for Collaborative Urban Planning".The sole responsibility of this publication lies with the author. The European Union is not responsible for any use that may be made of the information contained therein.

Institutional Review Board Statement: The study was conducted according to the guidelines of the Declaration of Helsinki, and approved by the Institutional Ethics Committee of POLITECNICO DI MILANO (protocol code 7/2019 on 18/03/2019, and 15/2020 on 15/07/2020).

Informed Consent Statement: Informed consent was obtained from all subjects involved in the study.

Data Availability Statement: Original data can be obtained by contacting the first author. Authorization for publication consent was obtained from all authors and involved partners.

Acknowledgments: The first draft of this paper was presented at the symposium Greening cities shaping cities on 12 and 13 October 2020, at the Politecnico di Milano; for more information visit www.greeningcities-shapingcities.polimi.it (accessed on 19 November 2021).

Conflicts of Interest: The authors declare no conflict of interest.

References

1. European Commission; Directorate-General for Research and Innovation. *Towards an EU Research and Innovation Policy Agenda for Nature-Based Solutions & Re-Naturing Cities: Final Report of the Horizon 2020 Expert Group on "Nature Based Solutions and Re Naturing Cities*; Publications Office of the European Union: Luxemburg, 2015; ISBN 978-92-79-46048-7.
2. Faivre, N.; Fritz, M.; Freitas, T.; de Boissezon, B.; Vandewoestijne, S. Nature-Based Solutions in the EU: Innovating with Nature to Address Social, Economic and Environmental Challenges. *Environ. Res.* **2017**, *159*, 509–518. [CrossRef]
3. Parenership for Environment and Disaster Risk Reduction; Friends of EbA. *Promoting Nature-Based Solutions in the Post-2020 Global Biodiversity Framework*; International Union for Conservation of Nature: Gland, Switzerland, 2020.
4. Nesshöver, C.; Assmuth, T.; Irvine, K.N.; Rusch, G.M.; Waylen, K.A.; Delbaere, B.; Haase, D.; Jones-Walters, L.; Keune, H.; Kovacs, E.; et al. The Science, Policy and Practice of Nature-Based Solutions: An Interdisciplinary Perspective. *Sci. Total Environ.* **2017**, *579*, 1215–1227. [CrossRef] [PubMed]
5. Bishop, I.D. Environmental Information and Technology: Is It All Too Much? *Landsc. Urban Plan.* **2000**, *47*, 111–114. [CrossRef]
6. Ferreira, V.; Barreira, A.P.; Loures, L.; Antunes, D.; Panagopoulos, T. Stakeholders' Engagement on Nature-Based Solutions: A Systematic Literature Review. *Sustainability* **2020**, *12*, 640. [CrossRef]

7. Rainisio, N.; Boffi, M.; Riva, E. Positive Change in Environment: Aesthetics, Environmental Flowability and Well-Being. In *Enabling Positive Change: Flow and Complexity in Daily Experience*; Inghilleri, P., Riva, G., Riva, E., Eds.; DE GRUYTER OPEN: Warsaw, Poland, 2015; ISBN 978-3-11-041024-2.
8. Joye, Y. Architectural Lessons from Environmental Psychology: The Case of Biophilic Architecture. *Rev. Gen. Psychol.* **2007**, *11*, 305–328. [CrossRef]
9. Kellert, S.R.; Heerwagen, J.H.; Mador, M.L. *Biophilic Design: The Theory, Science, and Practice of Bringing Buildings to Life*; Wiley: Hoboken, NJ, USA, 2008.
10. Kellert, S.; Wilson, E. *The Biophilia Hypothesis*; Island Press: Washington, DC, USA, 1993.
11. Xue, F.; Gou, Z.; Lau, S.S.-Y.; Lau, S.-K.; Chung, K.-H.; Zhang, J. From Biophilic Design to Biophilic Urbanism: Stakeholders' Perspectives. *J. Clean Prod.* **2019**, *211*, 1444–1452. [CrossRef]
12. Jiang, B.; Larsen, L.; Deal, B.; Sullivan, W.C. A Dose–Response Curve Describing the Relationship between Tree Cover Density and Landscape Preference. *Landsc. Urban Plan.* **2015**, *139*, 16–25. [CrossRef]
13. Kaplan, R.; Kaplan, S. *The Experience of Nature: A Psychological Perspective*; Cambridge University Press: Cambridge, UK, 1989; ISBN 978-0-521-34939-0.
14. White, E.V.; Gatersleben, B. Greenery on Residential Buildings: Does It Affect Preferences and Perceptions of Beauty? *J. Environ. Psychol.* **2011**, *31*, 89–98. [CrossRef]
15. Wolch, J.R.; Byrne, J.; Newell, J.P. Urban Green Space, Public Health, and Environmental Justice: The Challenge of Making Cities 'Just Green Enough. *Landsc. Urban Plan.* **2014**, *125*, 234–244. [CrossRef]
16. Ulrich, R.S.; Simons, R.F.; Losito, B.D.; Fiorito, E.; Miles, M.A.; Zelson, M. Stress Recovery during Exposure to Natural and Urban Environments. *J. Environ. Psychol.* **1991**, *11*, 201–230. [CrossRef]
17. Cousins, J.J. Justice in Nature-Based Solutions: Research and Pathways. *Ecol. Econ.* **2021**, *180*, 106874. [CrossRef]
18. Dale, A.; Newman, L.L. Sustainable Development for Some: Green Urban Development and Affordability. *Local Environ.* **2009**, *14*, 669–681. [CrossRef]
19. Haase, D.; Kabisch, S.; Haase, A.; Andersson, E.; Banzhaf, E.; Baró, F.; Brenck, M.; Fischer, L.K.; Frantzeskaki, N.; Kabisch, N.; et al. Greening Cities—To Be Socially Inclusive? *About the Alleged Paradox of Society and Ecology in Cities. Habitat Int.* **2017**, *64*, 41–48. [CrossRef]
20. Curran, W.; Hamilton, T. Just Green Enough: Contesting Environmental Gentrification in Greenpoint, Brooklyn. *Local Environ.* **2012**, *17*, 1027–1042. [CrossRef]
21. Conway, D.; Li, C.Q.; Wolch, J.; Kahle, C.; Jerrett, M. A Spatial Autocorrelation Approach for Examining the Effects of Urban Greenspace on Residential Property Values. *J. Real Estate Finance Econ.* **2010**, *41*, 150–169. [CrossRef]
22. Zhang, Y.; Dong, R. Impacts of Street-Visible Greenery on Housing Prices: Evidence from a Hedonic Price Model and a Massive Street View Image Dataset in Beijing. *ISPRS Int. J. Geo-Inf.* **2018**, *7*, 104. [CrossRef]
23. Zuniga-Teran, A.A.; Gerlak, A.K. A Multidisciplinary Approach to Analyzing Questions of Justice Issues in Urban Greenspace. *Sustainability* **2019**, *11*, 3055. [CrossRef]
24. Li, H.; Liu, Y. Neighborhood Socioeconomic Disadvantage and Urban Public Green Spaces Availability: A Localized Modeling Approach to Inform Land Use Policy. *Land Use Policy* **2016**, *57*, 470–478. [CrossRef]
25. Gould, K.A.; Lewis, T.L. *Green Gentrification: Urban Sustainability and the Struggle for Environmental Justice*; Routledge: London, UK, 2016; ISBN 978-1-317-41780-4.
26. Li, X.; Zhang, C.; Li, W.; Kuzovkina, Y.A.; Weiner, D. Who Lives in Greener Neighborhoods? The Distribution of Street Greenery and Its Association with Residents' Socioeconomic Conditions in Hartford, Connecticut, USA. *Urban For. Urban Green.* **2015**, *14*, 751–759. [CrossRef]
27. Sipilä, M.; Tyrväinen, L. Evaluation of Collaborative Urban Forest Planning in Helsinki, Finland. *Urban For. Urban Green.* **2005**, *4*, 1–12. [CrossRef]
28. Mahmoud, I.; Morello, E. Co-Creation Pathway as a Catalyst for Implementing Nature-Based Solution in Urban Regeneration Strategies Learning from CLEVER Cities Framework and Milano as Test-Bed. *Urban. Inf. Interruptions Intersect. Shar. Overlapping New Perspect. Territ.* **2018**, *25*, 204–210.
29. Kronenberg, J. Why Not to Green a City? Institutional Barriers to Preserving Urban Ecosystem Services. *Ecosyst. Serv.* **2015**, *12*, 218–227. [CrossRef]
30. Kirkpatrick, J.B.; Davison, A.; Daniels, G.D. Sinners, Scapegoats or Fashion Victims? Understanding the Deaths of Trees in the Green City. *Geoforum* **2013**, *48*, 165–176. [CrossRef]
31. Lyytimäki, J.; Sipilä, M. Hopping on One Leg—The Challenge of Ecosystem Disservices for Urban Green Management. *Urban For. Urban Green.* **2009**, *8*, 309–315. [CrossRef]
32. Kirkpatrick, J.B.; Davison, A.; Harwood, A. How Tree Professionals Perceive Trees and Conflicts about Trees in Australia's Urban Forest. *Landsc. Urban Plan.* **2013**, *119*, 124–130. [CrossRef]
33. Boffi, M.; Riva, E.; Rainisio, N. Positive Change and Political Participation: Well-Being as an Indicator of the Quality of Citizens' Engagement. In *Enabling Positive Change: Flow and Complexity in Daily Experience*; Inghilleri, P., Riva, G., Riva, E., Eds.; DE GRUYTER OPEN: Warsaw, Poland, 2015; ISBN 978-3-11-041024-2.
34. Evans-Cowley, J.; Hollander, J. The New Generation of Public Participation: Internet-Based Participation Tools. *Plan. Pract. Res.* **2010**, *25*, 397–408. [CrossRef]

35. Sæbø, Ø.; Rose, J.; Flak, L. The Shape of Eparticipation: Characterizing an Emerging Research Area. *Gov. Inf. Q.* **2008**, *25*, 400–428. [CrossRef]
36. Desouza, K.C.; Bhagwatwar, A. Technology-Enabled Participatory Platforms for Civic Engagement: The Case of U.S. *Cities. J. Urban Technol.* **2014**, *21*, 25–50. [CrossRef]
37. Picon, A. Smart Cities. A Spatialised Intelligence. *AD* **2015**, *49*, 5403. [CrossRef]
38. Boffi, M.; Pola, L.; Fumagalli, N.; Fermani, E.; Senes, G.; Inghilleri, P. Nature Experiences of Older People for Active Ageing: An Interdisciplinary Approach to the Co-Design of Community Gardens. *Front. Psychol.* **2021**, *12*, 4175. [CrossRef]
39. Fumagalli, N.; Fermani, E.; Senes, G.; Boffi, M.; Pola, L.; Inghilleri, P. Sustainable Co-Design with Older People: The Case of a Public Restorative Garden in Milan (Italy). *Sustainability* **2020**, *12*, 3166. [CrossRef]
40. Kanhere, S.S. Participatory Sensing: Crowdsourcing Data from Mobile Smartphones in Urban Spaces. In Proceedings of the ICDCIT 2013, Bhubaneswar, India, 5–7 February 2013; Hota, C., Srimani, P.K., Eds.; Springer: Berlin/Heidelberg, Germany, 2013; pp. 19–26.
41. Brooks, N. *Vulnerability, Risk and Adaptation: A Conceptual Framework*; Tyndall Centre for Climate Change Research: Norwich, UK, 2003; p. 20.
42. Bosselmann, P. *Urban Transformation: Understanding City Form and Design*; Island Press: Washington, DC, USA, 2012; ISBN 978-1-61091-149-8.
43. Ertiö, T.-P. Participatory Apps for Urban Planning—Space for Improvement. *Plan. Pract. Res.* **2015**, *30*, 303–321. [CrossRef]
44. Cardone, G.; Cirri, A.; Corradi, A.; Foschini, L.; Ianniello, R.; Montanari, R. Crowdsensing in Urban Areas for City-Scale Mass Gathering Management: Geofencing and Activity Recognition. *IEEE Sens. J.* **2014**, *14*, 4185–4195. [CrossRef]
45. Capineri, C.; Huang, H.; Gartner, G. Tracking Emotions in Urban Space. *Two Experiments in Vienna and Siena. Riv. Geogr. Ital.* **2018**, *125*, 273–288.
46. Hudson-Smith, A.; Evans, S.; Batty, M. Building the Virtual City: Public Participation through e-Democracy. *Knowl. Technol. Policy* **2005**, *18*, 62–85. [CrossRef]
47. Anagnostou, K.; Vlamos, P. Square AR: Using Augmented Reality for Urban Planning. In Proceedings of the 2011 Third International Conference on Games and Virtual Worlds for Serious Applications, Athens, Greece, 4–6 May 2011; pp. 128–131.
48. Bosché, F.; Tingdahl, D.; Carozza, L.; Gool, L.V. Markerless Vision-Based Augmented Reality for Enhanced Project Visualization. *Gerontechnology* **2012**, *11*, 69. [CrossRef]
49. Imottesjo, H.; Kain, J.-H. The Urban CoBuilder—A Mobile Augmented Reality Tool for Crowd-Sourced Simulation of Emergent Urban Development Patterns: Requirements, Prototyping and Assessment. *Comput. Environ. Urban Syst.* **2018**, *71*, 120–130. [CrossRef]
50. Rodríguez Bolívar, M.P.; Alcaide Muñoz, L. *E-Participation in Smart Cities: Technologies and Models of Governance for Citizen Engagement*; Public Administration and Information Technology; Springer: Cham, Switzerland, 2019; Volume 34, ISBN 978-3-319-89473-7.
51. Stratigea, A.; Papadopoulou, C.-A.; Panagiotopoulou, M. Tools and Technologies for Planning the Development of Smart Cities. *J. Urban Technol.* **2015**, *22*, 43–62. [CrossRef]
52. Appleyard, D. Understanding Professional Media. In *Human Behavior and Environment*; Altman, I., Wohlwill, J.F., Eds.; Plenum Press: New York, NY, USA, 1977; pp. 43–88. ISBN 978-1-4684-0810-2.
53. Bosselmann, P. *Representation of Places: Reality and Realism in City Design*; University of California Press: Oakland, CA, USA, 1998; ISBN 978-0-520-20658-8.
54. Piga, B.E.A.; Morello, E. Environmental Design Studies on Perception and Simulation: An Urban Design Approach. *Int. J. Sens. Environ. Archit. Urban Space* **2015**. [CrossRef]
55. Sheppard, S.R.J. *Visual Simulation: A User Guide for Architects, Engineers, and Planners*; Van Nostrand Reinhold: New York, NY, USA, 1989; ISBN 978-0-442-27827-4.
56. McKechnie, G.E. Simulation Techniques in Environmental Psychology. In *Perspectives on Environment and Behavior*; Stokols, D., Ed.; Plenum Press: London, UK, 1977; pp. 169–189, ISBN 13: 978-1-4684-2279-5.
57. Browning, M.H.E.M.; Saeidi-Rizi, F.; McAnirlin, O.; Yoon, H.; Pei, Y. The Role of Methodological Choices in the Effects of Experimental Exposure to Simulated Natural Landscapes on Human Health and Cognitive Performance: A Systematic Review. *Environ. Behav.* **2020**. [CrossRef]
58. Calogiuri, G.; Litleskare, S.; Fagerheim, K.A.; Rydgren, T.L.; Brambilla, E.; Thurston, M. Experiencing Nature through Immersive Virtual Environments: Environmental Perceptions, Physical Engagement, and Affective Responses during a Simulated Nature Walk. *Front. Psychol.* **2018**, *8*, 2321. [CrossRef]
59. Kahn, P.H.; Severson, R.L.; Ruckert, J.H. The Human Relation With Nature and Technological Nature. *Curr. Dir. Psychol. Sci.* **2009**, *18*, 37–42. [CrossRef]
60. Kjellgren, A.; Buhrkall, H. A Comparison of the Restorative Effect of a Natural Environment with That of a Simulated Natural Environment. *J. Environ. Psychol.* **2010**, *30*, 464–472. [CrossRef]
61. Lassonde, K.A.; Gloth, C.A.; Borchert, K. Windowless Classrooms or a Virtual Window World: Does a Creative Classroom Environment Help or Hinder Attention? *Teach. Psychol.* **2012**, *39*, 262–267. [CrossRef]
62. Loomis, J.M.; Blascovich, J.J.; Beall, A.C. Immersive Virtual Environment Technology as a Basic Research Tool in Psychology. *Behav. Res. Methods Instrum. Comput.* **1999**, *31*, 557–564. [CrossRef] [PubMed]
63. Dede, C. Immersive Interfaces for Engagement and Learning. *Science* **2009**, *323*, 66–69. [CrossRef] [PubMed]

64. Edler, D.; Keil, J.; Wiedenlübbert, T.; Sossna, M.; Kühne, O.; Dickmann, F. Immersive VR Experience of Redeveloped Post-Industrial Sites: The Example of "Zeche Holland" in Bochum-Wattenscheid. *KN-J. Cartogr. Geogr. Inf.* **2019**, *69*, 267–284. [CrossRef]

65. Loyola, M.; Rossi, B.; Montiel, C.; Daiber, M. Use of Virtual Reality in Participatory Design. In *Blucher Design Proceedings;* Editora Blucher: Porto, Portugal, 2019; pp. 449–454.

66. Williams, A.S.; Angelini, C.; Kress, M.; Ramos Vieira, E.; D'Souza, N.; Rishe, N.D.; Medina, J.; Özer, E.; Ortega, F. Augmented Reality for City Planning. In Proceedings of the Virtual, Augmented and Mixed Reality. Design and Interaction, Copenhagen, Denmark, 19–24 July 2020; Chen, J.Y.C., Fragomeni, G., Eds.; Springer: Cham, Switzerland, 2020; pp. 256–271.

67. Saßmannshausen, S.M.; Radtke, J.; Bohn, N.; Hussein, H.; Randall, D.; Pipek, V. Citizen-Centered Design in Urban Planning: How Augmented Reality Can Be Used in Citizen Participation Processes. In Proceedings of the Designing Interactive Systems Conference 2021, Virtual Event, USA, 28 June 2021; pp. 250–265.

68. Nitoslawski, S.A.; Galle, N.J.; Van Den Bosch, C.K.; Steenberg, J.W.N. Smarter Ecosystems for Smarter Cities? A Review of Trends, Technologies, and Turning Points for Smart Urban Forestry. *Sustain. Cities Soc.* **2019**, *51*, 101770. [CrossRef]

69. Frumkin, H.; Bratman, G.N.; Breslow, S.J.; Cochran, B.; Kahn, J.P.H.; Lawler, J.J.; Levin, P.S.; Tandon, P.S.; Varanasi, U.; Wolf, K.L.; et al. Nature Contact and Human Health: A Research Agenda. *Environ. Health Perspect.* **2017**, *125*, 075001. [CrossRef]

70. Velarde, M.D.; Fry, G.; Tveit, M. Health Effects of Viewing Landscapes—Landscape Types in Environmental Psychology. *Urban For. Urban Green.* **2007**, *6*, 199–212. [CrossRef]

71. Parsons, R.; Tassinary, L.G.; Ulrich, R.S.; Hebl, M.R.; Grossman-Alexander, M. The View From The Road: Implications for Stress Recovery and Immunization. *J. Environ. Psychol.* **1998**, *18*, 113–140. [CrossRef]

72. Ulrich, R.S. View through a Window May Influence Recovery from Surgery. *Science* **1984**, *224*, 420–421. [CrossRef]

73. Tennessen, C.M.; Cimprich, B. Views to Nature: Effects on Attention. *J. Environ. Psychol.* **1995**, *15*, 77–85. [CrossRef]

74. Berto, R. The Role of Nature in Coping with Psycho-Physiological Stress: A Literature Review on Restorativeness. *Behav. Sci.* **2014**, *4*, 394–409. [CrossRef]

75. Richardson, M.; McEwan, K.; Maratos, F.; Sheffield, D. Joy and Calm: How an Evolutionary Functional Model of Affect Regulation Informs Positive Emotions in Nature. *Evol. Psychol. Sci.* **2016**, *2*, 308–320. [CrossRef]

76. Dijkstra, K.; Pieterse, M.; Pruyn, A. Physical Environmental Stimuli That Turn Healthcare Facilities into Healing Environments through Psychologically Mediated Effects: Systematic Review. *J. Adv. Nurs.* **2006**, *56*, 166–181. [CrossRef]

77. Kline, G.A. Does A View of Nature Promote Relief From Acute Pain? *J. Holist. Nurs.* **2009**, *27*, 159–166. [CrossRef] [PubMed]

78. Berman, M.G.; Jonides, J.; Kaplan, S. The Cognitive Benefits of Interacting With Nature. *Psychol. Sci.* **2008**, *19*, 1207–1212. [CrossRef] [PubMed]

79. Kuo, F.E.; Sullivan, W.C. Aggression and Violence in the Inner City: Effects of Environment via Mental Fatigue. *Environ. Behav.* **2001**, *33*, 543–571. [CrossRef]

80. Sullivan, W.C.; Kuo, F.E.; Depooter, S.F. The Fruit of Urban Nature: Vital Neighborhood Spaces. *Environ. Behav.* **2004**, *36*, 678–700. [CrossRef]

81. Wells, N.M.; Evans, G.W. Nearby Nature: A Buffer of Life Stress among Rural Children. *Environ. Behav.* **2003**, *35*, 311–330. [CrossRef]

82. Engemann, K.; Svenning, J.-C.; Arge, L.; Brandt, J.; Geels, C.; Mortensen, P.B.; Plana-Ripoll, O.; Tsirogiannis, C.; Pedersen, C.B. Natural Surroundings in Childhood Are Associated with Lower Schizophrenia Rates. *Schizophr. Res.* **2020**, *216*, 488–495. [CrossRef]

83. Krabbendam, L.; van Os, J. Schizophrenia and Urbanicity: A Major Environmental Influence—Conditional on Genetic Risk. *Schizophr. Bull.* **2005**, *31*, 795–799. [CrossRef]

84. Kaplan, S. The Restorative Benefits of Nature: Toward an Integrative Framework. *J. Environ. Psychol.* **1995**, *15*, 169–182. [CrossRef]

85. Korpela, K.M. Adolescents' Favourite Places and Environmental Self-Regulation. *J. Environ. Psychol.* **1992**, *12*, 249–258. [CrossRef]

86. Korpela, K.M.; Hartig, T.; Kaiser, F.G.; Fuhrer, U. Restorative Experience and Self-Regulation in Favorite Places. *Environ. Behav.* **2001**, *33*, 572–589. [CrossRef]

87. Gerber, S.M.; Jeitziner, M.-M.; Wyss, P.; Chesham, A.; Urwyler, P.; Müri, R.M.; Jakob, S.M.; Nef, T. Visuo-Acoustic Stimulation That Helps You to Relax: A Virtual Reality Setup for Patients in the Intensive Care Unit. *Sci. Rep.* **2017**, *7*, 13228. [CrossRef] [PubMed]

88. Pasca, L.; Carrus, G.; Loureiro, A.; Navarro, Ó.; Panno, A.; Tapia Follen, C.; Aragonés, J.I. Connectedness and Well-Being in Simulated Nature. *Appl. Psychol. Health Well-Being* **2021**. [CrossRef]

89. Yeo, N.L.; White, M.P.; Alcock, I.; Garside, R.; Dean, S.G.; Smalley, A.J.; Gatersleben, B. What Is the Best Way of Delivering Virtual Nature for Improving Mood? An Experimental Comparison of High Definition TV, 360° Video, and Computer Generated Virtual Reality. *J. Environ. Psychol.* **2020**, *72*, 101500. [CrossRef] [PubMed]

90. Felnhofer, A.; Kothgassner, O.D.; Schmidt, M.; Heinzle, A.-K.; Beutl, L.; Hlavacs, H.; Kryspin-Exner, I. Is Virtual Reality Emotionally Arousing? Investigating Five Emotion Inducing Virtual Park Scenarios. *Int. J. Hum.-Comput. Stud.* **2015**, *82*, 48–56. [CrossRef]

91. De Kort, Y.A.W.; Meijnders, A.L.; Sponselee, A.A.G.; IJsselsteijn, W.A. What's Wrong with Virtual Trees? Restoring from Stress in a Mediated Environment. *J. Environ. Psychol.* **2006**, *26*, 309–320. [CrossRef]

92. Anderson, A.P.; Mayer, M.D.; Fellows, A.M.; Cowan, D.R.; Hegel, M.T.; Buckey, J.C. Relaxation with Immersive Natural Scenes Presented Using Virtual Reality. *Aerosp. Med. Hum. Perform.* **2017**, *88*, 520–526. [CrossRef]

93. Herman, L.M.; Sherman, J. Virtual Nature: A Psychologically Beneficial Experience. In *Virtual, Augmented and Mixed Reality. Multimodal Interaction*; Lecture Notes in Computer Science; Chen, J.Y.C., Fragomeni, G., Eds.; Springer: Cham, Switzerland, 2019; Volume 11574, pp. 441–449. ISBN 978-3-030-21606-1.

94. Wang, X.; Shi, Y.; Zhang, B.; Chiang, Y. The Influence of Forest Resting Environments on Stress Using Virtual Reality. *Int. J. Environ. Res. Public Health* **2019**, *16*, 3263. [CrossRef] [PubMed]

95. Gao, T.; Zhu, L.; Qiu, L.; Gao, T.; Gao, Y. Exploring Psychophysiological Restoration and Individual Preference in the Different Environments Based on Virtual Reality. *Int. J. Environ. Res. Public Health* **2019**, *16*, 3102. [CrossRef] [PubMed]

96. Elliot, A.J.; Maier, M.A. Color Psychology: Effects of Perceiving Color on Psychological Functioning in Humans. *Annu. Rev. Psychol.* **2014**, *65*, 95–120. [CrossRef]

97. Mehta, R.; Zhu, R. (Juliet) Blue or Red? Exploring the Effect of Color on Cognitive Task Performances. *Science* **2009**, *323*, 1226–1229. [CrossRef]

98. Lichtenfeld, S.; Elliot, A.J.; Maier, M.A.; Pekrun, R. Fertile Green: Green Facilitates Creative Performance. *Pers. Soc. Psychol. Bull.* **2012**, *38*, 784–797. [CrossRef]

99. Akers, A.; Barton, J.; Cossey, R.; Gainsford, P.; Griffin, M.; Micklewright, D. Visual Color Perception in Green Exercise: Positive Effects on Mood and Perceived Exertion. *Environ. Sci. Technol.* **2012**, *46*, 8661–8666. [CrossRef] [PubMed]

100. Clarke, T.; Costall, A. The Emotional Connotations of Color: A Qualitative Investigation. *Color Res. Appl.* **2008**, *33*, 406–410. [CrossRef]

101. Fetterman, A.K.; Robinson, M.D.; Meier, B.P. Anger as "Seeing Red": Evidence for a Perceptual Association. *Cogn. Emot.* **2012**, *26*, 1445–1458. [CrossRef]

102. Hill, R.A.; Barton, R.A. Red Enhances Human Performance in Contests. *Nature* **2005**, *435*, 293. [CrossRef]

103. Jadid, A.; Koplin, M.; Stephan, S.; Hering-Bertram, M.; Paelke, V.; Teschke, T.; Helmut, E. Express Yourself/City—Smart Participation Culture Technologies. In *Smart Cities in the Mediterranean. Coping with Sustainability. Objectives in Small and Medium-sized Cities and Island Communities*; Stratigea, A., Kyriakides, E., Nicolaides, C., Eds.; Springer: Berlin/Heidelberg, Germany, 2017.

104. Adams, F.M.; Osgood, C.E. A Cross-Cultural Study of the Affective Meanings of Color. *J. Cross Cult. Psychol.* **1973**, *4*, 135–156. [CrossRef]

105. Valdez, P.; Mehrabian, A. Effects of Color on Emotions. *J. Exp. Psychol. Gen.* **1994**, *123*, 394–409. [CrossRef] [PubMed]

106. Petit, A.; Siret, D.; Simonnot, N. Capturing Chromatic Effects in Urban Environment. In *Experiential Walks for Urban Design: Revealing, Representing, and Activating the Sensory Environment*; Piga, B.E.A., Siret, D., Thibaud, J.-P., Eds.; Springer Tracts in Civil Engineering; Springer International Publishing: Cham, Switzerland, 2021; pp. 207–222. ISBN 978-3-030-76694-8.

107. Manav, B. Color-Emotion Associations, Designing Color Schemes for Urban Environment-Architectural Settings. *Color Res. Appl.* **2017**, *42*, 631–640. [CrossRef]

108. Palmer, S.E.; Schloss, K.B. An Ecological Valence Theory of Human Color Preference. *Proc. Natl. Acad. Sci. USA* **2010**, *107*, 8877–8882. [CrossRef] [PubMed]

109. Armondi, S.; Bruzzese, A. Contemporary Production and Urban Change: The Case of Milan. *J. Urban Technol.* **2017**, *24*, 27–45. [CrossRef]

110. Boros, J.; Mahmoud, I. Urban Design and the Role of Placemaking in Mainstreaming Nature-Based Solutions. Learning From the Biblioteca Degli Alberi Case Study in Milan. *Front. Sustain. Cities* **2021**, *3*, 38. [CrossRef]

111. Denissen, J.J.A.; Butalid, L.; Penke, L.; van Aken, M.A.G. The Effects of Weather on Daily Mood: A Multilevel Approach. *Emotion* **2008**, *8*, 662–667. [CrossRef]

112. Höppe, P. Different Aspects of Assessing Indoor and Outdoor Thermal Comfort. *Energy Build.* **2002**, *34*, 661–665. [CrossRef]

113. Jendritzky, G.; de Dear, R.; Havenith, G. UTCI—Why Another Thermal Index? *Int. J. Biometeorol.* **2012**, *56*, 421–428. [CrossRef]

114. Rouse, J.W.; Haas, R.; Deering, D.; Schell, J.; Harlan, J. *Monitoring the Vernal Advancement and Retrogradation (Green Wave Effect) of Natural Vegetation*; ScienceOpen, Inc.: Washington, DC, USA, 1974.

115. Russell, J.A.; Pratt, G. A Description of the Affective Quality Attributed to Environments. *J. Pers. Soc. Psychol.* **1980**, *38*, 311–322. [CrossRef]

116. Benedikt, M.L. To Take Hold of Space: Isovists and Isovist Fields. *Environ. Plan. B Plan. Des.* **1979**, *6*, 47–65. [CrossRef]

117. Morariu, V.I.; Prasad, V.S.N.; Davis, L.S. Human Activity Understanding Using Visibility Context. *IEEERSJ IROS Workshop Sens. Hum. Spat. Concepts FS2HSC* **2007**, *9*, 1–8.

118. Birant, D.; Kut, A. ST-DBSCAN: An Algorithm for Clustering Spatial–Temporal Data. *Data Knowl. Eng.* **2007**, *60*, 208–221. [CrossRef]

119. Cheng, H.D.; Jiang, X.H.; Sun, Y.; Wang, J. Color Image Segmentation: Advances and Prospects. *Pattern Recognit.* **2001**, *34*, 2259–2281. [CrossRef]

120. Fairchild, M.D. Color Appearance Models. In *The Wiley-IS&T Series in Imaging Science and Technology*, 3rd ed.; John Wiley & Sons, Inc.: Chichester, West Sussex, UK, 2013; ISBN 978-1-118-65309-8.

121. Bunce, C. Correlation, Agreement, and Bland–Altman Analysis: Statistical Analysis of Method Comparison Studies. *Am. J. Ophthalmol.* **2009**, *148*, 4–6. [CrossRef]

122. Giavarina, D. Understanding Bland Altman Analysis. *Biochem. Medica* **2015**, *25*, 141–151. [CrossRef]

123. Elsadek, M.; Sayaka, S.; Fujii, E.; Koriesh, E.; Moghazy, E.; Fatah, Y.A.E. Human Emotional and Psycho-Physiological Responses to Plant Color Stimuli. *J. Food Agric. Environ.* **2013**, *11*, 1584–1591.

124. Elsadek, M.; Sun, M.; Fujii, E. Psycho-Physiological Responses to Plant Variegation as Measured through Eye Movement, Self-Reported Emotion and Cerebral Activity. *Indoor Built Environ.* **2017**, *26*, 758–770. [CrossRef]
125. Zhang, Y.; Hedo, R.; Rivera, A.; Rull, R.; Richardson, S.; Tu, X.M. Post Hoc Power Analysis: Is It an Informative and Meaningful Analysis? *Gen. Psychiatry* **2019**, *32*, e100069. [CrossRef]
126. Feldman Barrett, L.; Russell, J.A. Independence and Bipolarity in the Structure of Current Affect. *J. Pers. Soc. Psychol.* **1998**, *74*, 967–984. [CrossRef]
127. Gibson, J.J. The Theory of Affordances. In *Perceiving, Acting and Knowing*; Shaw, R., Bransford, J., Eds.; Lawrence Erlbaum Associates; distributed by the Halsted Press Division; Wiley: Hillsdale, NJ, USA; New York, NY, USA, 1977; ISBN 978-0-470-99014-8.
128. Roe, J.; Aspinall, P. The Emotional Affordances of Forest Settings: An Investigation in Boys with Extreme Behavioural Problems. *Landsc. Res.* **2011**, *36*, 535–552. [CrossRef]
129. Han, K.-T.; Ruan, L.-W. Effects of Indoor Plants on Self-Reported Perceptions: A Systemic Review. *Sustainability* **2019**, *11*, 4506. [CrossRef]
130. Bringslimark, T.; Hartig, T.; Patil, G.G. The Psychological Benefits of Indoor Plants: A Critical Review of the Experimental Literature. *J. Environ. Psychol.* **2009**, *29*, 422–433. [CrossRef]
131. Boffi, M.; Rainisio, N. To Be There, or Not to Be. Designing Subjective Urban Experiences. In *Urban Design and Representation*; Springer International Publishing: Cham, Switzerland, 2017; pp. 37–53.
132. Piga, B.E.A. The Combined Use of Environmental and Experiential Simulations to Design and Evaluate Urban Transformations. In *Quality of Life in Urban Landscapes*; Cocci Grifoni, R., D'Onofrio, R., Sargolini, M., Eds.; The Urban Book Series; Springer: Cham, Switzerland, 2018; pp. 357–364. ISBN 978-3-319-65580-2.
133. Sheppard, S.R.J. Guidance for Crystal Ball Gazers: Developing a Code of Ethics for Landscape Visualization. *Landsc. Urban Plan.* **2001**, *54*, 183–199. [CrossRef]
134. Sheppard, S.R.J. Validity, Reliability and Ethics in Visualization. In *Visualization in Landscape and Environmental Planning*; Bishop, I.D., Lange, E., Eds.; Routledge—Taylor and Francis Group: New York, NY, USA, 2005; pp. 79–97.
135. Boffi, M.; Rainisio, N.; Riva, E.; Inghilleri, P. Social Psychology of Flow: A Situated Framework for Optimal Experience. In *Flow Experience. Empirical Research and Applications*; Springer: Berlin, Germany, 2016.

sustainability

Article

The Improvement of User Satisfaction for Two Urban Parks in Dubai, UAE: Bay Avenue Park and Al Ittihad Park

Chuloh Jung *, Nahla Al Qassimi, Mohammad Arar and Jihad Awad

Department of Architecture, College of Architecture, Healthy and Sustainable Buildings Research Center, Ajman University, Ajman P.O. Box 346, United Arab Emirates; n.alqassimi@ajman.ac.ae (N.A.Q.); m.arar@ajman.ac.ae (M.A.); j.awad@ajman.ac.ae (J.A.)
* Correspondence: c.jung@ajman.ac.ae

Abstract: The population of Dubai has increased dramatically in the last 40 years. Along with social changes, neighborhood parks are becoming increasingly important for enhancing the residents' quality of life. This study aims to evaluate the physical environment of parks and investigate park users' satisfaction in neighborhood parks of Dubai. After defining the park and surrounding environment for access, a field survey was performed at Bay Avenue Park and Al Ittihad Park. The data for analysis were collected from the Department of Geographical Information System (GIS) Center at Dubai Municipality. The results show that the standard duration was 60–90 min, and the walking/driving time was 10–20 min. "Children Facility" and "Various Attractions" were low in both parks. The statistical results of multiple regression analysis of the derived factors and satisfaction show that Bay Avenue Park influenced satisfaction in the surrounding environment for access, pedestrian space, park facility, convenience and comfort of the park, and various attractions and activities. Furthermore, Al Ittihad Park influenced satisfaction in pedestrian space, green landscape, surrounding environment for access, park facilities, and safe access. Therefore, factors such as park facilities, surrounding environment for access, and pedestrian space were analyzed to affect satisfaction in both parks. The analysis of the surrounding environment for access factors using GIS would methodologically help determine priorities for future improvements around parks. However, this study is limited by the scope and investigation period of the target parks, and detailed factors related to the surrounding environment for access are also not evaluated.

Keywords: surrounding environment for access; neighborhood park; user satisfaction; park facility; Bay Avenue Park; Al Ittihad Park

Citation: Jung, C.; Al Qassimi, N.; Arar, M.; Awad, J. The Improvement of User Satisfaction for Two Urban Parks in Dubai, UAE: Bay Avenue Park and Al Ittihad Park. *Sustainability* **2022**, *14*, 3460. https://doi.org/10.3390/su14063460

Academic Editors: Israa H. Mahmoud, Eugenio Morello, Giuseppe Salvia and Emma Puerari

Received: 26 January 2022
Accepted: 7 March 2022
Published: 16 March 2022

Publisher's Note: MDPI stays neutral with regard to jurisdictional claims in published maps and institutional affiliations.

1. Introduction

Since the 1980s, Dubai has experienced rapid economic growth in the gross domestic product (GDP) from 29.57 billion USD in 1980 to 401.51 billion USD in 2020 [1,2]. Dubai's population expanded from 254,400 in 1980 to 2,921,376 in 2020 [3]. Such an unprecedented development and urban expansion have increased personal income and more comfortable lifestyles in tax-free cities [4,5]. However, the lack of urban open space with the green area, even though Dubai is a planned city from the initiation compared to other metropolitan cities in the Middle East, created the unprecedented unbalance in the life of Dubai residents. Thus, Dubai residents are looking for spaces that will allow them to engage in relaxing leisure activities [6]. Previously, most leisure activities took place in desert parks far away from residences and theme parks in the outskirts, so activities were concentrated on such facilities [7,8]. The high frequency of use resulted in great congestion and damage to the natural environment [9]. Currently, neighborhood parks have become necessary for Dubai residents to satisfy their resting and recharging needs [10].

The importance of parks within walking distance has been recognized in the past, and parks have been placed in easily accessible locations where residents can conveniently use them in real life [11–13]. Furthermore, efforts are being made to increase the park utilization rate by reflecting the diverse needs of the citizens in major global cities [14]. For example, 96% of New Yorkers enjoy the benefits of parks within a 10-min walk [15]. Despite its many urban problems, New York City has become one of the most desirable cities to live in [16]. Accordingly, the importance of neighborhood parks within walking or short driving distance is gradually increasing in Dubai as it improves the quality of life among urban residents [17,18]. Furthermore, in recent years, parks have served as a place to maintain a lasting relationship with society by providing a pleasant environment, revitalizing the residents' activity through walking, and promoting walking activities for the elderly [19]. However, even though neighborhood parks near their residence are helpful for the residents, it is challenging to expand new parks due to economic problems such as non-profitability and difficulty purchasing land [20].

Most of the current neighborhood parks, initially established as urban parks, within walking distances consist of spatial composition and facilities that provide green resting areas [21]. However, in reality, the utilization rate is decreasing because these parks do not reflect the actual needs of the modern citizens, such as sports-related facilities, walking/jogging tracks, and bicycle tracks [22]. The social function of parks needs to adapt to the changes in society and lifestyles. Evaluation of the users' satisfaction and needs is essential to improve parks for the use of residents and improve their quality of life through these parks [23]. In addition, unlike urban and regional parks, in the case of parks located within walking distances, the surrounding environment for access to the park can directly affect the park's utilization rate, considering that it is intended for use by foot [24,25].

Most of the plans focus on improving the environment inside the park, which can lead to a decreased actual utilization rate after the park is built [26,27]. In other words, it is crucial to improve the surrounding environment for access from the residence to the parks so that the people can recognize it as a suitable facility in their locality and use it to their satisfaction [28]. Therefore, to improve the satisfaction levels and increase the utilization rate of neighborhood parks, measures to enhance the internal environment and access to parks reflecting modern city residents' park usage patterns are recommended [29,30].

Various studies to improve parks have been continuously conducted by analyzing users' behavior and satisfaction [31,32]. Plunz et al. (2019) [33] argued that urban parks should be created as resident-friendly parks to increase their efficiency as a comfortable resting space for people using the city park. Turan et al. (2016) classified neighborhood parks in Rize, Turkey according to the type of facilities provided, and suggested improvements for each type. In addition, Zhang et al. (2018) [34] analyzed the satisfaction level of neighborhood parks in China by classifying them into the urban, river, and natural park types and found that users prioritized benefits over facilities or environmental conditions. Chan et al. (2018) [35] mentioned the necessity of creating urban parks using scrap land because residents of areas adjacent to the industrial complex in Hong Kong prefer small parks near their residences. Neckel et al. (2020) [36] suggested that park revitalization should be promoted in connection with the park's walking and climbing functions, as satisfaction in the use of regional parks increases through mental and physical stability and health promotion in the area. Gholipour et al. (2021) [37] analyzed facility use and satisfaction surveying urban park users. They pointed out that many improvements are needed in the facilities used and convenience facilities such as information facilities and parking lots. However, previous studies aimed to suggest improvement plans and their effects on parks as open spaces in the city, rather than focusing on the accessibility of these parks [38]. As such, there have been many studies on the level of satisfaction, problems, and space required by park visitors, but there are insufficient studies on user behavior and satisfaction of urban parks within walking or short driving distance for residents, and the necessary facilities and directions for improvement [39,40]. Ayala-Azcárraga et al. (2019) [41] analyzed the relationship between the perceived spatial characteristics such as size, number, and distance to the park, infrastructure, and environmental components of three size-categories of urban parks in Mexico City and their use. They analyzed the use of these spaces to promote well-being, considering the relationship with three dimensions (health, community, and satisfaction with life) to recover from stress and fatigue, encourage physical activity, and facilitate social contact. Anastasiou and Manika (2020) [42] identified the characteristics that affect the residential satisfaction of open urban space in a medium-sized Greek city and the complex patterns between the characteristics elements of these spaces and the visitors' satisfaction. Liu and Xiao (2021) [43] assigned the potential factors affecting people's perception of and satisfaction with urban parks based on the online comments data from Dianping and explored the relationship between these factors and people's satisfaction, and further identified the significant factors. For the next phase of research, after the analysis of user satisfaction by survey or online data, a stimulus-response theory which is an environmental, psychological theory of the process of perception and cognition by which humans respond to environmental stimuli such as Russell and Lanius (1984) [44] and Rapoport (2016) [45] can be used to build qualitative analysis data from Dubai residents.

Neighborhood urban parks have an attraction distance of fewer than 1000 m and a size of more than 30,000 m^2 [46,47]. It is for the use of those who live within walking or short driving distance [48]. Moreover, residents within the neighborhood should use it without significant restrictions in terms of time and space [49]. Although the impact of having proper accessibility from the residential areas to the park is essential, most park satisfaction studies are limited to the evaluation of the park's internal facilities and physical environment [50]. Improving Park utilization by enhancing accessibility has not been considered [51]. Accessibility considers the travel time and the characteristics of the destination, such as the presence or absence of sidewalks, provision of shade by trees, and pedestrian obstacles [52]. In particular, these characteristics can be considered essential variables that can affect the satisfaction level with a park accessible by foot [53].

This study investigates neighborhood parks within walking or short driving distances that are advantageous to residents. In addition, a satisfaction analysis is conducted regarding the accessibility factors that have not been dealt with thus far, and the difference in influence between internal and external environments is analyzed to provide a holistic review for satisfaction improvement. Finally, the study aims to provide data for park creation and reorganization to improve future utilization rates by statistically determining the effects of these internal and external environments on user satisfaction to the Dubai Municipality since there is no previous research for urban parks in metropolis with the similar development pace of urban expansion or similar hot desert climate urban context.

2. Materials and Methods

2.1. Target Sites

For the spatial scope of the study, parks satisfying each criterion were selected in the following order, as shown in Table 1. First, by referring to the park status of Dubai Municipality in 2021 [54], 20 neighborhood parks were identified (summarized in Figure 1) [55,56]).

Table 1. List of Neighborhood Parks in Dubai (* = Neighborhood Parks between 30,000 m^2 and 100,000 m^2).

#	Name	Location	Area	Features
1	Al Barsha Pond Park	Al Barsha	210,400 m^2	- Park House/Café - Kiosk/Toilets/Pond/Solar-powered boats
2	Al Ittihad Park *	Palm Jumeirah	98,200 m^2	- Shops/Restaurants - Playgrounds/Water features/Toilets
3	Al Khazzan Park	City Walk	15,200 m^2	- Shaded Playgrounds - Toilets
4	Al Sufouh Park	Al Sufouh	14,000 m^2	- Coffee shop - Sports facilities/Toilets
5	Al Wasl Park	Jumeriah 1	14,800 m^2	- Shaded Playgrounds - Toilets
6	Bay Avenue Park *	Business Bay	35,000 m^2	- Shaded Playgrounds/Skate Park - Toilets
7	Dubai Creek Park	Dubai Creek	960,000 m^2	- Barbeques/Picnic Area/Playgrounds - Coffee shop/Toilets
8	Dubai Hills Park	Dubai Hills	180,000 m^2	- Splash Park/Skate park/Ice rink/Dog park - Playgrounds/Toilets
9	Dubai Miracle Garden *	Al Barsha South	72,000 m^2	- Flower gardens/Restaurants - Toilets
10	JLT Park *	JLT	35,000 m^2	- Restaurants/Coffee shop - Toilets
11	Jumeirah Beach Park *	Umm Suqeim 2	35,100 m^2	- Barbeques/Food kiosks - Toilets
12	Love Lake Dubai	Al Qudra Lake	105,500 m^2	- Barbeques/Coffee shop - Toilets
13	Mamzar Beach Park	Al Mamzar	1,060,000 m^2	- Chalets/Swimming pool/Sports courts - Barbeque area/Restaurants/Toilets
14	Mushrif Park	Al Khawaneej	5,250,000 m^2	- Restaurants/Sports fields/Bike track - Toilets
15	Nad Al Sheba Cycle Park *	Nad Al Sheba	60,000 m^2	- Change Rooms - Toilets
16	Quranic Park	Al Khawaneej	600,000 m^2	- Shaded seating/Playgrounds - Toilets
17	Al Safa Park	Al Safa	640,000 m^2	- Cafeterias/Theater/Track and Field courts - Toilets
18	The Block	Dubai Creek	25,500 m^2	- Sports facilities/Climbing walls /Cafes - Toilets
19	Umm Suqeim Park	Umm Suqeim 1	28,000 m^2	- Cafes - Toilets
20	Zabeel Park	Zabeel	475,000 m^2	- Barbeques/Picnic Area/Playgrounds - Restaurants/Toilets

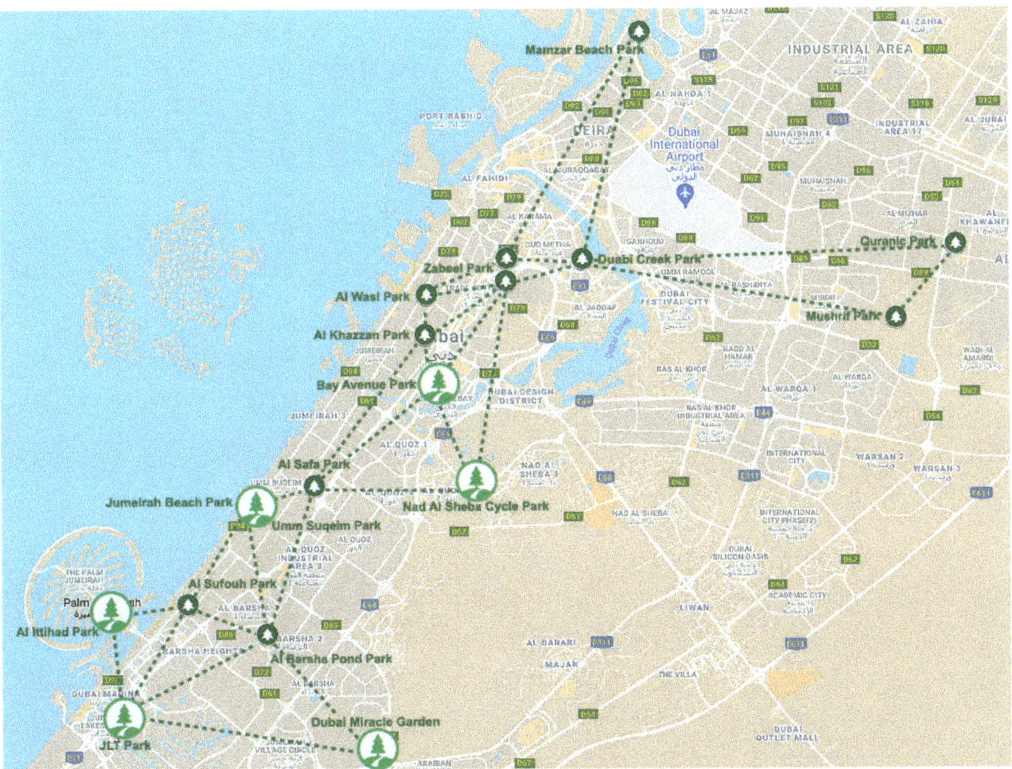

Figure 1. The Network of Neighborhood Park Locations in Dubai.

Next, neighborhood parks with an area of 30,000 m^2 or more, which is the legal standard for neighborhood parks within walking or short driving distance in Dubai Municipality regulation, and smaller than 100,000 m^2, the minimum area of an urban neighborhood park in Dubai Municipality regulation was selected. It was found that 6 out of 20 neighborhood parks met this requirement (Figure 1) [57].

Finally, a 1000 m park attraction distance was set from the park boundary to select a park with a high park utilization rate. Then, the area percentage of the high-rise residential areas in the zone was calculated for each park. The higher density within a 1000 m radius, the more residents use the parks. Compared to other park areas with low-rise villa and townhouses, the area percentage of the high-rise residential areas in the zone of Bay Avenue Park (12.8%) and Al Ittihad Park (9.6%) showed the highest ratios in that order since both parks are in high-density family-oriented apartment areas.

Bay Avenue Park's first target site is Al A'amal Street in Business Bay (Figure 2). It can be easily accessed from the Business Bay Metro Station by a short 15-min walk. A convenient Bay Avenue parking area is available for those who come by car [58]. The Business Bay district is known for its commercial development and high-rise office towers. However, it also contains urban parks, which offer a wide variety of physical activities. Dubai Properties developed Bay Avenue Park in 2014 as an addition to the Bay Avenue project, including a retail shopping center [59]. It covers an area of 35,000 m^2 and is in the heart of the Business Bay area. It features a large green area along with a jogging track (Figure 3), two outdoor gyms, and other recreational facilities, such as one of the best skate parks in Dubai and two playgrounds to entertain children [60].

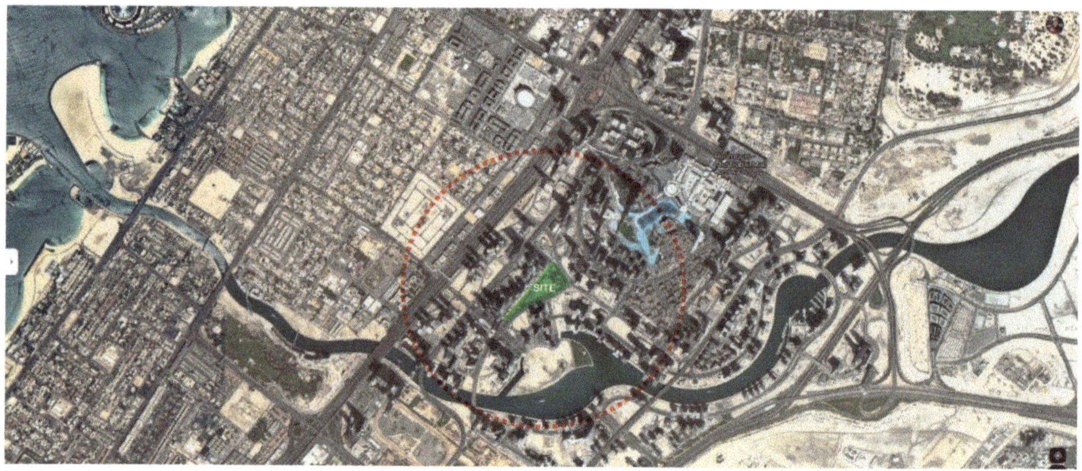

Figure 2. Location Map of Bay Avenue Park (Accessibility Radius 1 km).

Figure 3. Urban Context of Bay Avenue Park.

These amenities are ideal for the 191,000 residents of Business Bay, especially those living in the Executive Towers complex [61]. Bay Avenue Park also has coffee shops, a dancing fountain, and a jogging path that runs for 1 km around the park. Park users enjoy the area's landmarks, such as Burj Khalifa and Zaha Hadid's Opus building, while walking or jogging in the park.

The second target site, Al Ittihad Park, is located in Palm Jumeirah (Figure 4). It is easily accessible through the Palm Monorail (Al Ittihad Park station), and the lifted monorail track runs through the park [62]. It is a tranquil park spread across 98,200 m². The word Ittihad (union in Arabic) implies the ethos of living together in harmony [63]. The most iconic feature of this park is its range of trees and plants. Al Ittihad Park was inaugurated in 2012 on the UAE's 41st National Day. It is located adjacent to the Golden Mile Galleria Mall and Shoreline Apartments complex and provides residents with a walking distance place to spend their leisure time [64]. One of the main features is the 3.2-km jogging track, one of the best places in the city for an outdoor run, brisk walk, or a casual stroll in the park (Figure 5). It has 60 varieties of plants and trees indigenous to the region; it is home to over 600 palm trees planted on the edges of the running track. It also has a play area for children with swings, slides, and other fun rides.

Figure 4. Location Map of Al Ittihad Park (Accessibility Radius 1 km).

Figure 5. Urban Context of Al Ittihad Park.

2.2. Survey Method

The survey was divided into a preliminary and the main questionnaire to select the objective criteria. The questionnaires consisted of main categories on user behavior, satisfaction with the internal environment and accessibility to the park, and personal characteristics. Regarding accessibility from the residence to the park, based on the literature related to the pedestrian environment, an objective analysis was conducted to determine the subjective view of the difficulties in surrounding environment for access. Items related to satisfaction with the park environment were organized by extracting indices based on previous studies.

Among preliminary variables, a total of 25 critical variables was extracted via four zoom brainstorming sessions between 10 June 2021 and 24 July 2021 with nine professionals with more than 20 years of experience in their fields (2 urban design professors from the University of Sharjah and Ajman University, two landscape architecture professors from UAE University and Ajman University, two architecture professors from UAE University and Ajman University, two urban design department managers from Dubai Municipality, and one general manager from landscape design firm in Dubai) (Table 2).

Table 2. Survey Questionnaire Structure.

Main Categories		Questions	Evaluation
Park User Behavior		1. Frequency of Visit, 2. Duration of Visit, 3. Reason for Visit, 4. Transportation Mode	Multiple Choices Question
		1. Visiting Time, 2. Walking/Driving Time to Access	Short Answers Question
Satisfaction Level	Park Environments	1. The Size of the Park, 2. Green Area, 3. Shades, 4. Trail Length, 5. Trail Width, 6. Lightings, 7. Parking, 8. Resting Place, 9. Public Facility, 10. Children Facility, 11. Sports Facility, 12. Sports Court, 13. Safe Environment, 14. Attractions, 15. Scenery	5-Point Likert Scale Question
	Accessibility to the park	1. Safe Environment, 2. Scenery, 3. Attractions, 4. Tree Shades, 5. Connectivity of Walkway, 6. Level of Slopes, 7. Pedestrian/Car Separation, 8. Lightings, 9. Car Speed, 10. Hindrance	5-Point Likert Scale Question
Surveyor Information		1. Gender, 2. Age, 3. Occupation, 4. Monthly Income Range	Multiple Choices Question

A preliminary questionnaire was performed on 1 September 2021, to correct and supplement questions. The survey was conducted at Bay Avenue Park and Al Ittihad Park simultaneously from 7:00 AM to 8:00 PM for a total of 10 days, from 6–15 September 2021. The survey was conducted in a face-to-face method by graduate students, who were well aware of the purpose of the research, to ensure the reliability of the survey results. Excluding ten copies that were judged to be insincere, a total of 375 answered questionnaires, 189 from Bay Avenue Park and 186 from Al Ittihad Park, were used for the analysis.

2.3. Analysis Method

The flow of analysis is shown in Figure 6. The environment of each park was investigated through a field survey by graduate students to evaluate the physical environment related to park use. Physical accessibility items were selected based on previous studies related to walking [65]. The analysis data were collected from the Department of Geographical Information System (GIS) Center at Dubai Municipality [66]. Spatial data were constructed for the surrounding environment for access within a straight distance of 1000 m from the park boundary [67]. As for the analysis data, using the facility spatial information DB, GIS data for pedestrian-related sidewalks, pedestrian traffic lights, street trees, bus stops, traffic safety signs, and intersections were constructed and analyzed for each park.

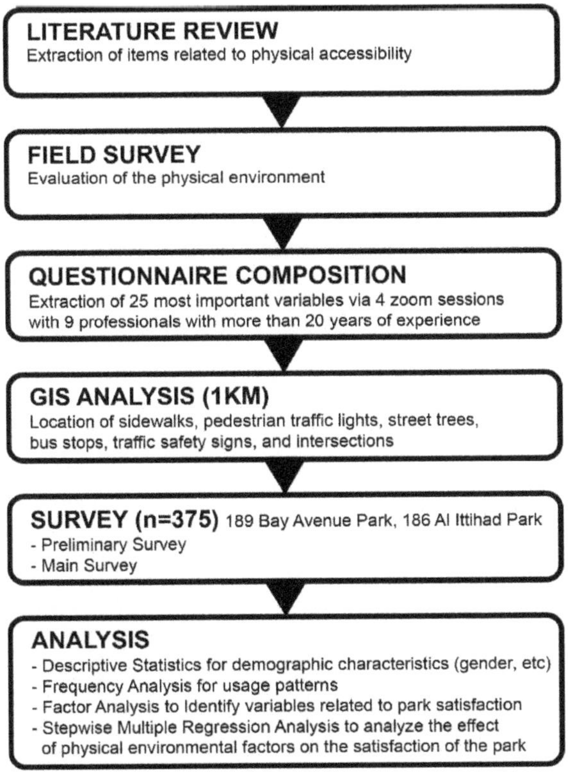

LITERATURE REVIEW
Extraction of items related to physical accessibility

FIELD SURVEY
Evaluation of the physical environment

QUESTIONNAIRE COMPOSITION
Extraction of 25 most important variables via 4 zoom sessions
with 9 professionals with more than 20 years of experience

GIS ANALYSIS (1KM)
Location of sidewalks, pedestrian traffic lights, street trees,
bus stops, traffic safety signs, and intersections

SURVEY (n=375) 189 Bay Avenue Park, 186 Al Ittihad Park
- Preliminary Survey
- Main Survey

ANALYSIS
- Descriptive Statistics for demographic characteristics (gender, etc)
- Frequency Analysis for usage patterns
- Factor Analysis to Identify variables related to park satisfaction
- Stepwise Multiple Regression Analysis to analyze the effect
  of physical environmental factors on the satisfaction of the park

Figure 6. The Flow of the Research Analysis.

The questionnaires were analyzed using the IBM SPSS Statistics 26 program. First, descriptive statistics and frequency analysis were conducted to understand the general characteristics and satisfaction of the survey subjects. Finally, factor analysis was conducted to identify the variables and factors related to the satisfaction level. Finally, stepwise multiple regression analysis was performed to analyze the effect of the physical environmental factors extracted through factor analysis on each park's satisfaction level.

3. Results

3.1. Physical Environment & Accessibility Assessment

By examining the status of facilities within the two parks using a field survey, as shown in Table 3, the types and number of landscaping, resting, and convenience facilities were similar. However, there were differences in exercise facilities. Bay Avenue Park has a promenade at its center; hence, many users engage in simple physical activities such as walking and jogging. In Al Ittihad Park, children-friendly facilities were installed in a multipurpose sports space, which many residents used. Further, in Bay Avenue Park, various outdoor exercise equipment was also installed. Still, Al Ittihad Park was found to be slightly deteriorated due to the facilities' lack of expansion and maintenance.

Table 3. Physical Environments of Bay Avenue Park and Al Ittihad Park.

Category	Bay Avenue Park	Al Ittihad Park
Physical Environment		
Size	35,000 m^2	98,200 m^2
Resting Places	Coffee shops, 46 Benches, 6 Outdoor cafés	Coffee shops, 118 Benches, 8 Outdoor cafés
Sports Facility	1.0 km jogging track, 2 Outdoor gyms,	3.2 km jogging track
Parking	Parking lot (975), Bicycle racks	Parking lot (1346), Bicycle racks
Landscape	Lawn, Bushes, Dates palm trees (10 varieties of indigenous plants and trees)	Lawn, Bushes, Dates palm trees (60 types of indigenous plants and trees)
Children Facility	Skate parks, 2 Playgrounds, Dancing fountain	Playground with swings, slides, and other fun rides
Commercial Activity	Bay Avenue Mall	Golden Mile Galleria Mall
Sidewalk Area	24,364 m^2	19,496 m^2
Pedestrian Traffic Lights	28	18
Traffic Safety Signs	84	68
Number of Trees	158	600

The data on the approach environment index was constructed by performing a 1 km radius buffer from the park boundary. Dubai residents mostly used the park within 1 km of the attraction distance. The park needs to be within 10 min walking or 5 min driving distance because it needs to be accessible to the residents. In evaluating the pedestrian networks, sidewalks for safety were vital as they affect safe passage from residential areas to parks. The analysis revealed that the sidewalk area of Bay Avenue Park was 24,364 m^2 and that of Al Ittihad Park was 19,496 m^2, indicating that the two parks were similar.

Moreover, there were 28 pedestrian traffic lights around Bay Avenue Park and 18 around Al Ittihad Park. Traffic safety signs were analyzed among the 84 around Bay Avenue Park and 68 around Al Ittihad Park. It was confirmed that there was no significant difference in pedestrian safety facilities around the park. However, it was found that the number of street trees and intersections differed within the two parks' induction zones. In the case of street trees, Al Ittihad Park has 442 more than Bay Avenue Park, despite the similar sidewalks within the shaded area. It means that Al Ittihad Park is relatively better than Bay Avenue Park in terms of comfort, such as green cover and shade provided by trees.

3.2. User Behavior and Satisfaction Assessment

As shown in Table 4, In terms of the demographic characteristics of park users, particularly gender distribution, the proportion of female users of Bay Avenue Park was more, at 52.4%. However, in Al Ittihad Park, male users were more at 56.5%. In terms of age, 21.7% in Bay Avenue Park were in their 40 s, and 28.5% were in their 40 s at Al Ittihad Park. By occupation, in Bay Avenue Park, housewives accounted for the most at 29.6%, and in Al Ittihad Park, housewives accounted for 34.4%, followed by office workers, self-employed, and students.

Table 4. Demographic Characteristics of Survey Participants.

Category		Bay Avenue Park		Al Ittihad Park	
		Number	Percentage	Number	Percentage
Gender	Male	90	47.6	105	56.5
	Female	99	52.4	81	43.5
	Total	189	100.0	186	100.0
Age Group	10–19	14	7.4	15	8.1
	20–29	18	9.5	33	17.7
	30–39	29	15.3	44	23.7
	40–49	41	21.7	53	28.5
	50–59	30	15.9	18	9.7
	60–69	34	18.0	14	7.5
	Above 70	23	12.2	9	4.8
	Total	189	100.0	186	100.0
Occupations	Students	30	15.9	23	12.4
	Housewives	56	29.6	64	34.4
	Office Workers	46	24.3	55	29.6
	Self-Employed	21	11.1	33	17.7
	Others	36	19.1	11	5.9
	Total	189	100.0	186	100.0
Average Monthly Income (AED)	≤10,000	19	10.2	13	7.0
	10,000–20,000	76	40.6	90	48.1
	20,000–30,000	44	22.5	48	25.9
	30,000–40,000	33	17.6	26	14.1
	≥40,000	17	9.1	9	4.9
	Total	189	100.0	186	100.0

In terms of usage patterns, Bay Avenue Park and Al Ittihad Park users were primarily similar in Figure 7. First, looking at the frequency of visits, one time per week was the highest in Al Ittihad Park (38.0%), and more than five times per week was the highest in Bay Avenue Park (27.4%).

Regarding the duration of the visit, the percentage of respondents who used 60 to 90 min for both Bay Avenue Park and Al Ittihad Park was the highest at 32.7% and 45.9%, respectively. This is consistent with research showing that residents spend a relatively shorter duration, amounting to less than two hours when using a nearby park [68].

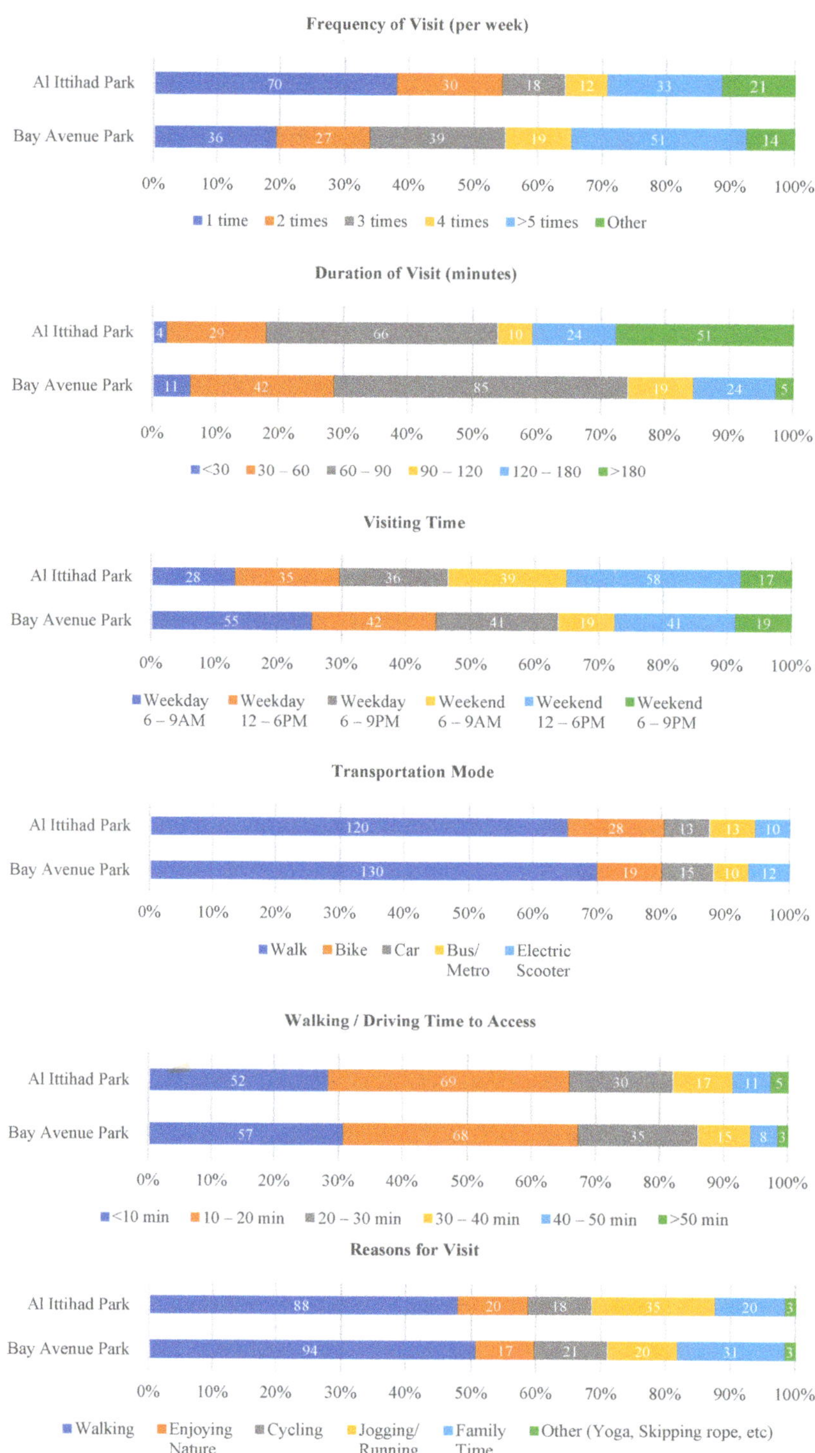

Figure 7. Analysis of the Usage Behaviors of Park Users.

A high walking rate was seen in terms of means of transportation from the residence to the park. Further, the respondents who said that the walking time was between 10 and 20 min were 37.4% in Bay Avenue Park and 38.1% in Al Ittihad Park. It means that parks are used more by the residents living in the vicinity than those from other areas. Furthermore, 43.4% of Bay Avenue Park users and 29.6% of Al Ittihad Park users mainly used the park for walking and jogging.

3.3. Park Physical Environment & Accessibility Satisfaction Assessment

Table 5 shows the results of analyzing the park users' satisfaction with the physical environment on a 5-point Likert scale. First, looking at the evaluation results of Bay Avenue Park users, satisfaction with most items were found to be higher than that of Al Ittihad Park. Among them, the "Safe Environment" item had the highest average value of 3.72 points, indicating a relatively high level of satisfaction with safety. It is because Bay Avenue Park has the characteristic of an open type, wherein the inside of the park can be seen clearly. This result can be attributed to the various facilities arranged around the linear promenade next to the Bay Avenue mall corridor and its spatial characteristics that enable natural monitoring. Next, satisfaction with the "Trail Width" was relatively high at 3.57 points. The circular walkway with a 5 m or more width inside Bay Avenue Park is highly satisfactory because it provides a smooth environment for users to walk and jog.

Table 5. Satisfaction of Park's Physical Environments ($p < 0.05$).

Category	Total		Bay Avenue Park		Al Ittihad Park		F Value
	Mean	Standard Deviation	Mean	Standard Deviation	Mean	Standard Deviation	
Shades	3.52	0.870	3.50	0.866	3.53	2.876	0.008
Green Area	3.54	0.785	3.53	0.820	3.54	0.749	2.042
The Size of the Park	3.48	0.813	3.48	0.834	3.47	0.792	0.479
Trail Width	3.49	0.790	3.57	0.757	3.42	0.816	0.919
Trail Length	3.40	0.837	3.47	0.815	3.31	0.853	0.118
Parking	2.92	0.880	3.15	0.814	3.32	0.885	3.960
Lightings	3.19	0.906	3.47	0.872	2.69	0.852	3.089
Resting Place	3.15	0.981	3.33	0.962	2.89	0.969	1.895
Public Facility	2.98	1.032	3.34	0.894	2.97	1.031	6.163
Children Facility	2.56	0.883	2.68	0.881	2.60	0.867	0.006
Sports Facility	3.22	0.869	3.48	0.795	2.41	0.865	1.701
Sports Court	3.08	0.902	3.21	0.864	2.95	0.922	0.679
Safe Environment	3.49	0.796	3.72	0.782	3.25	0.739	0.554
Attractions	2.79	0.846	2.90	0.842	2.68	0.837	1.967
Scenery	3.31	0.894	3.40	0.879	3.21	0.902	0.315

When looking at user satisfaction with the environment of Al Ittihad Park, the average value of satisfaction for "Green Area" was the highest at 3.54 points, followed by "Shades" at 3.53 points, "The Size of the Park area" at 3.47, "Trail Width" at 3.42, and "Trail Length" at 3.31. This is because Al Ittihad Park is rich in natural elements such as green space and trees, and the trail has more than 600 vernacular trees, so users can experience nature, which is rarely obtained in city life. However, satisfaction with "Public Facility", "Lightings" and "Parking" was low, which could be because most users have to commute on foot from their apartment in Palm Jumeirah due to limited parking space near the park. It is also because of the increasing demands of Dubai residents looking for parks that are easily accessible in their daily lives and provide various activities and facilities. Therefore, these factors should be reflected in the improvement of park facilities.

Table 6 shows the results of analyzing the satisfaction levels of park users for the items related to accessibility from their residence to the park. Looking at the satisfaction evaluation results, it was found that Bay Avenue Park users were generally more satisfied with the surrounding environment for access than Al Ittihad Park users. This is because Bay Avenue Park was better than Al Ittihad Park regarding the number of plantings and intersections related to comfort and accessibility among the pedestrian-related environments in the park's attraction area. However, users' satisfaction with "Attractions" was evaluated as the lowest among the park surrounding environment for access, with a score of 3.04 at Bay Avenue Park and 2.72 at Al Ittihad Park. This is due to the locational characteristics of the two parks, which are located inside a grid-type residential block composed of high-rise houses. It results from the lack of consideration for creating a street environment that can revitalize the individuality and culture of the region in urban planning; thus, the characteristic street landscape for each area has not been formed. Therefore, in future street environment improvement projects, it is necessary to change the landscape in consideration of regional characteristics through changes in street tree patterns. Thus, it is possible to improve the satisfaction with the access road to the park by adding exciting sights to the pedestrian environment and applying attractive environmental color schemes for buildings and outdoor advertisements.

Table 6. Satisfaction of Accessibility ($p < 0.05$).

Category	Total		Bay Avenue Park		Al Ittihad Park		*F* Value
	Mean	Standard Deviation	Mean	Standard Deviation	Mean	Standard Deviation	
Level of Slopes	3.78	0.888	3.92	0.933	3.64	0.819	0.153
Safe Environment	3.72	0.886	3.89	0.871	3.54	0.869	5.687
Scenery	3.53	0.896	3.73	0.857	3.33	0.893	0.391
Attractions	2.88	0.865	3.04	0.858	2.72	0.845	1.399
Tree Shades	3.44	0.934	3.46	0.913	3.43	0.957	0.356
Connectivity of Walkway	3.58	0.833	3.66	0.817	3.51	0.845	0.739
Pedestrian/Car Separation	3.63	0.868	3.76	0.890	3.50	0.826	0.151
Lightings	3.31	0.935	3.48	0.931	3.41	0.911	2.011
Car Speed	3.32	0.821	3.46	0.808	3.18	0.814	1.751
Hindrance	3.43	0.964	3.61	0.943	3.24	0.949	0.000

3.4. Factoring of Physical Environment Variables

Regarding Bay Avenue Park, factor analysis was conducted to identify the interrelationship among the 25 physical environmental variables related to park user satisfaction and the intrinsic factors. If the correlation between each variable and element is low, the reliability of the analysis may be reduced. Therefore, after factor analysis by principal component analysis and the varimax rotation method, insignificant variables were excluded through factor loadings.

All 25 variables showed a factor loading of ≥ 0.5, and the Kaiser-Meyer-Olkin (KMO) value was high at 0.912. Bartlett's sphericity test found no problem with fit within the significance level of 1%. There were five extracted factors, and the cumulative explanatory power was approximately 64.18% (Table A1).

The first factor included variables related to the safe environment of the access road, level of slopes, lighting facilities, car speed, hindrance, and surrounding environment. This was associated with the overall environment of the access road to the park; hence, it was named "surrounding environment for access". The second factor was the park's size, the length and width of the trail, and the scenery. Most of the variables referred to concepts related to the space for walking; thus, it was named "pedestrian space". The third factor was named "park facility" because it was related to its facilities, such as resting places, lighting facilities, public facilities, and individual sports facilities. The fourth factor included variables associated with the connectivity of walkways, pedestrian/car separation, the amount of green area within the park, and the amount of shade inside and outside the park. This concept encompassed the convenience of accessing the park and the comfort of green spaces inside and outside the park. Thus, it was named "convenience and comfort of the park". The fifth factor included variables such as children's and group sports facilities in the park and attractions inside and outside the park. It was named the "various attractions and activities" factor because it referred to users' demands for various interests and activities.

The results of the analysis of Al Ittihad Park show that factor analysis was conducted for a total of 25 variables, except for the factor loading of 0.5 or less in the "attractions on the access roads". The KMO value, which indicates the adequacy of the factors, was found to be as high as 0.885. The Bartlett sphericity test value satisfied the significance level of 0.01; therefore, the factor analysis result was statistically significant. There were a total of five extracted factors, the eigenvalue of each factor was over 1.0, and the cumulative explanatory power was approximately 63.09% (Table A2).

The first factor was analyzed to include children's facilities, individual sports facilities, sports courts, the scenery in the park, lighting facilities, parking, public facilities, and other places in the park, and was named "park facilities". The second factor was called "surrounding environment for access", It was related to the lighting facilities from the residence to the park, pedestrian/car separation, car speed, connectivity of pedestrian walkway, and the overall environment of the approach to the park as a hindrance. The third factor was named "pedestrian space" because it was related to the park's area and the length and width of the park trail. The fourth factor included the amount of shade in the park, access road, and beautiful scenery. It was named "green landscape" because it is related to the provision of shade and landscape creation by greenery. The fifth factor is associated with the access road's safety and stability from the residence to the park, including the slope level. It was named "safe accessibility."

3.5. Analysis of Factors Affecting User Satisfaction

Multiple regression analysis was performed using a stepwise selection method with the factors extracted by factor analysis as the independent variable and park use satisfaction as the dependent variable to determine the relationship between the physical environment and satisfaction related to park use. The tolerance values of all variables were above 0.1, and the regression model was found to be suitable (=99.566 (<0.01)). The five input factors affected the satisfaction of using Bay Avenue Park at a significance level of 1%. The contribution rate was 0.724, which explained 72.4% of the total variance (Table 7).

Table 7. Multiple Regression Analysis for Satisfaction with Bay Avenue Park (* = $p < 0.01$).

Factors	Unstandardized Coefficients		Standardized Coefficients	t	R^2	F
	B	**Standard Error**	**Beta**			
Factor 1. Surrounding environment for access	0.343	0.023	0.548	14.317 *		
Factor 2. Pedestrian Space	0.243	0.023	0.356	9.308 *		
Factor 3. Park Facility	0.221	0.023	0.354	9.251 *	0.723	99.564 *
Factor 4. Convenience/Comfort of the Park	0.214	0.023	0.342	8.927 *		
Factor 5. Various Attractions & Activities	0.154	0.023	0.244	6.391 *		

After examining standardized regression coefficient values, surrounding environment for access, pedestrian space, park facility, convenience and comfort of the park, and various attractions and activities appeared in that order. "Surrounding environment for access" was analyzed to have the most significant influence on the satisfaction of using Bay Avenue Park. The "surrounding environment for access" factor's relative importance was about 1.5 times higher than other factors. In the case of high satisfaction with the environment, such as Bay Avenue Park, the "surrounding environment for access" should be considered to improve user satisfaction.

The results of the factor analysis show that rather than the improvement of the park environment, if the safe environment, level of slopes, lightings, removal of hindrance, and car speed on the access roads included in "surrounding environment for access" is improved and supplemented, it will be able to enhance user satisfaction significantly. Next, "pedestrian space", "park facility" and "convenience and comfort of the park" were analyzed as critical environmental factors in using the park, in that order. Given that Bay Avenue Park users' satisfaction with related aspects was relatively high, it is judged that maintenance and management are necessary. In addition, in terms of "various attractions and activities", it is possible to improve user satisfaction by installing more facilities for children and organizing various events to reflect users' needs.

The results of the multiple regression analysis using each factor of Al Ittihad Park as the independent variables are as follows. It was found that there was no problem of multicollinearity, and the regression model was found to be suitable at 61.942. In addition, the contribution rate of all factors was 0.627, explaining 62.7% of the total variance (Table 8).

Table 8. Multiple Regression Analysis for Satisfaction with Al Ittihad Park (* = $p < 0.01$).

Factors	Unstandardized Coefficients		Standardized Coefficients	t	R^2	F
	B	**Standard Error**	**Beta**			
Factor 3. Pedestrian Space	0.248	0.026	0.413	9.122 *		
Factor 4. Green Landscape	0.226	0.026	0.376	8.336 *		
Factor 2. Surrounding environment for access	0.222	0.026	0.369	8.166 *	0.626	61.941 *
Factor 1. Park Facilities	0.210	0.026	0.349	7.736 *		
Factor 5. Safe Access	0.148	0.026	0.249	5.510 *		

The results of looking at the standardized regression coefficient values are as follows. It was found that pedestrian space, green landscape, surrounding environment for access, park facilities, and safe access affected satisfaction in that order. This is because Al Ittihad Park has the advantage of abundant natural greenery through the promenade of more than 600 trees. Consequently, it is evaluated that the "pedestrian space" and "green landscape" factors showed a high proportion as the determinants of satisfaction.

These results show that, in the case of urban parks with great natural environments as significant resources such as Al Ittihad Park, it is necessary to focus on the continuous maintenance and management of the natural environment inside and outside the park to satisfy the urban residents' desire to be in nature. Next, in terms of "surrounding environment for access", it is possible to increase user satisfaction by improving the lighting facility on the access roads, pedestrian/car separation, car speed on the access roads, and connectivity of walkway included in the environmental factors of Al Ittihad Park. In terms of "park facilities", user satisfaction was low. Therefore, the satisfaction level can be increased by replacing and maintaining facilities.

Factor and regression analyses were performed on Bay Avenue Park and Al Ittihad Park, and the results are as follows. The three factors of "park facilities", "surrounding environment for access" and "pedestrian space" appeared to be the common denominator factors affecting user satisfaction despite the differences in the types of the two parks.

This is an essential factor to consider, especially among various internal and external environments related to the satisfaction of using a neighborhood park within walking or short driving distance.

However, in Bay Avenue Park, "convenience and comfort of the park" and "various attractions and activities" factors affected satisfaction, and in Al Ittihad Park, "green landscape" and "safe access" factors affected satisfaction. This is considered the result of the differences in an urban context. In other words, it is possible to improve the park by its characteristics by enhancing the different environments for each park. In addition, it will contribute to the improvement of user satisfaction.

4. Discussion & Conclusions

This study targeted neighborhood parks located within walking distances that have characteristics that can be used conveniently in the daily life of Dubai residents. The results of this empirical study on the satisfaction of park users are summarized as follows.

First, as a result of analyzing the usage behavior of the park, the standard usage time for Bay Avenue Park and Al Ittihad Park was less than 60–90 min, and the walking/driving time to access was about 10–20 min. For use, walking and jogging showed the highest ratios. However, there was a difference in the frequency and duration of use.

Second, the results of the comparative analysis between the field survey on the park environment and the satisfaction analysis results are as follows. Although the number of facilities in the two parks was similar, Al Ittihad Park showed lower satisfaction with the facilities than Bay Avenue Park. Therefore, we evaluate that an overall overhaul of the Al Ittihad Park facility is required. Furthermore, users' satisfaction with natural elements such as green spaces, trees, and trails was high in Al Ittihad Park. Accordingly, it seems that continuous protection and management of natural environmental resources, such as green areas and abundant street trees, should be carried out. The items that showed low satisfaction in both parks were "children's facility" and "various attractions". Despite the recent increase in leisure activities for families with children in parks near their homes, the facilities for this social phenomenon are insufficient. Therefore, to increase the desire of Dubai residents to visit parks, this should be considered when improving neighborhood park facilities within walking distance.

Third, the results of the GIS analysis by building data on the external environment of the park are as follows. The number of pedestrian traffic lights and traffic safety signs related to pedestrian safety and the area of sidewalks within a distance of 1 km was similar between the two parks. However, in the pedestrian environment, the number of street trees related to comfort and intersections related to accessibility was higher in Bay Avenue Park than in Al Ittihad Park. The surrounding environment for access to the park was better at Bay Avenue Park. The survey's analysis of the surrounding environment for access shows that satisfaction with "Scenery" was the lowest in the two parks. It resulted from lacking considerations in urban planning for street environment creation that can preserve the region's individuality and culture; thus, a unique streetscape for each area has not formed.

Fourth, factor analysis was conducted for each environmental variable to analyze the determinants that affect the satisfaction level with parks within walking or short driving distance. The results of the multiple regression analysis of the derived factors and satisfaction are as follows. Bay Avenue Park affects satisfaction in the surrounding environment for access, pedestrian space, park facility, convenience and comfort of the park, and various attractions and activities. On the other hand, Al Ittihad Park affects satisfaction in pedestrian space, green landscape, surrounding environment for access, park facilities, and safe access. In particular, factors such as "park facilities", "surrounding environment for access" and "pedestrian space" were found to affect satisfaction in both parks commonly. It could confirm the importance of the surrounding environment for access within the lien and the facility aspect within the park.

Thus, this study investigates the user behavior of neighborhood parks within walking or short driving distance. Given the dramatic social changes from the unprecedented fast urban expansion in Dubai, neighborhood parks are becoming critically important in enhancing the quality of life among Dubai residents. This study has academic significance as it statistically clarifies that the surrounding environment for access is an essential factor influencing the park users' satisfaction. Since there are no similar cities with an unprecedented fast urban expansion like Dubai, Dubai has its problems. Therefore, this research may not be transferable to other international metropolises other than urban parks in Dubai and Abu Dhabi in U.A.E.

As described in above first to the fourth point, practical directions were presented on which areas should be improved, maintained, and managed to improve the satisfaction of park users such as (1) more walking and jogging tracks, (2) continuous protection and management of natural environmental resources, such as green areas and street trees, (3) more "children's facility" and "various attractions" to support leisure activities, and (4) the urban planning for a unique streetscape that can preserve the individuality and culture of the region.

The scope of the target parks limits the study, and thus it could not deal with the types and characteristics of various neighborhood parks within walking distance. In addition, the study had a limited investigation period, and detailed factors related to the surrounding environment for access were not evaluated. Moreover, the limitation of this study is only focusing on the users' satisfaction with the physical environments. Therefore, future study should focus more on the analysis of the relationship between physical characteristics such as spatial characteristics such as size, number, and distance to the park, infrastructure, and environmental components and qualitative factors such as health, community, and satisfaction with life to recover from stress and fatigue, encourage the physical activity, and facilitation of social contact. Besides the park survey, the analysis method could be enriched with the online comments data.

For the future planning of urban parks in Dubai, including the government policy, vision, and urban planning implementation, this research can be used as primary data for improving parks towards a more greening cities approach by conducting research on more diverse types of parks in the future, greening strategies, and detailing surrounding environment for access factors, such as the width of sidewalks and types of pavement materials.

Author Contributions: Conceptualization, C.J. and N.A.Q.; methodology, C.J., M.A. and J.A.; software, M.A. and C.J.; validation, C.J., J.A. and N.A.Q.; formal analysis, N.A.Q.; investigation, M.A., J.A. and N.A.Q.; resources, C.J., M.A. and N.A.Q.; data curation, C.J. and M.A.; writing—original draft preparation, C.J.; writing—review and editing, M.A., J.A. and N.A.Q.; visualization, C.J. and M.A.; supervision, N.A.Q. and J.A.; project administration, N.A.Q. All authors have read and agreed to the published version of the manuscript.

Funding: This research received no external funding.

Institutional Review Board Statement: The study was conducted according to the guidelines of Ajman University Research Ethics Committee.

Informed Consent Statement: Informed consent was obtained from all subjects involved in the study.

Data Availability Statement: New data were created or analyzed in this study. Data will be shared upon request and consideration of the authors.

Acknowledgments: The authors would like to express their gratitude to Ajman University for APC support and the Healthy and Sustainable Buildings Research Center at Ajman University for providing an excellent research environment.

Conflicts of Interest: The authors declare no conflict of interest.

Appendix A

Table A1. Result of Factor Analysis for Satisfaction with Bay Avenue Park.

Variables	Factors				
	Factor 1	Factor 2	Factor 3	Factor 4	Factor 5
Safe Environment	0.788	0.206	0.104	0.109	0.192
Level of Slopes	0.756	0.217	0.046	0.184	0.174
Lightings on the Access Roads	0.668	−0.021	0.381	0.283	0.076
The hindrance on the Access Roads	0.628	0.123	0.291	0.149	−0.011
Car Speed on the Access Roads	0.612	0.093	0.312	0.381	0.078
The scenery on the Access Roads	0.521	0.359	−0.120	0.220	0.344
The Size of the Park	0.128	0.818	0.179	0.115	0.076
Trail Length	0.283	0.747	0.278	0.087	0.031
Trail Width	0.105	0.742	0.044	0382	0.088
Parking	0.092	0.595	0.334	0.005	0.222
The scenery in the Park	0.172	0.545	0.215	0.383	0.354
Public Facility	0.082	0.182	0.688	0.297	0.163
Lighting Facility	0.316	0.223	0.675	0.183	−0.022
Children's Facility	0.170	0.331	0.616	0.021	0.265
Sports Facility	0.214	0.236	0.563	0.091	0.346
Safe Environment	0.342	0.182	0.555	0.293	0.199
Connectivity of Walkway	0.308	0.121	0.102	0.681	0.209
Green Area	0.135	0.381	0.205	0.679	0.000
Tree Shades in Sidewalk	0.349	0.080	0.132	0.666	0.242
Shades in the Park	0.261	0.347	0.337	0.657	0.013
Pedestrian/Car Separation	0.486	0.056	0.128	0.576	0.210
Sports Facility	0.044	0.324	0.186	0.060	0.685
Sports Court	0.218	0.067	0.502	0.074	0.608
Attractions on the Access Roads	0.253	−0.068	0.079	0.344	0.586
Attractions in the Park	0.114	0.455	0.323	0.201	0.521
Eigenvalue	4.309	3.591	3.262	3.258	2.262
Percentage of Variance (%)	16.573	13.813	12.547	12.537	8.703
Cumulative Variance (%)	16.573	30.388	42.937	55.476	64.181
Sampling Adequacy by KMO Measure			0.912		
Bartlett's Test of Sphericity			$p = 0.000$		

Table A2. Result of Factor Analysis for Satisfaction with Al Ittihad Park.

Variables	Factors				
	Factor 1	Factor 2	Factor 3	Factor 4	Factor 5
Children's Facility	0.818	−0.078	0.071	0.040	0.052
Sports Facility	0.699	0.130	0.073	0.307	−0.177
Sports Court	0.672	−0.002	0.204	0.214	0.033
Attractions in the Park	0.651	0.086	0.122	0.216	0.028
Lighting Facility	0.634	0.297	0.118	0.095	0.111
Parking	0.564	0.134	0.330	−0.264	0.138
Public Facility	0.558	0.043	0.204	0.175	0.398
Resting Place	0.527	0.118	0.452	0.134	0.234
Safe Environment	0.509	0.093	0.302	0.176	0.354
Lightings on the Access Roads	0.222	0.808	0.041	0.077	0.085
Pedestrian/Car Separation	−0.078	0.753	0.271	0.265	0.078
Car Speed on the Access Roads	0.125	0.698	0.095	0.094	0.280
Connectivity of Walkway	0.042	0.598	0.189	0.432	0.042
The hindrance on the Access Roads	0.080	0.532	0.354	0.053	0.452
The scenery on the Access Roads	0.053	0.510	0.324	0.124	0.453
The Size of the Park	0.249	0.161	0.796	0.203	0.092
Trail Length	0.265	0.217	0.765	0.164	0.015
Trail Width	0.296	0.261	0.757	0.214	0.038
Tree Shades in Sidewalk	0.182	0.292	0.053	0.723	0.193
Shades in the Park	0.205	0.102	0.346	0.631	0.304
Green Area	0.199	0.068	0.396	0.625	0.297
The scenery in the Park	0.425	0.114	0.244	0.605	0.004
The scenery on the Access Roads	0.242	0.378	0.076	0.535	0.292
Safe Environment	0.119	0.254	0.081	0.209	0.712
Level of Slopes	0.010	0.246	−0.097	0.424	0.668
Eigenvalue	4.286	3.302	3.041	2.948	2.186
Percentage of Variance (%)	17.151	13.211	12.173	11.801	8.749
Cumulative Variance (%)	17.151	30.363	42.538	54.338	63.091
Sampling Adequacy by KMO Measure	0.885				
Bartlett's Test of Sphericity	$p = 0.000$				

References

1. Jung, C.; Awad, J. The improvement of indoor air quality in residential buildings in Dubai, UAE. *Buildings* **2021**, *11*, 250. [CrossRef]
2. Knoema. United Arab Emirates—Gross Domestic Product in Current Prices. 2021. Available online: https://knoema.com/atlas/United-Arab-Emirates/GDP (accessed on 2 July 2021).
3. World Population Review. Dubai Population, 2021–2. 2021. Available online: https://worldpopulationreview.com/world-cities/dubai-population (accessed on 16 June 2021).
4. Giuffrida, N.; Le Pira, M.; Inturri, G.; Ignaccolo, M.; Calabrò, G.; Cuius, B.; Pluchino, A. On-demand flexible transit in fast-growing cities: The case of Dubai. *Sustainability* **2020**, *12*, 4455. [CrossRef]
5. Zaidan, E.; Kovacs, J.F. Resident attitudes towards tourists and tourism growth: A case study from the Middle East, Dubai in the United Arab Emirates. *Eur. J. Sustain. Dev.* **2017**, *6*, 291. [CrossRef]

6.   Ryan, C.; Ninov, I. Dimensions of destination images—The relationship between specific sites and overall perceptions of place: The example of Dubai Creek and "Greater Dubai". *J. Travel Tour. Mark.* **2011**, *28*, 751–764. [CrossRef]

7.   Bodolica, V.; Spraggon, M.; Saleh, N. Innovative leadership in leisure and entertainment industry: The case of the UAE as a global tourism hub. *Int. J. Islamic Middle East. Financ. Manag.* **2020**, *13*, 323–337. [CrossRef]

8.   Jung, C.; Awad, J.; Sami Abdelaziz Mahmoud, N.S.A.; Salameh, M. An analysis of indoor environment evaluation for the Springs development in Dubai, UAE. *Open House Int.* **2021**, *46*, 651–667. [CrossRef]

9.   Khan, M.S.; Woo, M.; Nam, K.; Chathoth, P.K. Smart city and smart tourism: A case of Dubai. *Sustainability* **2017**, *9*, 2279. [CrossRef]

10.   Aram, F.; Solgi, E.; Garcia, E.H.; Mosavi, A. Urban heat resilience at the time of global warming: Evaluating the impact of the urban parks on outdoor thermal comfort. *Environ. Sci. Eur.* **2020**, *32*, 117. [CrossRef]

11.   Lin, B.B.; Fuller, R.A.; Bush, R.; Gaston, K.J.; Shanahan, D.F. Opportunity or orientation? Who uses urban parks and why? *PLoS ONE* **2014**, *9*, e87422. [CrossRef]

12.   Rigolon, A. A complex landscape of inequity in access to urban parks: A literature review. *Landsc. Urban Plan.* **2016**, *153*, 160–169. [CrossRef]

13.   Tempesta, T. Benefits and costs of urban parks: A review. *Aestimum* **2015**, *67*, 127–143.

14.   Larson, L.R.; Jennings, V.; Cloutier, S.A. Public parks and wellbeing in urban areas of the United States. *PLoS ONE* **2016**, *11*, e0153211. [CrossRef] [PubMed]

15.   Loughran, K. Urban parks and urban problems: An historical perspective on green space development as a cultural fix. *Urban Stud.* **2020**, *57*, 2321–2338. [CrossRef]

16.   Rigolon, A.; Browning, M.; Jennings, V. Inequities in the quality of urban park systems: An environmental justice investigation of cities in the United States. *Landsc. Urban Plan.* **2018**, *178*, 156–169. [CrossRef]

17.   Alawadi, K.; Benkraouda, O. The debate over neighborhood density in Dubai: Between theory and practicality. *J. Plan. Educ. Res.* **2019**, *39*, 18–34. [CrossRef]

18.   Awad, J.; Jung, C. Extracting the Planning Elements for Sustainable Urban Regeneration in Dubai with AHP (Analytic Hierarchy Process). *Sustain. Cities Soc.* **2022**, *76*, 103496. [CrossRef]

19.   Elsheshtawy, Y. Where the sidewalk ends: Informal Street corner encounters in Dubai. *Cities* **2013**, *31*, 382–393. [CrossRef]

20.   Denley, D. Reinventing the public Park- The block in Dubai. *Landsc. Archit. Front.* **2019**, *7*, 134–146. [CrossRef]

21.   Jang, K.M.; Kim, J.; Lee, H.Y.; Cho, H.; Kim, Y. Urban green accessibility index: A measure of pedestrian-centered accessibility to every Green Point in an urban area. *ISPRS Int. J. Geoinf.* **2020**, *9*, 586. [CrossRef]

22.   Alipour, S.M.H.; Galal Ahmed, K.G. Assessing the effect of urban form on social sustainability: A proposed "Integrated Measuring Tools Method" for urban neighborhoods in Dubai. *City Territ. Archit.* **2021**, *8*, 1. [CrossRef]

23.   Alawadi, K. Place attachment as a motivation for community preservation: The demise of an old, bustling, Dubai community. *Urban Stud.* **2017**, *54*, 2973–2997. [CrossRef]

24.   Cohen, D.A.; Han, B.; Nagel, C.J.; Harnik, P.; McKenzie, T.L.; Evenson, K.R.; Katta, S. The first national study of neighborhood parks: Implications for physical activity. *Am. J. Prev. Med.* **2016**, *51*, 419–426. [CrossRef] [PubMed]

25.   Moulay, A.; Ujang, N.; Said, I. Legibility of neighborhood parks as a predicator for enhanced social interaction towards social sustainability. *Cities* **2017**, *61*, 58–64. [CrossRef]

26.   Furlan, R.; Sinclair, B.R. Planning for a neighborhood and city-scale green network system in Qatar: The case of MIA Park. *Environ. Dev. Sustain.* **2021**, *23*, 14933–14957. [CrossRef]

27.   Park, Y.; Rogers, G.O. Neighborhood planning theory, Guidelines, and Research [Guidelines], and research: Can area, population, and boundary guide conceptual framing? *J. Plan. Lit.* **2015**, *30*, 18–36. [CrossRef]

28.   Park, K. Park and neighborhood attributes associated with park use: An observational study using unmanned aerial vehicles. *Environ. Behav.* **2020**, *52*, 518–543. [CrossRef]

29.   Miyake, K.K.; Maroko, A.R.; Grady, K.L.; Maantay, J.A.; Arno, P.S. Not just a walk in the park: Methodological improvements for determining environmental justice implications of park access in New York City for the promotion of physical activity. *Cities Environ.* **2010**, *3*, 1–17. [CrossRef] [PubMed]

30.   Xiao, Y.; Wang, Z.; Li, Z.; Tang, Z. An assessment of urban park access in Shanghai–Implications for the social equity in urban China. *Landsc. Urban Plan.* **2017**, *157*, 383–393. [CrossRef]

31.   Roberts, H.; Kellar, I.; Conner, M.; Gidlow, C.; Kelly, B.; Nieuwenhuijsen, M.; McEachan, R. Associations between park features, park satisfaction and park use in a multi-ethnic deprived urban area. *Urban For. Urban Green.* **2019**, *46*, 126485. [CrossRef]

32.   Zengel, R.; Dogrusoy, İ.T. The importance of sociocultural habits in park design, leisure behaviour and user satisfaction. A Comparative Study on Two Parks in İzmir, Turkey. *J. Settl. Spat. Plan.* **2014**, *5*, 107–117.

33.   Plunz, R.A.; Zhou, Y.; Carrasco Vintimilla, M.I.C.; Mckeown, K.; Yu, T.; Uguccioni, L.; Sutto, M.P. Twitter sentiment in New York City parks as measure of well-being. *Landsc. Urban Plan.* **2019**, *189*, 235–246. [CrossRef]

34.   Zhang, Y.; Moyle, B.D.; Jin, X. Fostering visitors' pro-environmental behaviour in an urban park. *Asia Pac. J. Tour. Res.* **2018**, *23*, 691–702. [CrossRef]

35.   Chan, C.S.; Si, F.H.; Marafa, L.M. Indicator development for sustainable urban park management in Hong Kong. *Urban For. Urban Green.* **2018**, *31*, 1–14. [CrossRef]

36. Neckel, A.; Da Silva, J.L.; Saraiva, P.P.; Kujawa, H.A.; Araldi, J.; Paladini, E.P. Estimation of the economic value of urban parks in Brazil, the case of the City of Passo Fundo. *J. Clean. Prod.* **2020**, *264*, 121369. [CrossRef]

37. Gholipour, S.; MahdiNejad, J.E.D.; Saleh Sedghpour, B. Security and urban satisfaction: Developing a model based on safe urban park design components extracted from users' preferences. *Secur. J.* **2021**, 1–30. [CrossRef]

38. Alawadi, K.; Khaleel, S.; Benkraouda, O. Design and planning for accessibility: Lessons from Abu Dhabi and Dubai's neighborhoods. *J. Hous. Built Environ.* **2021**, *36*, 487–520. [CrossRef]

39. Wang, D.; Brown, G.; Liu, Y. The physical and non-physical factors that influence perceived access to urban parks. *Landsc. Urban Plan.* **2015**, *133*, 53–66. [CrossRef]

40. Zhai, Y.; Baran, P.K. Urban park pathway design characteristics and senior walking behavior. *Urban For. Urban Green.* **2017**, *21*, 60–73. [CrossRef]

41. Ayala-Azcárraga, C.; Diaz, D.; Zambrano, L. Characteristics of urban parks and their relation to user well-being. *Landsc. Urban Plan.* **2019**, *189*, 27–35. [CrossRef]

42. Anastasiou, E.; Manika, S. Perceptions, Determinants and Residential Satisfaction from Urban Open Spaces. *Open J. Soc. Sci.* **2020**, *8*, 1. [CrossRef]

43. Liu, R.; Xiao, J. Factors affecting users' satisfaction with urban parks through online comments data: Evidence from Shenzhen, China. *Int. J. Environ. Res. Public Health* **2021**, *18*, 253. [CrossRef] [PubMed]

44. Russell, J.A.; Lanius, U.F. Adaptation level and the affective appraisal of environments. *J. Environ. Psychol.* **1984**, *4*, 119–135. [CrossRef]

45. Rapoport, A. *Human Aspects of Urban Form: Towards a Man—Environment Approach to Urban Form and Design*; Elsevier: Amsterdam, The Netherlands, 2016; pp. 60–81.

46. Almeida, C.M.V.B.; Mariano, M.V.; Agostinho, F.; Liu, G.Y.; Giannetti, B.F. Exploring the potential of urban park size for the provision of ecosystem services to urban centres: A case study in São Paulo, Brazil. *Build. Environ.* **2018**, *144*, 450–458. [CrossRef]

47. Kim, H.S.; Lee, G.E.; Lee, J.S.; Choi, Y. Understanding the local impact of urban park plans and park typology on housing price: A case study of the Busan metropolitan region, Korea. *Landsc. Urban Plan.* **2019**, *184*, 1–11. [CrossRef]

48. Zhai, Y.; Baran, P.K. Do configurational attributes matter in context of urban parks? Park pathway configurational attributes and senior walking. *Landsc. Urban Plan.* **2016**, *148*, 188–202. [CrossRef]

49. Cho, G.H.; Rodriguez, D. Location or design? Associations between neighbourhood location, built environment and walking. *Urban Stud.* **2015**, *52*, 1434–1453. [CrossRef]

50. Giles-Corti, B.; Bull, F.; Knuiman, M.; McCormack, G.; Van Niel, K.; Timperio, A.; Boruff, B. The influence of urban design on neighbourhood walking following residential relocation: Longitudinal results from the RESIDE study. *Soc. Sci. Med.* **2013**, *77*, 20–30. [CrossRef]

51. Sarkar, C.; Webster, C.; Pryor, M.; Tang, D.; Melbourne, S.; Zhang, X.; Jianzheng, L. Exploring associations between urban green, street design and walking: Results from the Greater London boroughs. *Landsc. Urban Plan.* **2015**, *143*, 112–125. [CrossRef]

52. Wang, D.; Brown, G.; Liu, Y.; Mateo-Babiano, I. A comparison of perceived and geographic access to predict urban park use. *Cities* **2015**, *42*, 85–96. [CrossRef]

53. Tang, B.S.; Wong, K.K.; Tang, K.S.; Wai Wong, S. Walking accessibility to neighborhood open space in a multi-level urban environment of Hong Kong. *Environ. Plan. B Urban Anal. City Sci.* **2021**, *48*, 1340–1356. [CrossRef]

54. Dubai Municipality. GIS Services. 2021. Available online: https://www.dm.gov.ae/municipality-business/planning-and-construction/geographic-information-systems/gis-services/%20Dubai%20Municipality (accessed on 16 June 2021).

55. Dubai Travel Planner. 22 Beautiful Parks and Gardens in Dubai You Won't Believe! 2021. Available online: https://www.dubaitravelplanner.com/best-parks-in-dubai/ (accessed on 14 May 2021).

56. Time Out. 20 Family-Friendly Parks, Playgrounds and Picnic Spots in the UAE. 2020. Available online: https://www.timeoutdubai.com/kids/435378-20-family-friendly-parks-playgrounds-and-picnic-spots-in-the-uae (accessed on 18 July 2021).

57. Sugiyama, T.; Francis, J.; Middleton, N.J.; Owen, N.; Giles-Corti, B. Associations between recreational walking and attractiveness, size, and proximity of neighborhood open spaces. *Am. J. Public Health* **2020**, *100*, 1752–1757. [CrossRef] [PubMed]

58. Bayut. All about Al Ittihad Park in Palm Jumeirah. 2021. Available online: https://www.bayut.com/mybayut/al-ittihad-park-palm-jumeirah/ (accessed on 22 May 2021).

59. Bay Avenue. About Bay Avenue. 2020. Available online: https://www.bayavenue.ae/about-us/ (accessed on 28 May 2021).

60. Propsearch. Bay Avenue Mall and Park. 2021. Available online: https://propsearch.ae/dubai/bay-avenue-mall-park (accessed on 12 May 2021).

61. Dubai Properties. Executive Tower Business Bay. 2021. Available online: https://www.dp.ae/our-portfolio/homes-to-buy/13/the-executive-towers/ (accessed on 21 June 2021).

62. Nakheel. Al Ittihad Park: Nakheel. 2021. Available online: https://www.nakheel.com/our-development-al-ittihad-park.html (accessed on 8 April 2021).

63. Bayut. All about the Bay Avenue Park. 2021. Available online: https://www.bayut.com/mybayut/bay-avenue-park-business-bay/ (accessed on 12 June 2021).

64. DubaiCity. Golden Mile Galleria. 2021. Available online: https://www.dubaicity.com/golden-mile-galleria-in-palm-jumeirah/ (accessed on 16 April 2021).

65. Lotfi, S.; Koohsari, M.J. Measuring objective accessibility to neighborhood facilities in the city (A case study: Zone 6 in Tehran, Iran). *Cities* **2009**, *26*, 133–140. [CrossRef]
66. Dubai Municipality. Neighborhood Parks. 2021. Available online: https://www.dm.gov.ae/projects/neighborhood-parks/ (accessed on 2 July 2021).
67. Brown, G.; Schebella, M.F.; Weber, D. Using participatory GIS to measure physical activity and urban park benefits. *Landsc. Urban Plan.* **2014**, *121*, 34–44. [CrossRef]
68. Plunkett, D.; Fulthorp, K.; Paris, C.M. Examining the relationship between place attachment and behavioral loyalty in an urban park setting. *J. Outdoor Recreat. Tour.* **2019**, *25*, 36–44. [CrossRef]

 *sustainability*

Review

Green(er) Cities and Their Citizens: Insights from the Participatory Budget of Lisbon [†]

Roberto Falanga [1,*], Jessica Verheij [2] and Olivia Bina [1,3]

[1]   Instituto de Ciências Sociais, Universidade de Lisboa, Avenida Professor Aníbal de Bettencourt 9, 1600-189 Lisbon, Portugal; o.c.bina.92@cantab.net
[2]   Institute of Geography & Center for Regional Economic Development (CRED), University of Bern, 3012 Bern, Switzerland; jessica.verheij@giub.unibe.ch
[3]   Geography & Resource Management, The Chinese University of Hong Kong, Hong Kong, China
*   Correspondence: roberto.falanga@ics.ulisboa.pt
[†]   This article was presented at the International Symposium on Nature-Based Solutions "Greening Cities Shaping Cities", Milan, Italy, 12–13 October 2020.

Citation: Falanga, R.; Verheij, J.; Bina, O. Green(er) Cities and Their Citizens: Insights from the Participatory Budget of Lisbon. *Sustainability* **2021**, *13*, 8243. https://doi.org/10.3390/su13158243

Academic Editors: Israa H. Mahmoud, Eugenio Morello, Giuseppe Salvia and Emma Puerari

Received: 7 April 2021
Accepted: 19 July 2021
Published: 23 July 2021

Abstract: There is rising scholarly and political interest in participatory budgets and their potential to advance urban sustainability. This article aims to contribute to this field of study through the specific lens of the city of Lisbon's experience as an internationally acknowledged leader in participatory budgeting. To this end, the article critically examines the lessons and potential contribution of the Lisbon Participatory Budget through a multimethod approach. Emerging trends and variations of citizen proposals, projects, votes, and public funding are analysed in tandem with emerging key topics that show links and trade-offs between locally embedded participation and the international discourse on urban sustainability. Our analysis reveals three interconnected findings: first, the achievements of the Lisbon Participatory Budget show the potential to counteract the dominant engineered approach to urban sustainability; second, trends and variations of the achievements depend on both citizens' voice and the significant influence of the city council through policymaking; and, third, the shift towards a thematic Green Participatory Budget in 2020 was not driven by consolidated social and political awareness on the achievements, suggesting that more could be achieved through the 2021 urban sustainability oriented Participatory Budget. We conclude recommending that this kind of analysis should be systematically carried out and disseminated within city council departments, promoting much needed internal awareness of PBs' potential as drivers of urban sustainability. We also identify further research needed into the sustainability potential of green PBs.

Keywords: participatory budget; urban sustainability; European green capital; European green deal; Lisbon

1. Introduction

In the last few decades, spreading scepticism towards democratically elected governments and their institutions, along with a shared need to improve democratic decision-making, has convinced public authorities to promote participatory processes in policy making [1,2]. At the end of the 1980s, a particular category of participatory practices came to the fore: Participatory Budgets (PB hereafter). Initiated in Latin America, PBs provided new impetus to participatory processes by allocating a share of the local public budget to citizen-led initiatives. While aiming to get the most marginalised groups of civil society closer to democratic institutions and representatives, PBs were celebrated by movements and parties on the left of the political spectrum for their capacity to foster social justice, transparency, and accountability [3,4]. During the 1990s and 2000s, international agencies such as the World Bank, United Nations, Organisation for Economic Cooperation and Development, and European Union endorsed the PB for its potential to recover citizenry trust towards democracy. The magnitude of its dissemination reached the five continents, with higher rates in Latin America and Europe [5]. The disruption of the Covid-19 pandemic, however, has had major impacts over the implementation of participatory processes in general and PBs in particular, as public authorities were required to prioritise immediate responses to the global contagion of the coronavirus, at the expense of PB projects [6].

In 2019, Portugal was the European country with the highest rate of local PBs implemented out of a national legal framework [5]. The enthusiasm around the PB dates to the early 2000s and then grew exponentially after the Lisbon PB in 2008/2009 [7], which marked the first city-wide PB implemented by a European capital and introduced a new model of direct participation. Citizens have since been invited to submit and vote ideas for public funding in all policy domains, including urban sustainability within the thematic area "environment, green structure, and energy". However, only in 2020 did a more explicit and tangible link emerge between Lisbon's PB and urban sustainability. As Lisbon received the European Green Capital Award (EGCA hereafter) and subsequently announced the launch of a Green PB, it raised great interest worldwide. This promising turn of events and of policy attention towards more sustainable urban agendas was shaken by the global pandemic (Covid-19), yet, as we will see, the city council strove to keep its political promise to shift the thematic focus of the PB towards urban sustainability, incorporating the European Green Deal into the 2021 version of the Lisbon PB.

The recent global pandemic has added to the already abundant calls for greater attention to the health of ecosystems and to the negative impacts of rapid urbanisation and of cities on local and global sustainability [8,9]. Two notable trends seem relevant here: first, the growing role of cities in national and international agreements promoting sustainability (see Sustainable Development Goal 11, UN General Assembly 2015) and related agendas for climate mitigation and adaptation, and reducing biodiversity loss [10]; second, the growing demand for nature-based solutions which identifies the participation of citizens and businesses as key [11,12]. Hence, this article aims to contribute to the participatory dimension of urban sustainability agendas through the specific lens of participatory budgets, by asking what the lessons and potential contribution of the Lisbon PB experience are in advancing urban sustainability. To this end, we first describe the key concepts for our inquiry into participatory budgeting and urban sustainability, by then focusing on the case study of Lisbon's accumulated experience of 11 PB editions and its ongoing thematic shift towards a Green PB in 2020, followed by a "sustainability-oriented" PB in 2021. After presenting our mixed-method approach based on the analysis of grey literature, of statistical data on the 11 PB editions, and of four semi-structured interviews with key local actors, we discuss our findings by highlighting emerging trends and variations throughout Lisbon's PB experience. This analysis allows us to identify the most common urban sustainability topics proposed by citizens. We will show that the Lisbon PB has had remarkable success in attracting citizen proposals and funding projects in key areas of urban sustainability, covering topics that often go beyond the more technocratic focus on green spaces and infrastructure present in international agendas of climate change and biodiversity loss. We go on to argue that PB achievements are to be understood within the context and framework of the public agendas set by city councils, as these influence trends and variations within the PB program. We also found that the pursuit of pedagogical and instrumental purposes in shifting towards a thematic PB is based on a rather weak evidence-based awareness about the PB's contribution in this field. We conclude by recommending that this kind of analysis should be systematically carried out and disseminated within city council departments, promoting much needed internal awareness of PBs' potential as driver of urban sustainability. We also identify further research needs regarding the sustainability potential of green PBs.

2. Conceptual Framework: Urban Sustainability and Citizen Participation

Since the origins of the concept of sustainable development based on the balance between ecological sustainability and human needs, the urban dimension has been a significant part of the global debate. Already the Brundtland Report of 1987 referred to "the urban challenge" by addressing extreme forms of urbanisation as associated with growing social and environmental problems [13]. However, whereas initially cities tended to be considered mainly as hotspots of environmental hazards, and still are [14], the idea of urban sustainable development nonetheless gained momentum, as urban governments became increasingly seen as drivers of positive change. In the last few years, local governments have sought to implement policies to reduce their ecological footprint and gained international projection [15] for being considered an appropriate administrative level to enable changes in everyday structures and behaviours in the face of global climate change. Hence, sustainability processes are nowadays often analysed on city level, linking environmental improvements to local economic development [16,17].

This trend culminated in 2015 with the inclusion of a specific "urban" goal in the 17 Sustainable Development Goals [18], which re-energised the momentum initiated with the Rio Conference in 1992 [19], and the more recent UN New Urban Agenda in 2017 [20]. SDG 11 addresses the challenges faced within urban contexts, while going "far beyond the typical focus on housing and slum upgrading" ([21], p. 94) to also include targets related to climate resiliency, waste management, public space, mobility systems, and, ultimately, forms of participatory planning. Nonetheless, the concept of urban sustainability is often criticized for being an oxymoron [22], not addressing the inevitable trade-offs between the environmental, economic, and social dimensions of sustainability [23,24]. The essence of this critique lays in the fact that urban sustainability itself falls short to define what is to be sustained, by whom, and for whom [25].

While the understanding of urban sustainability can range from approaches focusing mostly on the environmental dimension towards others prioritising social justice and well-being [24], the human dimension has gained a more prevalent role during the 1990s. "Social sustainability" has since been evoked to highlight the interdependence between social, economic, and environmental goals [25,26] with a growing concern towards policies aiming to eradicate poverty and social exclusion [26]. In this regard, the need to mitigate environmental hazards is increasingly being linked to calls for wider citizen participation, in order to provide local communities with new conditions to improve their quality of life [27]. Experiments of co-design and co-production of knowledge [28] through transdisciplinary practices [29] have further allowed delving into the contested character of urban sustainability. In addition, international agencies have equally emphasised the role of public participation in creating a shared vision of cities' future by ensuring the inclusion of different social groups [30].

Scholars contend that the development of participatory processes with citizens provides evidence for effectively supporting environmental policies and reducing the risk of future backlashes [31]. As residents have a chance to shape the development of their neighbourhood or city, solutions are likely to be more aligned to local needs and aspirations [32], as shown in seminal experiences under the "Global Agenda 21" [19]. In some cases, participatory initiatives create new opportunities for partnerships in the management of specific projects [33–35] either led by public agencies or grassroots groups [36,37], and possibly develop into more extensive sustainability programmes [38]. When participation is prompted for the strategic planning of green infrastructures and services [35,39,40], new links can be created among ecological, economic, and societal elements of urban sustainability [41].

Acknowledging the potential of citizen participation for urban sustainability, Hayward [42] suggests looking at participatory processes through the lenses either a deliberative or participatory approach. The former aims at giving the opportunity to a selected poll of citizens to reflect on socio-ecological and human systems [43]. In this type of settings, citizens are asked to construct a new ecological rationality [44], possibly reliant upon multiple governance arrangements [45]. In some cases, deliberative initiatives also engage experts to foster the public debate with citizens, although this may raise concerns as to the growing distrust towards science and political power in green thinking [46,47]. Participatory approaches are rather concentrated on the sharing of power among multiple agents and are more frequently open to the participation of all citizens [48]. Along with participatory experiences in the field of urban planning [49], the PB is the most known participatory process that encompasses mechanisms aimed at collecting proposals from citizens and allocate public funding for their implementation [50]. However, while the uptake of PBs worldwide has been significant, little is known about its actual contribution to advance urban sustainability.

In a recent publication, Cabannes [51] notes that there are few internationally documented PBs that have put environmental issues at the centre of the public debate. Yet those that have done so have shown a promising potential to bridge public debates with concrete actions on climate change. Scholars hold, in fact, that funding community projects through the PB provides citizens with greater know-how in taking the lead of environmental projects [52], as well as higher degrees of control over service delivery and goals of redistributive justice [53]. Other scholars, however, cast light on potential risks, as PBs may incorporate the urban sustainability agenda because of budget cuts imposed on local authorities, particularly in the aftermath of the financial crisis [34], resonating with those who claim that the welfare state has been significantly narrowed, especially in the peripheral Eurozone [54]. The question on whether PBs can contribute to advance urban sustainability, thus, remains open, as evidence is needed to substantiate hopes and promises about its potential.

3. The Lisbon Participatory Budget

Portugal was one of the most affected countries of the 2008 financial crisis in Europe, eventually entering an international bailout provided by the International Monetary Fund together with the European Bank and the European Commission from 2011 to 2014. Like in other peripheral Eurozone member states, budgetary cuts imposed on local authorities were significant [55], which led policymakers to search for innovative and effective mechanisms of participatory governance. Likewise, the first Lisbon PB in 2008/2009 gave citizens the possibility to submit and vote project ideas across all policy domains. Over the last decade, the Lisbon PB was run based on a yearly cycle with a dedicated budget of 5 million euros in the first editions, later cut to 2.5 million in 2012, which, according to the city council, was a direct consequence of the austerity measures. The Lisbon PB was the first PB implemented on a municipal scale by a European capital, having a major impact over the dissemination of PBs in Portugal and reverberation at the international level [7,56].

Available data show that from 2008 to 2018 the PB has received a total number of 2073 proposals from citizens: 139 out of them were funded based on citizen voting, which covered slightly more than a 36-million-euro budget. Today, the Lisbon PB is one of the longest-running PBs in the country and has recently been awarded as the best 2020 participatory practice by the national network of participatory city councils in Portugal (More information at: https://op.lisboaparticipa.pt/, accessed on 12 June 2021). Some scholars have pointed out how the Lisbon PB has contributed to advance urban sustainability, as distinguished by van der Jagt and colleagues [34] in their extensive analysis of 20 cities in 14 EU-countries. Buijs and colleagues [57] similarly suggest that the Lisbon PB provided new conditions for citizens to influence the decision-making and improve their understanding of the implications of urban sustainability, based on the many sustainability-oriented projects funded through the PB, such as paths, parks, gardens, and ecological corridors. "The success in Lisbon is that citizen concern for their urban environment was captured by the L-PB [Lisbon PB] process. It may also be the case that the L-PB process increased awareness of the relevance of green to the public" [57] (p. 42).

In 2019, a collaboration between the Lisbon city council and the European agency Climate KIC led to a trial of a PB for schools, dedicated to encouraging the participation of youngsters in sustainability projects [51]. At the same time, the creation of a PB Green Seal aimed at recognising PB proposals with a focus on urban sustainability. These actions can be considered as a proxy to the forthcoming 2020 Green PB, announced by the Mayor in October 2019 in light of the 2020 European Green Capital Award (EGCA)—the first time a capital city from a Southern-European country received this award (The European

Green Capital Award is an initiative launched in 2008 by the European Commission to celebrate and recognize cities' achievements in terms of urban sustainability and to provide role-models for other European cities. Any city government can apply for the award by submitting an application, and every year a jury selects one city for its efforts and consistent record of high environmental standards. The award is meant to serve as a platform for policy-sharing and inspiration, as the European Green Capital shares its best-practices and experience with other cities. More information about the European Green Capital Award: https://ec.europa.eu/environment/europeangreencapital/winning-cities/2020-lisbon/, accessed on 5 May 2021). The Green PB was one of the various actions planned for 2020, together with tree-planting programs, the (re)opening of a (renewed) public green space each month of the year, and the implementation of a reduced-emissions-zone in Lisbon's downtown. The Green PB was understood as the first PB of its kind, despite other cities having invested in similar programs, as in the case of Lublin (Poland), Brussels (Belgium), Odemira (Portugal), and Matz and Bordeaux (France). However, neither embraced urban sustainability as an exclusive thematic focus or made available full information regarding procedures and outputs.

Since the Covid-19 pandemic eventually forced the Lisbon city council to suspend the Green PB in 2020, the 2021 edition went on to recover and enhance the intention to shift towards a thematic PB. This time, the Lisbon PB is oriented to promote urban sustainability as framed within the recently-issued European Green Deal and its several action areas. In addition, the 2021 PB added one specific thematic area concerning physical and mental health as well as sport, to encourage citizen proposals in response to the impacts of the pandemic and support the as agenda linked to the European Capital of Sport. Given its scope, the PB shows a significant capacity to readapt its design to dramatic contingencies and build on the international debate on sustainability. Despite the ongoing crisis, and going beyond the PB, the city intends to implement significant environmental changes to ensure sustainable land use, green infrastructure, and water management. As enshrined in the 2012 Lisbon master plan and later highlighted by the Lisbon's Action Plan for Biodiversity published in 2015, the city aims to expand its green infrastructure by over 200 hectares and create linkages between different green areas through so-called green corridors. In the last few years, the city has invested in developing a more sustainable mobility system based on public transport and cycle networks and has become involved in climate change adaptation and mitigation, culminating into the publication of its Strategy for Climate Change Adaptation in 2017 with a strong commitment to reducing greenhouse gas emissions.

4. Methodology

Our interest was raised by the combination of the international achievements of Lisbon in the fields of participatory budgeting and urban sustainability, which culminated in the announcement of the Green PB in 2020, followed by the 2021 sustainability-oriented PB edition framed within the European Green Deal. Our research asked what the lessons and potential contributions are of the Lisbon PB experience in advancing urban sustainability, and specifically seeks to understand its contribution throughout the past 11 editions. In addition, we ask whether its thematic shift towards sustainability relied upon a shared understanding of its achievements. To this end, we analysed the past 11 PB editions by means of a mixed-method approach, as synthetised in the Table 1 below.

Firstly, we conducted an analysis of the policy documents produced on the integration of urban sustainability agenda into the Lisbon PB. We took into account three primary sources: (i) the PB "Charter of Principles" issued in 2008, which has guided this process thus far; (ii) the yearly evaluation reports of each one of the 11 PB editions, available on the city council's website; and (iii) the Lisbon application to the EGCA, including its evaluation by the evaluation board. The analysis was conducted between April 2018 and September 2020 to provide a focussed overview of the main arguments posed by the official policy discourse.

Table 1. Methods and goals.

Methods	Goals
Analysis of grey literature.	Provide a comprehensive understanding of the official policy discourse on the Lisbon PB
Quantitative and qualitative data analysis of past PB editions (or statistical analysis)	Provide an evidence-based overview of the 11 PB editions
Expert interviews	Provide complementary data from situated knowledge of key actors

Secondly, we collected and analysed data on all citizen proposals selected for public voting and funded projects in the PB from 2008/2009 to 2018. We focussed on proposals and projects listed in one of the nine thematic areas defined by the PB, namely, "environment, green structure, and energy". This is due to the fact that, according to our conceptual framework, this theme condenses the main tenets of the urban sustainability agenda. Data collection was based on the systematic consultation of the PB website, oriented by a twofold purpose: on a quantitative stance, we aimed to draw a comprehensive picture of trends and variations of proposals submitted by citizens, citizen votes to the proposals, winning projects (i.e., proposals to be implemented as a result of citizens' voting), and public funding allocated for the implementation of projects. On a qualitative stance, we aimed to make sense of the thematic area by analysing the content of citizen proposals by identifying emerging key topics.

Thirdly, we interviewed four key actors to account for their personal perspectives on PB and the urban sustainability agenda in Lisbon. Interviews were conducted with the advisor in charge of the KIC Climate for the design of the Lisbon Green Seal and the PB for schools; the head of the municipal direction for participation; the PB coordinator; and an internal advisor to the municipal councillor for environment, green structure, climate, and energy who coordinated the Lisbon EGCA application.

5. Findings

Findings are presented below by following the same order of the section above. Accordingly, we first examine findings from the document analysis; secondly, we delve into the data retrieved from the consultation of the 11 PB editions; and, lastly, we highlight the main arguments provided by our four interviewees.

5.1. Document Analysis

The PB "Charter of Principles" is meant to guide the implementation of the PB according to eight propositions: (i) to promote participatory democracy as issued by the Portuguese Constitution; (ii) to foster the following goals: active citizenship, civic education, policy responsiveness, and transparency; (iii) to share power with citizens by giving them the opportunity to propose and vote new public measures; (iv) to multiply channels of participation, both face-to-face and online to ensure its reach; (v) to organise the PB through sequential steps on a yearly basis; (vi) to make available and accessible information related to the PB; (vii) to be accountable on the implementation of projects; (viii) to monitor and assess the PB in order to adapt the process to emerging needs. This chart provides the political and administrative framework to enhance the city council's capacity to learn from citizens' input and provide them with a voice into decision making through the PB. Considering these eight propositions, the analysis of the yearly PB reports shows the impacts of political decisions over the institutional design of this process, which demonstrates a great adaptative nature to the local sociopolitical setting.

The PB reports permit providing insights on the most remarkable changes, such as the reduction of the budget from 5 million to 2.5 million euros in 2012 (and still in place). While no public information is made available on the budgetary weight of the PB since 2008, it is possible to collect data from the 2019 PB, which represents around 0.4% of public investment in the city budget. Considering the remarkable cut operated in 2012 and the inflating economy of Lisbon in the last few years, these data are likely to have sensibly changed over time. Allegretti and Antunes [56] argue, in fact, that up to the 2012 cut, the Lisbon PB corresponded to 5.4% of the investment capacity of the municipality. Nevertheless, this change did not affect the participation of citizens in significant ways, as shown in the analysis described below. Moreover, in 2021, new criteria for the spatial distribution of citizen proposals were introduced. Proposals have since been divided into city-wide (up to 500.000 euros each) and neighbourhood-based proposals (up to 150.000 euros each), which is expected to better respond to citizens' needs according to the implementation scale of the projects. The analysis of the PB reports further informs on changes adopted in the channels of participation, as the first two PB editions were run online only, followed by a mixed face-to-face and online version from 2010 onwards. Provided with multiple channels, the PB was expected to improve its reach in the city and include citizens via local meetings, and online through the PB official website. These changes were in place until the 2018 PB edition.

The growing importance of the PB as a key space for citizen engagement motivated the city council to emphasize its potential in its application to the ECGA, stressing its role in five out of the twelve indicators of the application form: sustainable mobility, sustainable land use, noise, green growth and eco-innovation and governance(The complete list of topics is: climate change mitigation, climate change adaptation, sustainable urban mobility, sustainable land use, nature and biodiversity, air quality, noise, waste, water, green growth and eco-innovation, energy performance, and governance. The Lisbon application to EGCA was scored high in the indicators of sustainable mobility, sustainable land use and green growth, while it did not score particularly well on the indicators for governance (4 out of 12) and noise (6 out of 12). The PB was mostly presented as a flagship of public decision making, however, without providing further details on the sustainability-oriented projects funded through the PB—despite its clear contribution to the development of green spaces and cycle infrastructure. The evaluation board recognised the PB as a "very well established [practice], with over €31 million allocated over the past 10 years" [58] (p. 45). The board further appreciated the reference to citizen participation in eco-innovation, as the city owns a set of "[c]lear and concise policy driven plans involving citizens, universities, businesses and foreign partners" [58] (p. 43). Yet, the board contended that it is not always clear which citizens are involved and how, nor what plans are being prepared for the future.

5.2. Data Analysis

To better understand the potential disclosed by the PB in this field, we collected data on the thematic area "environment, green structure, and energy" from 2008 to 2018, and focussed on four items: citizens' proposals; votes; projects; and public funding. The figure below puts together findings from the analysis made per PB edition (for more detailed information, see Table A1 in Appendix A). An overview of the emerging trends shows that 38% of citizen proposals were submitted in the selected thematic area, while around 27% of the projects were eventually funded in this thematic area. Data about the volume of votes show that around 18% were made by citizens for projects in this thematic area ($n = 54863$), which corresponded to slightly more than 16 million euros allocated for their implementation (around 44% of the overall budget, slightly less than 36 million euros).

Figure 1 elucidates the relatively large amount of budget allocated to this field, with a considerable peak at the very beginning of the PB in the city, as the first PB holds a primacy in all rates. The visible drop down in 2012 may be explained by the remarkable changes made in that year, as a response to the financial crisis. Nevertheless, the drop contrasts with a relatively high rate of proposals (36%) pointing to the significant citizens' interest in the implementation of projects within the thematic area despite the budgetary cut operated in that year. The increase of projects and funding allocated to their implementation in 2016 opposes to the decrease in the number of proposals, which has persisted in 2017 and 2018. To provide a more substantive explanation of these trends and variations, we analysed the 785 citizen proposals selected from 2008 to 2018 in the thematic area. Our goal was to better understand the emerging topics related to urban sustainability by disaggregating the official classification of the thematic area. Accordingly, we identified eight key topics:

- Topic 1 'city/neighbourhood': promotion of green initiatives at the neighbourhood and/or city level;
- Topic 2 'recovery': regeneration and green recovery of streets, squares, and public spaces;
- Topic 3 'gardens': improvement and/or creation of new parks and gardens for all and/or for children;
- Topic 4 'farming': creation of new urban farming and gardening;
- Topic 5 'mobility': promotion of sustainable mobility;
- Topic 6 'animals': animal protection;
- Topic 7 'energy': sustainable energy (e.g., street led lighting);
- Topic 8 'pollution': combat air and noise pollution.

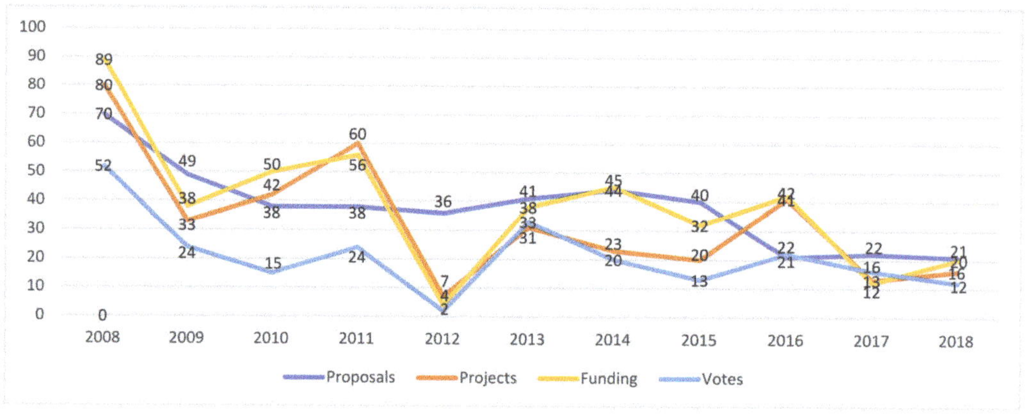

Figure 1. Percentages of proposals, votes, projects, and funding in the "environment, green structure, and energy" thematic area of the Lisbon PB from 2008 to 2018. Data source: Lisbon city council (http://op.lisboaparticipa.pt, accessed on 12 June 2021).

The eight key topics help characterise both qualitatively and quantitatively the most frequent aspects considered by citizens. Figure 2 below shows the rates of citizen proposals per key topic, while extended information about proposals and projects can be found in Table A2 in Appendix B.

The content analysis of citizens' proposals suggests that some topics hold greater consistency, despite the rise and decline of others, as in the case of the first five topics and topic 7 'energy'. Topics 1 'city/neighbourhood', 2 'recovery', and 3 'gardens' are significant throughout the PB editions, the latter peaking in 2016. In contrast, topics 7 'energy' and 8 'pollution' show the lowest rates overall. The inversely proportional relation between topic 3 'gardens' and topic 2 'recovery' is substantiated by the former taking over from the decrease of the latter. Moreover, topics 6 'animals' and 8 'pollution' increased considerably in the last PB editions, whereas topics 5 'mobility' (absent in 2017 and 2018 editions), 4 'farming' and 7 'energy' (absent in the 2018 edition) decreased significantly. To better understand the magnitude of each key topic, we also collected data about the votes made by citizens and the allocated public funding for the implementation of projects from 2008 to 2018 (Table 2).

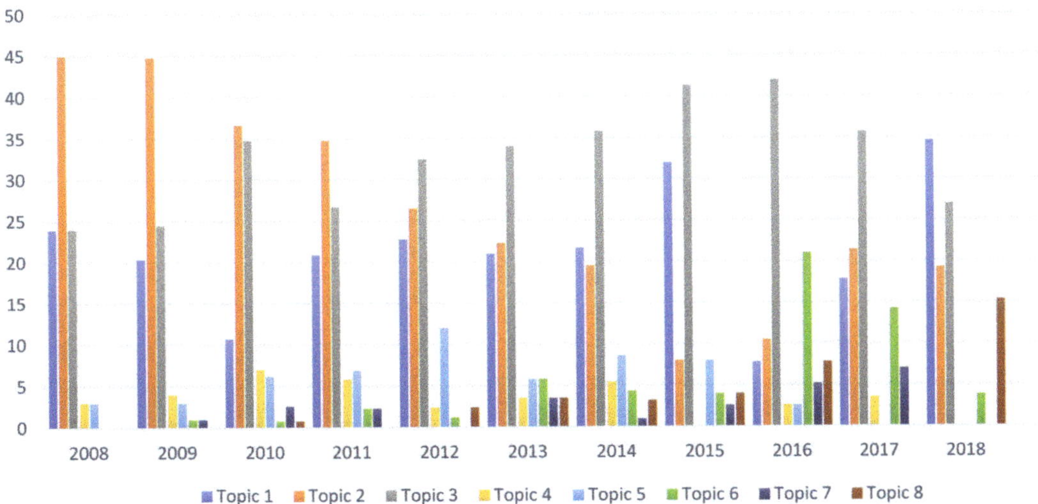

Figure 2. Percentages of citizen proposals per key topic in the "environment, green structure, and energy" thematic area of the Lisbon PB from 2008 to 2018. Data source: Lisbon city council (http://op.lisboaparticipa.pt, accessed on 12 June 2021).

Table 2. Votes and public funding allocated per key topic in the "environment, green structure, and energy" thematic area of the Lisbon PB from 2008 to 2018.

Key Topic	Proposals (2008–2018)	Projects (2008–2018)	Votes (2008–2018)	Funding (2008–2018)
1 'city/neighbourhood'	163	3	5298	1,800,000 euros
2 'recovery'	223	9	6494	3,250,000 euros
3 'gardens'	255	16	31,351	6,340,000 euros
4 'farming'	31	1	346	150,000 euros
5 'mobility'	48	2	4832	3,100,000 euros
6 'animals'	30	5	6382	825,000 euros
7 'energy'	16	1	89	500,000 euros
8 'pollution'	19	1	71	50,000 euros

Source: Lisbon city council (http://op.lisboaparticipa.pt, accessed on 12 June 2021).

Highlights from the table above show that topic 3 'gardens' holds the record with more than 30,000 votes received throughout the PB editions and the largest amount of funding, which is consistent with the number of projects implemented throughout the 11 PB editions ($n = 16$). The high rate of votes builds on—at least—a couple of landmark projects that mobilised great support from citizens: the restoration of the Botanic Garden in 2013, and the "Caracol" garden in 2016, the latter gathering 9447 votes, the highest ever in the Lisbon PB. Funding is relatively high in topics 2 'recovery' and 5 'mobility', as well. The latter confirms that, despite the few approved projects ($n = 2$), these can have a significant impact on public spending for demanding a wider approach to urban intervention in the green infrastructure.

5.3. Interviews

Exploratory interviews helped us to draw a clearer picture of emerging trends, variations, and topics throughout the 11 PB editions. Accordingly, we interviewed four key actors: the advisor in charge of the KIC Climate for the design of the Lisbon Green Seal and the PB for schools; the head of the municipal direction for participation; the PB coordinator; and the politically appointed advisor to the municipal councillor for environment, green structure, climate, and energy. Interviews took around one hour each and were based on a similar semi-structured approach.

The KIC Climate advisor was interviewed in July 2020 to discuss the progressive tendency of the PB towards urban sustainability themes, with a focus on the Green Seal launched in 2019, which, according to our interviewee, was designed to raise citizens' awareness. The Seal was also conceived as a proxy to the 2020 Green PB through a step-by-step strategy aimed to learn from the past editions and prepare the future of this participatory process. Likewise, the piloting of the PB for schools—focused on themes of sustainability—was thought of as an opportunity for young students to take decisions over a more sustainable future. In September 2020, we interviewed the municipal director of citizen participation and the PB coordinator at the city council. The former emphasised the growing political commitment towards urban sustainability, which found in the PB a potential ally. As the deputy mayor in charge of the PB encouraged a more inclusive and face-to-face approach to participation, the reduction in number of citizens' proposals selected for public voting aimed to ease the internal coordination for the implementation of the projects. In 2018, new "pedagogical goals" mobilised the department of participation that soon embraced the political intention to promote a Green PB in 2020 to encourage behavioural changes in the city. This effort, she contended, did not entail only the incorporation of urban sustainability themes, but rather the capacity to make its values accessible to all through a common language shared between the city council and citizens. In this way, the PB could be approached not as a mere financial tool for the implementation of projects, but rather as a generator of new bridges between the city council and citizens.

Along with pedagogical goals, the PB coordinator reinforced the instrumental purpose of the envisioned transition towards the 2020 Green PB. The thematic focus on urban sustainability implied, according to our interviewee, the opportunity to keep high the international interest on the PB by building upon the connection with the European agency KIC Climate. Most importantly, this shift implied the opportunity to solve the pitfalls of this process regarding the time-demanding analysis of citizens' proposals in all policy areas and implementation of projects. Both stages are coordinated by the PB team with other municipal departments, which ends up increasing the red tape and slowing project implementation (see also [59]), with implications over the credibility of the PB with citizens.

Lastly, the interview with the internal advisor to the councillor responsible for Lisbon's application to the European Green Capital Award allowed us to better position the PB within the urban sustainability agenda pursued by the city council. While the interview focused mainly on the development of Lisbon's Green Infrastructure from an urban planning perspective, it gave important insights into the city's journey towards being acknowledged with the title of 2020 European Green Capital. The winning application submitted in 2017 was the third application submitted by Lisbon, having reached second place the previous year. According to our interviewee, the EGCA communicated in 2018 had an additional impact over the EGC programme: based on the city's track record with participatory processes, the EGC assessment formula for indicator 12 ("Governance") was changed slightly to include an explicit reference to "Partnerships and Public Involvement". As a result, from 2019 onwards, applicant cities are required to describe their strategies for citizen participation, which should be considered, according to our interviewee, an important contribution of Lisbon's application.

6. Discussion

The inclusion of a thematic focus on urban sustainability, which was planned for the 2020 PB edition and then took place in 2021 by incorporating the European Green Deal action areas, is a turning point in the history of the Lisbon PB. Scholars contend that citizen participation can effectively contribute to advance urban sustainability [42,51], and the Lisbon PB seems to prove efficacy in allocating public funding to local issues perceived as most pressing or urgent in this field [26]. This process further shows that, despite the considerable cuts induced on the available budget in 2012, public engagement has been relatively high throughout the 11 editions, with a more visible decrease in numbers over the last few editions. Bearing this in mind, and relying upon evidence retrieved from our study, our discussion develops around three main findings.

Firstly, data show that the Lisbon PB aligns with urban sustainability goals and seems to counteract the dominant engineered approach towards sustainability, focused on green growth and innovation. As Gulsrud and colleagues [60] stress, the international agenda tends to undermine 'ecological green resources' (e.g., parks, trees, grassland, water bodies, etc.) by prioritising short-term deliverables (e.g., cycle paths, storm-water channels, footpaths and access routes, energy-efficient built environment) in detriment of biophysical achievements. In Lisbon, the authors (ibidem) contend that some of these trends reverberate in the unbalanced commitment to planning public spaces and nature-based solutions beyond the city boundaries, while additional criticisms are raised as to the capacity of the city to address rising real estate prices and social exclusion (cf. [61]). Against this background, our findings show that the PB contrasts with the otherwise dominant focus on green growth and eco-innovation of several other policy programmes, including the ECGA [62]. Despite the slow decline over the last years, findings show that around 33% of citizen proposals and projects, 18% of votes, and 44% of public funding are dedicated to the "environment, green structure, and energy" thematic area of the PB. Interestingly, these outcomes align only partially to the emphasis put by international agendas on urban infrastructure. Instead, other topics, often excluded from the dominant agenda on sustainability, are prevalent in the PB, as is the case of animal protection. We therefore note the potential of a green PB to promote a more balanced approach to socio-ecological challenges, while counteracting an engineering-led discourse of urban sustainability present in, among others, the EGCA. Scholars have argued that this approach reduces the opportunity to promote equitable, inclusive urban transitions [62]. Hence, given Portugal's challenging economic context of the last decade and the correlation between national wealth and the performance of Europe's capital cities in terms of sustainability [16], this contribution of the Lisbon PB towards urban sustainability is particularly valuable.

At the time of writing, the 2021 Lisbon PB is being implemented with a 2.5-million-euro budget. Preliminary data show a sensible decrease of citizen participation at the stage of proposals' submission (=251), and only 75 out of them considered for public voting. An overview on the distribution of proposals in the new eight thematic areas shows a preponderance of the mental and physical health and sport theme ($n = 75$), followed by smart and sustainable mobility ($n = 45$), adaptation and mitigation of climate changes ($n = 43$), and reduction of pollution and promotion of biodiversity ($n = 41$). Circular economy and efficient building collected the same number of proposals ($n = 16$), followed by the promotion of just food systems ($n = 10$), and only 5 proposals for clean and renewable energy. The decrease in numbers of proposals should be analysed attentively in the future, taking into consideration the impacts of the pandemic in the field of citizen participation [6]. At the moment, we can only speculate about possible reasons behind this decline, as these data can either reveal a process of cultural accommodation to the thematic shift, or the adoption of online tools only due to the restrictions of the Covid-19 pandemic. These data may also indicate a mounting disillusionment regarding the PB as a policy mechanism, out of any direct correlation with urban sustainability themes.

Secondly, PB achievements should be understood in light of the local agenda and public action of the city council. The 2012 PB edition represented a turning point due to cuts in the allocated budget from 5 to 2.5 million euros and the distribution of the budget into city-wide projects and smaller interventions at the neighbourhood level, which all seemed to influence the decrease in votes, projects, and allocated budget, but not proposals. As citizen interest around urban sustainability remained high up to 2016, the last few PB editions showed a decrease in proposals, contrasting with a relatively high number of votes, projects, and allocated funding. This variation can be traced back to changes prompted under the deputy mayor in charge of the PB, who aimed to reduce the number of citizen proposals selected for public voting to ease the red tape and, later in 2018, to boost a new pedagogical ethos to prepare the shift towards the 2020 Green PB. Also in 2016, the 'Caracol' garden, one of the most emblematic PB projects for urban sustainability, was funded with a record number of votes. This project relied upon an unprecedented citizen mobilisation that reverberated citizen support for the restauration of the Botanic garden in 2013, which attracted one of the highest rates of votes ever. As pointed out by Allegretti and Antunes [55], the social dynamics behind the submission of citizen proposals heavily rely upon pre-existing networks and interest groups, which often explain the success of specific projects in the Lisbon PB.

The impact of the two garden projects can be understood in the framework of the identified key topics, where topic 3 'gardens' emerges as the most frequent. Its progressive growth throughout the 11 PB editions emphasises the decrease of topic 2 'recovery'. This inverse relation may relate to a broader shift in the public discourse and citizen understanding of urban sustainability in Lisbon, as the focus on the quality of green public spaces seems to prevail over the more infrastructure-based interventions aimed at improving streets and squares, as argued above. This hypothesis is also substantiated by empirical knowledge about the city council's action in this field, which plays a key role in managing significant trade-offs of urban development [23], with implications on governance tools, as in the case of the PB. For example, the lack of proposals in Topics 4 'farming' and 7 'energy' in the last edition, as well as the absence of proposals in Topic 5 'mobility' in the last couple of editions, can be traced back to the mainstreaming of urban policies in these fields, which reduced public demand when it comes to PB processes.

Thirdly, our interviews suggest that the shift towards a thematic PB in 2020 was primarily seen as an opportunity to pursue pedagogical purposes and reduce the red tape of the PB. On the one hand, the PB was expected to foster a common language on, and understanding of, sustainability between the city council and citizens. On the other, this shift was expected to ensure a more straightforward analysis of citizen proposals and higher implementation rates. The streamlining of this process, therefore, was less a result of a shared understanding of its achievements, as—despite the step-by-step process towards urban sustainability theme via the Green Seal and the PB for schools—the announcement of the 2020 Green PB relied upon low evidence-based awareness of its contribution. These limits may have lowered the political ownership paired by an equally low rate of citizen participation in the first urban-sustainability-oriented PB in 2021. The systematisation of knowledge on the contribution of the PB to urban sustainability provided in this study, therefore, offers a review of the PB's achievements and contributes to our understanding and awareness of the reasons why, and the ways in which, such specific participatory arrangements can make a difference in citizen-centred sustainable policy-making. It thus translates the insights from Lisbon's case study into lessons for the improvement and advancement of PB as an additional mechanism that may respond to the global and local calls for change, and greater focus on sustainable urban futures.

7. Conclusions

Our study aimed to contribute to the participatory dimension of urban sustainability agendas through the specific lens of participatory budgets, by asking what are the lessons and potential contributions of the Lisbon PB experience towards urban sustainability, and analysing in detail Lisbon's developments in this field, keeping in mind the failed attempt to implement a Green PB in 2020, and the ongoing sustainability-oriented PB in 2021. Accordingly, this study presents and discusses the Lisbon PB experience and its achievements with a focus on the "environment, green structure and energy" thematic area, with the aim to understand its achievements and substantiate the potential to advance the city's urban sustainability agenda.

Our analysis reveals three interconnected findings. First, the Lisbon PB demonstrated considerable achievements in the "environment, green structure and energy" thematic area, corroborating the idea that citizens are indeed interested in promoting urban sustainability. Proposals and projects cover a wide range of topics, some of them showing the potential to go beyond an engineering-based vision of green spaces and infrastructure, as well as one less aligned with international agendas of climate change and biodiversity loss. Second, the achievements heavily depended on public action by the city council in deciding to invest in specific policies, with reverberations over the types of citizen demands to the PB. Finally, the shift towards a thematic PB in 2020 was not driven by social and political awareness of the PB achievements in terms of urban sustainability, but rather by the pursuit of pedagogical and instrumental (efficiency-led) purposes. The implementation of the 2021 sustainability-oriented PB, however, widens its scope and is likely to hold a potential that may be further analysed in the future. We recommend that this kind of data collection and analysis should be systematically carried out and disseminated within city council departments, promoting much needed internal awareness of PBs' potential as drivers of urban sustainability. We also identify further research needs into the sustainability potential of green PBs.

At the time of this study, the 2021 PB edition is in progress and it is impossible to retrieve significant up-to-date insights from its outcomes that ought to be addressed in future research. However, we may interpret the low participation rate in the 2021 PB as the need to accommodate society to the idea that now the PB only funds citizen proposals in the field of urban sustainability. It might also relate to the reduced capacity of the city council to capitalise knowledge from past editions and raise awareness about the opportunity to reorientate the PB towards urban sustainability. Last, these data may also mean that the public attitude, interest, and trust towards the PB as a policy mechanism per se is declining. Likewise, we acknowledge three main limitations in our study that can hopefully inspire the scholarly debate. Firstly, the focus on one city (albeit a leader in this field) is inevitably limited, and international comparative analyses on the contribution of the PB in this field will enhance our understanding of the potential of citizen participation to advance urban sustainability. Secondly, the focus on the Lisbon PB thematic area "environment, green structure and energy" may have sidelined insights from other thematic areas, which could contribute to other aspects of urban sustainability, thus, future research should take into consideration citizen proposals and projects funded in all PB thematic areas for a more comprehensive picture of its achievements. Thirdly, an in-depth investigation into citizens' views of urban sustainability would make an important addition to this initial overview of Lisbon's experience.

Despite the limitations inherent to our exploratory study, our findings offer useful insights into the potential of participatory budgets in contributing towards more sustainable cities. As we enter the third decade of the 21st century, the pressure and hopes around the role of cities to fight climate change and contribute to reduce biodiversity loss is growing. Agendas, including those based on nature-based solutions, are building momentum and expectations. We believe that a better understanding of the PB potential can make an important contribution towards more sustainable futures, promoting greater awareness and commitment of both citizens and local governments.

Author Contributions: Conceptualization, R.F., J.V., and O.B.; methodology, R.F.; validation, R.F., J.V. and O.B.; formal analysis, R.F.; investigation, R.F., J.V., and O.B.; resources, R.F.; data curation, R.F.; writing—original draft preparation, R.F., J.V., and O.B.; writing—review and editing, R.F., J.V., and O.B.; visualization, R.F.; supervision, R.F.; project administration, R.F.; funding acquisition, R.F. and O.B. All authors have read and agreed to the published version of the manuscript.

Funding: This research was funded by Fundação para a Ciência e Tecnologia, grant number SFRH/BPD/109406/2015, and by CONEXUS, European Union Horizon 2020 Grant Agreement 867564.

Institutional Review Board Statement: Not applicable.

Informed Consent Statement: Not applicable.

Data Availability Statement: Data analysed in this article with reference to the Lisbon Participatory Budget can be found at: https://op.lisboaparticipa.pt/, accessed on the 12 June 2021.

Conflicts of Interest: The authors declare no conflict of interest.

Appendix A

Table A1. Citizen proposals, votes, winning projects, and public funding in the Lisbon PB.

PB Editions	Total of Citizen Proposals (No.)	Citizen Proposals in the Thematic Area (No.)	%	Total of Votes (No.)	Votes in the Thematic Area (No.)	%	Total of Winning Projects (No.)	Winning Projects in the Thematic Area (No.)	%	Total of Public Funding (€)	Public Funding in the Thematic Area (€)	%
2008	89	62	70%	1101	572	52%	5	4	80%	5,130,176	4,550,000	89%
2009	200	98	49%	4719	1148	24%	12	4	33%	4,817,492	1,825,000	38%
2010	291	112	38%	11,570	1722	15%	7	3	42%	4,500,000	2,250,000	50%
2011	228	86	38%	17,887	4265	24%	5	3	60%	4,600,000	2,600,000	56%
2012	231	83	36%	29,911	620	2%	15	1	7%	2,375,000	100,000	4%
2013	208	85	41%	35,922	11,788	33%	16	5	31%	2,475,000	940,000	38%
2014	211	92	44%	36,032	7386	20%	13	3	23%	2,428,000	1,100,000	45%
2015	189	75	40%	42,130	5644	13%	15	3	20%	2,500,000	800,000	32%
2016	182	38	21%	51,591	11,595	22%	17	7	41%	2,480,000	1,050,000	42%
2017	128	28	22%	37,673	5898	16%	15	2	13%	2,513,000	300,000	12%
2018	122	26	21%	34,672	4225		19	3	16%	2,505,000	500,000	20%
TOTAL	2073	785	38%			18%	139	38	27%	36,441,176	16,015,000	44%

Appendix B

Table A2. Proposals from the emerging topics in the thematic area "environment, green structure, and energy" of the Lisbon PB.

PB Editions	Topic 1 City/Neighbourhood		Topic 2 Recovery		Topic 3 Gardens		Topic 4 Farming		Topic 5 Mobility		Topic 6 Animals		Topic 7 Energy		Topic 8 Pollution	
	No.	%	No.	%	No.	%	No.	%	No.	%	No.	%	No.	%	No.	%
2008	15	24%	28	45%	15	24%	2	3%	2	3%	0	0%	0	0%	0	0%
2009	20	20%	44	45%	25	24%	4	4%	3	3%	1	1%	1	1%	0	0%
2010	12	11%	41	37%	39	35%	8	7%	7	6%	1	1%	3	3%	1	1%
2011	18	21%	30	35%	23	27%	5	6%	6	7%	2	2%	2	2%	0	0%
2012	19	23%	22	26%	27	32%	2	2%	10	12%	1	1%	0	0%	2	2%
2013	18	21%	19	22%	29	34%	3	3%	5	6%	5	6%	3	3%	3	3%
2014	20	22%	18	19%	33	36%	5	5%	8	9%	4	4%	1	1%	3	3%
2015	24	32%	6	8%	31	41%	0	0%	6	8%	3	4%	2	3%	3	4%
2016	3	8%	4	10%	16	42%	1	3%	1	3%	8	21%	2	5%	3	8%
2017	5	18%	6	21%	10	36%	1	3%	0	0%	4	14%	2	7%	0	0%
2018	9	35%	5	19%	7	27%	0	0%	0	0%	1	4%	0	0%	4	15%
TOTAL	163	21%	223	28%	255	32%	31	4%	48	6%	30	4%	16	2%	19	2%

References

1. Pateman, C. *Participation and Democratic Theory*, 1st ed.; Cambridge University Press: Cambridge, UK, 1970.
2. Fung, A. Putting the Public Back into Governance: The Challenges of Citizen Participation and Its Future. *Public Adm. Rev.* **2015**, *75*, 513–522. [CrossRef]
3. Avritzer, L. New Public Spheres in Brazil: Local Democracy and Deliberative Politics. *Int. J. Urban. Reg. Res.* **2006**, *30*, 623–637. [CrossRef]
4. Baiocchi, G. *Militants and Citizens: The Politics of Participatory Democracy in Porto Alegre*; Stanford University Press: Stanford, CA, USA, 2005.
5. Dias, N.; Enríquez, S.; Júlio, S. *The Participatory Budgeting World Atlas*; Epopeia and Oficina: Faro, Portugal, 2019.
6. Falanga, R. Citizen Participation during the COVID-19 pandemic: Insights from Local Practices in European Cities. 2020. Available online: http://library.fes.de/pdf-files/bueros/lissabon/17148.pdf (accessed on 30 December 2020).
7. Falanga, R.; Lüchmann, L. Participatory Budgets in Brazil and Portugal: Comparing Patterns of Dissemination. *Policy Stud.* **2020**, *41*, 603–622. [CrossRef]
8. Grimm, N.; Schindler, S. Nature of Cities and Nature in Cities: Prospects for Conservation and Design of Urban Nature in Human Habitat. In *Rethinking Environmentalism: Linking Justice, Sustainability, and Diversity*; Sharachchandra, L., Brondizio, E.S., Byrne, J., Mace, G.M., Martinez-Alier, J., Eds.; MIT Press: Cambridge, UK, 2018; pp. 99–125.
9. Simon, D.; Arano, A.; Cammisa, M.; Perry, B.; Pettersson, S.; Riise, J.; Valencia, S.; Oloko, M.; Sharma, T.; Vora, Y.; et al. Cities coping with COVID-19: Comparative perspectives. *City* **2021**, *25*, 1–42. [CrossRef]
10. Elmqvist, T.; Fragkias, M.; Goodness, J.; Güneralp, B.; Marcotullio, P.J.; McDonald, R.I.; Parnell, S.; Schewenius, M.; Sendstad, M.; Seto, K.C.; et al. *Urbanization, Biodiversity and Ecosystem Services: Challenges and Opportunities*; Springer: Berlin/Heidelberg, Germany, 2013.
11. Dorst, H.; van der Jagt, A.; Raven, R.; Runhaar, H. Urban Greening through Nature-Based Solutions—Key Characteristics of An Emerging Concept. *Sustain. Cities Soc.* **2019**, *49*, 101620. [CrossRef]
12. McCormick, K. *Cities, Nature and Innovation: New Directions*; Lund University: Lund, Sweden, 2020.
13. Brundtland Commission. *Our Common Future. Report of the World Commission on Environment and Development*; Oxford University Press: Oxford, UK, 1987.

14. IPBES. *Workshop Report on Biodiversity and Pandemics of the Intergovernmental Platform on Biodiversity and Ecosystem Services*; IPBES: Bonn, Germany, 2020.

15. Rudd, A.; Simon, D.; Cardama, M.; Birch, E.L.; Revi, A. The UN, the Urban Sustainable Development Goal, and the New Urban Agenda. In *Urban Planet: Knowledge towards Sustainable Cities*; Griffith, C., Maddox, D., Simon, D., Watkins, M., Frantzeskaki, N., Romero-Lankao, P., Parnell, S., Elmqvist, T., McPhearson, T., Bai, X., Eds.; Cambridge University Press: Cambridge, UK, 2018; pp. 180–196.

16. Akande, A.; Cabral, P.; Gomes, P.; Casteleyn, S. The Lisbon Ranking for Smart Sustainable Cities in Europe. *Sustain. Cities Soc.* **2019**, *44*, 475–487. [CrossRef]

17. Bulkeley, H. *Cities and Climate Change*; Routledge: New York, NY, USA, 2013.

18. UNGA. *Transforming Our World: The 2030 Agenda for Sustainable Development, UN General Assembly, Resolution Adopted by the General Assembly on 25 September 2015, Geneva*; UN: Geneva, Switzerland, 2015.

19. UNCED. Agenda 21, United Nations Conference on Environment and Development, United Nations General Assembly. 1992. Available online: http://www.unorg/esa/sustdev/agenda21texthtm13/12/02 (accessed on 23 March 2020).

20. UN. New Urban Agenda, Habitat III. 2017. Available online: https://unhabitat.org/sites/default/files/2019/05/nua-english.pdf (accessed on 15 July 2020).

21. Klopp, J.M.; Petretta, D.L. The Urban Sustainable Development Goal: Indicators, COMPLEXITY and the politics of Measuring Cities. *Cities* **2017**, *63*, 92–97. [CrossRef]

22. Rees, W.E. Is 'Sustainable City' An Oxymoron? *Local Environ.* **1997**, *2*, 303–310. [CrossRef]

23. Campbell, S. Green Cities, Growing Cities, Just Cities?: Urban Planning and the Contradictions of Sustainable Development. *J. Am. Plan. Assoc.* **1996**, *62*, 296–312. [CrossRef]

24. Connelly, S. Mapping Sustainable Development as A Contested Concept. *Local Environ.* **2007**, *12*, 259–278. [CrossRef]

25. Mitlin, D.; Satterthwaite, D. Sustainable Development and Cities. In *Sustainability, the Environment and Urbanization*; Pugh, C., Ed.; Earthscan, London and Sterling: London, UK, 1996.

26. Dempsey, N.; Brown, C.; Bramley, G. The Key to Sustainable Urban Development in UK Cities? The Influence of Density on Social Sustainability. *Prog. Plan.* **2012**, *77*, 89–141. [CrossRef]

27. McKenzie, S. *Social Sustainability: Towards Some Definitions*; Hawke Research Institute Working Paper Series, nr 27; University of South Australia: Adelaide, Australia, 2004.

28. Simon, D.; Palmer, H.; Rise, J. *Comparative Urban Research from Theory to Practice: Co-Production for Sustainability*; Policy Press: Bristol, UK, 2020.

29. Fokdal, J.; Bina, O.; Chiles, P.; Omajäe, L.; Paadam, K. *Enabling the City: Inter- and Transdisciplinary Encounters*; Routledge: New York, NY, USA, 2021.

30. Leipzig Charter. Leipzig Charter on Sustainable European Cities. 2007. Available online: https://english.leipzig.de/construction-and-residence/urban-development/leipzig-2020-integrated-city-development-concept-seko/implementation/leipzig-charter/ (accessed on 2 May 2007).

31. Kinzer, K. How Can We Help? An Exploration of the Public's Role in Overcoming Barriers to Urban Sustainability Plan Implementation. *Sustain. Cities Soc.* **2018**, *39*, 719–728. [CrossRef]

32. Trudeau, D. Integrating Social Equity in Sustainable Development Practice: Institutional Commitments and Patient Capital. *Sustain. Cities Soc.* **2018**, *41*, 601–610. [CrossRef]

33. Van der Steen, M.; Van Twist, M.; Chin-A-Fat, N.; Kwakkelstein, T. Pop-Up Public Value: Public Governance in the Context of Civic Self Organisation. 2013. Available online: http://www.nsob.nl/wp-content/uploads/NSOB_Denktank_Pop-up-UK-DEF-web.pdf (accessed on 28 February 2020).

34. Van der Jagt, A.P.N.; Elands, B.H.M.; Ambrose-Oji, B.; Gerőházi, E.; Steen Møller, M.; Buizer, M. Participatory Governance of Urban Green Spaces: Trends and Practices in the EU. *Nord. J. Archit. Res.* **2016**, *3*, 11–40.

35. Lieberherr, E.; Odom Green, O. Green Infrastructure through Citizen Stormwater Management: Policy Instruments, Participation and Engagement. *Sustainability* **2018**, *10*, 2099. [CrossRef]

36. Mayer, A.L.; Shuster, W.D.; Beaulieu, J.J.; Hopton, M.E.; Rhea, L.K.; Roy, A.H.; Thurston, H.W. Building Green Infrastructure via Citizen Participation: A Six-Year Study in the Shepherd Creek (Ohio). *Environ. Pract.* **2012**, *14*, 57–67. [CrossRef]

37. Verheij, J.; Nunes, M. Justice and Power Relations in Urban Greening: Can Lisbon's Urban Greening Strategies Lead to More Environmental Justice? *Local Environ.* **2020**, *26*, 329–346. [CrossRef]

38. Wilker, J.; Rusche, K.; Rymsa-Fitschen, C. Improving Participation in Green Infrastructure Planning. *Plan. Pract. Res.* **2016**, *31*, 229–249. [CrossRef]

39. Hysing, E. Representative Democracy, Empowered Experts, and Citizen Participation: Visions of Green Governing. *Environ. Polit.* **2013**, *22*, 955–974. [CrossRef]

40. Portney, K.E.; Berry, J.M. Participation and the Pursuit of Sustainability in US Cities. *Urban Aff. Rev.* **2010**, *46*, 119–139. [CrossRef]

41. Polasky, S.; Caldarone, G.; Duarte, T.K.; Goldstein, J.; Hannahs, N.; Ricketts, T.; Tallis, H. Putting Ecosystem Service Models to Work: Conservation, Management, and Trade-offs. In *Natural Capital: Theory and Practice of Mapping Ecosystem Services*; Kareiva, P., Tallis, H., Ricketts, T.H., Daily, G.C., Polasky, S., Eds.; Oxford University Press: New York, NY, USA, 2011; pp. 249–263.

42. Hayward, B.M. The Greening of Participatory Democracy: A Reconsideration of Theory. *Environ. Polit.* **1995**, *4*, 215–236. [CrossRef]

43. Dryzek, J.S.; Pickering, J. Deliberation as A Catalyst for Reflexive Environmental Governance. *Ecol. Econ.* **2017**, *131*, 353–360. [CrossRef]
44. Dryzek, J.S. *Rational Ecology: Environment and Political Economy*; Basil Blackwell: Oxford, UK, 1987.
45. Stevenson, H.; Dryzek, J.S. The Discursive Democratisation of Global Climate Governance. *Environ. Polit.* **2012**, *21*, 189–210. [CrossRef]
46. Yearley, S. Green Ambivalence about Science: Legal-Rational Authority and the Scientific Legitimation of Social Movement. *Br. J. Sociol.* **1992**, *43*, 511–532. [CrossRef]
47. Paehlke, R. Democracy and Environmentalism: Opening A Door to the Administrative State. In *Managing Leviathan. Environmental Politics and the Administrative State*; Paehlke, R., Torgerson, D., Eds.; Broadview Press: Peterborough, ON, Canada, 2005; pp. 25–43.
48. Smith, G. *Democratic Innovations. Designing Institutions for Citizen Participation*; Cambridge University Press: Cambridge, UK, 2009.
49. Healey, P. *Collaborative Planning: Shaping Places in Fragmented Societies*; UBC Press: Newcastle, UK, 1997.
50. Sintomer, Y.; Herzberg, C.; Röcke, A.; Allegretti, G. Transnational Models of Citizen Participation: The Case of Participatory Budgeting. *J. Public Delib.* **2012**, *8*, 70–116. [CrossRef]
51. Cabannes, Y. *Contributions of Participatory Budgeting to Climate Change Adaptation and Mitigation. Current Local Practices around the World & Lessons from the Field*; IOPD: Barcelona, Spain, 2020.
52. Maksymiuk, G.; Kimic, K. 'Green Projects' in Participatory Budgets Inclusive Initiatives for Creating City's Top Quality Public Spaces. Warsaw Case Study. Warsaw University of Life Sciences—SGGW. 2016. Available online: https://core.ac.uk/download/pdf/84251405.pdf (accessed on 30 March 2017).
53. Epting, S. Participatory Budgeting for Environmental Justice. *Ethics Policy Environ.* **2020**, *23*, 22–36. [CrossRef]
54. Balbona, D.L.; Bebega, S.G. Austerity and Welfare Reform in South-Western Europe: A Farewell to Corporatism in Italy, Spain and Portugal? *Eur. J. Soc. Secur.* **2015**, *17*, 271–291. [CrossRef]
55. Teles, F. Local Government and the Bailout: Reform Singularities in Portugal. *Eur. Urban Reg. Stud.* **2016**, *23*, 455–467. [CrossRef]
56. Allegretti, G.; Antunes, S. The Lisbon Participatory Budget: Results and Perspectives on An Experience in Slow but Continuous Transformation. *Field Actions Sci. Rep.* **2014**, *11*.
57. Buijs, A.E.; Elands, B.H.M.; Havik, G.; Ambrose-Oji, B.; Gerőházi, E.; Jagt van der, A.; Mattijssen, T.J.M.; Steen Møller, M.; Vierikko, K. Innovative Governance of Urban. Green Spaces: Learning from 18 Innovative Examples across Europe. Project 'Green Surge' Research Report. 2016. Available online: https://research.wur.nl/en/publications/innovative-governance-of-urban-green-spaces-learning-from-18-inno (accessed on 31 January 2016).
58. EGC Evaluation Board. Expert Panel Technical Assessment Synopsis Report European Green Capital Award 2020. Available online: https://ec.europa.eu/environment/europeangreencapital/press-communications/egca-publications/ (accessed on 27 January 2020).
59. Graça, M. Participação Pública: Mecanismos e Práticas No Contexto da Administração Pública e o Caso do Orçamento Participativo de Lisboa. Master's Thesis, ISCTE-Lisbon University Institute, Lisbon, Portugal, December 2018.
60. Gulsrud, N.M.; Ostoić, S.K.; Faehnle, M.; Maric, B.; Paloniemi, R.; Pearlmutter, D.; Simson, A.J. Challenges to Governing Urban Green Infrastructure in Europe—The Case of the European Green Capital Award. In *The Urban Forest—Cultivating Green Infrastructure for People and the Environment*; Pearlmutter, D., Calfapietra, C., O'Brien, L., Samson, R., Sanesi, G., Ostoic, S.K., del Amo, R.A., Eds.; Springer International Publishing: Berlin/Heidelberg, Germany, 2017.
61. Garcia-Lamarca, M.; Anguelovski, I.; Cole, H.; Connolly, J.T.; Arguelles, L.; Baro, F.; Loveless, S.; Pérez, C.; Shokry, G. Urban Green Boosterism and City Affordability: For Whom Is the 'Branded' Green City? *Urban Stud.* **2019**, *58*, 90–112. [CrossRef]
62. Sareen, S.; Grandin, J. European Green Capitals: Branding, Spatial Dislocation or Catalysts for Change? *Geogr. Ann. Ser. B Hum. Geogr.* **2019**, *102*, 101–117. [CrossRef]

MDPI
St. Alban-Anlage 66
4052 Basel
Switzerland
Tel. +41 61 683 77 34
Fax +41 61 302 89 18
www.mdpi.com

Sustainability Editorial Office
E-mail: sustainability@mdpi.com
www.mdpi.com/journal/sustainability